Techniques in Protein Chemistry

Techniques in Protein Chemistry

Edited by

Tony E. Hugli
Department of Immunology
Research Institute of Scripps Clinic
La Jolla, California

ACADEMIC PRESS, INC.
Harcourt Brace Jovanovich, Publishers
San Diego New York Berkeley
Boston London Sydney Tokyo Toronto

Academic Press Rapid Manuscript Reproduction

COVER: Computer graphic representation of a 21-residue peptide fragment of human C3a anaphylatoxin. The original print was provided with the assistance of Dr. Arthur J. Olson, Department of Molecular Biology, Research Institute of Scripps Clinic. The program utilized GRAMPS computer graphic language developed by T.J. O'Donnell and A.J. Olson.

Academic Press, Inc.
San Diego, California 92101

United Kingdom Edition published by
Academic Press Limited
24–28 Oval Road, London NW1 7DX

Library of Congress Cataloging-in-Publication Data

Techniques in protein chemistry / edited by Tony E. Hugli.
p. cm.
A collection of reports of new methods and techniques resulting from the second annual meeting of the Protein Society held on Aug. 13–17, 1988, in San Diego.
Includes index.
ISBN 0-12-682000-7 (hardcover)(alk. paper)
ISBN 0-12-682001-5 (paperback)(alk. paper)
1. Proteins--Analysis--Congresses. 2. Amino acid sequence--Congresses. I. Hugli, T. E. (Tony E.) II. Protein Society.
QP551.T333 1989
547.7'5046--dc19 89-80
CIP

Printed in the United States of America
89 90 91 92 9 8 7 6 5 4 3 2 1

Contents

Section III
Amino Acid Analysis

Section IV
Applications of HPLC

Section V
Protein Chemistry and Analyses

Section VI
Standard Test Peptide

Foreword

At the first symposium of the Protein Society in August 1987, one of the exciting aspects of the meeting was the profusion of excellent posters and abstracts. How could all this current, fascinating information be brought rapidly to a larger audience? The new publications committee, under the chairmanship of George Rose considered this question, and the outcome is this volume of *Techniques in Protein Chemistry*.

The present volume is based on expanded abstracts from the second symposium of the society, held in August 1988, as well as several invited contributions. It represents only a fraction of the excellent work that was discussed at the symposium and is itself an experiment. There were more than 1000 participants in the symposium, and over 300 abstracts. If this volume of *Techniques in Protein Chemistry* proves useful, the series may be expanded to feature a larger fraction of the contributions at the 1989 symposium.

This first volume represents the vision and hard work of Tony Hugli. Tony welded the agreement between the society, Academic Press, the authors, and the industrial sponsors that has made rapid publication possible at a reasonable price. Then he selected the abstracts, identified associate editors, coordinated communication between the authors and the associate editors, and edited the volume. To the extent that the volume and the series proves useful, much of the credit belongs to Tony and the associate editors who gave of their time to complete the volume on schedule.

David Eisenberg
President
The Protein Society

Preface

The Protein Society held its first annual meeting in San Diego in August of 1987. A commitment to excellence was apparent from the beginning, and the efforts of an enthusiastic membership have produced a highly successful organization. Meetings of the society have provided a forum for reporting major advances in the analysis of various aspects of protein architecture and function. The scientific programs at each meeting have featured dynamic leaders in the field, and attendance has always been greater than expected. The second annual meeting, in 1988, saw a marked expansion in size and sophistication over the inaugural meeting.

By 1988 it was clear that the society would continue to grow and prosper. Unfortunately, no formal mechanism existed to report details from the proceedings of these meetings to the society's membership and to scientists at large. The scientific sessions were already too extensive to be published in their entirety. It was suggested that a collection of reports of new methods and techniques, condensed into a single volume for rapid publication, could capture a significant aspect of the meeting for later reference. This idea was proposed to the publications committee and accepted on an experimental basis for the second annual meeting. For this first volume of *Techniques in Protein Chemistry*, topics were chosen that have both general interest and practical value to protein chemists; they emphasize new methods and applications in protein sequencing, highlight advanced applications of mass spectrometry and nuclear magnetic resonance technology, present a status report on limitations of amino acid microanalysis, and update advances in high-performance liquid chromatography. One chapter is also devoted to reports of general protein chemistry. As a bonus, participants in a special workshop organized to determine the structure of synthetic test peptide-3 (STP-3), a peptide designed to test the analytical limits of current technology in the field, reported their strategies in solving the structure of the "mystery" peptide.

Under the guidance of the eight associate editors, the six selected topics were integrated into sections that summarize recent and significant technical advances in protein chemistry. This information is timely and will be useful both to investigators actively involved in protein chemistry and to those just wishing to be informed of current trends in the field. Taken together, the articles in this volume provide an excellent review of the analytical techniques currently available for the molecular characterization of proteins. We have attempted to select reports that reflect limitations as well as identify new procedures and advances in protein analysis. In this regard, the volume serves as an informative progress report of selected sessions from the annual meeting of the society. A small number of articles were also solicited from members not presenting abstracts in order to obtain a more general overview of specific areas of expertise. These articles provide the continuity needed for a more coherent documentation of each topic.

This volume is a direct result of the willingness of researchers who attended the second annual meeting of the Protein Society on August 13–17, 1988, to share their reports. It is a strong statement in support of the volume that so many participants were willing and even anxious to contribute articles to *Techniques in Protein Chemistry*. I thank all the authors for their contributions and the cooperative spirit in which they met the challenge of our rigorous deadlines.

Tony E. Hugli

Acknowledgments

The following organizations contributed to the production of this volume:

AAA LABORATORY
6206 89th Avenue, S.E,.Mercer Island, Washington 98040-4599, (206) 364-3446

ADVANCED CHEMTECH
P.O. Box 1403, Louisville, Kentucky 40201, (502) 636-0000

AMERSHAM CORPORATION
2636 South Clearbrook Drive, Arlington Heights, Illinois 60005-4692, (312) 593-6300

APPLIED BIOSYSTEMS
850 Lincoln Center Drive, Foster City, California 94404, (415) 570-6667

BECKMAN
Bioanalytical Systems Group, 2500 Harbor Boulevard, Fullerton, California 92634, (714) 773-6646

BIONOVUS, INC.
344 Lakeside Drive, Foster City, California 94404, (415) 578-8700

BIOSYM TECHNOLOGIES, INC.
10065 Barnes Canyon Road, Suite A San Diego, California 92121, (619) 458-9990

DIONEX CORPORATION
1228 Titon Way, P.O. Box 3603, Sunnyvale, California 94088-3603, (408) 737-0700

HEWLETT-PACKARD GMBH
Waldronn Analytical Division, Hewlett-Packard Strasse, 7517 Waldbronn 2,West Germany, 0 72 43/602-0

IMMUNO-DYNAMICS, INC.
P.O. Box 766, La Jolla, California 92038, (619) 452-1270

MILLGEN/BIOSEARCH
Division of Millipore, 186 Middlesex Turnpike, Burlington, Massachusetts 01803, (800) 872-0071

MULTIPLE PEPTIDE SYSTEMS, INC.
10955 John Jay Hopkins Drive, San Diego, California 92121, (619) 455-3710

PHARMACIA LKB TECHNOLOGY, INC.
800 Centennial Avenue, Piscataway, New Jersey 08854, (201) 457-8000

PORTON INSTRUMENTS, INC.
18740 Oxnard Street, Suite 303, Tarzana, California 91350, (800) 822-6333

RAININ INSTRUMENTS COMPANY, INC.
Mack Road, Woburn, Massachusetts 01801, (617) 935-3050

TOSO HAAS
Rohm & Haas Building, Independence Mall West Philadelphia, Pennsylvania 19105, (215) 592-3100

VYDAC
17434 Mojave Street, Box 867, Hesperia, California 92345, (619) 244-6107

Section I

Introduction

CHEMICAL MICROSEQUENCING TECHNIQUES

John E. Shively and David W. Speicher

Beckman Research Institute of the City of Hope, Duarte, CA, and Wistar Institute, Philadelphia, PA.

Keeping up to date on the latest techniques in protein chemistry is of prime interest to the protein chemist. The myriad problems encountered during the purification and structural analysis of minute amounts of protein of great biological interest continue to challenge protein chemists. Few current reference works exist in these areas despite continuing rapid changes and improvements to the methodology. The compilation of new or improved techniques in this section represent the current state of the art, and as such, will be open to close scrutiny and testing in many laboratories. The techniques which survive this scrutiny or go on to further iterations of improvement will certainly benefit the scientific community, since the need for improved and more sensitive methods of sequence analysis is apparent.

Protein sequencing techniques and the requisite sophisticated equipment are constantly evolving. The major driving force toward improvements in this field has historically been a continuing need for improved sensitivity and several articles in this section are focused on raising the sequencing sensitivity level beyond its already impressive low picomole level. In this context it is also critical for both sequence practitioners and investigators supplying proteins for analysis to have realistic concepts of the current capacities of available techniques and sequencing facilities. The article by Niece et al is an especially important contribution that describes the results of a well designed survey of 40 microchemistry core facilities. The data described can be used by participants to evaluate how their facility compares to other facilities. Other protein chemists can derive interesting suggestions concerning the relative capacities of different instruments as well as the need for cautious application of computerized sequence interpretation. Also, these data provide the investigator, that utilizes a core facility, with an accurate indication of the performance level of the average facility with a "real" unknown. In addition to the survey of core facilities, articles in this section cover most areas of recent progress in protein chemistry techniques as summarized below.

I. AMINO-TERMINAL SEQUENCING FROM SDS GELS

An extremely popular technique highlighted in this section by several articles is the separation of proteins by SDS gel electrophoresis, electrophoretic transfer of stained bands to polyvinylidene (PVDF) membranes, and subsequent sequence analysis. The technique hails from the pioneering work of Aebersold et al (1) using electrotransfer of SDS gel separated proteins to glass fiber papers, to the vastly improved method of Matsudaira (2) using PVDF membranes. The popularity of the methods rests with the fact that SDS gels can quickly separate small amounts of complex protein mixtures and the fact that most protein chemists use SDS gels as the routine method to analyze the purity of protein samples. The problems with use of gels to isolate small amounts of sequencable protein are manifold and have limited the use of this approach until recently. The reader will be pleased to see a selection of articles describing improvements and variations of the methodology including efficient transfer of proteins from one and two dimensional (2D) gels.

The extremely high resolving power of 2D gels is usually sufficient to warrant using this method as the only purification step, but in the case of Parmelee and coworkers, minor components of a crude tissue extract were first enriched by affinity chromatography. This approach was necessary to circumvent the limited load capacity of 2D gels with the resulting extremely low amount of protein present in a single spot. In this report, initial sequence yields after transfer to PVDF were in the 10 to 20 picomole range which corresponds to about 1 to 2 μg per spot for a 100 kDa protein. The power of this technique is emphasized by the fact that no serious alternate methods exist for the facile isolation of a few micrograms of a large protein from complex mixtures.

Several problems associated with use of PVDF membranes are addressed in the article by Speicher including: nonlinear transfer of different molecular weight proteins, losses during staining/destaining, and variable recoveries from cycle-to-cycle in the gas phase sequencer. This latter problem is accentuated if a glass fiber filter is not used, but substantial time savings and improved background can result by elimination of the filter. Modifications to the sequencer program have minimized this problem and the sequence analysis of as much as 22 residues from 3.3 picomoles (initial cycle yield) of a 90 kDa peptide transferred from a 2D gel is reported. Direct adsorption of purified proteins in solution to PVDF is also presented as an alternate, possibly superior, loading method to direct spotting of proteins onto PVDF.

The problem of blocked amino-terminals is addressed by Hulmes et al who treat samples on the PVDF membranes with acid to obtain internal cleavages at serines or at amino-terminal acetyl groups. This method would be expected to be most successful for the more acid labile asp-pro bonds. This group also demonstrates sensitive amino acid analysis on samples spotted on PVDF. This method may yield critical information on blocked samples. Other methods of dealing with blocked samples on

PVDF include chemical cleavage with: CNBr (cleavage at met), o-iodosobenzoic acid (cleavage at trp), or hydroxylamine (cleavage at asn). Samples may also be digested with proteases after transfer to nitrocellulose as an alternative approach (3).

It is often important to recover activity of a protein after purification even if the principle use of the purified protein is for sequence analysis. Recovery of activity is an especially formidable task when SDS gels are used for protein isolation. Fischer and coworkers describe a urea extraction method that gave activity for various peptides in the FSH assay. The samples were transferred to PVDF and sequenced to determine their relatedness to known peptide hormones.

II. OTHER IMPROVEMENTS IN AMINO-TERMINAL SEQUENCING

Currently amino-terminal sequencing by Edman chemistry is limited to the 10 to 20 picomole range, although occasionally samples down to the 1 to 2 picomole range have been reported to be sequenced. Commercial instruments usually certify a more conservative 100 to 200 picomole level. Despite the broad range of sensitivities obtained in different laboratories and situations, the ultimate sensitivity is limited by the detection methodology for the PTH-amino acid derivatives. Horn and coworkers describe a potentially new method for increasing the sensitivity of the Edman chemistry by using fluorescent or chromophoric PTC amino acid esters. This report should be taken as a first step in the development of the method since no data is reported for peptides or proteins. Instead the authors demonstrate the kinetics and yields for producing PTC amino acid esters versus the production of PTH derivatives. Overall, the method shows some promise for the use of hindered fluorescent or chromophoric alcohols in the conversion reaction. This approach may stimulate renewed interest in improving the types of derivatives formed in the Edman chemistry.

Haniu and Shively present an updated manual Edman protocol which uses a simple apparatus capable of sequencing six to ten samples simultaneously in the 100 to 200 picomole range. The advantage of this approach is simplicity and economy. In the past, workers avoided manual Edman chemistry because of the skill and time required. As a result, little effort has been made to improve the methodology over the last ten years. The authors describe a gas phase sequencing apparatus employing disposable flow reactors and commercially available Hamilton valves to control the delivery of reagents and solvents. With this method, one may sequence an entire peptide map within 2 to 3 days, or perform amino-terminal sequence analysis on 10 to 20 samples per day. This report may encourage others to attempt manual methods, and thereby greatly increase productive output.

III. CARBOXY-TERMINAL SEQUENCING METHODS

There has been a growing need for improved methods for determining the carboxy-terminal sequence of a protein. Hawke and coworkers present a new version of the strategy for end labeling a protein, digesting with a protease, isolating the end labeled peptide, and performing conventional amino-terminal sequence analysis. Although this strategy is somewhat involved, it relies on established techniques and, at least in theory, will lead to the desired information. In this report the investigators have established the methodology using a test peptide. It remains to be seen how well the method will work for proteins. An interesting aspect of this work is the use of carboxypeptidase P, which has a very broad specificity for C-terminal amino acids, and may not suffer from the usual problems encountered with carboxypeptidase digests.

Miller and coworkers have tackled the extremely difficult problem of direct carboxy-terminal sequencing by chemical means. The work described includes an analysis of the chemical reactions required for degradations with trimethylsilylisothiocyanate to give amino acid thiohydantions. Progress in this area is limited to two or three cycles at the 1 to 2 nmol level. The authors are optimistic concerning the potential for higher sensitivity and longer sequencer runs. Major problems yet to be solved are associated with particular amino acid residues such as proline, aspartic acid, and glutamine.

IV. COMPUTER ANALYSIS OF AMINO ACID SEQUENCES

One of the problems encountered with the acquisition of large amounts of data in any field is compilation and comparative analysis. At present, most protein chemists use the commercial gene and protein databases for simple sequence homology searches. A discussion of more advanced topics including searching for structural motifs is given by Gribskov and Eisenberg. The authors describe probes which are sensitive to three dimensional motifs, profile analysis with assigned probes, and normalization techniques to remove variable sequence lengths within a profile. These types of analysis will help identify, in a new sequence, motifs that appear in other proteins with similar function.

V. REFERENCES

1. Aebersold, R., Teplow, D.B., Hood, L.E., and Kent, S.B. (1986) J. Biol. Chem. 261:4229-4238.

2. Matsudalra, P. (1987) J. Biol. Chem. 262:10035-10038.

3. Aebersold, R., Leavitt, J., Saavedra, R.A., Hood, L.E., and Kent, S.B.H. (1987) Proc. Natl. Acad. Sci. USA 84:6970-6974.

1

STRATEGIES FOR MICROCHARACTERIZATION OF PROTEINS USING DIRECT CHEMISTRY ON SEQUENCER SUPPORTS

Jeffrey D. Hulmes, May C. Miedel and Yu-Ching E. Pan

Department of Protein Biochemistry, Roche Research Center
Hoffmann-La Roche Inc., Nutley, NJ 07110

I. INTRODUCTION

Purification of a sufficient quantity of protein for structural analysis is always a difficult task with the goal being to get the most information from the least amount of sample. With the increased use of protein electroblotting from SDS-PAGE for sequence determination (1-4), more protein chemistry is being done directly on samples immobilized on polybrene treated glass fiber filters (GFF) and polyvinylidene difluoride (PVDF) membranes. Blotted proteins are directly sequenced in the hope that they are pure and sequencable, but this is not always the case. This article describes three routine procedures developed in our laboratory for direct analysis of protein or peptide samples immobilized on sequencer supports. They are: 1) the use of gas phase trifluoroacetic acid [TFA(g)] for specific cleavage and/or possible deblocking of polypeptides on GFF or PVDF; 2) the use of o-phthalaldehyde (OPA) as a blocking reagent for obtaining sequence from peptide mixtures and reducing sequencing background on the gas-phase sequencer; 3) amino acid analysis (AAA) of proteins on PVDF for composition and quantitation after acid hydrolysis using a postcolumn fluorescamine amino acid analyzer.

II. MATERIALS AND EQUIPMENT

Polybrene and 6N HCl was from Pierce (Rockford, IL). Thioglycolic acid was from Sigma (St. Louis, MO). Fluorescamine, recombinant proteins and synthetic peptides were obtained from Hoffmann-La Roche (Nutley, NJ). Polyvinylidene difluoride (PVDF, Immobilon) was from Millipore (Milford, MA). Trifluoroacetic acid (TFA, R3) and other sequencer reagents were from Applied Biosystems (ABI, Foster City, CA).

Amino acid sequence analyses, using automated Edman degradation, were performed using an ABI model 470A gas phase sequencer (5). Phenylthiohydantoin (PTH) amino acids were identified either "off-line" with a Waters HPLC system using a Beckman/Altex Ultrasphere 250 x 4.6 mm C-18 column (6) or "on-line" with an ABI model 120A PTH analyzer.

Amino acid analyses were performed on an instrument using a cation exchange column (Aminex HP-C, 4 x 150 mm, BioRad, Richmond, CA) and detected by postcolumn reaction with fluorescamine (7).

III. TREATMENT OF POLYPEPTIDE SAMPLES ON SEQUENCER SUPPORTS

A. TFA(g) cleavage and deblocking of immobilized polypeptides

During protein sequencing it is not uncommon to find that a protein is blocked at its amino terminus. Usually this fact is discovered only after the sample has been committed to sequencing and it is now attached to a sequencer support (filter). Simpson and Nice (8) developed a procedure which allows cyanogen bromide (CNBr) cleavage of proteins on polybrene treated GFF. It has proven useful for verifying the presence of blocked protein samples and also releases peptides that may yield some sequence data. We have developed a simple method that cleaves polypeptides at the amino terminal side of Ser or Thr residues (Ser-N, Thr-N) and can also deblock peptides containing acetyl-Ser, a common blocked amino terminus. The method is applicable to proteins attached to both GFF and PVDF membranes[1]. It was successfully used to deblock and sequence the amino terminal of the "proteolipid" subunit of bovine chromaffin granule ATPase (9).

Immobilized, blocked peptide and protein samples can be chemically fragmented and possibly deblocked by treatment with TFA(g). Cleavage is performed on the immobilized sample (on GFF or PVDF) in a reaction tube (15 ml polypropylene) by adding TFA(g) via the sequencer's cartridge inlet line with the R3 reservoir (TFA) turned on for ≥ 5 minutes. The tube is capped so that samples are in the presence of TFA and left at room temperature (22°C) for 6-15 days or heated in an oven at 45°C for 2-3 days.

As an example of this cleavage and deblocking procedure, the blocked (acetylated) synthetic peptide, Ac-PT12Y, was sequenced before and after treatment with TFA(g) for 2 days at 45°C. The results are shown in figure 1. The cleavage occurred at the three Ser residues thus releasing three sequencable peptides including the deblocked N-terminus. Cleavage efficiency was between 20 and 35% under these conditions.

---

[1]Hulmes, et al., manuscript in preparation

The time and temperature to be used for the TFA(g) cleavage depends on the information sought. To test if a protein is blocked, 2-3 days at 45°C is sufficient. In order to obtain sequence data from an immobilized, blocked protein, cleavage at room temperature for 6-14 days usually works better because of increased specificity and reduced background. As little as 200 picomoles of blocked peptide has been tested with sequence yields following cleavage approaching 50%. Some modification of Lys and Tyr residues has been observed while deamidation of Asn and Gln is generally insignificant. TFA(g) is more specific than liquid TFA and has less chance of modifying amino acids or introducing contaminants[1]. This method will also work for unbound peptides in a reaction tube.

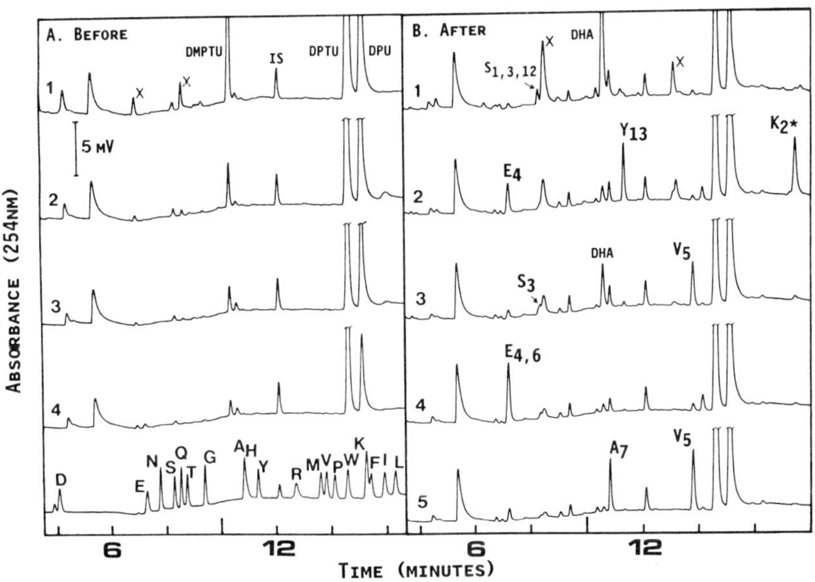

Fig. 1. Sequence analyses of Ac-PT12Y on GFF before and after treatment with TFA(g). 400 pmols of sample was loaded and sequenced (A). The GFF containing the blocked peptide was treated with TFA(g) for 2 days at 45°C and resequenced with the first five cycles shown (B). PTH amino acids from the three major peptides released by cleavage at Ser-N are indicated. A modified derivative of Lys (K_2*) was present after treatment. Unknown junk peaks are denoted by X. Ac-PT12Y has the following sequence and cleavage sites (↑).

5 10
Ac-S-K-S-E-V-E-A-A-A-E-L-S-Y
↑ ↑ ↑

DHA, dehydroalanine; DMPTU, N′,N-dimethyl-N′-phenylthiourea; DPTU, N′,N-diphenylthiourea; DPU, N′,N-diphenylurea; Ac, N-acetyl group.

The method of gas phase TFA cleavage described here offers a suitable alternative to CNBr cleavage. It can easily prove that a protein sample is blocked in a simple and less toxic manner. For small polypeptides (<100 residues) the TFA cleavage can be used for cleavage and possible deblocking that may lead to sequence data.

B. OPA blocking of polypeptide mixtures on the gas-phase sequencer

OPA, a reagent which reacts irreversibly with primary amines, has been used on the spinning cup and gas-phase sequencers to obtain sequence of proline containing peptides from protein/peptide mixtures, to reduce sequencer background and to extend sequence runs (10,11). When applied to any cycle which contains a proline, OPA will react with and block all unwanted sequences thus allowing unambiguous assignment of the proline containing peptides. These type of peptide mixtures are often generated by TFA(g) cleavage as well as by CNBr cleavage which has been used with OPA blocking (12). We have developed a method for OPA blocking on the gas-phase sequencer similar to the one described by Gross et al. (11) and a sequencer program and an example of its use is described here.

An initial sequence run of a peptide mixture or a portion of a sequencer filter containing cleaved immobilized peptides is necessary to determine the position of proline residues. The remainder of the sample is sequenced with a modified sequencer cycle (5PROPA, Table I) inserted when a proline is the next residue to be reacted. With ABI's model 470A microsequencer, Solvent 1 (S1, n-heptane) is replaced with a solution containing 0.02 mg OPA/ml n-butyl chloride (S3). The S1 wash step must be deleted from all sequencer cycles when OPA is used. This has no effect on sequencing. The OPA concentration is important because greater amounts introduce several large junk peaks while lower amounts are insufficient for complete blocking. The two 7 second deliveries of OPA are good for up to about 5 nanomoles of protein. OPA is reacted under basic conditions with R2 (TMA) and excess reagent is washed out with S3 and S2. Extra cleavage with R3 (TFA(g)) is incorporated into the program because the PTC derivative of proline is more difficult to cleave.

OPA is soluble and very stable in butyl chloride. The OPA blocked peptides are also stable under sequencer conditions (10) so background remains low for many cycles following application.

An example of the application of this cycle on a microsequencer is shown in figure 2. An equimolar mixture of three synthetic peptides plus recombinant alpha interferon was applied to GFF and loaded onto the sequencer with the 5PROPA cycle incorporated into the program for OPA reaction prior to residue #4. All the peptides were blocked except the proline containing interferon which continued to sequence. OPA blocking can be used on proteins with heterogeneous N-termini as well as on peptide mixtures and impure samples.

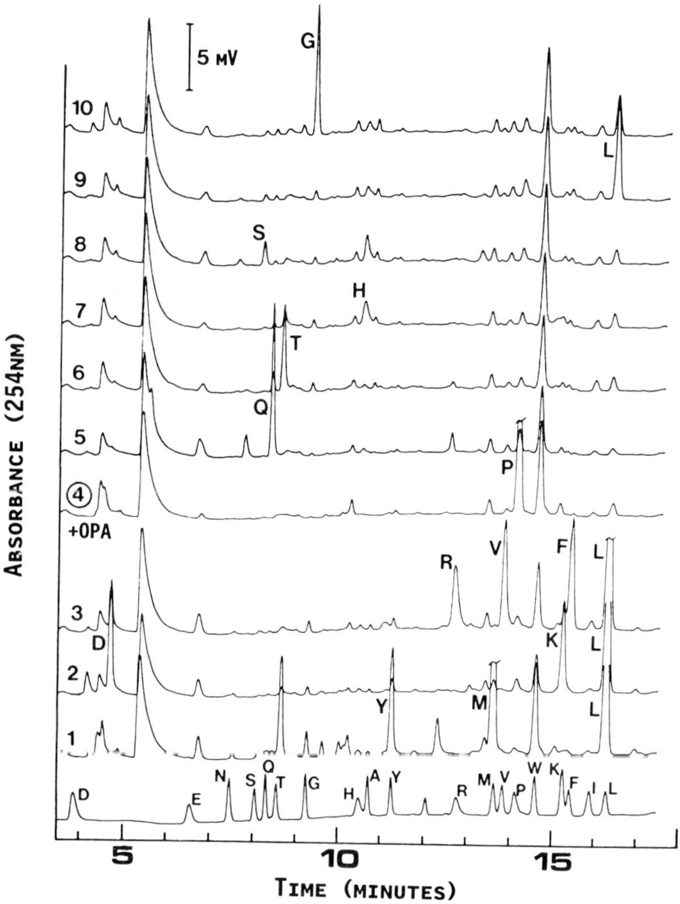

Fig. 2. Application of OPA using the 5PROPA cycle on a mixture of polypeptides. A mix of 3 synthetic peptides (12-27 aa's) plus rIFN-α was sequenced with a program containing the 5PROPA cycle at cycle 4. Three of the peptides were blocked while the rIFN-α containing P_4 continued to sequence. 300-450 picomoles of each polypeptide was loaded.

C. Amino acid analysis (AAA) of polypeptides on PVDF

Protein samples which have been immobilized onto PVDF by either spotting or electroblotting (2) can be hydrolyzed and subjected to AAA (4,13). Occasionally, it is difficult to determine if a sample on the PVDF membrane is blocked or there is too little to sequence. Direct AAA using a postcolumn amino acid analyzer offers a way to

TABLE I. Sequencer Program For OPA Blocking With Gas-Phase Sequencer

SEQUENCER CYCLE: 5PROPA[+]

STEP	CARTRIDGE FUNCTION	FLASK FUNCTION	TIME	STEP	CARTRIDGE FUNCTION	FLASK FUNCTION	TIME
1	deliver R2	pause	120	28	deliver R2	pause	400
2	prep S1	pause	6	29	argon dry	pause	60
3	deliver S1*	pause	7	30	deliver S2	pause	240
4	argon dry	pause	20	31	argon dry	pause	120
5	deliver R2	pause	300	32	deliver R3	argon dry	700
6	deliver S1*	pause	7	33	deliver R3	deliver S4	14
7	argon dry	argon dry	20	34	deliver R3	argon dry	8
8	deliver R2	pause	300	35	deliver R3	pause	540
9	argon dry	pause	100	36	deliver R3	collect	20
10	prep S3	pause	6	37	deliver R3	deliver S4	4
11	deliver S3	pause	50	38	deliver R3	argon dry	8
12	argon dry	pause	50	39	deliver R3	collect	30
13	deliver S2	pause	240	40	deliver R3	deliver S4	27
14	argon dry	argon dry	120	41	deliver R3	argon dry	5
15	deliver R2	argon dry	120	42	argon dry	increment FC	2
16	prep R1	argon dry	6	43	argon dry	empty	40
17	deliver R1	argon dry	3	44	transfer w/ Ar	pause	30
18	argon dry	argon dry	50	45	prep transfer	pause	6
19	deliver R2	argon dry	60	46	transfer w/ S3	pause	11
20	deliver R2	deliver R4	4	47	pause	deliver R4	5
21	deliver R2	argon dry	4	48	pause	argon dry	4
22	deliver R2	pause	330	49	pause	pause	10
23	deliver R1	pause	3	50	transfer w/ S3	pause	50
24	argon dry	pause	50	51	pause	argon dry	20
25	deliver R2	pause	400	52	transfer w/ Ar	pause	30
26	deliver R1	pause	3	53	deliver S3	argon dry	30
27	argon dry	pause	50	54	argon dry	argon dry	120
					TOTAL TIME FOR CYCLE: 82 MINUTES 3 SECONDS		

[+] for "off-line" ABI model 470A gas-phase sequencer. Program can be modified for "on-line" machines.

* ABI S1 (n-heptane) replaced with 0.02mg OPA/ml butyl chloride

TABLE II. AMINO ACID COMPOSITION OF POLYPEPTIDES ON PVDF

	rIL-2R++	rIFN-α	rHGH	rIL-1α	rIL-2 (A)	rIL-2 (B)	rIL-2 (C)
Asx B	13.2(11)	12.8(12)	20.8(20)	20.2(20)	11.8	11.4	12.1 (12)
Thr T	25.0(25)	10.5(10)	10.2(10)	10.4(12)	11.7	12.0	12.1 (13)
Ser S	16.3(17)	11.1(14)	16.9(18)	8.1(8)	8.6	8.5	6.7 (8)
Glx Z	37.4(41)	28.0(26)	27.8(27)	15.7(16)	16.9	18.3	18.4 (18)
Pro P	ND (18)	ND (5)	ND (8)	ND (6)	4.4	3.2	7.0 (5)
Gly G	15.8(12)	7.5(5)	8.9(8)	6.4(5)	3.3	2.8	2.6 (2)
Ala A	14.7(13)	9.3(8)	8.0(7)	13.7(13)	5.1	5.4	5.1 (5)
Cys C	ND (11)	ND (4)	*4.2(4)	ND (1)	+2.6	+2.5	+2.4 (3)
Val V	10.5(10)	6.8(7)	6.8(7)	6.0(7)	3.5	3.8	3.0 (4)
Met M	6.6(10)	5.6(5)	3.7(4)	3.0(4)	4.1	4.4	4.8 (5)
Ile I	5.8(6)	7.3(8)	7.1(8)	12.0(13)	7.3	7.7	7.5 (9)
Leu L	11.2(9)	21.1(21)	24.6(26)	16.3(15)	23.8	24.7	23.3 (22)
Tyr Y	7.8(7)	5.0(5)	8.1(8)	6.4(7)	4.0	3.4	4.3 (3)
Phe F	6.9(6)	10.4(10)	12.2(12)	10.1(9)	6.2	6.4	6.4 (6)
His H	4.4(6)	2.6(3)	3.3(3)	3.3(3)	2.6	2.3	2.9 (3)
Lys K	7.7(10)	9.4(11)	8.9(9)	10.6(11)	10.9	9.2	10.9 (11)
Arg R	11.6(11)	7.3(9)	11.1(11)	3.9(3)	4.1	3.5	3.9 (4)
Trp W	NE (3)	NE (2)	NE (1)	NE (2)	NE	NE	NE (1)
μg	0.24	0.15	0.96	1.30	0.87	0.75	0.93**
pmoles	10	8	46	76	58	50	62

++rIL-2R, recombinant IL-2 receptor; rIFN-α, recombinant alpha interferon; rHGH, recombinant human growth hormone; rIL-1α, recombinant interleukin-1α; rIL-2, recombinant interleukin-2

(A) - in solution, (B) - spotted onto PVDF, (C) - blotted onto PVDF.

** 4μg of rIL-2 was loaded onto gel and 0.93μg was blotted to PVDF.

* Sample was carboxymethylated (CM) prior to blotting and Cys was determined as CM-Cys.

+ Cys was detected as cysteic acid after performic acid oxidation of sample on PVDF membrane or in solution.

ND - Not done; NE - Not enough for detection

effectively quantitate the amount of sample present on the PVDF membrane as well as to acquire reliable compositional data.

PVDF membrane containing sample is placed in a hydrolysis tube and 200 µl of 6N HCl containing 4% thioglycolic acid is added. The tube is sealed <u>in vacuo</u> and the sample is hydrolyzed for 16-24 hours at 110°C. The liquid containing the amino acids is then extracted from the tube and the PVDF membrane is rinsed with an additional 50-100 µl of 6N HCl. This rinse is combined with the extracted HCl and dried. The dried sample is ready for amino acid analysis. Cysteine can be determined as cysteic acid by oxidizing the sample immobilized on the PVDF membrane prior to acid hydrolysis following the general procedure described by Hirs (14). Briefly, 50 µl of performic acid is added to the hydrolysis tube containing the immobilized sample and allowed to react for 1 - 2 hours. Then the performic acid is evaporated and the sample is hydrolyzed as described.

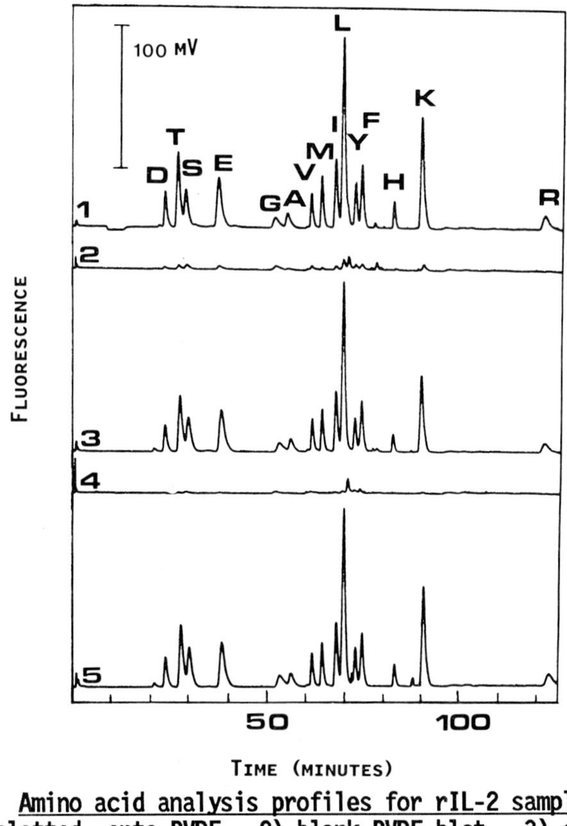

Fig. 3. <u>Amino acid analysis profiles for rIL-2 samples.</u> 1) electroblotted onto PVDF 2) blank PVDF blot 3) spotted onto PVDF 4) blank PVDF 5) in solution. Refer to Table II for composition.

The amino acid compositions acquired from several samples blotted onto PVDF membrane were close to the expected values and are presented in Table II. In addition, the recovery and composition of a protein in solution versus spotted to PVDF, and blotted to PVDF was examined (Table II). All three compositions were close to theoretical values with minimal background (fig. 3). The recovery of amino acids from the sample spotted onto PVDF (.75 μg) is approximately the same as that in solution (.87 μg). The amount in the blotted sample is much less than what was loaded on the gel probably due to sample lost during the blotting. The AAA of proteins on PVDF yields both accurate compositional and reliable quantitative information.

IV. STRATEGIES/CONCLUSION

Several possibilities exist for combining the methods described here along with others in order to obtain structural information of proteins or peptides immobilized to sequencer supports. Some of these possibilities are outlined in the following scheme:

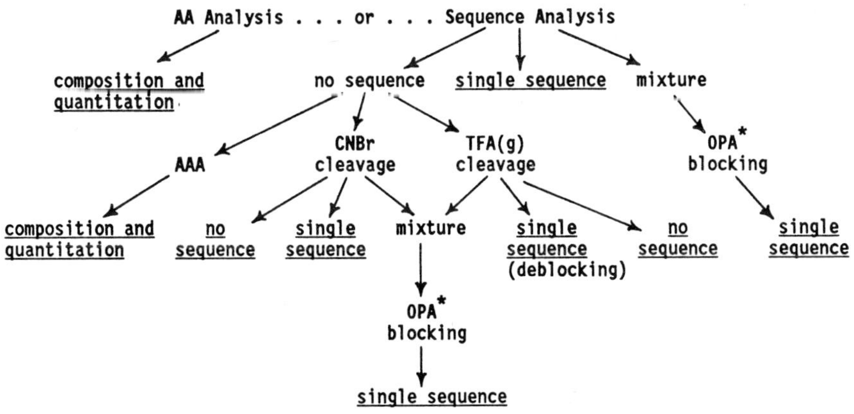

* note: in order to do OPA blocking it is necessary to divide the sample for a preliminary sequence run.

Other methods such as enzymatic digestions of immobilized proteins followed by microbore HPLC mapping (4) may be incorporated into this scheme. With the methods described in this article and the scheme outlined above it should be possible to extract structural data from immobilized protein and peptide samples.

Acknowledgments - The authors wish to thank all colleagues for providing synthetic peptides and recombinant proteins. We also thank Dr. C.-Y. Lai for review of the manuscript and Cathy Michaud for her assistance in its preparation.

V. REFERENCES

1. Aebersold, R.H., Teplow, D.B., Hood, L.E. and Kent, B.H. (1986) J. Biol. Chem. 261, 4229-4238.
2. Matsudaira, P. (1987) J. Biol. Chem. 262, 10035-10038.
3. Xu, Q.-Y. and Shively, J.E. (1988) Anal. Biochem. 170, 19-30.
4. Yuen, S.W., Chui, A.H., Wilson, K.J. and Yuan, P.M. (1988) ABI User Bulletin, Issue No. 36.
5. Hewick, R.M., Hunkapiller, M.W., Hood, L.E. and Dreyer, W.J. (1981) J. Biol. Chem. 256, 7990-7997.
6. Hawke, D., Yaun, P.-M. and Shively, J.E. (1982) Anal. Biochem. 120, 302-311.
7. Pan, Y.-C.E. and Stein, S. (1986) in Methods of Protein Microcharacterization (Shively, J.E., ed.), pp. 105-119, The Humana Press, Clifton, NJ.
8. Simpson, R.J. and Nice, E.C. (1984) Biochem. International 8, 787-791.
9. Mandel, M., Moriyama, Y., Hulmes, J.D., Pan, Y.-C.E., Nelson, H. and Nelson, N. (1988) Proc. Natl. Acad. Sci. USA 85, 5521-5524.
10. Brauer, A.W., Oman, C.L. and Margolies, M.N. (1984) Anal. Biochem. 137, 134-142.
11. Gross, J., Brauer, A.W., Bringhurst, R.F., Corbett, C. and Margolies, M.N. (1985) Proc. Natl. Acad. Sci. USA 82, 5627-5631.
12. Ramachandran, J., Rodriguez, H., Henzel, W. and Tsubokawa, M. (1986) in Methods in Protein Sequence Analysis (Walsh, K.A., Ed.), pp. 157-169, Humana Press, Clifton, NJ.
13. LeGendre, N. and Matsudaira, P. (1988) BioTechniques 6, 154-59.
14. Hirs, C.H.W. (1967) Methods Enzymol. 11, 197-199.

2

MANUAL GAS-PHASE MICROSEQUENCING OF PEPTIDES AND PROTEINS[1]

Mitsuru Haniu
John E. Shively

Department of Immunology
Beckman Research Institute of the City of Hope
Duarte, Ca 91010

I. INTRODUCTION

Since the establishment of automated Edman Chemistry for protein and peptide sequencing (1), significant improvements of the automated instrument have been made by many workers (2-4). Although less appreciated, manual Edman chemistry has played an important role in protein sequencing methodology. Fraenkel-Conrat was the first investigator to perform manual gas-phase Edman chemistry using filter paper (5-6). This method was adapted to automated gas-phase chemistry by Hewick et al (3). Manual protein sequencing by Edman chemistry is still widely used because of its application to handling multiple samples and the low cost of the apparatus (7-8). Disadvantages of liquid phase manual sequencing has been high sample wash-out (low repetitive yields) and high background peaks on HPLC analysis. Sample washout for small peptides was solved by utilizing Polybrene as a carrier (7,9,10) whereas the problem with background peaks remains unresolved. Polybrene has been successfully applied to several analyses in a spinning cup sequencer by Hunkapiller and Hood (2) and in gas-phase sequencing on a glass fiber filter by Hewick et al. (3)and Hawke et al. (4)· Despite the advantages of automated gas-phase sequencing, instruments are costly and can process only one sample at a time. The cycle time for these instruments ranges from 40-70 min (3, 4), but can be as short as 10-15 min for manual sequence analysis (11). Because of the potential advantages of manual sequence analysis, we have recently begun to apply more modern approaches to its improvement. Our first attempt was to apply the gas-phase Edman reaction to the manual method (12); however, this method used small glass fiber filter tips which give high sample washout presumably due to the low mass of supporting material. Recently, we have reported the use of a continuous flow reactor (CFR) containing Polybrene-coated silica gel for automated sequencing (13). These studies encouraged us to apply the CFR to manual gas-phase sequencing.

[1]This research was partly supported by the National Institutes of Health grants GM34426 and CA33572 from the National Cancer Institute to the City of Hope Cancer Center.

II. SEQUENCING METHODS

A. Apparatus.

A schematic of the manual sequencing apparatus is shown in Figure 1. Key components include the reagent-gas deliver system, the continuous flow reactors (CFR), a heating unit, and sample collector. The CFR is constructed from a 2 cm length of Teflon tubing (0.125 in ID or 0.0625 in. ID) containing Polybrene-coated spherical, porous silica gel (100-200 μm, particle size), as described by Shively et al. (13). The CFR is tightly connected to Teflon tubing (0.125 in. OD) at the bottom, and the top of the CFR is connected to a pipet tip (CY-3, Cole Scientific). Argon gas is distributed to several vials or bottles, including bottle 1 (argon), R2 (5% TEA), R3 (anhydrous TFA), and bottle 5 (argon). The compositions of the reagents and solvents are essentially the same as described by Hawke et al. (4). The reagent delivery is controlled by 3-way valves (Hamilton Valve, part No. 86777, Nevada). For six sample analyses, twelve valves are necessary. The pressure of the argon gas is adjusted to 3-5 psi and a flow rate 2ml/min. The CFR can be easily disconnected by a special adaptor as shown in Figure 2. The TFA adaptor must be released during the coupling reaction. Adaptors for R2 (TEA) and R3(TFA) are attached with individual argon lines. Bottles for Ar gas and for TFA were constructed from 5 ml glass vials (Pierce Chemical Co.) with Teflon cap seals. The 0.0625 in. OD teflon tubing is inserted into the vials through a "pull-through" fitting (13). Reagent 2 for 5% TEA-water is in a 100 ml reagent bottle (Pyrex glass bottle with Teflon cap seal). Bottle 2 is a trap to eliminate liquid TFA from the gas phase. Solvent washes are performed with a 1 ml Hamilton syringe fitted with a shortened pipet tip (made of polypropylene, length; 2 cm). The heating block is prepared from an aluminum block which fits a "Reacti-Thermo" heating block (Pierce Chemical Co.). Six CFRs can be set in the heating block (50° C) which has six separate holes of 5 mm diameter. The heating block was tilted at 90°or a vertical arrangement of CFRs. The ATZ amino acid derivatives are collected into 250 μl polypropylene microcentrifuge tubes.

B. Sample Application .

The Teflon tubing adaptor (pipet tip 1)on the CFR is disconnected (Figure 2) and the sample (3-5 μl) is applied in 25% TFA-1-propanol (1:1, v/v) or 10% TEA-1-propanol (1:1, v/v) using a Hamilton syringe. The CFR is connected to the original adaptor, and is dried with argon for 5 min. An alternative method of sample application is also used: the sample is injected directly onto silica, followed by adding a layer of new silica (2-3 mm thickness) and a Zitex disk.

C. Operation for Edman Reaction.

The operational steps for the gas-phase Edman reaction are shown in Table I. Although single coupling and single cleavage reactions are used for peptide samples, double coupling and double cleavage reactions are recommended for hydrophobic proteins because of difficulty in solubility. The cycle time for single coupling and single cleavage is 15-20 min, and for double coupling and double cleavage it is 40 min. A 5 μl aliquot of PITC solution is injected into the CFR with Hamilton syringe A (25 μl) by disconnecting the adaptor. The adaptor is

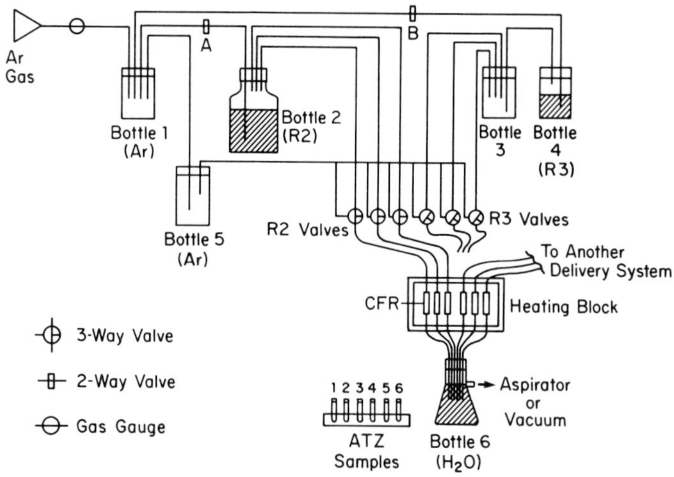

Figure 1. Overall schematic of manual gas-phase sequencing for multiple samples. The flow rate of argon gas (3-5 psi) was adjusted to 2 ml/min by inlet valve. Each CFR is connected to either the TEA-or TFA-gas delivery system. For six sample analyses, a second identical delivery system is required (not shown).

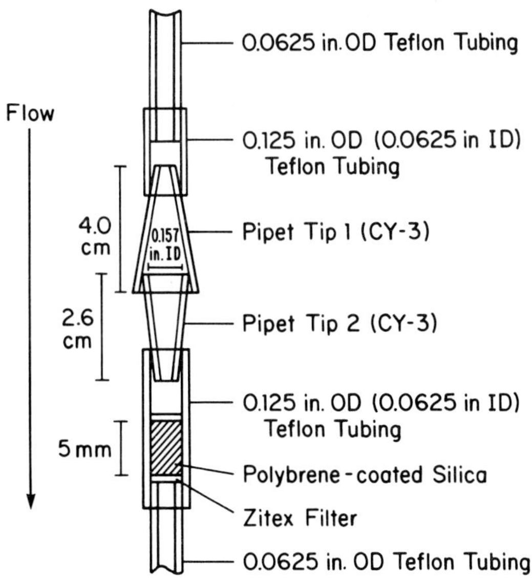

Figure 2. Diagram of a silica continuous flow reactor (CFR) and adaptor. A Polybrene-coated silica is supported by two Zitex filter disks in the Teflon tube which is connected to the pipet tip cut to fit.

immediately reconnected, and the CFR flushed with argon gas (2 ml/min) for 30 seconds. For the coupling reaction, a 2-way valve (valve A) is initially opened in order to pressurize the TEA bottle (bottle 3, Figure 1) for 5 seconds, then the TEA valve is opened to allow the flow of TEA gas for 5-10 minutes (2 ml/min). After the coupling reaction, valve A is closed, and the TEA valve is switched to Ar gas to dry the sample for 1 min. The solvent wash (0.5 ml) is performed with ethyl acetate (14) at 50° C by manually pressurizing with a Hamilton syringe (syringe B) over 1 min. The CFR is dried with argon for 0.5 min. For the cleavage reaction, the TFA vial (bottle 4) is pressurized by opening the 2-way valve (valve B). The TFA valve is opened for gas delivery for 3 min. To dry the CFR, valve B is closed, followed by opening the Ar gas line attached to the same valve (TFA valve). The drying step requires at least 1 min with a flow rate of 2 ml/min. The ATZ extraction is repeatedly performed with 100 μl of BuCl at 50°C. The ATZ residue is delivered into a microcentrifuge tube (250 μl) by pressurizing the CFR with a slow flow rate of argon. A second extraction with 100 μl of BuCl is necessary to completely remove the ATZ from the CFR.

TABLE 1. PROCEDURE FOR MANUAL EDMAN REACTI ONS

STEP	FUNCTION	OPERATION TIME	DEVICE
1. Sample	Dry	5 Min	Syringe A
2. Coupling	PITC	-	Syringe A
3. Coupling	Dry	0.5 Min	
4. Coupling	TEA gas	5-10 Min	
5. Coupling	Ethyl acetate Wash	1 Min	Syringe B
7. Coupling	Dry	0.5 Min	
8. Cleavage	TFA gas	3 Min	
9. Cleavage	Dry	1 Min	
10.Extraction	BuCl	1 Min	Syringe B
11.Extraction	BuCl	1 Min	Syringe B
12.Extraction	Dry	1 Min	
TOTAL TIME PER CYCLE		15-20 MIN	

D. **Manual Conversion Procedure and PTH Analysis.**

The ATZ-BuCl solution is dried in a vacuum centrifuge for 5 min, followed by conversion at 65^{0} C for 10 min in 20 μl of 25% TFA containing DTT (2-3 mg/1) under argon gas. The system used for PTH analysis has been published previously from our laboratory (13). An isocratic HPLC analysis was also used as described previously (15).

III. RESULTS AND DISCUSSION

A. Sequence Analyses of Peptides.

Typical chromatograms for the sequence analysis of 99 pmol of the small peptide (F-N-G-L-R-G-CONH2) are shown in Figure 3. The sample was applied to CFRs containing Polybrene-coated porous spherical silica. In the analysis of 99 pmol of sample, the initial yield was 73%, and the repetitive yield 89%. The major background peaks were the PTC derivative of diethylamine (PTU) and diphenylthiourea (DPTU). A small quantity of PTH-Gly was often observed as a background peak. The size of background peaks was controlled when the optimal flow rate of the ethyl acetate wash which was 0.5 ml/min or less. The sequence analyses for a second known peptide (G-V-H-L-A-A-F-E-E-E-R-G-G) were performed. The initial yields for analysis af the 90 and 230 pmol level were 65% and 54%, and for the repetitive yields 96% and 89%, respectively. A comparison of these results to the other manual sequencing using liquid phase chemistry (11) demonstrated an improvement in the repetitive yields and a slight decrease in the initial yields.

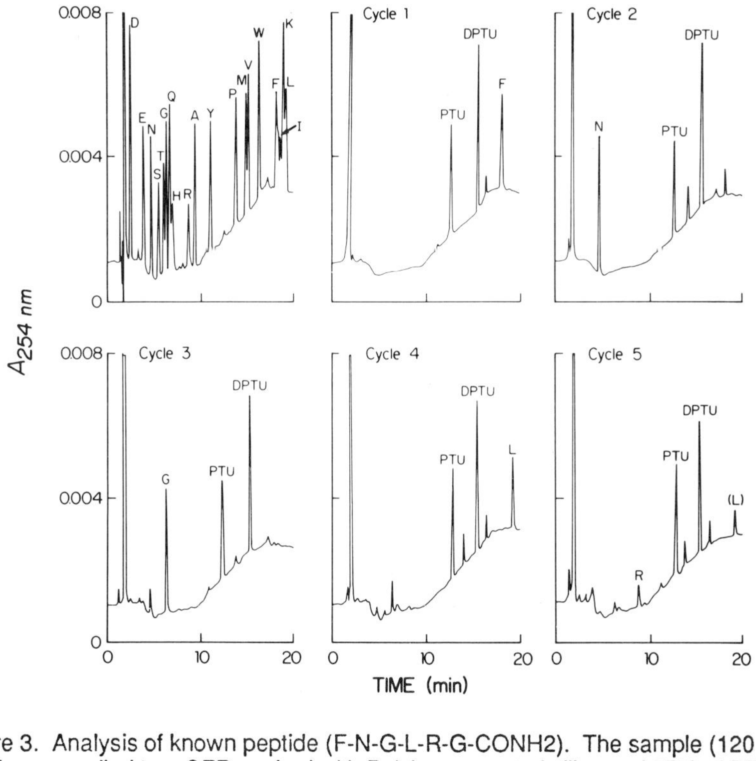

Figure 3. Analysis of known peptide (F-N-G-L-R-G-CONH2). The sample (120 pmol) was applied to a CFR packed with Polybrene coated silica and 35% of PTH at each cycle was analyzed by HPLC. Two major background peaks (PTU and DPTU) were observed eluting near at PTH-Trp and PTH-Val. These peaks could be reduced by carefully extending the ethyl acetate wash.

B. Sequence Analysis of Myoglobin.

For myoglobin analysis, a 10% PITC solution was used, followed by double coupling and double cleavage reactions. Thus, the total cycle time was 40 min. These procedures are required particularly for insoluble membrane proteins. PTH amino acids were analyzed by the isocratic HPLC method. The initial yield for analysis of the myoglobin was 72% and the repetitive yield from cycle 1 to cycle 10 was 90%. Thirteen cycles of myoglobin were analyzed by HPLC as shown in Figure 4. The sequence analysis of the protein generally gave a slightly higher background compared with the peptide analysis. The low recovery of PTH-Trp observed was probably due to oxidation.

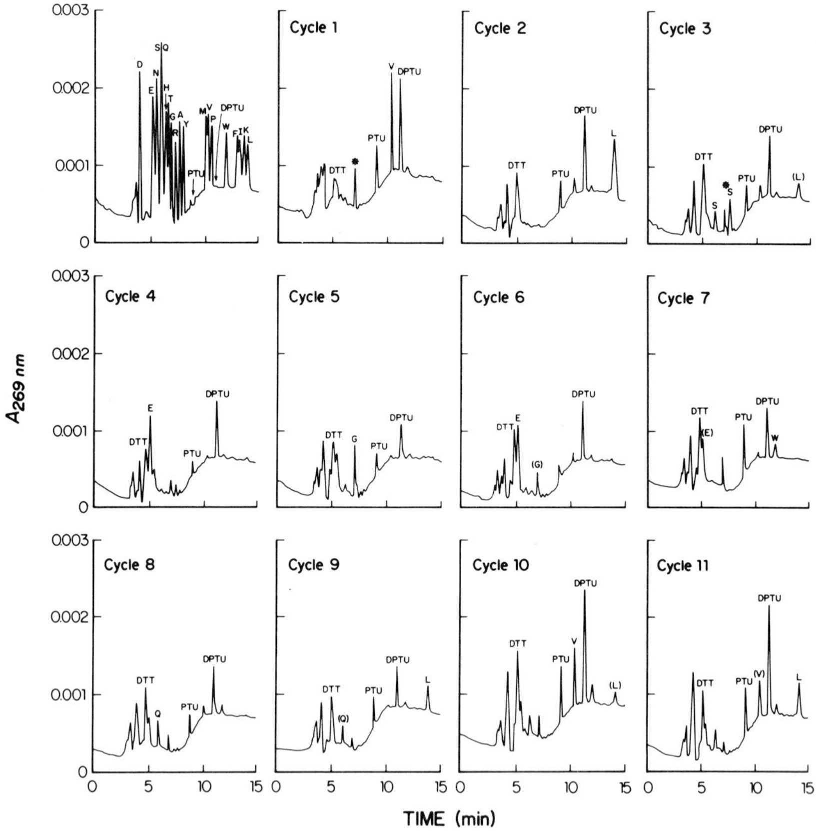

Figure 4. Sequence analysis of myoglobin.
Myoglobin (180 pmol) was analyzed on a CFR packed with Polybrene-coated silica. PTH amino acids were analyzed by the isocratic method. Two major backgrounds (PTU and DPTU) are indicated. The asterisk denotes the trace of PTH-Gly at cycle 1 and the DTT adduct of PTH-Ser at cycle 3.

IV. SUMMARY AND CONCLUSION

The manual Edman method described here can be used to determine the sequences of samples for which a small quantity of peptides and proteins (100-200 pmol) are available. Compared to other manual methods (9, 10), the sensitivity of this analysis was clearly improved. This method was simple, requiring less skillful sample manipulation and reaction procedures. The sample wash-out problem often encountered in manual Edman Chemistry was diminished by the gas-phase chemistry. The advantages of this system over previous manual methods are the capability for multiple sample analyses (six to ten samples) and the low cost of materials.

V. REFERENCES

1. Edman, P., and Begg, G. (1967) Eur. J. Biochem. 1, 80-91.

2. Hunkapiller, M. and Hood, L.E. (1978) Biochemistry, 17, 2124- 2133.

3. Hewick, R.M., Hunkapillar, M.W., Hood, L.E., and Dryer, W. J. (1981) J. Biol. Chem., 256, 7990-7997.

4. Hawke, D.H., Harris, D.C., and Shively, J.E. (1985) Anal. Biochem. 147, 315-330.

5. Fraenkel-Conrat (1954) J. Amer. Chem. Soc., 76, 3606-3607.

6. Schroeder, W.A. (1967) Method Enzymol., 11, 298-313.

7. Tarr, G.E., Beecher, J.F., Bell, M., and McKean, D.J. (1978) Anal. Biochem., 84, 622-627.

8. Chang, J.Y. (1983) Method Enzymol., 91, 455-466.

9. Klapper, D.G., Wilde, C.E., III., and Capra, J.D. (1978) Anal. Biochem., 85, 126-131.

10. Tarr, G.E. (1977) Methods Enzymol., 47, 335-357.

11. Tarr, G.E. (1986) in 'Methods of Protein Microcharacterization. (Shively, J.E., ed) pp. 115-193. Humana Press, NJ.

12. Mehl, F., Haniu, M., and Shively, J.E. (1985) Symposium of American Protein Chemists (San Diego, CA) abstract 112.

13. Shively, J.E., Miller, P., and Ronk, M. (1987) Anal. Biochem., 163, 517-529.

14. Shively, J.E., Hawke, D., and Jones, B.N. (1982) Anal. Biochem., 120, 312-322.

15. Tarr, G.E. (1981) Anal. Biochem., 111, 27-32

3

MICROSEQUENCING WITH PVDF MEMBRANES: EFFICIENT ELECTROBLOTTING, DIRECT PROTEIN ADSORPTION AND SEQUENCER PROGRAM MODIFICATIONS[1]

David W. Speicher

Wistar Institute
Philadelphia, PA 19104

I. INTRODUCTION

Substantial advances in automated Edman microsequencing over the past ten years have continually increased sensitivity. Currently, many protein chemistry laboratories can routinely obtain useful sequence information on 10 pmol or less of a protein or peptide. While further advances in instrumentation or chemistry may increase sensitivity even further, a frequent limitation for many projects is the inability to purify low picomole quantities of protein in a form suitable for sequencing.

The high resolving power of 1D and 2D polyacrylamide gels on a micro scale is well known, but difficulties with electroelution of low picomole quantities of protein have previously limited the value of this approach. Recently, many of these problems were minimized by development of glass fiber based membranes suitable for electroblotting of proteins from gels followed by direct sequencing of the adsorbed protein (1,2). Unfortunately, the performance of these membranes varied widely between laboratories. More recently, polyvinylidene difluoride (PVDF) membranes have been used for electroblotting followed by direct sequencing (3,4). This membrane has rapidly emerged as the preferred substrate for electroblotting/sequencing and improvements to the original methods have recently been reported (5,6).

In this report several problems commonly encountered with electroblotting and sequencing using PVDF membranes are examined including: 1) quantitative electrotransfer of proteins with a wide range of molecular weights at reasonable pH's, 2) direct adsorption of proteins in solution to PVDF as an alternate sample loading/ sample cleanup procedure,

---

[1]Supported in part by National Institutes of Health grant HL38794.

3) sequencer program changes which decrease variability without the need for polybrene or glass filters, 4) modification of stain/destain solutions to increase sequence yield.

II. METHODS

A. Electrophoresis and Electroblotting

Electrophoresis reagents were purchased from Bio-Rad and were used without further purification. Standard sized slab gels (12 cm x 16 cm x 1.5 mm) were prepared using the buffer system described by Laemmli (7) with modifications described by Hunkapiller and Hood (8). Multiple separating gel concentrations were tested in different experiments and ranged from as low as 5% to 15% depending upon the molecular weights of sample proteins. Also, in some experiments gradient gels were utilized to accommodate a wider size range. Specific precautions used were: the entire gel, including stacking gel, was cast 1-3 days before use, sucrose was used in the sample buffer instead of urea, samples were generally heated at 37°C for 10 min before loading, and 0.1 mM thioglycolate (Sigma) was used in the top electrode buffer. Electrophoresis was performed either overnight at about 80 V (r.t.) or at 350 V with cooling to 12°C (Bio-Rad Protean II unit) for approximately 3 hr.

Two-dimensional gels were prepared as described by O'Farrell (9) with modifications as previously described (10). Generally the second dimension was a 10% to 15% Laemmli gradient employing precautions described above. In some experiments the relative amounts of ampholines were adjusted to produce different pH ranges appropriate to different samples.

Electroblotting was performed using a transfer apparatus (Model G201 purchased from the Yale University Medical Instrumentation Facility) which produced a voltage gradient between the top and bottom of the gel as described by Gershoni et al. (11). In later experiments this device was modified by using larger capacity resistors (2 watts) to permit higher currents and faster transfer speeds. Tris-glycine buffers as described by Towbin (12) or 3'-[cyclohexylamino]-1-propanesulfonic acid (CAPS) buffer described by Matsudaira (3) were used. Different methanol concentrations were tested and generally 10% methanol was used. PVDF membranes were wetted briefly with methanol prior to preparing the blotting sandwich. Two sheets of PVDF were used in most experiments; protein passing through the first sheet was at least partially retained on the second sheet.

B. Direct Adsorption of Proteins to PVDF Membranes

Conditions for direct adsorption of proteins to PVDF were explored as an alternative to direct spotting of solution samples. For most experiments, either ß-lactoglobulin A (Applied Biosystems, Inc.) or horse apomyoglobin (Sigma) was incubated at 4°C in a siliconized glass test tube with a 5 mm x

9 mm piece of PVDF membrane that was prewetted by immersion in methanol. In some experiments constant agitation with a rotary shaker (Tekmar, VXR-S10) at 200 rpm was used. The quantity of protein bound was calculated based on duplicate amino acid analyses of the unbound supernatant after incubation. For most subsequent experiments, standard conditions were: 4°C, agitation, 24 hr., 200 μl final volume of 10 mM sodium phosphate, pH 7.5, 5 mm x 9 mm PVDF.

C. Sequence Analysis

An Applied Biosystems Model 470A gas phase sequencer with online PTH amino acid analysis (Model 120A) and a Nelson Analytical data system were used. Several changes to the standard Applied Biosystems sequencer programs and reagents were used for all sequence runs. Specifically, a 2.1 x 250 mm Supelcosil C18-DB column was used on the PTH analyzer and the terminal portion of the gradient was modified to resolve DPU from Trp. S4 was prepared from Baker acetonitrile and MilliQ water and PTH-Nleu was added to a final concentration of 50 pmol/120 μl; a fresh solution was prepared biweekly. Other changes that were developed specifically for gas phase sequencing of samples on PVDF membranes, without the need for glass fiber filters, are described in RESULTS.

III. RESULTS

A. Efficiency of Electrotransfer onto PVDF membranes

Ideally, electrotransfer for sequencing would result in quantitative transfer of all polypeptides without chemical modification of the sample and without retention of contaminants that could interfere with sequence analysis. The most persistent problem is the nonuniform transfer rate based on size which results in either incomplete transfer of large peptides or loss of smaller peptides from the blot on extended transfers. Two alternate solutions to this problem would be to either irreversibly bind the protein to the membrane or to produce a stronger electrophoretic field at the top of the gel to speed the elution of higher molecular weight components. This latter approach has been previously described (11) and was further evaluated here.

Initial experiments used an unmodified gradient voltage transfer apparatus which gave a 4:1 voltage ratio (top-to-bottom; five independent electrodes) in high ionic strength buffers such as Tris-glycine (11). Since the electrodes were spread over a height of about 26 cm and standard sized separating gels were only about 10 to 14 cm high, an actual 2:1 voltage gradient between the top and bottom of the gel was available. This gradient was clearly beneficial as indicated by an increased recovery of bands from the top third of the gel on the primary membrane, although a steeper voltage gradient would be preferred. Another limitation of the blotting apparatus was that it was designed to operate only at low power. We were able to increase the current capacity, and hence the speed of transfer, simply by increasing resistor capacities. Clearly the concept of voltage gradients is a promising approach to quantitative transfer, but this device only partially solved the problem. Further modifications are being investigated.

Tris-glycine buffers at 1X (25 mM Tris, 192 mM glycine, pH 8.3), 0.5X and 0.25X concentrations as well as 10 mM CAPS, pH 10 and 11, were tested for efficiency of transfer using a wide range of protein standards and experimental proteins/peptides. Both types of buffers produced reasonable transfers if transfer times were empirically optimized. No dramatic overall differences in uniformity of transfer or recovery of sequencable protein were noted between the two buffers; although specific proteins occasionally exhibited either poor elution out of the gel or poorer retention on the primary membrane in one of the buffer systems. For example, ß-lactoglobulin A had a higher affinity for PVDF in CAPS than in Tris-glycine buffers, but this seemed to be an exception rather than a general trend, i.e. some proteins showed higher retention with Tris-glycine buffers. Transfer times were always shorter for CAPS buffer apparently because it was a low ionic strength buffer and transfer can be performed at higher voltages. Similarly, lower ionic strength Tris-glycine buffers could be used at higher voltages without overheating, resulting in shorter transfer times. In later experiments, either 0.5X or 0.25X Tris-glycine was routinely used with optimal transfer times for most proteins between 2 to 4 hr depending upon gel concentration. Using these conditions and the modified blotting apparatus, transfer of proteins to the primary membrane was routinely nearly quantitative for most proteins, except for proteins in the top 15% of the gel which were generally >75% transferred.

Speed of transfer from the gel and affinity to the membrane were also dependent upon the methanol concentration. Concentrations between 0 and 20% were tested as well as the effect of adding low levels of SDS (0.01% or 0.001%) to the transfer buffer. For most proteins, addition of SDS or methanol concentrations of less than 10% decreased binding to PVDF; proteins in the lower half of the gel were recovered in low yield on either the primary or secondary membrane because most of the protein transferred through both membranes into the buffer. However, we have encountered at least four membrane proteins with apparently low affinity for SDS. These proteins were either not transferred at all or were only partially transferred using 10% methanol. Nearly quantitative transfer was obtained for three of these proteins by decreasing the methanol to 0% and maintaining the normal transfer time. However as indicated above, the standard proteins on this gel were recovered in poor yield. Therefore, 0.5X Tris-glycine buffer with 10% methanol has become the routine method and the occasional protein that remains in the gel is recovered in subsequent experiments by transfer without methanol.

Methanol can have multiple effects during protein transfer. It favors dissociation of SDS from the protein, it causes the gel matrix to shrink, it may decrease solubility of the protein in the transfer buffer, and it could decrease affinity of the protein for the hydrophobic PVDF membrane. In the 0 to 20% range normally used in transfers the major effect of methanol appears to be upon the dissociation of SDS from the protein. Since bound SDS provides the driving force for transfer, premature removal of SDS may cause the protein to remain in the gel while too tight a SDS-protein association will favor continued migration of the protein from the primary membrane onto the secondary membrane and eventually into the buffer.

B. Direct Adsorption of Protein to PVDF Membranes

Purified proteins including sequencer standards have been loaded on prewetted PVDF membranes by direct spotting in small volumes and subsequently sequenced with good yield (3). However, initial experiments in our laboratory suggested that protein binding to the membrane was quite variable using this method. Specifically, initial coupling and repetitive yields ranged widely for duplicate samples if sequenced without a polybrene coated filter. In addition, it was noted that duplicate spots on a single strip of membrane would occasionally adsorb incompletely since spotted standards sometimes exhibited streaking when the strip was stained with Coomassie blue. This variability suggested that the protein was sometimes incompletely adsorbed, with part of the sample dried on the membrane, but not tightly bound. The tendency toward incomplete binding was greatest at higher protein loads although the highest loads used, about 20 $\mu g/cm^2$ were well below the capacity reported by the manufacturer. A likely explanation for these observations is that protein adsorption may take longer than the drying time of the limited volume of solution that can be applied to a small area of PVDF membrane. If adsorption is not sufficiently rapid, the degree of adsorption may be greatly influenced by subtle differences in solvent composition, humidity, and protein concentration.

A number of direct adsorption experiments were performed to test these possibilities. Results are summarized in Table I. In the absence of agitation, adsorption was very slow requiring on the order of 24 hr to reach maximum binding. Also, the amount adsorbed after 24 hr decreased as the volume increased. In contrast, much more rapid adsorption occurred with agitation and the degree of adsorption was relatively independent of total sample volume. These results imply that diffusion of protein to the membrane is limiting in static samples although it is surprising that diffusion in volumes as small as 200 μl would be so slow.

It is important to note that even with agitation, binding was not complete at 1 hr at 4°C. Therefore, as hypothesized above, binding to PVDF membranes takes a significant amount of time and complete adsorption may not always occur for spotted samples before the solution dries, possibly trapping some of the protein in an unbound state. This unbound protein could be extracted during initial sequence cycles unless a polybrene coated glass fiber is placed below the membrane. In contrast, samples loaded on PVDF by solution adsorption consistently gave high, consistent initial couplings and repetitive yields as indicated in Table II without the need for a polybrene coated filter.

TABLE I. Direct adsorption of apomyoglobin to PVDF.

Experiment	Protein (pmol)	Sample Vol(ul)	Incubation Time(hr)	Agitate	Percent Bound
Time course	300	200	1	no	47
	300	200	2	no	62
	300	200	4	no	65
	300	200	24	no	91
	300	200	1	yes	90
	300	200	2	yes	88
	300	200	4	yes	93
	300	200	24	yes	98
Volume effect	300	200	24	no	91
	300	1000	24	no	62
	300	200	24	yes	98
	300	1000	24	yes	92
	300	2000	24	yes	91
Total protein	300	200	24	no	91
	600	200	24	no	85
	1200	200	24	no	74
Buffers/Detergents					
Isotonic KCl	300	200	24	yes	98
PBS	300	200	24	yes	98
200mM Tris,pH 8	300	200	24	yes	84
0.01% SDS	300	200	24	yes	98
0.1% SDS	300	200	24	yes	30
0.1% Octylglucoside	300	200	24	yes	99

All incubations were at 4°C in siliconized test tubes. PBS, phosphate buffered saline, pH 7.4. The percent bound was estimated based on duplicate amino acid analysis of the supernatant solution after incubation.

Direct adsorption is also an appropriate method for loading large amounts of protein onto PVDF for sequencing. As indicated in Table I, at least a nmole of horse apomyoglobin can be adsorbed with high efficiency. In addition, the protein was tightly adsorbed at this higher loading since the 1.2 nmole sample indicated in Table I sequenced with high initial coupling and repetitive yield (Table II). Similarly a number of common buffers including high concentrations of Tris did not interfere with adsorption or sequence results at least at neutral pH. Not surprisingly, high amounts of detergents tended to interfere with adsorption although moderate quantities (0.01% SDS or 0.1% octylglucoside) did not significantly deter adsorption (Table I) or sequencer performance (Table II).

TABLE II. Sequence results for standard proteins and typical high sensitivity experimental samples from 1D and 2D gels.

A. Horse apomyoglobin adsorbed to PVDF for 24 hours as indicated in Table I. Standard conditions are defined in METHODS.

Adsorption conditions	Init. Coupl.	Repet. yield
standard conditions, 300 pmol	60	93.6
"	48	92.6
"	58	93.1
standard conditions, 1200 pmol	77	94.6
Isotonic KCl, 300 pmol	62	93.0
Tris, 200 mM, 300 pmol	60	93.6
Octylglucoside, 0.1%, 300 pmol	66	93.0
None, glass filter with polybrene	62	91.4

B. Experimental samples from 1D and 2D gels.

Samples	M.W.	Init. Coupl.	Repet. yield	# of cycles
2D-spectrin BIV	T48K	18 pmol	92.0	19/19+
2D-spectrin AII	T25K	16 pmol	90.7	20/21+
2D-spectrin AII	T23K	2.5 pmol	n.d.	22/23
2D-ankyrin 30b	C30K	18 pmol	92.8	29/29
2D-ankyrin 90	C90K	3.3 pmol	94.4	16/16
1D-ankyrin	220K	8 pmol	92.5	20/20
1D-Protein 4.1	80K	9 pmol	91.3	21/22

The letter immediately before the molecular weight designates the type of enzyme, trypsin or chymotrypsin, used to prepare the indicated domain peptide. + indicates that the sequence could have been continued beyond the number of cycles run.

C. Modifications to the Gas Phase Sequencer

1. Identification of the source of variation

As noted above, a major problem associated with the use of PVDF membranes as sequencer supports is the large cycle-to-cycle variation in PTH amino acid recovery. This problem could be minimized by placing a glass fiber filter either above or below the PVDF membrane. However, if the glass fiber is not coated with polybrene, His and Arg yields are usually reduced to undetectable levels, especially when sequencing <10 picomoles. Since polybrene and/or glass fibers are not needed to retain well-adsorbed protein on PVDF membranes during gas phase sequencing, our objective was to reduce the variation without resorting to a glass filter or polybrene.

In initial experiments, PTH-Nleu was added to S4 to monitor recovery of sample from the flask stage forward. Surprisingly, PTH-Nleu was very rapidly degraded in the lot of S4 from Applied Biosystems in use at that time (approximately 1% loss per hour) with the appearance of a new peak eluting slightly before Pro. When 20% acetonitrile was prepared using Baker HPLC grade acetonitrile and MilliQ water, PTH-Nleu was substantially more stable (about 1% loss per day) and this manually formulated S4 with an internal standard was used for all subsequent sequences. It was immediately obvious that Nleu recoveries were excellent and that the variation associated with sequencing on PVDF was associated with events in the cartridge rather than in the flask.

For some sequences the yield of PTH's would suddenly drop or jump to a higher level. Sequences that showed great variability did not exhibit any greater carryover than sequences with more stable yields implying that extractions, rather than coupling or cleavage reactions, were responsible for the observed fluctuations. It was subsequently confirmed that the major cause of cycle-to-cycle variation in the gas phase sequencer was nonreproducible flow patterns during solvent deliveries, especially S3 and S2. The wettability of dry PVDF membranes was observed to be S1 > S2 > S3 and this resistance to wetting coupled with the poor porosity of PVDF and its hydrophobic nature tended to produce a random flow pattern of solvent that could sometimes miss the membrane or poorly wet it. Solvent flow patterns were further distorted by argon bubbles that would sometimes form and grow above the membrane. The likelihood of gas bubble accumulation above the membrane was the inverse of the wettability, i.e. S3 was the worst. Argon bubbles, once formed above the membrane, tended to remain in place for the rest of the delivery. Even when bubbles did not get trapped above the membrane, S3 and S2 would sometimes flow exclusively down one side of the chamber and could miss the sample, especially very small pieces such as those typically obtained from 2D gels. Consistent with this idea, samples on a single small piece of membrane (about 2 x 3 mm) showed the greatest variation. Additional variation could be caused by membrane pieces stacking up on each other followed by poor flow between the membrane pieces.

2. Sample loading for best results

The optimal sample arrangement is to load membrane so that the flow path is not obstructed and with minimal membrane overlap. The ideal membrane size is about 4 x 9 mm which is approximately the size of a band on a standard sized gel. Two pieces of this size can be loaded side-by-side by trimming opposing edges to fit the circle of the cartridge. Before loading into the cartridge, the edges are feathered by cutting the membrane part way thru every 1 to 2 mm being careful to not cut entirely through the entire width.

A second teflon seal above the PVDF membrane may also prevent extensive movement or stacking of multiple membrane pieces. Initial experiments to test the value of this approach were inconclusive. Certainly addition of the second membrane was not detrimental but it was not clearly beneficial.

3. Sequencer modifications

After identifying the variability problem as variable extraction with S3 and to a lesser extent S2, we initially changed the major S3 delivery (50 sec) to four 20 sec deliveries interspersed with 20 sec pauses. This change alone dramatically decreased the frequency and severity of yield variation, but it did not totally eliminate the problem. Further improvement was achieved by adding a 2 sec transfer with argon immediately before transfers with S3; this cleared the cartridge and line and displaced any accumulated trapped bubbles. **It is critical that argon transfer be kept to 2 sec** since longer times dry the membrane and dramatically decrease extraction. By keeping deliveries short and interspersing deliveries with pauses and argon delivery, the tendency to form and trap gas bubbles with subsequent channelling of solvent around the membrane was minimized. The cartridge was also inverted for PVDF runs to decrease the space above the membrane, thereby minimizing the size of argon bubbles that could form. Although variability in S2 extraction (indicated by reagent peak sizes) was less significant than S3, we found that multiple short deliveries of S2 interspersed with 2 sec argon dry steps dramatically increased extraction efficiency allowing substantial reduction in the total volume of S2 needed (reduced to a total of 78 sec compared with 240 sec in the standard program). Finally, a 2 sec S2 delivery was added before S1 delivery to purge argon trapped in the E block between the S1 and S2 valves. This "slug" of argon was especially troublesome during S2 delivery if it was not eliminated. The current sequencer program is described in Table III.

The modified program also performed well with polybrene coated glass filters and preliminary evidence suggests that the multiple S3 extractions may especially increase Arg recovery from both PVDF membranes and from glass filters as well.

TABLE III. The modified program - The standard cycle: 03CPTH was replaced with the cycle: 01CPVD listed below.

STEP	CARTRIDGE	FLASK	TIME	STEP	CARTRIDGE	FLASK	TIME
Steps 1-15 are unmodified				51	trans w/S3	pause	25
16	deliver S2	pause	3	52	pause	argon dry	10
17	deliver S1	argon dry	30	53	trans w/Ar	pause	2
18	deliver S2	argon dry	25	54	trans w/S3	pause	25
19	argon dry	argon dry	2	55	pause	argon dry	10
20	deliver S2	argon dry	25	56	trans w/Ar	pause	2
21	argon dry	argon dry	2	57	trans w/S3	pause	25
22	deliver S2	argon dry	25	58	pause	argon dry	10
23	argon dry	argon dry	375	59	trans w/Ar	pause	30
Steps 24-47 = original 19-42				60	deliver S3	argon dry	30
48	trans w/S3	pause	15	61	argon dry	argon dry	120
49	pause	argon dry	10				
50	trans w/Ar	pause	2		CYCLE TIME: 53 min 49 sec		

D. Losses upon staining and destaining

A moderate, but significant loss in sequencable protein was observed when adsorbed standard proteins were stained with Coomassie blue as compared with unstained samples. Since the most successful methods of eluting proteins from PVDF have proven to be aqueous acid/organic solvent mixtures, the possibility exists for significant protein losses during staining and destaining if either the stain or the destain contains commonly used 10% acetic acid/50% methanol mixtures.

To identify the source of losses on staining, a large number of duplicate samples of horse apomyoglobin were adsorbed to PVDF as described above. Since a known, consistent quantity of protein was tightly adsorbed to duplicate membranes, effects of staining could be reliably compared to unstained samples. Also, to accentuate any potential deleterious effects, incubation times for staining or destaining were increased from the normal 5 min to 30 min in most experiments in this series. With various conditions of staining and destaining, no significant change in background, repetitive yield or carryover were observed, but decreased initial coupling was noted in several cases compared to unstained samples. The results can be summarized as follows: 50% methanol alone had no effect, 10% acetic acid alone showed a slight reduction in initial yield, and 50% methanol with 10% acetic acid showed a dramatic reduction (30 - 50% decrease). Presence or absence of Coomassie blue appeared to have no effect. Therefore, the following staining and destaining solutions were tested and subsequently used routinely since no decrease in initial coupling was observed with this solution: stain - 0.5% Coomassie blue in 40% methanol; destain - 50% methanol. The staining sensitivity with these solutions was similar to that obtained with solutions containing acetic acid but stained bands were a lighter shade of blue. The only disadvantage appears to be poorer quality photographs due to the decreased contrast.

IV. DISCUSSION

In summary, Tris-glycine transfer buffers were preferred over CAPS largely since there were no clear advantages to CAPS buffers and the high pH of CAPS buffers could result in cyclization of amino-terminal glutamines or deamination of sensitive amines. The presence of high concentrations of Tris and glycine was not a problem as long as the membrane was extensively washed with water prior to staining - typically <5 pmol of glycine background was observed in cycle 1. No special problems were encountered in transfers from 2D gels. Elution from the gel and affinity to the PVDF membrane vary somewhat for individual proteins; while the majority of proteins behave uniformly, an occasional protein is encountered whose behavior is dependent upon pH, methanol concentration, or transfer time. With the exception of the rare protein that does not elute from the gel under normal conditions, most other nonideal behavior could be minimized by using a transfer apparatus with a higher field strength at the top of the gel. Alternatively, precoating the membrane with polybrene was recently reported to increase protein affinity to PVDF during transfer (5) and this approach deserves further investigation.

Direct adsorption by incubating protein solutions with a piece of PVDF membrane overnight in the cold room is superior to direct spotting. This method could also be used as a final cleanup step for purified proteins that are presented for sequencing in large volumes and/or in buffers that interfere with sequencing. The adsorption rates observed at reduced temperature suggest that direct spotting of small volume samples may not always give significant time for complete adsorption before the sample dries. The relatively slow speed of adsorption and the dramatic change in speed of adsorption with agitation using 200 μl volumes were somewhat surprising. While the interaction of proteins with PVDF is generally regarded to be essentially a hydrophobic interaction, the actual mechanism is not known. Tight adsorption may actually occur through multiple steps. Since most hydrophobic residues are buried in native proteins, initial adsorption may be through limited interactions with moderately hydrophobic regions on the surface of the protein or dipole interactions followed by denaturation of the protein to expose regions with greater hydrophobicity leading to a high affinity adsorption.

Our approach to sequence analysis on PVDF membranes was focused on the elimination of the polybrene - glass fiber to save the time associated with precycling and to obtain a cleaner background. Apparently the only problems associated with gas phase sequencing on PVDF result from poor solvent extraction. Since the cartridge was designed for a glass filter and not PVDF, it is not surprising that flow problems exist. The remedy described here is adequate; more than 50 sequences have been run with the modified program with only extremely rare, and minor fluctuations in yield. However, a better approach may be to use another reactor. It should be noted that Xu and Shively (5) recently reported that they could eliminate the cycle-to-cycle variability with PVDF membranes by using their continuous flow reactor (13). Another approach would be a reactor specifically designed for PVDF membranes such as that designed by Milligen for their solid phase sequencer.

Overall, PVDF membranes represent a dramatic advance in the field of electroblotting proteins and peptides from polyacrylamide gels for sequence analysis. In addition these membranes may be slightly better sequencing matrices than glass filters for purified proteins in solution. By incorporating a few simple programing changes, polybrene can be eliminated with gas phase sequencing. The resulting time savings largely compensates for the longer cycle time relative to pulsed liquid sequencing for many applications.

V. REFERENCES

1. R.H. Aebersold, D.B. Teplow, L.E. Hood and S.B.H. Kent (1986) J. Biol. Chem. 261:4229-4238.

2. J. Vandekerckhove, G. Bauw, M. Puype, J. Van Damme and M. Van Montagu (1985) Eur. J. Biochem. 152:9-19.

3. P. Matsudaira (1987) J. Biol. Chem. 261:10035-10038.

4. G. Bauw, M. De Loose, D. Inze, M. Van Montagu, and J. Vandekerckhove (1987) Proc. Natl. Acad. Sci. USA 84:4806-4810.

5. Q. Xu and J.E. Shively (1988) Anal. Biochem. 170:19-30.

6. M. Moos, Jr., N. Y. Nguyen, and T. Liu (1988) J. Biol. Chem. 263:6005-6008.

7. U. K. Laemmli (1970) Nature 227:680-685.

8. M.W. Hunkapiller and L.E. Hood (1983) Methods Enzymol. 91:486- 494.

9. P.H. O'Farrell (1975) J. Biol. Chem. 250, 4007-4021.

10. D.W. Speicher, J.S. Morrow, W.J. Knowles and V.T. Marchesi (1982) J. Biol. Chem. 257:9093 9101.

11. J.M. Gershoni, F.E. Davis, and G.E. Palade (1985) Anal. Biochem. 144:32-40.

12. H. Towbin, T. Sttaehelin, and J. Gorden (1979) Proc. Natl. Acad. Sci. 76:4350-4353.

13. J. E. Shively, P. Miller, and M. Ronk (1987) Anal. Biochem. 163:517-529.

4

RECOVERY OF BIOLOGICAL ACTIVITY, BLOTTING AND SEQUENCE ANALYSIS OF PROTEIN HORMONES AFTER SDS-PAGE[1]

Wolfgang H. Fischer, Joan Vaughan, David Karr, Richard McClintock, Joachim Spiess, Jean Rivier, and Wylie Vale

The Clayton Foundation Laboratories for Peptide Biology
The Salk Institute
La Jolla, California

I. INTRODUCTION

Sodium dodecyl sulfate polyacrylamide gel electrophoresis (SDS-PAGE) is a highly resolving separation technique for proteins. Until recently, it was not extensively used for preparative purposes due to two reasons: the limited capacity (low microgram amounts of protein) and the difficulties in recovering pure protein after separation. With the advent of electrophoretic transfer (blotting) procedures and the higher sensitivities offered by modern protein sequencers, SDS-PAGE has become a powerful tool for final purification of proteins to be sequenced. Especially, polyvinylidene difluoride (PVDF, Immobilon) membranes have proven to be extremely useful for electrophoretic transfer and subsequent sequence analysis. Problems arise when SDS-PAGE reveals the presence of several proteins and the position of the protein under investigation is unknown. For our continuing effort to identify follicle stimulating hormone (FSH) release modulating proteins from gonadal sources, we were therefore interested in identifying proteins that modulate FSH release after separation by SDS-PAGE.

II. EXPERIMENTAL PROCEDURES

Inhibin was isolated from ovine rete testis fluid using immunoaffinity chromatography (with antisera directed against the N-terminus of the

[1] Supported by NIH grant HD 13527 and contracts HD 32826 (CDB/NICHD) and B 87.14C/ICCR (Population Council). Research conducted in part by Clayton Foundation, California Division. W.V. is a Clayton Foundation Investigator.

alpha subunit), FPLC gel permeation chromatography and preparative reversed-phase high performance liquid chromatography (HPLC) (1). Recombinant human activin was provided by Genentech and was shown to have activity similar to activin we isolated from porcine follicular fluid as previously described (2). Follistatin (3) was isolated from porcine follicular fluid using immunoaffinity chromatography, Mono S cation exchange chromatography and semi-preparative reversed-phase HPLC (4).

Proteins (3 µg) were subjected to SDS-PAGE according to Laemmli (5) in a Bio-Rad Mini-Protean II electrophoresis cell (gel thickness, 0.5 mm; 12.5 % acrylamide). Samples were incubated in non reducing sample buffer (2% SDS, 20% glycerol, 62 mM Tris-HCl, pH 6.8) for 15 min at room temperature prior to application to the gel. After electrophoresis (200 V constant), the lanes were cut into 1-2 mm slices and extracted with 0.2 ml aqueous SDS (0.5 %) at 4° C for 16 h. Precipitation of SDS did not occur under these conditions. Extracts were saturated with urea resulting in a final volume of 0.32 ml and 40 µg of bovine serum albumin (BSA) were added. Sephadex G-10 equilibrated with 0.1 M TES (pH 8, NaOH) was used as the gel filtration medium in disposable columns (Evergreen, gel bed: 8x45 mm). Aqueous urea (8M, 0.1 ml) was applied to the column before the sample was loaded. The main protein fraction (0.2ml) of each column was collected. The elution volume of protein was determined in parallel experiments with BSA using the Bio-Rad Protein Assay Kit. The eluants were added in triplicate to enzymatically dispersed and plated rat anterior pituitary cells (50 ml eluant/well) (6). Some samples were assayed at a five-fold dilution, also. After 72 hours, medium was removed from the cells and tested for FSH using a radioimmunoassay kit provided by the National Hormone and Pituitary Program of NIADDK.

Electroblotting to Immobilon (Millipore, Bedford, Mass.) membranes (7) was performed in a Bio-Rad Mini Trans-Blot electrophoretic transfer cell at constant 0.2 A for 10 minutes. Proteins were visualized with 0.1% Ponceau S in water (containing 10% methanol, 1% acetic acid, 1.5% trichloroacetic acid, and 1.5% sulfosalicilic acid). The excised protein bands were subjected to sequence analysis in an Applied Biosystems protein sequencer 470A equipped with an on-line PTH Analyzer 120A from the same manufacturer. The standard sequencing program O3RPTH was used.

III. RESULTS

Three proteins previously isolated and shown to modify FSH release in the rat anterior pituitary cell in vitro bioassay (6) were used in this study: Inhibin $\alpha\beta_A$ heterodimer (1), activin $\beta_A\beta_A$ homodimer (2) and follistatin monomer (3). The preparations had different degrees of purity ranging from 70 to >95 %. Proteins were eluted from gel slices after SDS-PAGE with dilute aqueous SDS (0.5 %). A guide lane of the gel was stained with Coomassie blue. The extracts were saturated with urea and then subjected to gel filtration for the removal of low molecular weight materials. The protein fractions of the gel filtration were then tested for FSH-release modulating activity in an anterior pituitary primary cell culture assay (6). The results thereof are shown in Fig. 1. Recoveries of activity were estimated to be approximately 5% and are based on potency to modulate FSH secretion in the assay system described in Experimental Procedures. The amount of protein recovered was not determined. For all three preparations tested, the activity was recovered from a slice of the gel corresponding to a protein band on the stained guide lane. In cases where maximal effects were observed for several fractions, diluted samples were assayed as well (the data points are shown as inserts in Fig. 1) in order to determine the relative potency of these fractions. In subsequent experiments, we were thus able to identify these proteins on Immobilon electroblots. The blotting procedure essentially followed the protocol described by Matsudaira (7). Two sheets of Immobilon membrane were used and in the case of inhibin and activin, the second sheet was subjected to immunostaining (8) to further confirm the nature of the proteins. The sequence data obtained by sequencing the electrophoretically transferred proteins are summarized in Table 1.

	Sequence
	231 … 240 … 250
I	S T A P L P W P W S P A A L R L L Q R P
II	S T P P L P W P W S P A A L R L L Q (R) P
	309 … 320
III	G L E C D G K V N I C C K K Q F F V S F
IV	G L E *C* D G K V N I *C* *C* K (K) Q F (F) V
V	G L E *C* D G K V N I *C* *C* K K Q F F V S
	10 … 20
VI	G N C W L R Q A K N G R C Q V L Y K T E L S K
VII	G N *C* W L R Q A K N G R *C* Q V L Y K T (E) L S (K)

Table 1 N-terminal sequence data. **I**, Inhibin α chain (Mason et al. (9) and Mayo et al. (10), deduced from porcine ovarian cDNA); **II**, 18kd material from ovine testicular fluid, FSH release suppressing activity before reduction; **III**, Inhibin β chain (Mason et al. (9)); **IV**, 14kd material from ovine testicular fluid, FSH release suppressing activity before reduction; **V**, 28kd recombinant material, FSH release stimulating activity, molecular weight after reduction 14kd; **VI**, Follistatin (Esch et al. (3), deduced from ovarian cDNA); **VII**, 43kd material from porcine

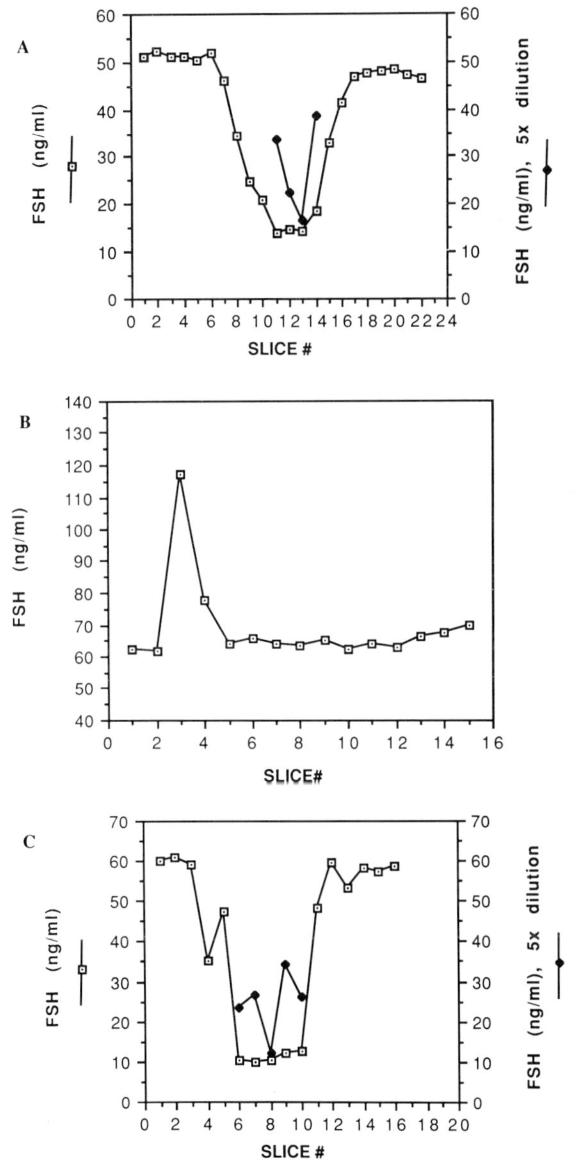

Fig. 1 Modulation of FSH release from rat anterior pituitary cells in vitro (6) by material recovered from SDS polyacrylamide gels. The data points shown as inserts were taken at a five fold dilution (see Experimental Procedures). Panel A: inhibin; Panel B: activin, Panel C: follistatin.

follicular fluid, FSH suppressing activity. (*C:* The identity of cystein residues was deduced from the presence of the phenylthiohydantoin derivative of dehydroalanine - dithiothreitol adduct.)

IV. DISCUSSION

The recovery of proteins from SDS polyacrylamide gels can be divided into two stages: 1) the elution of the protein from the gel and 2) the removal of SDS and gel artifacts that interfere with biological activity testing. For the elution of protein one has the choice between electroelution (11) and elution by diffusion (12). As the former requires special equipment, the number of samples that can be handled is limited. Elution by diffusion is less efficient, especially if the elution volume is limited, but many samples can be handled at the same time.

For recovery of proteins from SDS-containing solutions, several procedures have been described including ion exchange chromatography, protein precipitation, high performance liquid chromatography (HPLC). Ion exchange procedures tend to give variable results (12). Precipitation with trichloroacetic acid can destroy activity in sensitive proteins and requires an additional renaturation step (12). The most promising HPLC procedure is the "inverse gradient" mode described by Simpson et al. (13). Recovery of biological activity has not been described for this method.

For simultaneous processing of many samples, we investigated the use of gel filtration for removal of SDS and other low molecular weight contaminants. A large excess of urea is added before gel filtration. It is speculated that urea competes for binding to the protein and thus facilitates the removal of SDS. The urea and SDS are both separated from the protein in the gel filtration step. Bovine serum albumin is added to prevent nonspecific binding of proteins to the gel filtration matrix. As described in this study, the biological activity of inhibin, activin and follistatin can be restored in a cell culture assay that is very sensitive to the presence of denaturing agents and detergents. The recovery of approximately 5% of bioactivity reflects losses of material in the extraction and gel filtration step as well as incomplete renaturation after removal of denaturing agents. Considering the sensitivity of biological assay systems, this 5% recovery allows identification of low microgram quantities of material (3 μg for the hormones in this study). The procedure described can be useful in the final stages of isolation of novel bioactive proteins when SDS-PAGE reveals heterogeneities after extensive purification protocols. The previously uncharacterized follistatin preparation used in this study (4) revealed the presence of four protein bands after SDS-PAGE. The material exhibiting the highest potency in the bioassay showed the N-terminal amino acid sequence expected for follistatin (3). In a different study (14), enzymatic activity could be recovered employing the same procedure for glutaminyl cyclase (15).

V. LITERATURE

1. Rivier, J., Spiess, J., McClintock, R., Vaughan, J., Vale, W. (1985). Biochem. Biophys. Res. Commun. 133, 120.
2. Vale, W., Rivier, J., Vaughan, J., McClintock, R., Corrigan, A., Woo, W., Karr, D., and Spiess, J. (1986). Nature 321, 776.
3. Esch, F., Shimasaki, S., Mercado, M., Cooksey, K., Ling, N., Ying, S., Ueno, N., and Guillemin, R. (1987). Mol. Endocrinol. 1, 849.
4. Vaughan, J.M. unpublished results.
5. Laemmli, U.K. (1970). Nature 227, 680.
6. Vale, W., Vaughan, J., Yamamoto, G., Bruhn, T., Douglas, C., Dalton, D., Rivier, C., and Rivier, J. (1983). Methods Enzymol. 103, 565.
7. Matsudaira, P. (1987). J. Biol. Chem. 262, 10035.
8. Vaughan, J.M., Rivier, J., Corrigan, A.Z., McClintock, R., Campen, C.A., Jolley, D., Vogelmayr, J.K., Bardin, C.W., Rivier, C, and Vale, W. (1988). Methods Enzymol. in press
9. Mason, A.J., Hayflick, J.S., Ling, N., Esch, F., Ueno, N., Ying, S.Y., Guillemin, R., Niall, H., and Seeburg, P.H. (1985). Nature 318, 659.
10. Mayo, K.E., Cerelli, G.M., Spiess, J., Rivier, J., Rosenfeld, M.G., Evans, R.M., and Vale, W. (1986). Proc Natl. Acad. Sci. USA 83, 5849.
11. Hunkapillar, M.W., Lujan, E., Ostrander, F., and Hood, L. (1983). Methods in Enzymology 91, 227.
12. Hager, D.A. and Burgess, R.R. (1980). Anal. Biochem. 109, 76.
13 Simpson, R.J., Moritz, R.L., Nice, E.E., and Grego, B. (1987). Eur. J. Biochem. 165, 21.
14. Fischer, W.H. et al. unpublished results
15. Fischer, W.H. and Spiess, J. (1987). Proc. Natl. Acad. Sci. USA 84, 3628.

5

DIRECT SEQUENCE ANALYSIS OF RAT LIVER MEMBRANE GLYCOPROTEINS SEPARATED BY 2D-PAGE

David C. Parmelee, Timothy Benjamin, Chien-Hua Niu, and Snorri S. Thorgeirsson

National Cancer Institute, NIH, Bldg. 37, Room 3C-28,
Bethesda, MD 20892
USA

I. INTRODUCTION

The high resolution achieved by 2D-PAGE[1] has led to the extensive use of this technique in analyzing complex protein mixtures when studying the changes of cellular proteins during complicated biological processes such as differentiation, development and neoplastic transformation (1). Recently, computer based analysis of 2D-gels (2) has enabled the simultaneous observation of changes affecting numerous proteins following different treatments in biological studies. This type of analysis is currently the most powerful method for examining multiple cellular attributes to detect differences between specific cell populations.

The primary obstacle in effectively linking the 2D-PAGE technique to molecular and cellular biology is the difficulty in obtaining amino acid sequence data directly from the same polypeptides observed in the gels.

---

1 *2D-PAGE, two dimensional polyacrylamide gel electrophoresis; CHAPS, 3-[(cholamidopropyl)-dimethylammonio]-1-propanesulfonate; Con A, Concanavalin A; HPLC, high performance liquid chromatography; NP-40, Nonidet-P40, octylphenoxy polyethoxyethanol; PTH, phenylthiohydantoin; SDS, sodium dodecyl sulfate; TFA, trifluoroacetic acid; TRIS-HCl, tris (hydroxymethyl) aminomethane hydrochloride.*

However, the use of 1D-PAGE, accompanied by diffusion or electrodialysis procedures, has often been utilized to purify proteins for subsequent amino acid sequence analysis (3-12). Unfortunately these techniques are of limited value with 2D-gels, since incomplete recovery and NH_2-terminal blockage are more significant due to the minute amounts of individual proteins normally being separated by this low capacity system. Recently, numerous electroblotting techniques for transferring proteins onto supports compatible with direct gas-phase sequence analysis have been developed (13,14,15,16). This has allowed 2D-PAGE to be utilized as a preparative procedure for simultaneously purifying numerous proteins from relatively complex mixtures (13,14,15).

Our laboratory has utilized 2D-PAGE to detect a number of previously unidentified proteins whose regulation is markedly altered during the course of chemically-induced hepatocarcinogenesis (17). However, these individual proteins are usually present in minute amounts on a single 2D-gel because of sample complexity and the limited amount of material that can be applied per experiment. Normally, only the most abundant proteins could be directly analyzed using the current electroblotting and sequencing methodologies. However, we have successfully utilized a single 2D-gel to obtain sufficient amounts of four relatively minor membrane glycoproteins from a crude rat liver membrane preparation to determine their N-terminal structures. This was accomplished by partially purifying the proteins of interest before attempting the electrophoresis.

II. METHODS

A crude rat liver membrane extract was prepared and partially purified by affinity chromatography using Concanavalin A (Con A) and FPLC with Superose-12 to obtain an enriched mixture of glycoproteins having a specific range of molecular weights. This fraction was then subjected to 2D-PAGE using pH 4-6 Ampholytes in the first dimension and 7.5% SDS gels for the second (17,18). The solutions of acrylamide, bis-acrylamide and urea were pretreated with mixed bed ion exchange resin to remove undesirable ions and to avoid polymerization during storage (5). After electrophoresis, the proteins were electroblotted to Immobilon-P membranes (15) and stained with Coomassie Brilliant Blue R-250. Each sample was then prepared for amino acid sequence analysis by excision of the spot from the membrane and cutting it into smaller fragments (1 x 5 mm) which were then placed between the Zitex support and a TFA treated glass cartridge filter . The analyses were done using a 470A Applied Biosystems Gas Phase Protein/Peptide Sequencer connected to a 900A Data Module. The phenylthiohydantoin (PTH) amino acids were identified on-line with a 120A Applied Biosystems PTH Analyzer by reverse-phase HPLC using a Brownlee PTH-C-18 cartridge (2.1 x 220 mm). The performance of the sequencer was always checked by including 10-25 pmoles of beta-lactoglobulin as an internal standard with each unknown sample. If necessary, samples may be analyzed again without an internal standard to verify the sequence and to obtain quantitative information for any cycles that appeared to yield the same PTH derivative as that of the beta-lactoglobulin.

III. RESULTS

Partial purification of the crude rat liver membrane extract before 2D-PAGE circumvents many of the difficulties arising from the minute amounts of individual glycoproteins present in this complex sample. Such a fractionation increases the relative amount of the component present in a particular spot of the 2D-gel. An initial purification step utilizing affinity chromatography with lectins is limited to certain saccharide-containing proteins, but this procedure has several advantages over other separation techniques. Many types of immobilized lectin materials are commercially available and the specifically adsorbed glycoproteins can be eluted with various types of monosaccharides using mild conditions of pH and ionic strength. In addition, the binding of glycoproteins to most lectins are not significantly affected by up to 2.5% NP-40, a nonionic detergent commonly employed in the purification of solubilized integral membrane components (19).

The many glycoproteins present in the sample obtained by affinity chromatography using Con A is demonstrated in Fig. 1. The various proteins were separated into four fractions by FPLC using a gel filtration column containing Superose-12. Two dimensional-PAGE was then utilized to further purify the components in the highest molecular weight pool containing the glycoproteins which show the most pronounced modulation during hepatocarcinogenesis (unpublished results). The 2D-PAGE separation was optimized by analytically establishing the overall pH profile using an Ampholine ampholyte gradient from pH 3 to 10. The proteins of interest (differing only slightly in their isoelectric points) were then purified with a narrower pH range of ampholytes as shown in Fig. 2.

The glycoproteins labeled 1 thru 9 (Fig. 2) were selected for analysis by amino acid sequencing. Three of these demonstrated altered levels of expression in liver tumors (unpublished results). A single electroblotted 2D-gel yielded sufficient quantities of the various glycoproteins for direct amino acid sequencing and the N-terminal structure could be determined for four of them (see Fig. 3). The remaining five proteins of interest were not sequenceable in this manner, presumably because they had blocked N-termini. In these cases, the equipment was operating properly since the sequence was observed for the internal standard. The repetitive yields were normal in three of the experiments, but these

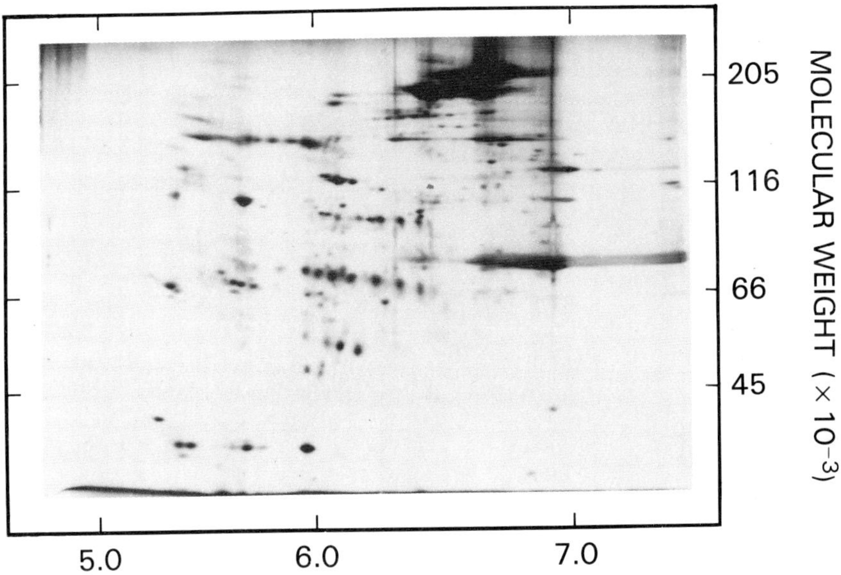

Figure 1: Analytical two dimensional-PAGE separation (pH 5-8) of plasma membrane glycoproteins (0.100 mg) from normal rat liver obtained by affinity chromatography using Concanavalin A. (Figures and Table I submitted elsewhere for publication)

values are sometimes relatively meaningless for certain sequences of amino acids or when analyzing small amounts of sample. For example, the repetitive yield of 99.8% obtained for RMGP-3 (see Fig. 3) is due to the fact that the first five amino acid residues sequenced are subject to significant errors in quantification. This results from insufficient resolution and instability of PTH standards (PTH-Lys) and variable recoveries from the sequencer or the immobilon membrane (PTH-Asp and -Glu). The apparent high recoveries of PTH-Thr and -Ser (seen for all RMGP peptides shown in Fig. 3) may be explained by the instability of these residues in the mixture of PTH standards.

IV. DISCUSSION

The excellent resolving power of 2D-PAGE has led to its widespread use as a tool to analyze proteins and to evaluate the various steps required to purify these materials using conventional procedures. Until

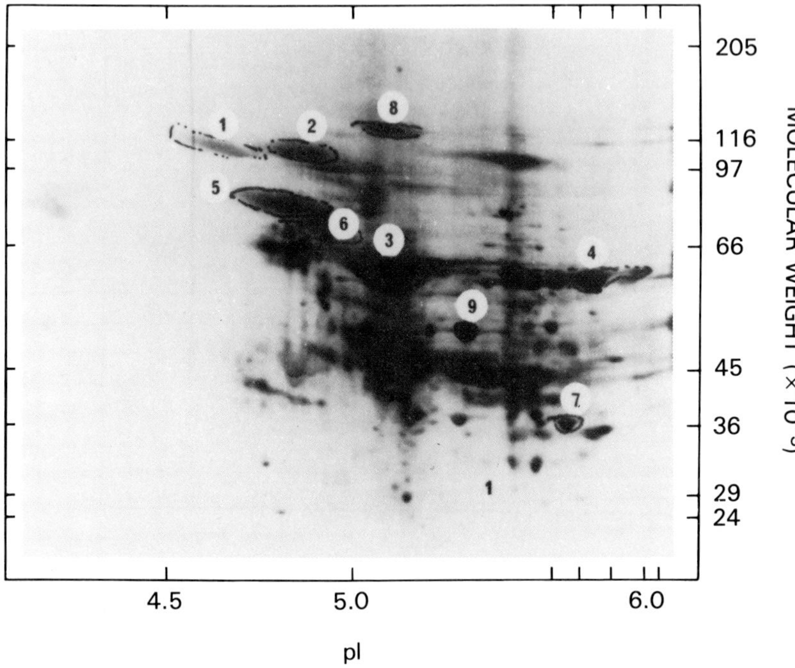

Figure 2: Preparative two dimensional-PAGE separation (pH 4-6) of the glycoproteins (about 0.350 mg) from the highest molecular weight fraction obtained by FPLC using Superose-12. The numbered spots indicate samples that were subjected to amino acid sequence analysis.

recently, this technique could not be utilized directly for amino acid sequence analysis of specific proteins from complex mixtures because only a limited amount of total material (usually about 0.100 mg) can be applied if maximum resolution is required. One solution to this problem has been to combine the purified proteins obtained from a large number of 2D-PAGE experiments.

We have described an alternative approach based on partially purifying the proteins of interest before attempting 2D-PAGE. This method is demonstrated by specifically isolating and analyzing several rat liver membrane glycoproteins. These components, characteristically difficult to purify by classical procedures, were present at relatively minor concentrations in the starting material. A solution enriched with the specific glycoproteins of interest was first obtained. It was then possible to apply this material to a single 2D-gel and electrophoretically purify sufficient quantities of the proteins for amino acid sequence analysis. This general procedure may be applied to other types of proteins by carefully chosing preliminary purification steps.

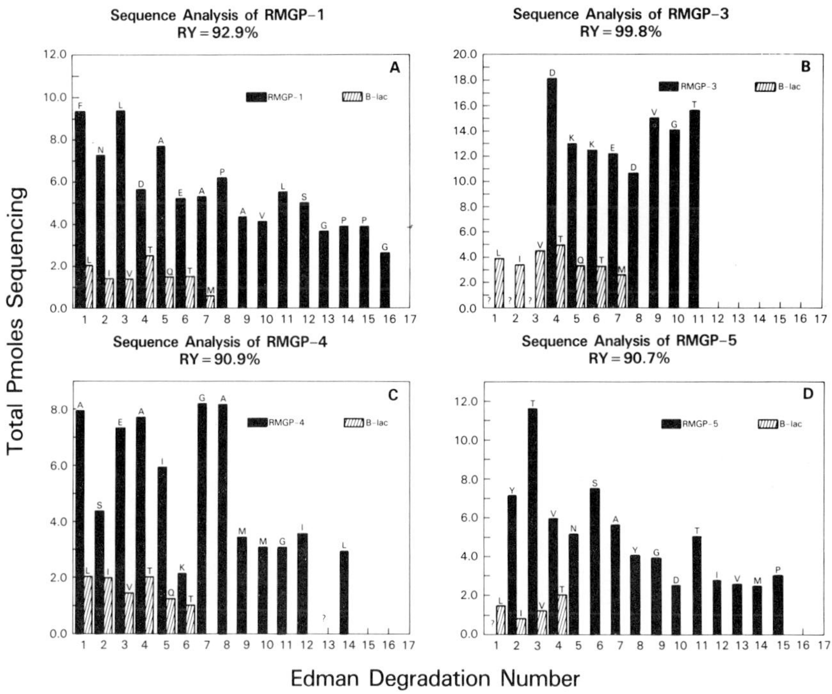

Figure 3: The yields of PTH-amino acids from the sequence analyses of the labeled spots excised from the electroblot as shown in Fig. 2. Panels A-D refer to spots 1, 3, 4, and 5, respectively. The shorter striped bars at the beginning of each experiment are the results obtained for the internal standard, beta-lactoglobulin. The yields of total PTH amino acids being sequenced were determined by subtracting the background amount of the PTH residue seen in the previous cycle and then correcting this value for the percentage of the sample analyzed by HPLC. The RY value refers to the repetitive yield, calculated by linear regression analysis, observed for each experiment.

We have successfully isolated and sequenced the N-terminal regions of four membrane glycoproteins from normal rat liver. Table I summarizes these results. No significant sequence similarities to other known proteins were observed using the ALIGN program of Dayhoff et al. (20). The glycoprotein designated as RLGP-1 is especially interesting since it is the major component that decreased during chemical hepatocarcinogenesis (unpublished results). Several of the proteins apparently had blocked N-terminal amino acids in their native states, since they were isolated from the same slab gel that yielded other

Table I:

N-Terminal Sequences of Electroblotted Rat Liver Membrane Glycoproteins

RLMP-	1 5 10 15	MW x 10^{-3}/pI
1[ab]	NH_2-PHE-ASN-LEU-ASP-ALA-GLU-ALA-PRO-ALA-VAL-LEU-SER-GLY-PRO-PRO-GLY-	145 / 4.5-4.6
2	(No sequence observed)[c]	143 / 4.7-4.8
3	NH_2--X[d]--X---X--ASP-LYS-LYS--X--ASP-VAL-GLY-THR-	70 / 5.0-5.1
4	NH_2-ALA-SER-GLU-ALA-ILE-LYS-GLY-ALA-MET-MET-GLY-ILE--X--LEU-	68 / 5.5-5.8
5	NH_2--X--TYR-THR-VAL-ASN-SER-ALA-TYR-GLY-ASP-THR-ILE-VAL-MET-PRO-	95 / 4.6-4.8
6	(No sequence observed)	83 / 4.8-4.9
7	(No sequence observed)	40 / 5.7
8	(No sequence observed)	155 / 5.0-5.1
9	(No sequence observed)	56 / 5.3

a *Proteins are denoted (RLMP-1 - 9) as shown in Fig. 2.*
b *The amino acid residues are given in the three letter notation.*
c *Mechanical malfunctions were not responsible when a lack of sequence was observed, since the 5 pmoles of internal standard was being properly analyzed (see Methods).*
d *X indicates no residue was identifiable by the sequence procedure, suggesting that the amino acid may be glycosylated or that it was a cysteine or tryptophan.*

sequenceable material and the proper operation of the sequencer was verified by an internal standard.

Performing sequence analysis directly on samples electroblotted from 2D-PAGE gels effectively breaks the barrier linking this powerful technique with that of protein chemistry to obtain structural information from biologically important materials. In addition, it now becomes feasible to directly confirm the identities of proteins being studied and/or compared in different laboratories, even when the only common data available consist of coordinates (MW/pI) obtained from 2D-PAGE. Making these comparisons will become easier as electrophoretic and sequence methodologies are improved to decrease analytical backgrounds, increase initial and repetitive yields, and to enhance the sensitivity of detecting the resulting amino acid derivatives.

V. REFERENCES

1. Celis, J. E., and Bravo, R., Eds. (1984) Two-dimensional gel electrophoresis of proteins. Methods and applications , pp. 1-487, Academic Press, Orlando, FL.

2. Miller, M. J. (1986) Expl. Biol. Med. 11,235-260.

3. Weiner, A. M., Platt, T., and Weber, K. (1972) J. Biol. Chem. 247, 3242-51.

4. Bhown, A. S., Mole, J. E., Hunter, F., and Bennett, J. E. (1980) Anal. Biochem. 103, 184-190.

5. Hunkapiller, M. W., Lujan, E., Ostrander, F., and Hood, L. E. (1983) *in* Methods in Enzymology (Hirs, C. H. W., and Timasheff, S. N., Eds.), Vol. 91, pp. 227-236, Academic Press, Orlando, FL.

6. Boguslaw, S. and Summers, D. F. (1988) Anal. Biochem. 168, 48-53.

7. Xu, Q. and Shively, J. E. (1988) Anal. Biochem. 170, 19-30.

8. Aebersold, R. H., Leavitt, J., Saavedra, R. A., Hood, L. E. and Kent, S. B. H. (1987) Proc. Natl. Acad. Sci. USA 84, 6970-6974.

9. Salinovich, O. and Montelaro, R. C. (1986) Anal. Biochem. 156, 341-347.

10. Hsieh, J., Lin, F. and Tam, M. F. (1988) Anal. Biochem. 170, 1-9.

11. Moos, M., Nguyen, N. Y. and Liu, T. (1988) J. Biol. Chem. 263, 6005-6008.

12. Yuen, S. W., Chui, S. H., Wilson, K. J., and Yuan, P. M. (1988) Applied Biosystems User Bulletin-Protein Sequencer. 36, 1-17.

13. Bauw, G., De Loose, M., Inze', D., Van Montagu, M., and Vandekerckhove, J. (1987) Proc. Natl. Acad. Sci. USA 84, 4806-4810.

14. Aebersold, R., Teplow, D. B., Hood, L. E., and Kent, S. B. (1986) J. Biol. Chem. 261, 4229-38.

15. Matsudaira, P. (1987) J. Biol. Chem. 262, 10035-38.

16. Vandekerckhove, J., Bauw, G., Puype, M., Van Damme, J., and Van Montagu, M. (1985) Eur. J. Biochem. 152, 9-19.

17. Wirth, P. J., Benjamin, T., Schwartz, D. M., and Thorgeirsson, S. S. (1986) Cancer Res. 46, 400-413.

18. O'Farrell, P.H. (1975) J. Biol. Chem. 250, 4007-4021.

19. Lotan, R., and Nicolson, G.L. (1979), Biochim. Biophys. Acta. 559, 329-376.

20. Dayhoff, M. O., Barker, W. C., and Hunt, L. T. (1983) *in* Methods in Enzymology (Hirs, C. H. W., and Timasheff, S. N., Eds.), Vol. 91, pp. 524-45, Academic Press, Orlando, FL.

6

THE USE OF PHENYLTHIOCARBAMYL AMINO ACID ESTERS FOR SENSITIVITY ENHANCEMENT IN EDMAN DEGRADATION

Marcus J. Horn
Sherrell L. Early
Sheila G. Magil

Applied Protein Technologies, Inc.
Cambridge, Massachusetts

I. INTRODUCTION

Recent efforts have focused upon improving the sensitivity of peptide and protein sequence analysis. A significant aspect of sensitivity enhancement is improvement of the detectability of the resulting cleaved amino acid derivatives. Much of the previous work on increasing this detectability has concentrated on modification of the standard Edman reagent, phenylisothiocyanate (PITC). These approaches have shown varied degrees of success. Our observations on the course of the methanolic-HCl conversion of anilinothiazolinone (ATZ) -amino acids to the corresponding phenylthiohydantoin (PTH) derivatives has led to the development of an alternative approach to detectability enhancement. Changes in temperature, acid concentration, and substitution of alcohols other than methanol were found to affect the ratio of the PTH to the intermediate phenylthiocarbamyl (PTC) amino acid ester. Modification of the conditions for methanolic-HCl conversion has allowed the successful production of high-yield stable chromophoric and fluorophoric esters (PTCE's). These derivatives are highly colored or fluorescent and may provide a 10-100 fold increase in signal to noise ratio over current methods.

II. BACKGROUND

The question of sensitivity in peptide and protein sequence analysis is continually being raised. We have previously examined some factors within the sequencing methodology which contribute to or detract from sensitivity enhancement (1). One aspect of any improvement in sensitivity involves the actual sample "handling" during the sequential degradation. Another equally significant aspect is the detectability of the resulting cleaved amino acid derivatives. This report presents some of our recent progress in the latter area - improvements in the detection of the derivatives of sequence analysis.

Previous work on increasing this detectability has focused upon modifications of the standard Edman reagent, phenylisothiocyanate (PITC). The use of radiolabelled PITC has been described (2) as a means of increasing sensitivity. Others have used chromophoric isothiocyanate analogs such as phenylazophenyl (3) or dimethylaminoazobenzene (4) isothiocyanate to enchance detectability. Fluorophoric Edman reagents (5), as well as compounds having "cryptic functionality" (6) (isothiocyanates requiring subsequent reaction with chromophores or fluorophores) have also been reported.

While showing varied degrees of success (much of it owing to low reagent reactivity), all of the above approaches possess an additional inherent drawback. Detectability, although an absolute measure, must be coupled with a high signal-to-noise ratio to achieve sensitivity increases. Since many of the by-products (phenylthiourea, diphenylthiourea, etc.) of the Edman degradation arise from breakdown of the isothiocyanate component, increases in detectability of the amino acid derivative are accompanied by simultaneous increases in detectability of by-products and thus no improvement in the signal-to-noise ratio.

An alternative is to convert one of the intermediate Edman compounds into a more easily detected derivative. Tsugita _et al_ (7), have recently adopted this approach by exposing the ATZ-amino acid to a radiolabelled primary amine to yield a labelled PTC-amino acid derivative. This method, as well as a similar method described by Appella _et al_ (8), requires the complete elimination of the acidic Edman cleavage reaction prior to introduction of the labelled amine. Additionally, the PTC-amino acid amides produced are prone to cyclization by the residual acid carrried over in the ATZ extraction solvent.

III. EXPERIMENTAL

The Edman degradation (Figure 1) consists of three distinct segments: coupling, cleavage and conversion. Strong acid cleavage of the PTC-peptide formed during the coupling step results in the formation of a 2-anilino-5-thiazolinone derivative (in the ammonium form). The classical conversion reaction, producing a PTH amino acid, takes place in aqueous acid, and actually consists of two reactions: hydrolysis of the thiazolinone to the PTC-amino acid and cyclization of the latter to the PTH amino acid. The first reaction is much faster than the second. We have characterized the cleavage reaction of seven peptides, each with a different PTC-amino acid terminus. Analysis of the products of the cleavage reaction by HPLC indicates that the vast majority of the cleavage product may be identified as the ATZ prior to conversion. Typically, ATZ-leucine is found in greater than 80% yield, while ATZ-glycine routinely appears in yields of about 90%. Cleavage conditions and dry-down conditions of the cleavage reaction affect ATZ-amino acid yields. ATZ-leucine has been previously synthesized preparatively in high yield (7).

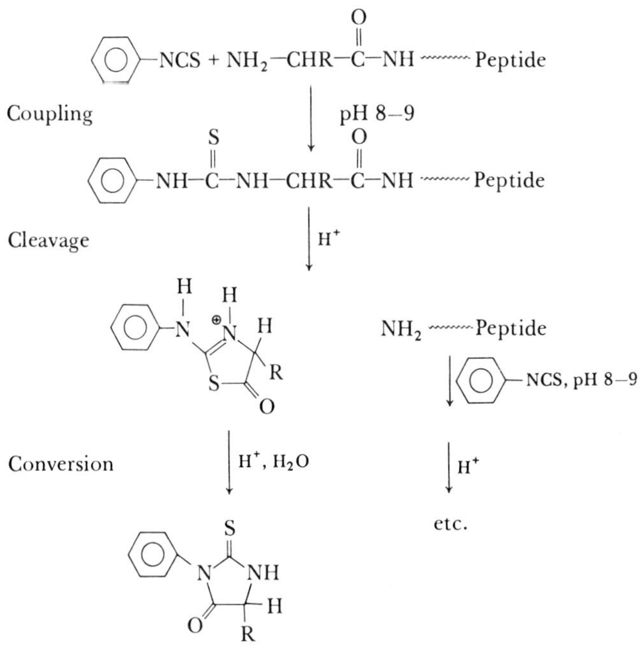

Figure 1. The Edman Degradation

Conversion of ATZ amino acids to the corresponding PTH derivatives may also be effected using methanolic-HCl (2) proceeding analogously through the intermediate PTC amino acid methyl ester (Figure 2). From our observations of the course of the methanolic-HCl conversion of ATZ-amino acids (9) (Figure 3), it appeared that changes in temperature, acid concentration, and substitution of alcohols other than methanol could affect the ratio of the PTH to the intermediate PTC amino acid ester. Under identical temperature and acidity conditions, methanol, 1-propanol and 2-propanol display significantly different rates of conversion to the PTH derivative (Figures 4a and 4b). Differences from unity, in each case, are accounted for by the corresponding PTC amino acid ester derivative.

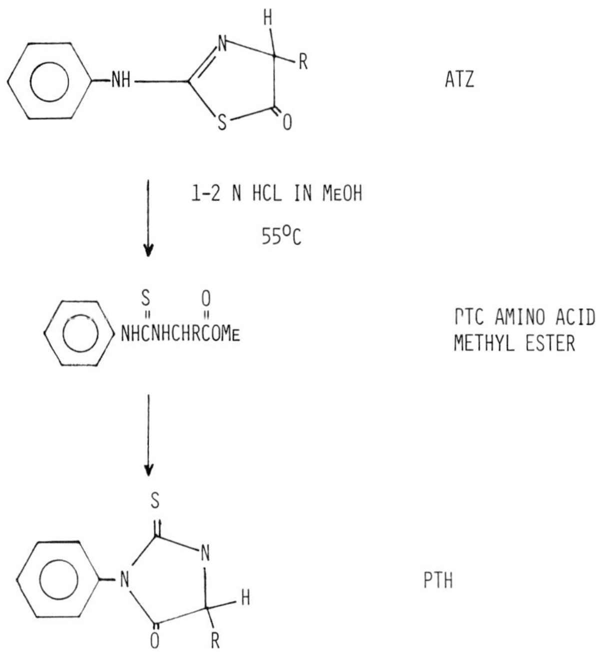

Figure 2. ATZ to PTH conversion using methanolic HCl.

When similar conditions are applied to other, more complex alcohols (Figure 5), the PTC amino acid esters become the principal products, rather than the PTH derivatives (Figures 6a and 6b). Conversion of the ATZ slowed down significantly, with the product mixture including ATZ, PTH and PTC amino acid ester derivative. Use of anhydrous acid conditions (Figure 7) allows production of high-yield amounts of the PTC amino acid esters from the ATZ's.

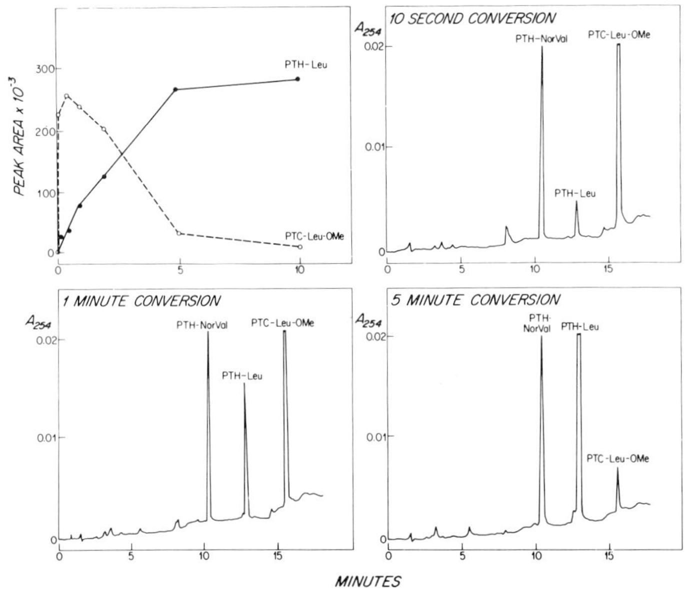

Figure 3. ATZ- to PTH-leucine conversion using 1N HCl in methanol, 65. Upper left: relative amounts of PTC-leucine methyl ester and PTH-leucine formed at various conversion times. ATZ was not detected. Other panels show HPLC chromatograms for certain time intervals. PTH-norvaline was used as internal standard. (Figure from ref.9)

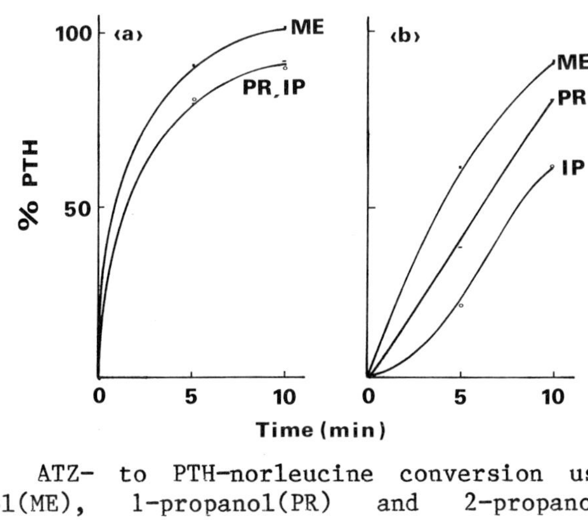

Figure 4. ATZ- to PTH-norleucine conversion using HCl in meth- nol(ME), 1-propanol(PR) and 2-propanol(IP), 60. Percentage PTH indicated; remainder is PTC ester. (a) 1.0N HCl. (b) 0.5N HCl.

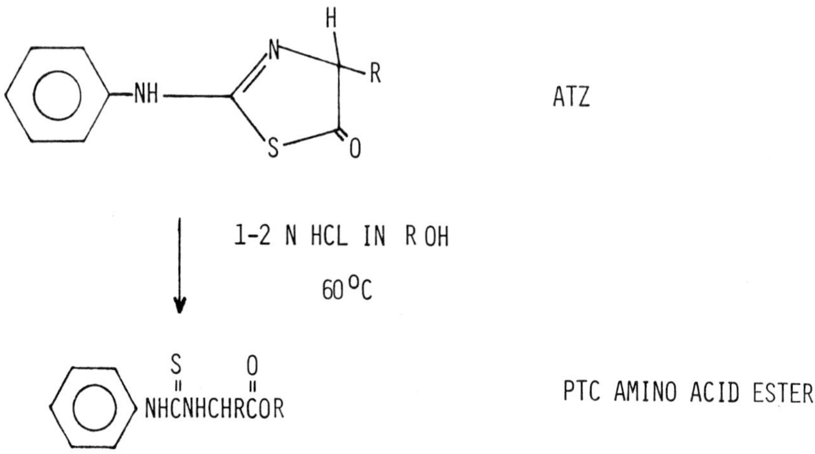

Figure 5. Conversion of ATZ amino acid to PTC amino acid ester.

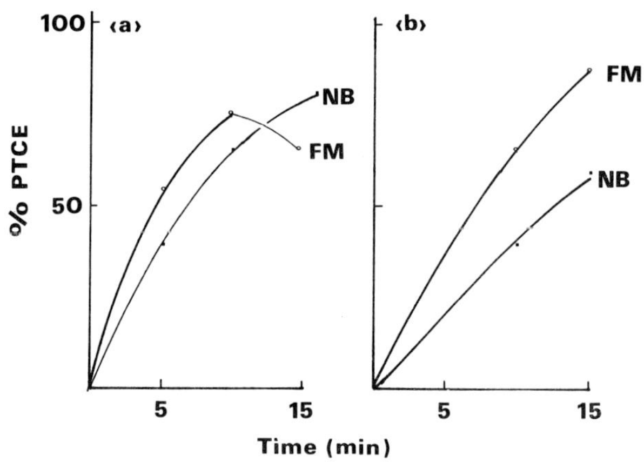

Figure 6. Conversion of ATZ-norleucine to PTC-norleucine ester using HCl in dioxane and 10 equivalents of nitrobenzyl alcohol (NB) and fluoenylmethyl alcohol (FM), 60. Percentage PTC-ester (versus theoretical) indicated. (a) 1.0N HCl. (b) 0.5N HCl.

A range of alcohols and their reaction with ATZ amino acids have been investigated. These derivatives (Figure 8) are highly colored or highly fluorescent and are readily separable from their starting alcohol by reversed phase HPLC. Resolution of the individual PTC amino acid esters (arising from a specific alcohol) is achievable on C8 and C18 columns with 10-100 fold increases in sensitivity over conventional PTh methodologies. Complete separation systems for the various ester groups are under development.

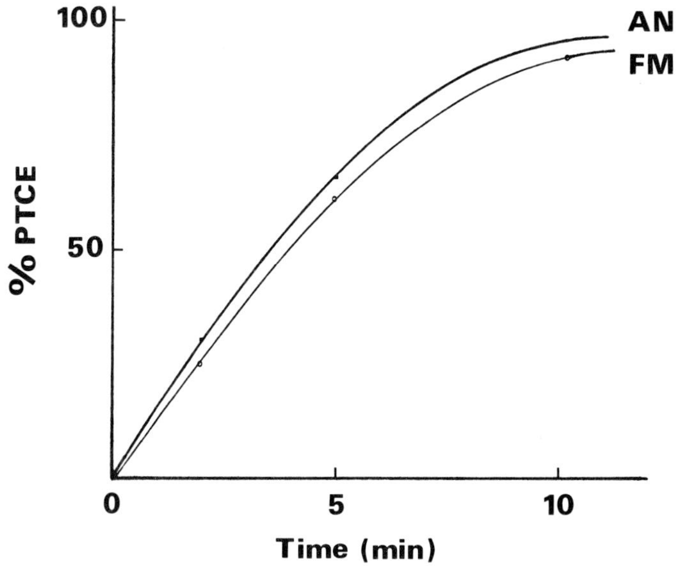

Figure 7. Conversion of ATZ-norleucine to PTC-norleucine ester using 1.0N HCl in anhydrous benzene and 10 equivalents of fluorenylmethyl alcohol (FM) and 9-anthracenemethanol (AN), 60. Percentage PTC-ester (versus theoretical) indicated.

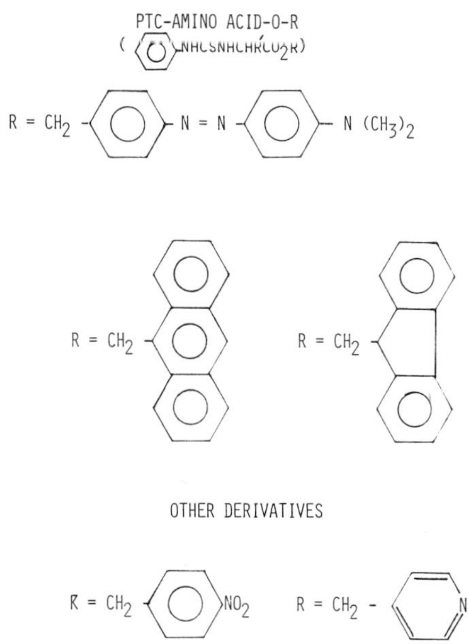

Figure 8. PTC-amino acid esters with enhanced detectability.

IV. CONCLUSIONS

ATZ-amino acids may be derivatized with a variety of alcohols to produce PTC-amino acid ester derivatives. Since the reaction environment required for production of the PTC-amino acid ester is identical to that necessary to perform the prior Edman step, no stringent acid removal is required. Steric hinderance associated with the sensitivity-enhancing alcohols actually promote the desired reaction by disfavoring the recyclization to the PTH derivative. The method gives an improved signal-to-noise ratio since the product's absorbance characteristics differ significantly from Edman by-products.

REFERENCES

1. Horn, M.J. and Bonner, A.G. (1982) in Methods in Protein Sequence Analysis, Elzinga, M., ed., Humana Press, Clifton, p.159.
2. Niall, H.D., *et. al.* (1974) Proc. Nat. Acad. Sci. USA *71*, 384.
3. Horn, M.J. and Bonner, A.G. (1977) in Solid Phase Methods in Protein Sequence Analysis, Proceedings of the 2nd International Conference, Previero, A. and Coletti-Previero, M.A., eds., Elsevier/North Holland, Amsterdam, p.163.
4. Chang, J.Y., Knecht, R. and Braun, D.G. (1982) in Methods in Protein Sequence Analysis, Elzinga, M., ed., Humana Press, Clifton (NJ), p.113.
5. Hirano, H. and Wittmann-Liebold, B. (1987) Abst. JASPEC 1987, p.128
6. L'Italien, J.J. and Kent. S.B.H. (1982) J. Chromatogr. *283*, 149.
7. Tsugita, A., Arai, I., Kamo, M., and Jones, C.S. (1988) J. Biochem. *103*, 399.
8. Appella, E., Inman, J.K. and Dubois, G.C. (1977) in Solid Phase Methods in Protein Sequence Analysis, Proceedings of the 2nd International Conference, Previero, A. and Coletti-Previero, M.A., eds., Elsevier/North Holland, Amsterdam, p.121.
9. Margolies, M., *et. al.* (1982) in Methods in Protein Sequence Analysis, Elzinga, M., ed., Humana Press, Clifton (NJ), p.189.

7

STUDIES ON C-TERMINAL ANALYSIS

David H. Hawke
Suzanne M. Meister
Pau-Miau Yuan
Michael W. Hunkapiller

Research and Development
Applied Biosystems, Inc.
Foster City, California 94404

John H. Richards

Department of Chemistry
California Institute of Technology
Pasadena, California 91125

I. INTRODUCTION

Despite the fact that end labelling has been widely used for DNA analyses, and in combination with an appropriate separation technique is central to all methods for DNA sequencing (1), reports of end-labelling of proteins are quite rare. The intrinsically higher chemical diversity of proteins (twenty different amino acids compared with the four bases in DNA) with a wide range of chemical reactivity has impeded the development of such techniques for protein chemists.

Such procedures would be useful in both finding the C-terminal fragment for amino terminal sequence analysis, as well as making alignments from partial cleavage maps. This could increase the analytical efficiency of sequencing projects tremendously by allowing workers to concentrate on peptides of particular interest. Selective isolation of the C-terminal peptide followed by amino terminal sequencing is also a potential alternative to direct chemical sequencing (2) from the carboxy terminus as a method for obtaining C-terminal sequence information.

A. *Selective Isolation of Carboxy Terminal Fragments.*

Several approaches to this problem have been discussed. Some of these are mentioned below. For a more complete list, please see reference 3. An earlier approach (4) to finding the C-terminal peptide involved chemical modification of all carboxyl groups as the first step of a series of manipulations. The final separation used ion exchange chromatography and was based on the charge differences; all but the carboxy terminal peptide should have a free carboxyl group. Later refinements also use carboxyl modification, either via water-soluble carbodiimide (WSC) couplings (5) or methyl ester formation (3). These continue to rely on charge differences to distinguish the C-terminal peptide, and their behavior under partial cleavage conditions has not been well established.

An interesting "reverse labeling" approach (6) has appeared in which all of the non-C-terminal peptides become ^{18}O labeled during enzymatic cleavage. The peptide whose carboxylate does *not* incorporate label is the C-terminal one. The determination requires a mass spectrometer, and there is some concern that unwanted carboxyl exchange may occur. This method should work with incomplete fragmentation, if the carboxylate exchange is controlled but subject to the mass limitation of the mass spectrometer available.

B. *Protein End Labeling*

There are two recent examples of attempts to end label proteins. An amino terminal labelling procedure (7) employed thioacylation to block lysyl side chains, then cleavage of the amino terminal amino acid reveals a free amino group (of the next residue). These workers also demonstrated its utility in peptide mapping. The amino terminal residue itself is lost, but presumably the derivative could be identified, if desired.

A carboxy terminal labeling approach (8) used carboxypeptidase to incorporate radiolabeled Phe-amide at the C-terminal end of a protein. This is a possibility which is suggested by the successful (9) use of CPY in forming peptide bonds (the formal reverse reaction of hydrolysis). CPY is somewhat difficult to control, however, and as expected, left a "ragged end" due to partial hydrolysis at the C-terminus.

C. *Chemico-Enzymatic Labeling*

We have devised a "chemico-enzymatic" approach to the problem (scheme 1). The general concept is to modify (block) all carboxyl groups, followed by a selective reaction to modify the C-terminal in preference to the side chain modified carboxyl groups. Water-soluble carbodiimide is used to amidate the carboxyls, then the "correct" carboxyl is selectively deamidated enzymatically.

NH_2...D...E...COOH ---> NH_2...Dx...Ex...COx --->

NH_2...Dx...Ex...COy ---> NH_2...Dx...Ex...COz

Scheme 1. General Modification Scheme. Carboxyls are blocked with "x", the terminal "x" is transformed to "y", which is then modified to "z".

Subsequent chemical modification of that residue can then place the desired label on the only free carboxyl group available: at the C-terminus. The selectivity of the deamidation step is crucial. If the side chain "protecting groups" react significantly, they will become available for the subsequent chemical modification. Several peptides, not just the C-terminal, would be labelled and the desired discrimination would be lost.

We have found that very high side-chain discrimination can be achieved in a model system. A preliminary report of this work has been made (10).

II. MATERIALS AND METHODS

General. Reagents were obtained from ABI, Aldrich or Sigma. Carboxypeptidase P was purchased from Boehringer-Mannheim and Sigma. Milli-Q water was used throughout.

HPLC. Reverse phase HPLC was performed with an ABI Model 120A microbore HPLC system on Aquapore RP-300 cartridge columns, usually 2.1 x 30 mm. Buffers: A: 0.1% aq. TFA, B: 0.085% TFA in 70% acetonitrile/water. Typical gradient: 0-60% B over 15 minutes followed by a rapid ramp to 90% B. The usual flow rate was 200 microliters/min at room temperature. Peptides were monitored at 214 nm. For preparative runs fractions were carefully hand-collected from a short length of stainless steel capillary tubing attached directly to the outlet of the UV-flowcell.

Peptide Synthesis. The peptide VLAEGYVGLRY was synthesized on an ABI Model 430A Peptide Synthesizer on PAM resin using the standard ABI protocols and reagents. It was cleaved with HF and purified by reverse phase HPLC (TFA/Acetonitrile/Water). The structure of the purified product was verified by amino acid analysis, sequencing, and FAB-MS.

Sequence Analysis. Sequencing was performed on a Liquid-Updated 475A Sequencing System (470A/120A/900A, ABI) using standard chemicals and only slightly modified cycles. For some runs the injection protocol was modified to allow 100 of 150 microliters to be injected on the Model 120A Online PTH Analyzer. PTH amino acids were resolved by the usual method. Data was collected and analyzed by the 900A.

Amino Acid Analysis (AAA). Composition data was acquired by the standard ABI tube hydrolysis procedure using 6 N HCl at 160 C for 1 hour in a sealed chamber. Hydrolysed amino acids were derivatized with PITC and analyzed on a Model 420A Derivatizer.

FAB-MS. Static Fast Atom Bombardment Mass Spectrometry was performed on a Jeol HX100 double focussing instrument operating at full accelerating potential.

Chemical Modifications. Chemical carboxyl group modifications were carried out essentially as described by Hoare and Koshland (11) in guanidinium HCl. However, for this study glycyl-L-leucinamide was used instead of glycinamide.

Enzymatic Deamidation Reactions. Deamidation was effected by Carboxypeptidase P (CPP) in 0.1 M TAPS buffer, pH 7.0. Typically 1-20 nmoles of peptide was reacted with 0.1-0.3 μg of enzyme in about 50 μL. For convenience the reactions were automated on a Gilson Model 231 autosampler. The instrument was programmed to perform the transfer and mixing steps in a clean target vial, then make injections onto the HPLC at appropriate intervals.

III. RESULTS AND DISCUSSION

The model peptide I (Figure 1), VLAEGYVGLRY, was synthesized containing one free side-chain carboxyl and a free carboxyl terminus. This was deemed small enough for it and all products to be thoroughly characterized. Carboxyl groups were activated with dimethylaminoethyl propyl carbodiimide,and coupled with excess G-L-NH_2. This reaction gave a single major product if allowed to proceed overnight, but appeared to be complicated if monitored earlier. We found no indication of significant racemization under these conditions, but our selection of G-L-NH_2 as the blocking group should reduce the likelihood of racemization interfering with the enzymatic deamidation because of the distance from the scissile bond.

I. **V-L-A-E-G-Y-V-G-L-R-Y**

II. **V-L-A-E-G-Y*-V-G-L-R-Y*-G-L-NH_2**
|
G-L-NH_2

III. **V-L-A-E-G-Y*-V-G-L-R-Y*-G-L-OH**
|
G-L-NH_2

IV. **V-L-A-E-G-Y*-V-G-L-R-Y*-G-L-G-L-NH_2**
|
G-L-NH_2

Figure 1. Peptide Structures. I: Starting peptide. II: Peptide I after the first modification with G-L-NH_2. III: Carboxy-terminally deamidated peptide II. IV: Peptide III after final modification with G-L-NH_2.

Amino acid composition, sequence analysis, and FAB-MS of the modified product are all consistent with structure II (Figure 1). The modified Glu sequences without difficulty, giving a PTH derivative, E', which almost coelutes with Ser', between PTH-Ala and PTH-Tyr (Figure 3). The carbodiimide adds to the Tyrosine hydroxyl group to form a relatively stable O-isourea (12), preventing normal recoveries during AAA and sequencing. This modification is indicated as Y*, and can be hydrolysed under mild conditions (not shown). Although the Y* does not appear to interfere with the reactions, it does increase the mass of the MH+ ion in the mass spectra of affected molecules.

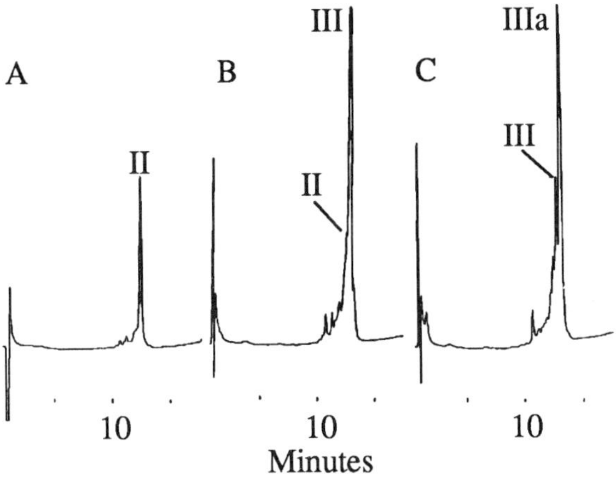

Figure 2. Enzymatic Deamidation Experiment. Left trace: Blocked peptide (II, about 200 pmol). Middle: Reaction mixture at one minute (mostly III). Right: Mixture after 45 minutes (mostly IIIa). These aliquots contained about 400 pmol of peptides. Scale: 0.2 AUFS at 214 nm.

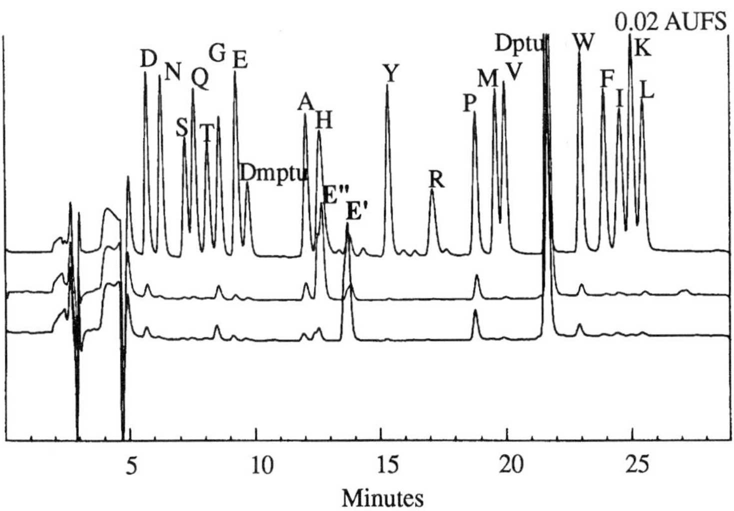

Figure 3. Sequencing Results: the Modified Glu Cycle. Lower Trace: III, side chain intact, C-terminal deamidated. Middle: IIIa, both side-chain and C-terminal deamidated. Upper: PTH Standards.

Deamidation of the modified peptide, II, is quite rapid by CPP (Figure 2). The reaction is essentially complete in less than one minute. A single major product , III, is observed (>90%). After extended time the side chain is significantly deamidated (IIIa), but further hydrolysis is not yet a problem.

The initial deamidation product is identified as the desired isomer, III, for several reasons. Its HPLC migration at pH 2 is consistent with amide hydrolysis. The sequencing results were indistinguishable from sequencing the starting modified peptide: the same peak was observed at the modified Glu position, hence it was intact, and the C-terminal Leu was easily assigned, ruling out further digestion. Also, the MH^+ is one AMU higher than for II, as expected from the hydrolysis of an amide. Since the side-chain is intact (Figure 3, Lower) the terminal Leu-NH_2 must be the one which deamidated.

The product of extended deamidation, IIIa, is hydrolysed at both amides. Again, the HPLC retention is consistent with further hydrolysis, but this time sequencing gives a different peak for the modified Glu at cycle 4 (Figure 3, Middle): E", which as expected elutes earlier than E'. The terminal Leu is still present, and the MH^+ has increased by another AMU, confirming that another amide has hydrolysed.

The third step of the labelling method is the further modification of the modified and C-terminally deamidated peptide (III) with the desired label. For this experiment we used G-L-NH_2 again. Alternative nucleophiles could be chosen to impart desirable characteristics (enhanced detectability or isolability, etc.) to the C-terminus. The final product, IV, is retained another minute on RPLC (not shown), and the sequence analysis confirms both the intact side chain (Figure 4a), and the additional G-L at the C-terminus (Figure 4b).

IV. CONCLUSIONS

We have demonstrated the feasibility in a model system of using a combination of chemical and enzymatic techniques to selectively label the carboxyl terminus of a peptide without interference from the side-chain carboxyl group. The key step in this process is the selective deamidation with carboxypeptidase (II --> III). Seryl carboxypeptidases are known to react with carboxamides (13), in contrast with most metallo-carboxypeptidases, but the scope of the reaction has not been well defined. We have found that carboxypeptidase P (from *Penicillium Janthinellum*) shows high selectivity for deamidating the C-terminus of the modified peptide while sparing the modified side-chain. High "side-chain discrimination" is crucial for the procedure to be generally applicable.

The advantages of this method over a direct C-terminal sequencing method are that sequencing of the labelled fragment can be performed in well-established, highly efficient and sensitive systems already available, and that additional information could be acquired through partial cleavage maps with a suitable label. The primary disadvantage is that the procedure requires additional sample manipulations and separations.

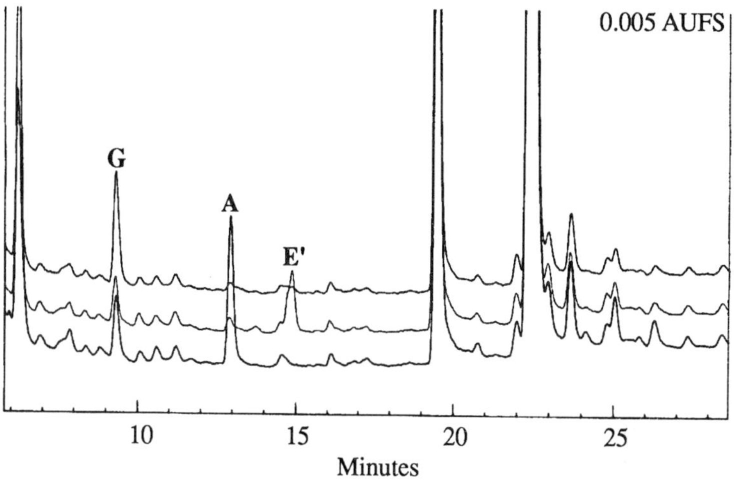

Figure 4a. Amino terminal sequencing of peptide IV. Raw data from cycles 3, 4, and 5 (lower to upper chromatogram), or A, E', and G, respectively are comparison plotted. Since E', not E", is found in cycle 4, the side chain remains stably blocked.

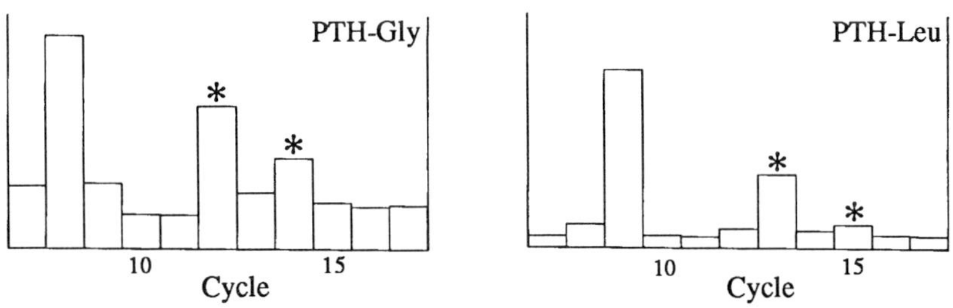

Figure 4b. Amino terminal sequencing of peptide IV. Partial amino acid histograms of yields (in pmol) for PTH-Gly and PTH-Leu. Full scale is 6.0 pmol for each. The assignments for cycles 12-15 are G-L-G-L, confirming the second modification reaction.

V. ACKNOWLEDGMENTS

The authors gratefully acknowledge Anita Hong for peptide synthesis, Pam Keim for amino acid analysis, Terry Lee for mass spectrometry, and Alex Glazer for stimulating discussions.

8

STUDIES ON THE USE OF SILYL COMPOUNDS FOR PROTEIN CARBOXY-TERMINAL SEQUENCE ANALYSIS[1]

Chad G. Miller, Cheng-te Kong, and John E. Shively

Immunology Division, Beckman Research Institute of the City of Hope, Duarte, CA.

I. INTRODUCTION TO THE CHEMICAL CHALLENGE.

The successful development of chemical strategies for the sequential degradation of proteins from the carboxyl-terminus has certainly not achieved a state of refinement comparable to the methodologies for amino-terminal sequencing (1). The chemical challenge arises from the very nature of the reactants themselves. The inherent lack of chemical reactivity of a protein terminal carboxylate coupled with the obligatory high reactivity of an appropriate sequencing reagent have rendered difficult any approach comparable to that of the relatively mild reaction conditions employed in Edman chemistry.

A potent nucleophile such as a protein alpha-amino group and a relatively stable yet efficient coupling reagent such as phenylisothiocyanate do not facilely translate into corresponding counterparts for a conceptually similar reaction imposed at the protein carboxyl-terminus. The documented chemical strategies consequently invoke methods to transform a nucleophilically unreactive carboxylate into a highly reactive electrophile and its subsequent reaction with nucleophiles. The current literature describes three principle chemical routes for degradative cycles releasing three alternative modifications of the carboxyl-terminal amino acid residues. The successive amino acids are identified as their thiohydantoin (2), iminohydantoin (3), or aldehyde (4) derivatives depending on the chemical approach selected. Each of these routes, along with interesting variations, has been intensively investigated and yet none has thus far demonstrated an efficiency, reproducibility, and sensitivity required to be routinely implemented as a carboxyl-terminal sequencing technique. We have concentrated our studies on degradative routes to the thiohydantoin amino acid derivative owing in part to the chemical relationship it bears to the phenylthiohydantoin product of Edman chemistry from which we may guide reaction conditons and analytical methods.

[1] Supported by NSF grant BBS-8714266.

II. THE COUPLING REACTION.

The chemical transformation of a protein carboxyl-terminus into a proteinylthiohydantoin by reaction with acetic anhydride and ammonium thiocyanate, reported by Stark (5) at the micromole level, at once demonstrated the potential efficacy of such a reaction. The precursary formation of a protein mixed anhydride utilizing acetic anhydride satisfied the stratagem of boosting the electrophilicity of the carboxyl-terminus allowing for nucleophilic attack by thiocyanate at a reasonable rate. Alternate versions of this chemistry employing alkali metal salts of thiocyanate or thiocyanic acid directly have been reported and may offer distinctive advantages, yet substantial progress boasting any one particular coupling reagent has been considerably limited. The method employing thiocyanic acid, developed by Gurd and coworkers (6), has proved to be one of the more successful approaches but suffers from the thermal instability of the reagent.

The use of trimethylsilylisothiocyanate (TMSITC) as a coupling reagent for protein carboxylates was described by Hawke et al (7). The reagent exhibited a greater stability than the thiocyanate salts and proved easier to manipulate in manually derived degradative cycles on nanomolar amounts of peptide and protein. The carboxyl-terminus is necessarily converted to the mixed acetyl anhydride by preactivation with acetic anhydride for five minutes at 55°C. The formation of a peptidylisothiocyanate, i.e. an acylisothiocyanate, was implicated as an obligatory intermediate enroute to the cyclized acylthiohydantoin as suggested by reasonable reaction mechanisms for the overall reaction, although no direct evidence to confirm this was available. A suggested chemical mechanism for the formation of the acylisothiocyanate is shown in Fig. 1. Direct reaction of the TMSITC molecule with the protein acetyl anhydride, or its azlactone, seems unlikely, despite a coupling temperature of 55°C for 30 minutes, due to the inherently low nucleophilic strength of the N-substituted isothiocyanate. It would be more tempting to propose the generation of acetylisothiocyanate in the complex reaction equilibrium mixture which may react with equilibrium amounts of protein carboxylates thus forming protein acetyl anhydrides and "free" thiocyanate. The thiocyanate, generated at the carboxyl terminus and associated with the anhydride, could readily transform the anhydride into an acylisothiocyanate. In fact, preliminary data demonstrate that apomyoglobin, pretreated with triethylamine, reacts

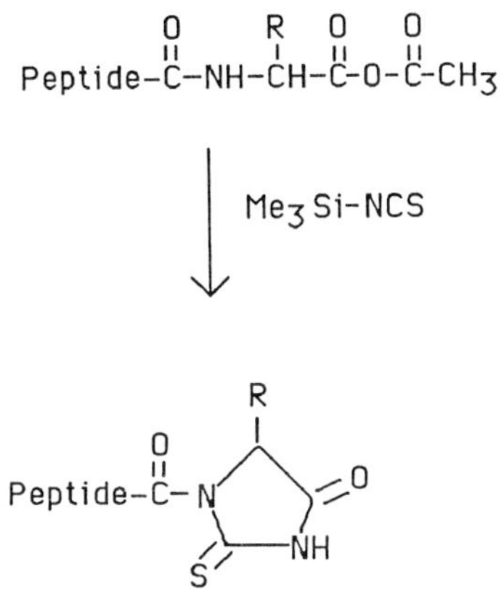

Fig. 1. Formation of the peptidylthiohydantoin from the anhydride.

with a premixed solution of acetic anhydride and TMSITC resulting in the C-terminal glycylthiohydantoin, after cyclization and cleavage chemistry. Indeed, the nucleophilic nature of the thiocyanate anion would presuppose a product distribution between acylthiocyanate and acylisothiocyanate isomers, the latter being that form which can cyclize to the thiohydantoin. It is not unrealistic to assume that in the presence of thiocyanate anion both isomers would form although the concentration of acylthiocyanate would diminish as isomerization to the more stable acylisothiocyanate proceeds as indicated in Fig. 2.

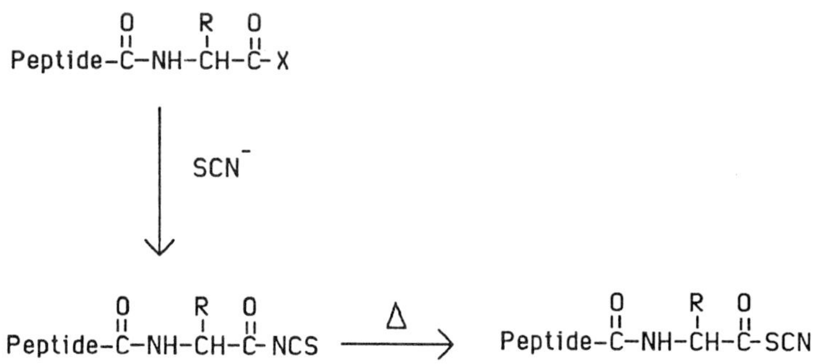

Fig. 2. Isomerization of an acylisothiocyanate to acylthiocyanate.

III. THE CYCLIZATION REACTION.

The cyclization of the acylisothiocyanate function at the carboxyl-terminus of the coupled protein to the acylthiohydantoin is formally the second reaction step in the two-step mechanism for acylthiohydantoin formation. However, its contribution to the overall degradative reaction rate seems overshadowed by the coupling reaction and the subsequent cleavage (or hydrolysis) reaction. The coupling-cyclization steps can be considered as two distinct reactions promoted under one set of reaction conditions. The subsequent cleavage reaction occurs under a different reaction condition. The advantage to resolving the coupling from the cyclization chemistry may indeed be more than academic. The cyclization step is reliant on the alpha-amide nitrogen to react in nucleophilic fashion at the thiocarbonyl moiety of the acylisothiocyanate group. The elevated reaction temperature of 55°C may in part be imparting sufficient energy for the carboxyl-terminal acylisothiocyanate to assume a correct conformation in which the intramolecular ring-closure event may ensue. Hawke and coworkers (7) had found a marked catalysis in the rate of acylthiohydantoin formation by addition of pyridine to the coupling-cyclization reaction medium. Although one may discount a rate enhancement solely attributable to general base catalysis due to the vast difference between the pKa of pyridine and that of an amide proton, it is possible to postulate a nucleophilic catalytic component wherein pyridine reacts with the acylisothiocyanate to form a more highly reactive electrophile assisting in the ring-closure. A similar proposition for nucleophilic catalysis by tertiary amines in the formation of phenylthiocarbamoyl derivatives in the Edman coupling reaction had been offered by Tarr (8). The overall reaction scheme for the cyclization to the thiohydantoin ring is illustrated in Fig. 3.

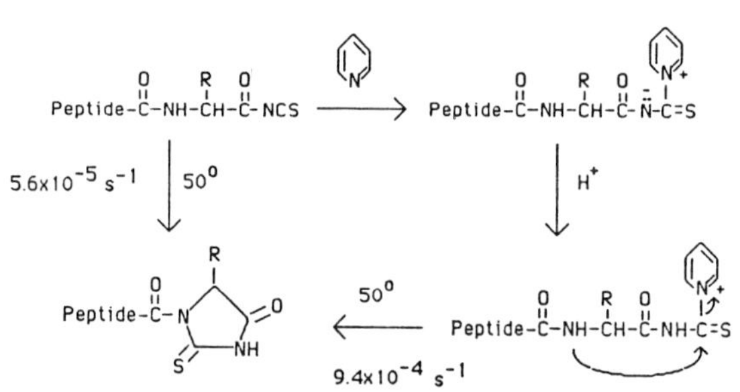

Fig. 3. Thiohydantoin ring formation.

IV. THE CLEAVAGE REACTION.

Successful coupling and cyclization chemistry results in the formation of the protein acylisothiocyanate which must be subjected to a cleavage reaction to generate the C-terminal amino acid thiohydantoin derivative. A number of different cleavage reaction conditions have been explored by earlier workers at the submicromole level of peptidyl or proteinyl-thiohydantoin. The broad spectrum of conditions ranged from harsh acid (12N HCl) and concentrated base (saturated aqueous triethylamine) to mild reaction with acetohydroxamic acid in slightly basic medium (9). Yields of recovered thiohydantoin derivatives fluctuated based on the approach used and were generallyin the 30 to 40 percent range. A cleavage reaction of a millimole amount of a dipeptidyl-thiohydantoin in dilute alkali resulted in a high yield of recovered thiohydantoin despite the concommitant hydrolytic ring-opening side-reaction (10). A brief survey of the conditions reported reveals the inadequacy of any one approach and clearly reflects the complexity of the problem. The difficulty encountered in the cleavage reaction is akin to that present in the coupling-cyclization steps in that the impediment to a smooth and facile reaction lies within the nature of the reactant. The acylthiohydantoin ring system presents four sites for reaction with nucleophiles or bases. The desired reaction at the N-1 exocyclic acyl function to liberate the thiohydantoin amino acid is competitive with ring oponing at either the thiocarbonyl or internal carbonyl functions along with a simple proton abstraction from the internal ring imide group. In addition, desulfuration of the internal thiocarbonyl function may occur rendering the conversion to a hydantoin derivative.

Our previously reported sequencing strategy employed the acidic conditions of 12N HCl to promote the cleavage reaction (7). Thiohydantoin derivatives of the first three C-terminal amino acids of apomyoglobin (3 nmoles immobilized to activated controlled pore glass beads) were recovered in fair yield (60 to 70%). Milder conditions for cleavage chemistry were obviously warranted and subsequently studied.

A cleavage reaction protocol employing a basic pH must take into account the base lability of the imide proton which, when ionized, drastically reduces the hydrolytic capacity of the N-1 acyl function. Cleavage results with imidazole suggest the optimal pH to lie in a range which favors protonation of the N-acylthiohydantoin (

pK_a=~7) and yet allows for a satisfactory equilibrium amount of ionized cleavage reagent. A cleavage reaction utilizing acetohydroxamic acid in aqueous pyridine at pH=8 compromises the extents of ionization of the acylthiohydantoin imide proton and the acetohydroxamic acid (pK_a=~9) which reacts preferentially as the hydroxamate oxyanion in accord with the known reactive species of hydroxyamine as depicted in Fig. 4. Our results with acetohydroxamate as a cleavage reagent for nanomolar amounts of peptidylthiohydantoins and proteinylthiohydantoins (immobilized on PVDF, Immobilon affinity membrane) proved its practicality by affording good recovery yields (60 to 80 %) of C-terminal thiohydantoin amino acids.

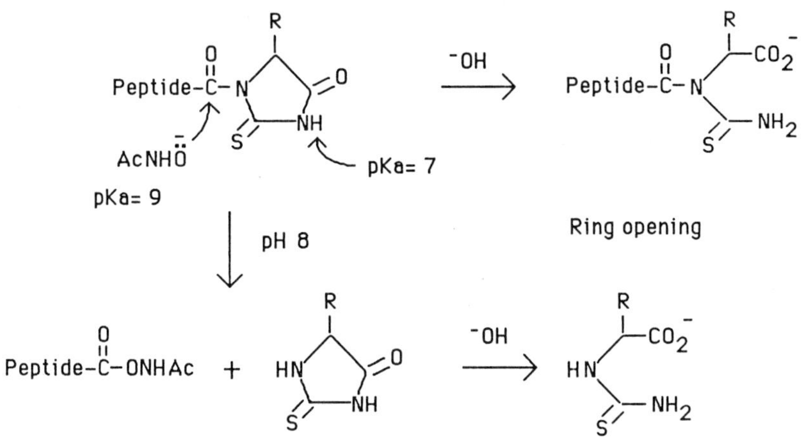

Fig. 4. Hydrolysis of peptidylthiohydantoins.

The base-catalyzed hydrolysis of the thiohydantoin ring, either as the acylthiohydantoin or the released thiohydantoin, would reduce the yield of the C-terminal thiohydantoin derivative. Oxidative desulfuration of the thiohydantoin to the hydantoin would also contribute to a diminished yield. The stability of thiohydantoin amino acids during the cleavage reaction was assessed by incubation of leucylthiohydantoin in 0.1M acetohydroxamate in aqueous pyridine at pH 8.5 for a typical time course of 30 minutes at 50°C. Approximately 70% of the thiohydantoin had decomposed as analyzed by the corresponding loss in detectable peak height of a reversed-phase HPLC trace of the cleavage mixture. The decomposition products were evident from a distribution of several new chromatographic peaks but were not yet characterized. The addition of dithiothreitol significantly prevented the decomposition of the thiohydantoin under the same model cleavage reaction conditions. The multipath mechanism for thiohydantoin decomposition is postulated to involve hydrolytic reactions and desulfuration of the ring system. The mode of dithiothreitol protection is unknown but may involve the association of the mercaptan with the thiohydantoin generating a rapidly equilibrating ensemble of unstable covalent adducts, which effectively competes with what otherwise is an unobstructed hydrolytic or oxidative pathway. Similarly, the lability of leucylthiohydantoin was greatly reduced in the presence of N-acetylcysteine. The addition of thioanisole surprisingly offered a significant degree of protection against thiohydantoin decomposition, which acknowledges the existence of perhaps additional modes of decomposition capable of being suppressed by thioethers.

The cleavage chemistry must regenerate unmodified protein C-terminal carboxylate for the degradation to be repetitive. The reaction of the protein acylthiohydantoin with acetohydroxamate presumably affords an O-acylhydroxamate intermediate according to proposed mechanism. The opportunity for blocking the C-terminus may potentially occur at the stage of the activated protein O-acylhydroxamate ester. The decomposition of this activated ester to the C-terminal azlactone would result in lower yields of subsequent thiohydantoin derivatives if an unmodified carboxyl-terminus could not be regenerated. The back hydrolysis of any azlactone product may require a longer exposure to the cleavage reaction medium than is necessary for the actual cleavage reaction itself. The presence of mercaptan such as dithiothreitol, to preserve the released thiohydantoin amino acid, may also react nucleophilically with the active ester, or serve as a competent cleavage reagent itself, forming initially a thiolester which can readily isomerize to the 2- or 3-hydroxy ester through a five-membered ring intermediate. The modification of a cleaved peptide fragment from the peptidyllthiohydantoin in an acetohydroxamate cleavage solution containing dithiothreitol was evidenced by the appearance of a FAB mass spectral molecular ion consistent with the structure of the dithiothreitol adduct to the peptide less the C-terminal residue (see below). A consequence of this type of reaction would be a reasonable recovery of the C-terminal thiohydantoin amino acid, or a good initial yield, but a very poor, or in the extreme non-existent, repetitive yield.

VI. REFERENCES

1. Smith LM Sanders JZ, Kaiser RJ, Hughes P, Dodd C, Connell CR, Heiner C, Kent SBH, Hood LE, (1986) Nature 321, 674
2. Hawke DH, Lahm HW, Shively JE, Todd CE (1987) Anal. Biochem. 166, 298
3. Furka A, Dibo G, Kovacs J, Sebestyen, F (1983) Anal. Biochem. 129, 14
4. Hargrave PA, Wold F (1973) Int. J. Peptide and Protein Res. 5, 85
5. Johnson L, Tarr G (1987) in Walsh K (ed) Methods in Protein Sequence Analysis 1986, p. 351, Humana Press, Clifton, NJ, USA
6. Rose K, Savoy LA, Simona MG, Offord RE, Wingfield P (1988) Biochem J. 250, 253
7. Jue PA, Doolittle RF (1985) Biochem 24, 162
8. Carles C, Gueguen P, Dumas BR (1987) FEBS Letters 212, 163
9. Breddam K, Johansen JT, Ottesen M, (1984) Carlsberg Res. Comm. 49, 457.
10. Hawke DH, Meister SM, Yuan PM, Hunkapiller MW (1988) in press.
11. Hoare DG, Koshland DE (1967) J. Biol. Chem. 242, 2447
12. Carraway KL, Koshland DE (1968) Biochim. Biophys. Acta 160, 212
13. Hayashi R (1988) in Bhown AS (ed) Protein/Peptide Sequence Analysis: Current Methodologies, p. 145, CRC Press, Baton Rouge, LA, USA

V. SAMPLE IMMOBILIZATION.

The initial studies with TMSITC as a coupling reagent and 12N HCl as cleavage reagent were performed on either peptide (bradykinin) in solution or protein (apomyoglobin) immobilized to activated controlled pore glass beads. The experiments described in this account used either synthetic peptides in solution or proteins immobilized to an activated polyvinylidine difluoride membrane (PVDF, Immobilon affinity membrane, Millipore) as shown in Fig. 5. The protein coated membrane

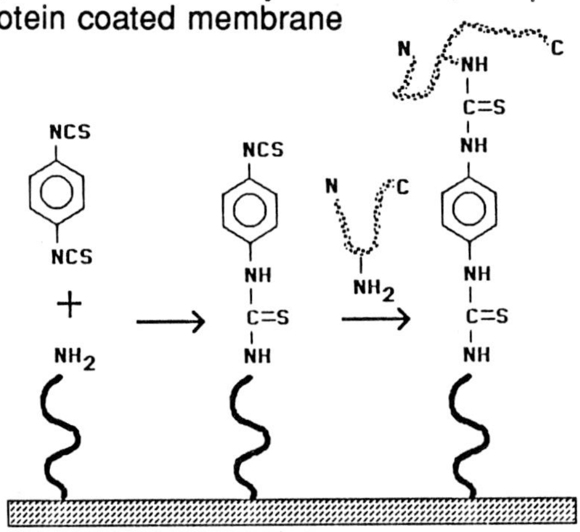

Fig. 5. Sample immobilization on derivatized PVDF or silica.

(about 1mm X 4mm) was inserted into a continuous flow reactor of the type described by Shively (11) fitted with inlet and outlet teflon narrow bore tubing. The peptide studies allowed for the isolation and characterization of reaction intermediates in the coupling-cyclization and cleavage steps. The utility of the solid-phase for protein samples is demonstrated by sequence analysis on either apomyoglobin or bovine serum albumin. A degradation cycle consisted of a series of distinct steps including preactivation with acetic anhydride, solvent wash, reaction with TMSITC, solvent wash, reaction with acetohydroxamate, solvent extraction of cleaved thiohydantoin amino acid, and a final solvent wash. The immobilization yields of protein on the activated PVDF membranes, as assessed by amino acid hydrolyses of the protein coated membrane, were greater than 90% with a similar retention yield after three cycles of degradation chemistry. The retention of protein by the membrane under the conditions of either 12N HCl cleavage or saturated aqueous triethylamine were dramatically lower to the extent of 20 - 50%.

VI. SPECIFIC APPLICATIONS.

The introduction of acetohydroxamate as a cleavage reagent into our methods necessitated its comparison with 12N HCl as an efficient protocol. The coupling-cyclization reaction was performed on acetic anhydride preactivated protein samples (50°C for 5 min) followed by treatment with TMSITC at 50°C for 30 min.The first round

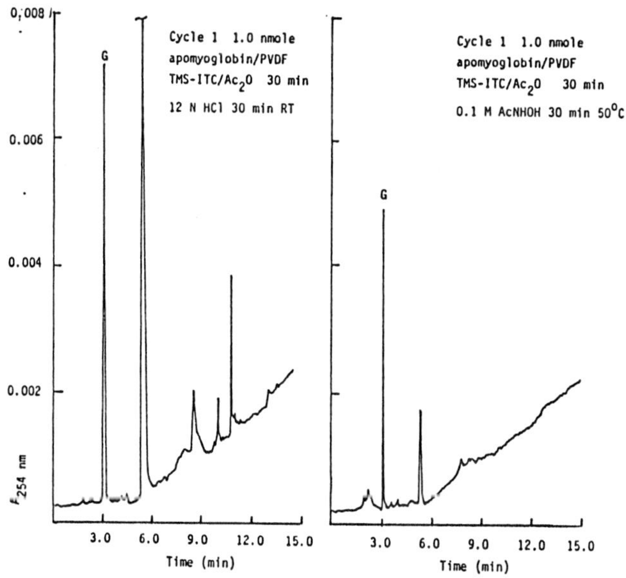

Fig. 6. Apomyoglobin cleaved with 12 N HCl or acetohydroxamate.

degradation cycles of 1 nmole of apomyoglobin, immobilized to activated-PVDF, employing either 0.1M acetohydroxamic acid in aqueous pyridine, pH 8.5, containing dithiothreitol at 50°C for 30 min, or 12N HCl at room temperature for 30 min each resulted in reasonable yields of glycylthiohydantoin as shown in Fig. 6. The calculated yields of glycylthiohydantoin are 80% and 50% for the two cleavage conditions. The solid-phase technique allows for a substantial reduction in chromatographic background by enabling extensive solvent rinses to be introduced into the step-wise cycle. The cleavage reagent, unless volatile, will be extracted along with the thiohydantoin and thus will contribute to background absorbance. This is the case with residue contaminants in 12N HCl and acetohydroxamate which both may be detected at 254 nm. The repetitive yields for either cleavage method are poor as subsequent thiohydantoin yields, namely glutamine-thiohydantoin, decreased abruptly. The diminution in thiohydantoin recovery may in part be due to the presence of a glutamine residue at the penultimate C-terminal position. There exists the potential for cyclization to an imide during the coupling-cyclization steps which would render the modified carboxy-terminus inert to further degradation. The fact that subtilisin, which has a glutamine residue at the carboxy-terminus, afforded no thiohydantoin after one complete cycle of degradation is compelling evidence that glutamyl, and presumably asparaginyl, residues are problematic. Three cycles of degradation of 1 nmole of bovine serum albumin, immobilized on activated PVDF, resulted in a fair recovery of alanylthiohydantoin (the C-terminal residue) after the first cycle, a similar recovery of leucylthiohydantoin at cycle two, and no alanylthiohydantoin detectable above background carryover from cycle-1, as shown in Fig. 7. Coupled with the various factors which may contribute to a reduced repetitive yield described above, the presence of a subterminal glutamine at residue position five may impede the degradative yield of successive C-terminal residues.

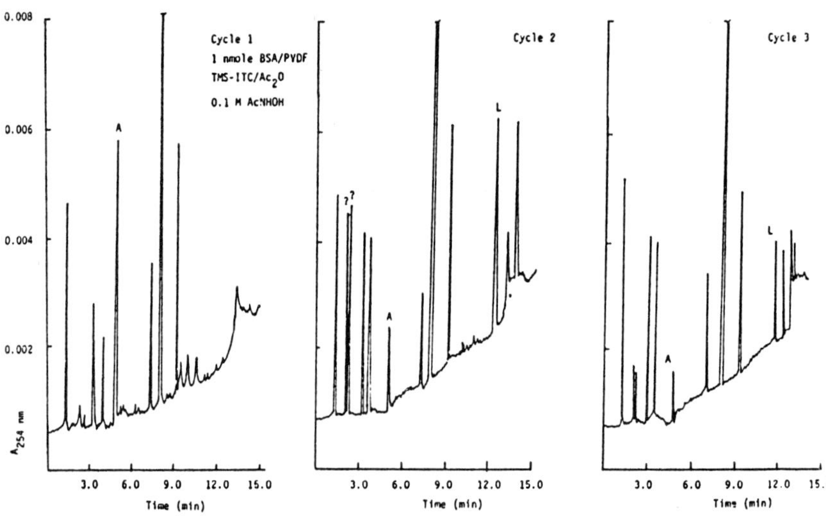

Fig 7. Three cycles of degradation of BSA immobilized on PVDF.

A synthetic heptapeptide peptide having the sequence Leu-Arg-Arg-Ala-Ala-Leu-Gly(CO_2H) was preactivated with acetic anhydride at 50°C for 5 min and treated with TMSITC at 50°C for 30 min. Aliquots of the coupling-cyclization mixture containing the peptidylthiohydantoin were exposed to alternate cleavage reaction conditions and analyzed by reversed-phase HPLC and FAB-MS. The peptide less the C-terminal glycine residue was identified under the various conditions of 12N HCl, 0.1M acetohydroxamic acid in aqueous pyridine at pH 8, and 0.1M acetohydroxamic acid in aqueous pyrindine at pH 8 containing dithiothreitol. Only in the last case was a molecular ion observed which corresponds to a dithiothreitol adduct to the resulting hexapeptide. This result clearly implicates the mercaptan as a source of side-reaction which reduces the effective amount of unmodified protein carboxyl-terminus available for degradation chemistry. We are currently exploring a number of alternative reductants for use in the cleavage chemistry.

VII. SUMMARY.

We have described in this account our continuing studies on the application of trimethylsilylisothiocyanate to the carboxy-terminal degradation of peptides and proteins. The reagent demonstrates a greater stability and more efficient reactivity than thiocyanate salts and thiocyanic acid. The problems associated with the sequencing chemistry are not solely dependent on the coupling reagent but, as described, are integral to the nature of the reaction paths any generic coupling reagent would have to promote. We have evidence that the cyclization step to form the thiohydantoin ring plays a primary role in the overall energetics of the degradation cycle. The cleavage step can be carried out under mild conditions with acetohydroxamate. The addition of dithiothreitol in the cleavage reaction preserves the intact thiohydantoin against decomposition yet may, as we have demonstrated, further react with the carboxyl-terminus generating a blocked functionality for successive cycles. A thiolester first formed as the dithiothreitol adduct might rearrange to the more hydrolytically stable oxygen ester. The side chain residues of aspartate and glutamate have been previously implicated as problematic for thiocyanate degradation chemistry due to anhydride ring formation (4). We have encountered similar impedence to successful reaction with glutamine as a C-terminal residue. The difficulties in chemistry owing to the individual amino acid side chains must be addressed in order for a method of degradation to become practical and generalized. Once these major impediments have been chemically solved, the degradation chemistry should be efficient enough to proceed through at least ten cycles and be amenable to automation.

VIII. REFERENCES.

1. M. Rangarajan, in " Protein/Peptide Sequence Analysis: Current Methodologies ", p.135, A.S. Bhown, Ed., CRC Press, Boca Raton, FA. (1988).

2. L.D. Cromwell and G.R. Stark, Biochemistry,**8**,4735 (1969).

3. G.E. Tarr, in " Solid-Phase Methods in Protein Sequence Analysis", p.139 R.A. Laursen, Ed., Pierce Chem. Co.,Rockford,IL (1975).

4. M.E. Parham and G.M. Loudon,Biochim. Biophys. Res. Commun.,**80**,7 (1978).

5. G.R. Stark,Biochemistry,**7**,1796 (1968).

6. F.E. Dwulet and F.N.R. Gurd, Int. J. Pept. Protein Res.,**13**,122 1979).

7. D.H. Hawke, H.W. Lahm, J.E. Shively, and C.W.Todd, Anal. Biochem.,**166**,298 (1987).

8. G. E. Tarr, in " Methods of Protein Microcharacterization ", p.177, J.E. Shively,Ed. Humana Press, Clifton, N.J. (1986).

9. J.L. Meuth, D.E. Harris, F.E. Dwulet, M.L. Crowl-Powers, and F.R.N. Gurd, Biochemistry, **16**,3750 (1982).

10. G.W. Kenner, H.G. Khorana, and R.J. Stedman,J. Chem. Soc.,673 (1953).

11. J.E. Shively,P. Miller, and M. Ronk, Anal. Biochem.,**163**,517 (1987).

9

AN IMPROVED INJECTION SYSTEM FOR ON-LINE LIQUID CHROMATOGRAPHS: APPLICATION TO HIGH SENSITIVITY PHENYLTHIOHYDANTOIN AMINO ACID ANALYSIS

GEOFFREY S. BEGG
RICHARD J. SIMPSON

Joint Protein Structure Laboratory,
Ludwig Institute for Cancer Research (Melbourne Branch)
and
The Walter and Eliza Hall Institute of Medical Research
PO Royal Melbourne Hospital, Parkville, Victoria 3050, Australia

I. INTRODUCTION

Twenty one years have elapsed since the first instrument for performing automated amino acid sequence analysis was described by Edman and Begg [1]. During the intervening period we have witnessed many improvements in both instrumentation [2] and identification of phenylthiohydantoin (PTH)-amino acids using HPLC [3] which now permit amino acid sequence information to be obtained at low picomole levels. For instance in 1967, approximately 10 nanomole of PTH-amino acid derivative was required for unambiguous identification whereas today this task can be performed using low picomole quantities (approximately 5 picomole). This has been largely due to the advent of microbore column (2.1 mm i.d.) HPLC.

One of the shortcomings of the first commercial automated sequencers introduced in 1970 and indeed in the more recent "gas-phase" sequencer described by Hewick <u>et al</u>. [2] in 1981 was the failure to appreciate the extent of heat destruction of labile PTH-amino acids (eg. carboxymethyl-Cys, -Lys, -Asp, -Met, -Trp, -Gln, -Thr and -Ser). Considerable improvements in this area were made with on-line PTH-amino acid analysis [4]. In addition to a more efficient PTH-amino acid analysis this on-line approach has resulted in improved yields of the more labile PTH-amino acid derivatives.

Abbreviations used: HPLC, high performance liquid chromatography; i.d., internal diameter; o.d., outer diameter; PTFE, polytetrafluoroethylene; PTH-, phenylthiohydantoin.

In spite of the advantages of on-line PTH-amino acid analysis several shortcomings exist in the presently available commercial sequencers (Applied Biosystems Inc. models 470A and 477A equipped with model 120A PTH-amino acid analyser). One is the constraint on the percentage of sample applied to the on-line PTH-amino acid analyser. With the present configuration less than half of the PTH sample is injected onto the HPLC column while the remainder is transferred to a non-refrigerated fraction collector where the levels of the more labile PTH-amino acids decrease due to heat destruction. A second shortcoming is the difficulty in maintaining those parameters (flow rate, time) which control the transfer of liquid from the conversion flask of the sequencer into the sample loop of the HPLC injection valve.

This report describes an improvement to the existing commercially available on-line injection system. We have designed a simple, low-cost device constructed from commercially available components which is easily incorporated into existing Applied Biosystems sequencers equipped with on-line PTH-amino acid analysers. The device permits the quantitative injection of the total PTH-amino acid sample and in contrast to the Applied Biosystems system is not dependant upon critical timing for efficient transfer of PTH-amino acids from the conversion flask to the sample loop of the injection valve.

II. MATERIALS AND METHODS

Figure 1a is a diagram of the injection system device. A length of large-bore PTFE tubing (1.5 mm o.d. x 0.8 mm, i.d., #243201, Beckman Instruments Inc., Altex Division, San Ramon, CA., USA). **A** is connected between the outlet port of the Rheodyne (model no. 7125) injection valve **B** of the HPLC and a solenoid valve (Miniature TFE Isolation Solenoid Valve, #190224 S30, Angar Scientific Inc., Cedar Knolls, NJ, USA) **C**. The outlet from the solenoid valve is connected to waste.

Variation of the amount of fluid transferred into the sample-loop of the injection valve is done by simply varying the length of PTFE tubing **A**. For instance, total sample transfer from the conversion flask of sequencers (either model 470A or 477A) is achieved by adjusting the length of PTFE tubing to approximately 1500 mm.

Alternatively the amount of sample transferred to the sample-loop can be varied by introducing a 500-µl gas-tight syringe (Scientific Glass Engineering Pty. Ltd., Ringwood, Victoria, Australia, #500A-RN-GSG) **D**, in-line, between the injector valve **B** and the solenoid valve **C** (see Fig. 1b). In this case the stainless-steel plunger of the gas-tight syringe is replaced with a hollow PTFE plunger **E** (a suitable length of PTFE tubing, 3.0 mm o.d. x 1.5 mm i.d.; #243202, Beckman Instruments Inc. with a smaller dimension PTFE tubing, 1.5 mm o.d. x 0.8 mm i.d inserted into its full length and which connects to the solenoid valve **C**) so that a flow-through design results. Variation in volume of fluid transferred to the sample-loop is achieved by simply altering the length (by sliding) of the PTFE-plunger **E** in the syringe-barrel.

III. RESULTS AND DISCUSSION

The commercial on-line injection system as supplied by Applied Biosytems with the model 120A PTH-amino acid analyser connected either to the model 470A or 477A sequencer relies on the large difference in viscosity of the sample solvent compared to that of argon gas. As shown schematically in Figure 1c, relatively large-bore tubing **F** leads from the conversion flask **G** of the sequencer to the 50-μl sample-loop of the Rheodyne injection valve **B** in the model 120A PTH-amino acid analyser. Attached to the exit port of the injection valve is a section of capillary tubing leading to a waste container. Dissolved PTH-amino acid derivatives are forced by gas pressure from the conversion flask into the injection valve and into the sample-loop. Liquid flow is relatively fast since at this stage only low-viscosity argon gas is passing through the fine capillary tubing **H** and out to waste. Once the sample liquid has filled the loop of the injection valve and starts to flow through the capillary tubing the rate of flow markedly decreases. The program of the sequencer must be adjusted accurately in order to minimise the further flow of liquid through the loop and into the capillary tubing since any liquid which has entered the capillary tubing is routed to the waste container and consequently lost.

In the injection system we have designed (Fig. 1a) the capillary tubing situated on the exit port of the Rheodyne injection valve in the commercial instrument (**H** in Fig. 1c) is replaced by a length of large-volume, large-bore tubing **A** which is closed gas-tight by an "isolation" solenoid valve **C**. When a transfer function of the Sequencer is activated, and the solenoid valve **C** is shut-off, liquid positioned in the conversion flask will flow into tubing **F** (and **A**) until the pressure ahead equals the applied pressure. When the pressures are equal the liquid stops and stays in this position indefinitely. The position where the sample stops may be adjusted by altering either the gas pressure or the volume of tubing **A**. This allows precision adjustment of the flow of the sample (if its volume equals the volume of the sample loop) into the loop of the injection valve. Since movement of the liquid comes to a complete stop, no critical timing is necessary. The transfer time in the sequencer programs (for both models 470A and 477A) is adjusted to 100 s (twice the time necessary). Once the liquid has come to rest in the sample loop the sequencer can proceed with its own management and the sample in the sample loop will inject onto the column by activation of the injection valve. When the PTH-amino acid analyser cycle has been completed the injection valve is returned to the "Load" position with the sample loop containing solvent, and the connecting tubing (**F** and **A** in Fig. 1a) under transfer pressure. Before the PTH-amino acid from the next Edman degradation cycle is dissolved, the isolation solenoid valve (**C** in Fig. 1a) is opened to waste. This allows the release of pressure and a further gas flow through the tubing (**F** and **A**) transferring liquids to waste. This gas flow is continued (200 s) until the whole system is completely empty. The solenoid valve **C** is again closed before the next PTH-amino acid derivative is dissolved and transferred from the conversion flask to the sample-loop of the injection valve.

Control of solenoid valve **C** is by a simple electronic timer (based on the 555 integrated circuit) started at the correct time by a function of the sequencer. For the 470A sequencer the "LC Start" function which closes an external contact connected to a terminal strip outside the back panel of the sequencer is used to start the timer. Diode gating is used so that the model 120A PTH-amino acid analyser may still be connected to the same external contact. A circuit diagram for installing the injection device is shown in Figure 2. With the model 477A sequencer, no suitable external contact is provided so that the function "Block Flush" is used. The function "Block Flush" is unique in that it is the only function where valves 14 and 20 operate at the same time. Connection to both valves 14 and 20 via plug J1 in the 477A sequencer is used to start the timer controlling the injector solenoid valve **C**. Use is made of the optocouplers (4N28 Integrated circuits) so that disturbance to the electronics of the sequencer does not take place. The optocouplers are connected so as to cause the timer to start only when both valves 14 and 20 operate

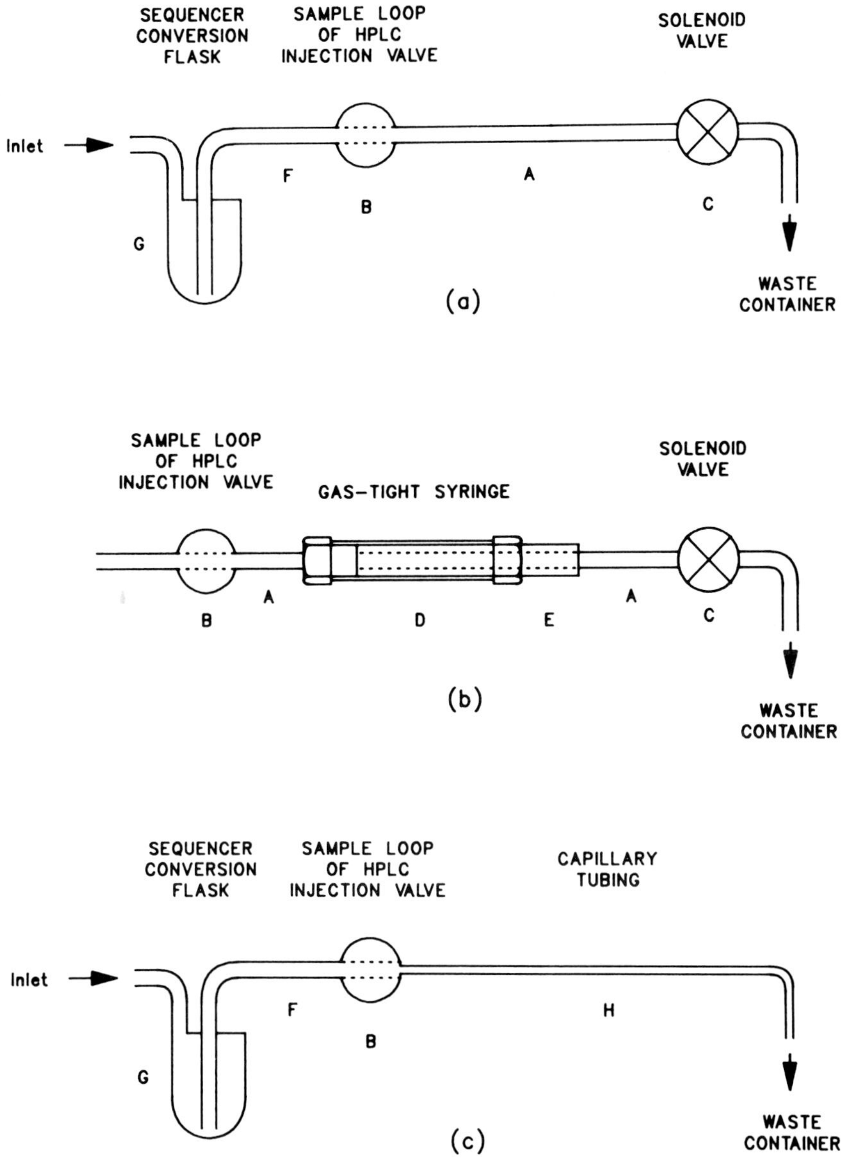

Fig. 1. Schematic diagram of on-line injection systems. (a) Improved injection system described in this report. (b) Optional variable volume device for use in Fig. 1a. (c) Injection system utilized in Applied Biosystems sequencers (models 470A and 477A) equipped with on-line model 120A PTH-amino acid analysers.

together. The function "Block Flush" is used just prior to "Clear Inj to Waste" function so that operation of the injection solenoid valve will occur at the correct time. The function "Block Flush" is also used at other times during the conversion cycle but because of the timing, no programming problem occurs. With normal conversion cycle programming used with the 477A sequencer the timer may be set anywhere between 206 and 416 s (or 210 and 574 s for the model 470A sequencer), hence by using a set time of 300 s allows for any minor drift in the timer. Modifications to the existing operational programs for Applied Biosystems sequencers (models 470A and 477A) required for installation of this injection device are listed in Table 1.

When using a gas pressure of 3.5 psi the volume of the PTFE tubing connecting the injection valve to the solenoid valve (**A** in Fig. 1a) should be four times the volume of the tubing connecting the conversion flask to the injection valve **F** plus the volume of the sample loop. Variation in tubing dimensions (length, internal diameter) **F** in any individual sequencers can be easily compensated for by adjustment of the length of PTFE tubing **A** (in Fig. 1a). Typical length of PTFE tubing **A** (0.8 mm i.d.) required for 100% sample injection was 1650 mm.

By introducing a gas-tight syringe into the system (**D** in Fig. 1b) the volume of fluid transferred to the sample-loop can be varied with ease by adjusting the PTFE-plunger **E**. If the desired volume to be injected is less than the total volume of the sample loop, gas will also be injected onto the HPLC column. In our experience using the model 120A PTH-amino acid analyser this situation presents no chromatographic problems. For 100% sample injection it is necessary to adjust the volume of solvent S4 used to dissolve the PTH-amino acid derivative to match the volume of the injection valve sample-loop (50 μl). This was achieved by - (i) programming for one addition only of solvent S4 and - (ii) selecting a length of PTFE tubing (500 mm of 0.3 mm i.d.) suitable for the solvent measuring loop in the sequencers (models 470A and 477A) but the length should be adjusted for possible variations of tubing diameter.

If necessary, the original commercial injection device can be readily restored by simply changing one connection at the outlet port of the HPLC injection valve back to the original capillary tubing.

A typical sequence analysis performed on Applied Biosystems sequencer (model 470A plus model 120A PTH-amino acid analyser) equipped with our improved injection system is shown in Figure 3. A more than two-fold increase in

TABLE I. MODIFIED CONVERSION CYCLE PROGRAMS FOR OPERATION OF INJECTION DEVICE, WITH APPLIED BIOSYSTEMS SEQUENCERS EQUIPPED WITH ON-LINE MODEL 120A PTH-AMINO ACID ANALYSERS

A. <u>Model 477A Sequencer</u>

Step	Function	Fxn#	Value	Elapsed Time
10	Block Flush	23	6	32 min 15 sec
11	Clear Inj to Waste	14	200	35 min 35 sec
12	Load S4	12	8	35 min 43 sec
13	Argon Dry	22	4	35 min 47 sec
14	Pause	25	198	39 min 5 sec
15	Load Injector	15	100	40 min 45 sec
16	Inject	16	1	40 min 46 sec

B. <u>Model 470A Sequencer</u>

Step	Cartridge Function	Flask Function	Time
20	Deliver R3	LC Start	10
21	Deliver R3	Collect	200
22	Deliver R3	Load S4	10
23	Deliver R3	Argon Dry	4
24	Deliver R3	Pause	350
25	Deliver R3	Collect	100

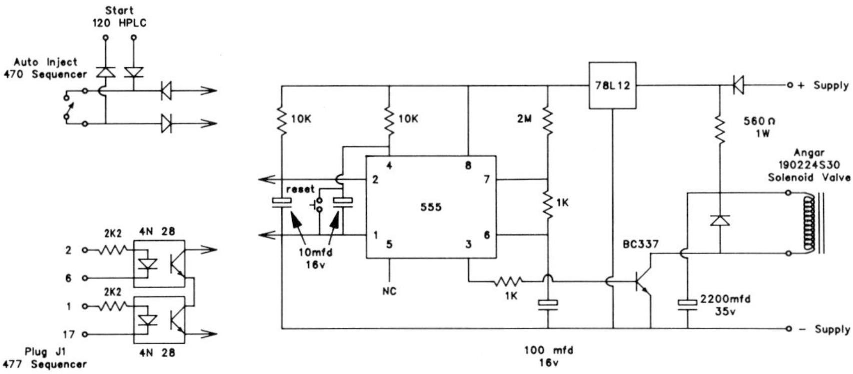

Fig. 2. <u>Circuit diagram of electronic timer controlling the isolation solenoid valve **C** in Fig. 1a</u>. Power to operate the timer is obtained from the power supply of the sequencer.

yield of PTH-amino acids is routinely achieved when compared with the commercial systems.

These results demonstrate the utility of this improved injection system for high sensitivity protein microsequencing in situations where analysis of the total PTH-amino acid derivative is advantageous.

Although other methods for high percentage transfer of dissolved PTH-amino acid derivatives to an on-line HPLC for analysis have been described [5], such a system requires substantive changes to the sequencer architecture. Since installation of the transfer device described in this report does not involve architectural modifications to the sequencer, any future updates to either software or hardware can be readily incorporated to instruments fitted with this device.

The injection device described here has been in constant use in our laboratory on both the model 470A and 477A Applied Biosystems sequencers for over 6 months, approximately 4000 injections, and has proved invaluable in low level sequence analyses.

IV. SUMMARY

A simple, low-cost device for the quantitative transfer of samples to on-line liquid chromatographs is described. The system is assembled from commercially-available components and comprises a length of PTFE tubing, a solenoid valve and a simple timer connected to the exit port of the injector valve. The key advantage of this system for protein microsequencing is that the total phenylthiohydantoin amino acid derivative generated from each Edman degradation cycle of automated protein sequencers is available for identification and quantification by HPLC analysis. In contrast to existing commercial instruments, no critical timing is required for the transfer of sample from the sequencer conversion flask to the sample loop of the HPLC injection valve. This device is easily adapted for use with Applied Biosystems sequencers (either model 470A or 477A) equipped with on-line phenylthiohydantoin amino acid analysers (model 120A) and requires no architectural changes.

V. ACKNOWLEDGMENT

We thank Michael R. Rubira for his assistance in assessing the utility of this injection system.

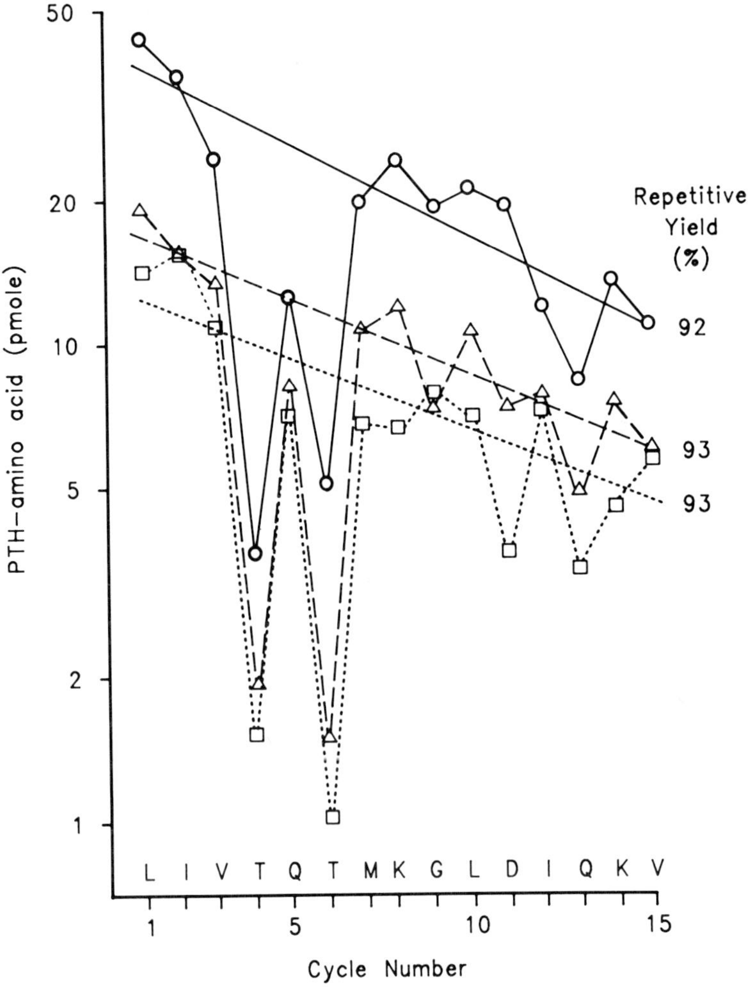

Fig. 3. Sequence analysis of β-lactoglobulin using model 470A Applied Biosystems sequencer equipped with an on-line model 120A PTH-amino acid analyser installed with an improved injection device described in Fig. 1a. Sample (100 picomoles, β-lactoglobulin) and reagents/solvents were from Applied Biosystems. (0-0), data obtained using injection system described in text (Fig. 1a). (Δ-Δ), data obtained using commercial injection system (Fig. 1c); portion of PTH-amino acid derivative which was routed into the fraction collector was immediately dried after the sequencer run, reconstituted in S4 (20% aqueous acetonitrile) and manually analysed using the model 120 PTH-amino acid analyser (-). One-letter notation has been used for amino acids.

VI. REFERENCES

1. Edman, P., and Begg, G. (1967) Eur. J. Biochem. 1, 80.
2. Hewick, R.M., Hunkapiller, M.W., Hood, L.E., and Dreyer, W.J. (1981) J. Biol. Chem. 256, 7990.
3. Zimmerman, C.L., Apella, E., and Pisano, J.J. (1977) Anal. Biochem. 77, 569.
4. Machleidt, W., and Hofner, H. (1980) In "Methods in Peptide and Protein Sequence Analysis" (C.H.R. Birr, ed.), p. 35. Elsevier Press.
5. Rodriguez, H. (1985) J. Chromatogr. 350, 217.

10

A SYNTHETIC PEPTIDE FOR EVALUATING PROTEIN SEQUENCER AND AMINO ACID ANALYZER PERFORMANCE IN CORE FACILITIES: DESIGN AND RESULTS

Ronald L. Niece[1], Kenneth R. Williams[2], Cynthia L. Wadsworth[1], James Elliott[2], Kathryn L. Stone[2], Walter J. McMurray[3], Audree Fowler[4], Donna Atherton[5], Rusty Kutny[6], Alan J. Smith[7]

[1]University of Wisconsin Biotechnology Center
Protein/DNA Sequence/Synthesis Facility
Madison, WI 53705

[2]Yale University School of Medicine
Protein and Nucleic Acid Chemistry Facility
New Haven, CT 06510

[3]Yale University Comprehensive Cancer Center
New Haven, CT 06510

[4]Dept. of Biological Chemistry
and Molecular Biology Institute
UCLA School of Medicine
Los Angeles, CA 90024

[5]The Rockefeller University
New York, NY 10021

[6]Eastman Pharmaceuticals
Great Valley, PA 19355

[7]Dept. of Biological Chemistry
University of California
Davis, CA 95616

I. INTRODUCTION

Recent advances in the biotechnology of protein and nucleic acid sequencing and synthesis have fostered the development of more than 100 core facilities in the U.S. whose primary function is to bring these techniques within reach of a wide spectrum of investigators that includes not only specialists in these areas but also cellular and molecular biologists in general. Because much of the needed equipment is expensive and requires considerable expertise as well as a large number of samples to keep it operating continuously at peak efficiency, core facilities staffed by professional personnel represent an effective and economical means of bringing these technologies to bear on problems related to biochemical research. In order to efficiently utilize a core facility, a prospective user must have realistic expectations concerning the technical capabilities of these facilities. A recent survey of forty core facilities clearly demonstrates that the relatively isolated instances where extremely high sensitivity analyses have been reported are not representative of what can be expected when a typical "unknown" protein sample, prepared by the investigator, is submitted to a well-run core facility, whether it is located in a university, government, or industrial laboratory (1). As part of our plans to provide the sequencing community with control materials, the Research Resource Facilities Group designed and synthesized a 40-residue peptide with characteristics suited for evaluation of protein/peptide sequencers and amino acid analyzers. The peptide was initially released to 103 core facilities as an "unknown" sample to provide data concerning the protein sequencing and amino acid analysis capabilities of existing core facilities. Prior to revealing the actual sequence and composition of these standard peptides, STD-1 and STD-2, at the Second Symposium of the Protein Society, sixty sets of data had been returned which together provide the basis for this report.[1]

[1]Samples of STD-1 and STD-2 are available from KRW.

II. MATERIALS AND METHODS

A. Design of the STD-1 and STD-2 Peptides

The forty-residue STD-1 and STD-2 peptides designed at the University of Wisconsin Biotechnology Center (CLW and RLN) were sufficiently long to challenge the capabilities of automated protein sequencing instruments and with suitable compositions to challenge amino acid analyzer capabilities. The standard peptides were identical except that the one designed for amino acid analysis (STD-2) had the amino-terminus acetylated. The amino acid sequence of these two peptides is shown in Fig. 2.

For protein sequencing the amino-terminal portion of the peptide (STD-1) was designed to provide data reflecting sequencer performance. Proline (which frequently results in incomplete cleavage and excessive lag) and serine (which may cause some blockage of the amino-terminus) were avoided. The sequencer performance could be assessed quickly on an overnight run for repetitive yield using alanine residues at positions 4 and 10 and for initial yield using valine, tyrosine, and alanine at positions 1, 2, and 4. Alanine residues were regularly spaced throughout the sequence to permit an accurate calculation of repetitive yields over various sequences within the peptide. Many less well recovered amino acids such as aspartate, glutamate, arginine, tryptophan, histidine, and cysteine were present early in the sequence to facilitate their identification and accurate quantitation. The middle portion of the sequence was designed to provide sequence interpretation challenges due to increasing lag before washout became significant. Proline was inserted at positions 18 and 25 for this purpose. The carboxy-terminal third of the peptide was designed to present sequencing challenges in the face of increasing washout. Arginine residues at positions 38 and 40 were present to limit washout. The two serines at residues 31 and 32 followed by glycine were expected to be difficult to interpret. With the limited amount of material present, it was expected that only a few picomoles (pmol) of material would be sequencing after 30 cycles of degradation.

In addition to the features relating to protein sequencer performance, consideration was also given to HPLC identification and quantitation of the resulting PTH-derivatives. Aspartic acid was placed early in the sequence to evaluate its separation from the ammonia-PITC artifact which is prominent in early cycles. Tryptophan was also placed early to permit observation of its oxidation products and its separation from DPU (diphenylurea). Histidine was positioned in the sequence between alanine and cysteine because PTH-histidine is usually eluted between PTH-alanine and PTH-dehydroalanine (formed from unmodified cysteine during sequencing). Similarly, arginine was positioned in the sequence between alanine and tyrosine. Histidine and arginine appeared throughout the sequence to permit monitoring the constancy of their elution times throughout the course of the run.

In terms of amino acid composition, the standard peptide was designed so that after acid hydrolysis all 18 of the commonly observed amino acids would be present. To monitor the extent of hydrolysis, a sequence consisting of isoleucyl-isoleucyl-valine was present. The ratios of isoleucine to leucine and of serine to threonine were both 3 to 1 to monitor the degree of resolution of these pairs of amino acids that are frequently incompletely resolved by ion exchange chromatography.

General sequence design considerations included ease of peptide synthesis, stability, and solubility. The peptide was short enough and of a suitable sequence to synthesize chemically and to permit rigorous characterization. Bonds potentially unstable at the extremes of pH found in the sequencer, specifically ASP-PRO and ASN-GLY, were also avoided. Tryptophan is unstable and was placed late in the peptide synthesis to minimize its destruction. A sufficient number of amino acids charged at different pH ranges were included to permit solubility in a broad range of HPLC solvent systems.

Finally the peptide was designed to be useful in monitoring a variety of other techniques typically used in a core facility. It was large enough so that it could be run on SDS polyacrylamide gels and electroblotted. Chemical and enzymatic cleavage sites were incorporated so that the resulting fragments could be sorted according to different characteristics. For example, the extent of cyanogen bromide (CNBr) cleavage could be monitored at the two methionines neither of which was followed by serine or threonine (which can reduce the yield of CNBr cleavage). One CNBr cleavage site was closely followed by proline so that the mixture of peptides produced during _in situ_ CNBr cleavage could be sorted out by blocking non-proline terminated peptides at the appropriate sequencing cycle. Several different enzymatic cleavage sites were incorporated. The potential fragments differ in hydrophobicity and/or charges to permit HPLC separation.

B. Synthesis, Characterization and Distribution of STD-1 and STD-2 Peptides

The STD-1 peptide was synthesized at Yale University (JE) on an Applied Biosystems Model 430A Solid Phase Peptide Synthesizer using standard techniques. After synthesis and removal of the NH_2-terminal tBOC, one half of the resin was acetylated with acetic anhydride to make STD-2. Following HF cleavage, the peptide/resin mixture was extracted with 40 ml 7M guanidine hydrochloride and the peptides purified on two Vydac C-18 columns connected in series. The sequence of

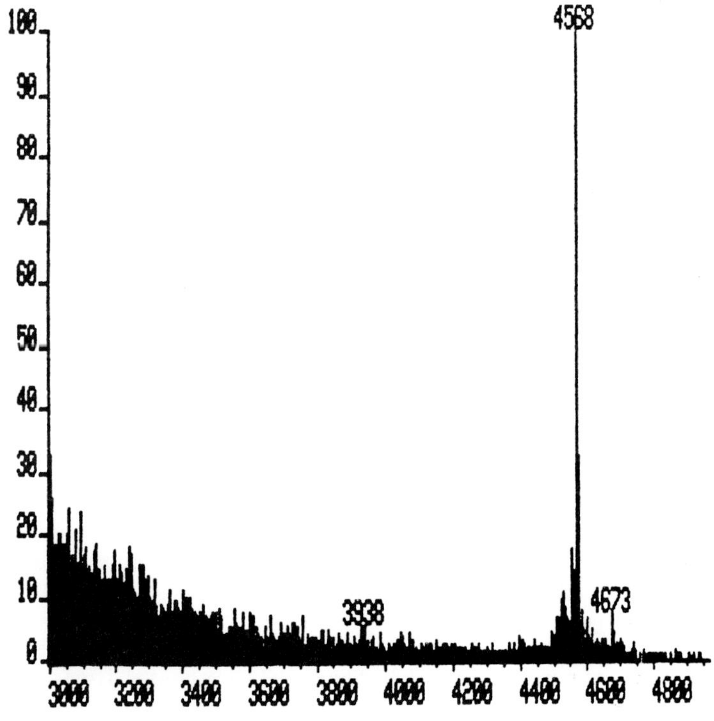

Fig. 1 FAB positive ion mass spectrometry of STD-2. Based on this analysis, the protonated molecular ion of STD-2 has a mass of 4568.22 daltons (predicted value is 4568.31 daltons). Sample was recorded using a thioglycerol matrix and a xenon gun operated at 8 keV.

STD-1 was confirmed by direct sequencing of 5 nanomoles (nmol); its purity was at least 98% with 0.9% preview. The molecular weights were determined by FAB mass spectrometry (WM). As shown in Fig. 1, the observed and predicted protonated molecular weights for STD-2 were within 0.1 atomic mass units of each other.

Aliquots of the standard peptides (100 pmol STD-1 and 1.09 nmol STD-2) were dried in 1.5 ml Eppendorf tubes that had been pre-washed in 0.05% trifluoroacetic acid, 50% acetonitrile and then distributed to the 103 core facilities on a mailing list that had been compiled during the previous two years by the Research Resource Facilities Group. Specific instructions for re-solubilizing both the STD-1 and STD-2 peptides were included with these samples. Extensive testing of randomly selected aliquots indicated that by following the instructions it would be possible to load approximately 75 pmol of STD-1 onto the sequencer and to hydrolyze either 0.5 μg or 4.5 μg STD-2. The instructions specified that for high sensitivity analyzers 10% (i.e. 0.5 μg) and that for less sensitive analyzers 90% (i.e. 4.5 μg) of STD-2 should be taken for hydrolysis and amino acid analysis, respectively. To guarantee the confidentiality of the resulting data, the responses were returned to a third party who removed any postmarks or other identification relating to the originating laboratory prior to forwarding the data to the authors.

III. RESULTS AND DISCUSSION

A. Sequencing of STD-1

A total of 54 responses were received among which two instrument failures were reported. This indicates that there is about a 4% chance of instrument failure when unknown samples are run. Fig. 2 provides a graphical representation of the 54 sets of STD-1 sequencing data. The sequence of the peptide is given at the bottom. The figure indicates the overall number of correct, incorrect, and tentative calls at each position in the sequence. The same data are grouped and averaged in Table I according to instrument model. Based on the data in Table I, Applied Biosystems instruments clearly give superior results with this peptide compared to those obtained with the Beckman 890 "spinning cup" instrument or a manual approach. Because the peptide was designed to be used with high sensitivity instrumentation, only a limited amount was provided accounting for the relatively poor showing of

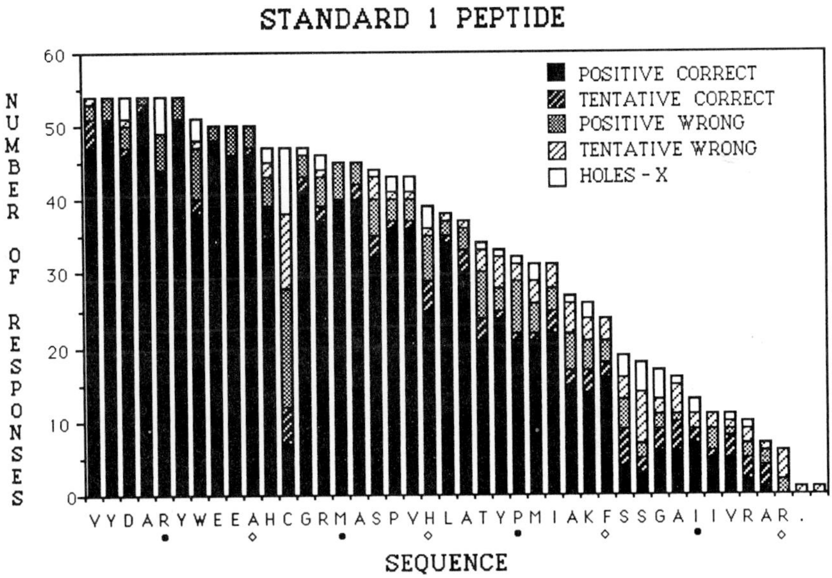

Fig. 2 Distribution of calls for each cycle. Any indication of qualification or uncertainty by the operator led to placing that amino acid in the tentative identification category. When more than one residue was listed in a cycle it was plotted as wrong. Cycles for which no residue was called and cycles labelled X were plotted as holes. The sequence is listed across the bottom with the amino terminus at the left.

older instrumentation and methods. Similarly, the sequence was selected to evaluate performance of modern instrumentation. It is also clear that the addition of the Model 120 on-line PTH-HPLC to the Model 470 sequencer approximately doubled the number of residues that can be sequenced. In contrast, there does not seem to be any significant difference in either the number of residues sequenced, the repetitive yields, or the accuracy of the three models of HPLC equipped instruments. For these three instruments the overall average number of correctly identified residues was 23.9 and the accuracy of sequence calls was 94.6% for those cycles that were "positively identified". The accuracy for the 93 residues which were only tentatively identified was 48%; this indicates a very good ability on the part of the personnel who interpreted the data to discriminate between a confident and tentative call.

TABLE I. Sequencing of STD-1

Instrument	Number	Coupling yield[a] (picomole)			Repetitive yield[b] (%)			Number of correct[c] Residues			Accuracy[c] (%)		
		Mean	Range	Std. Dev.	Mean	Range	Std. dev.	Mean	Range	Std. dev.	Mean	Range	Std. dev.
Applied Biosystems													
470A	4	44.0	--	--	90.6	87.7-93.9	2.5	12.2	6-17	4.2	92.4	75.0-100	10.3
Applied Biosystems													
477A/120	7	77.8	--	--	89.0	81.7-92.2	3.6	25.6	14-35	7.0	91.6	64.5-100	11.5
475A/120	5	50.6	37.6-64.3	10.9	86.7	76.3-93.0	5.6	20.8	9-32	9.5	94.9	90.0-100	4.3
470A/120	26	50.9	24.1-81.7	15.4	87.1	79.6-92.7	3.4	24.0	10-37	6.4	95.4	79.3-100	5.5
Total	38	52.5	24.1-81.7	15.6	87.5	76.3-92.7	3.9	23.8	9-37	7.1	94.6	64.5-100	7.0
Beckman													
890	3	--	--	--	--	--	--	2.0	0-3	1.4	30.9	0-75.0	32.0
Manual													
	2	--	--	--	--	--	--	2.5	0-5	2.5	41.7	0-83.3	41.7

[a] Coupling yield from the data sets in which the recovery of valine in cycle 1 could be calculated unambiguously. The total number of such cases was 477/120 (1), 475/120 (3), 470/120 (12) and 470 (1).

[b] Based on the background corrected yield of alanine 4 and alanine 16 or, if the sequence did not extend this far, on the yield of alanine 4 and alanine 10.

[c] Based on positively identified cycles only; those cycles that were left blank or that had either multiple or tentative assignments were not scored.

In addition to the sequencing data above, results were also obtained from three laboratories that specialize in extended sequencing runs on low picomole amounts of samples. As shown in Table II, these laboratories correctly identified an average of 36.5 residues with an accuracy of 99%. The average repetitive yield was greater than 91%. The instruments used were Models 470A and 477A equipped with the Model 120 on-line PTH-HPLC.

Five of the twelve respondents with Model 475A and 477A sequencers included the data generated from the computer assisted automatic sequence identification. Although the version of software was not specified, the runs were initiated at least 2 months after the release of version 1.30 for 900A and 1.50 for 477A in January 1988. Sequence calling routines were identical to those in the previous versions (1.10 and 1.20 for 900A and 477A, respectively) which were released in May and June 1987. Comparison of the number of errors in the sequence identified by the operator with those in the computer generated sequence to the end of confident operator calls, a total of 97, indicates that unreviewed use of the

TABLE II. Results of STD-1 sequencing on three "high performance" instruments

	470A/120	477A/120	477A/120
Positively identified residues			
correct	38	34	37
incorrect	0	1	0
% accuracy	100	97	100
Tentatively identified residues			
correct	--	3	--
incorrect	--	1	--
Data system calls through last manual call			
correct	--	34	30
incorrect	--	5	9
holes	--	1	0
% accuracy	--	87	77
Data system calls through last picomole ratio greater than 10			
correct	--	30	29
incorrect	--	2	1
holes	--	1	0
% accuracy	--	94	97
Repetitive yield	92	91	91
First cycle yield	47	57	77

automatic sequence identification would produce sequences of consistently lower accuracy, only 89% compared to 95%. More errors occurred when the picomole ratio (a software defined statistic relating amount of the identified amino acid to its background amount) was small. If automatic calls only to the last cycle with the picomole ratio greater than 20 (rather than through the last operator call) were considered, the accuracy was the same as confident operator calls (95%), but only on 56 compared to 97 calls. If data through the last cycle with picomole ratio greater than 10 were considered, the accuracy was 90% of 72 calls and can be defined as the end of the confident computer calls. Within the data set of confident computer calls for picomole ratios between 11-20 there were 12 correct out of 16 calls; for picomole ratios between 5-10 there were 10 correct out of 12 calls; and for picomole ratios less than 5 there were 2 correct out of 5 calls. Results from labs specializing in extended sequencing were similar and are included in Table II. Clearly operators identify sequences farther and more accurately than the computer; however, this analysis suggests that when the operator and the computer calls differ within the limit of confident computer calls the cycle merits re-evaluation by the operator.

B. Amino acid analysis of STD-2

A total of 48 responses included amino acid composition data, and of these, three (6%) reported instrument malfunctions. Table III summarizes the other 45 compositions. The data were standardized by calculating the mole percent of each amino acid multiplied by 40 (the length of the peptide) to obtain the residues per mole of peptide. The data were divided into two groups based on the two most commonly used chemistries, post-column ninhydrin and pre-column phenylisothiocyanate derivatization. Columns 4-6 of Table III detail the results from samples analyzed using ninhydrin after hydrolysis of 90% (900 pmol) of STD-2. When 10% (100 pmol) of the sample was used for the ninhydrin method, the results were too variable to include here. The PTC analysis data, shown in columns 7-9, resulted from the hydrolysis of only 10% of the sample. Too few results of fluorescent detection methods (OPA, Fmoc) were received to permit analysis.

With these data, we can determine the accuracy of an amino acid analysis of an unknown sample submitted to a core facility. In 27 of the 48 compositions (58%) submitted, the error was greater than 10%, i.e., more than 4 residues per mole were incorrect (cysteine and tryptophan were excluded given the recognized difficulties of hydrolysis). Only eight of the data sets were better than 95% correct (2 or fewer incorrect residues). It is obvious that significant problems exist with amino acid analysis.

Interpretation of the amino acid analysis data and assessment of the possible contributing factors to these poor results was complicated by the many variables involved. Factors affecting amino acid analysis include the possible introduction of contaminants, hydrolytic destruction or incomplete hydrolysis, type of chemistry used, and instrument performance. We had attempted to eliminate hydrolysis as a variable; however, a significant fraction of analysts chose to use their own hydrolysis procedure rather than the requested protocol. After grouping the samples among all the

Table III. Amino acid analysis of STD-2

	All responses (n=45)		Ninhydrin responses (n=10)			PTC responses (n=19)			Actual
Amino acid	Mean	Std dev	Mean	Std dev	Range[a]	Mean	Std dev	Range[a]	
ALA	6.98	1.24	6.7	1.14	5.2-8.0	7.0	0.99	5.4-8.1	7
ARG	4.01	0.95	3.6	0.34	3.6-4.4	4.3	0.87	2.6-4.9	4
ASX	1.44	0.55	1.5	0.18	1.1-1.9	1.3	0.26	1.0-1.5	1
CYS[b]	0.54	0.29			Not calculated				1
GLX	2.65	0.81	2.7	0.18	2.3-3.3	2.5	0.65	1.8-3.6	2
GLY	3.04	0.93	3.2	0.50	2.6-4.2	2.7	0.37	2.1-3.7	2
HIS	1.89	0.45	1.8	0.05	1.7-2.1	1.8	0.12	1.3-2.1	2
ILE	2.27	0.49	2.5	0.26	2.0-2.9	2.2	0.24	1.4-3.0	3
LEU	1.37	0.47	1.3	0.09	1.1-1.5	1.4	0.34	1.1-1.8	1
LYS	1.26	0.31	1.2	0.10	0.9-1.6	1.2	0.05	1.0-1.4	1
MET	1.49	0.48	1.3	0.46	0.5-2.0	1.5	0.18	0.8-1.8	2
PHE	1.11	0.17	1.0	0.01	0.9-1.1	1.1	0.05	1.0-1.4	1
PRO	1.98	0.51	2.4	0.29	1.7-2.3	2.1	0.07	1.9-2.5	2
SER	3.24	0.44	2.9	0.17	2.5-3.3	3.3	0.18	2.8-3.8	3
THR	1.28	0.45	1.4	0.10	1.0-1.7	1.3	0.08	1.0-1.5	1
TYR	2.60	0.69	2.7	0.07	2.5-3.0	2.7	0.33	2.1-3.1	3
VAL	2.88	0.44	2.8	0.17	2.3-3.0	3.0	0.09	2.7-3.5	3
TRP[c]	1.10	0.64			Not calculated				1

[a]Highest and lowest values were excluded from range.
[b] n=11
[c] n=4

various categories, the sample size per group was too small to permit adequate comparisons. However, an examination of the eight "acceptable" compositions with fewer than two incorrect residues revealed no correlation with instrument type, chemistry, amount of sample, or hydrolysis conditions. Contaminants introduced during handling of the peptide could explain the poor compositional data obtained. Some facilities appear not to have sufficient controls to detect suboptimal instrument performance or the presence of contaminants; nine of the 48 compositions incorrectly quantitated 12 or more residues. Further work is needed to establish good standards and controls encompassing all phases of the amino acid analysis process used in core facilities.

IV. CONCLUSIONS

Based on the results obtained on STD-1, it appears reasonable to expect an average facility equipped with an Applied Biosystems protein/peptide sequencer with an on-line HPLC to correctly sequence approximately 24 residues of an unknown peptide, if the peptide is at least 95% pure and at least 75 pmol is loaded onto the instrument and it is free of other primary amines and interfering substances. Under these conditions the overall accuracy of operator sequence identification is about 95% which is above the accuracy of automatic sequence identification that 6 facilities obtained with the current version of software available on the Model 900 Data System. Automatic calls extending as far as operator calls were only 89% accurate while automatic calls only to where the picomole ratio fell below 10 were 90% accurate. Although three laboratories that specialize in obtaining extended sequences from picomole amounts of peptides and proteins were able to accurately call an average of 12 additional residues of sequence, this potential cannot in general be routinely reached in a typical, high throughput core facility.

It is clear that amino acid analysis at moderate to high sensitivities lacks precision and consistency. Few laboratories can routinely provide compositions which are at least 90% correct with 0.1 to 1 nanomole of peptide.

V. ACKNOWLEDGEMENTS

The cooperation of all of the anonymous laboratories that graciously used their resources to analyze the samples and provide the data described here was essential to making this project possible. Their support is most gratefully acknowledged. The three high performance laboratories that were not a part the regular anonymous sample are thanked for contributing their data. The assistance of Dr. Bernard Forget, Associate Dean, Yale University in providing the anonymity of the respondents is appreciated. Directors of protein and/or nucleic acid chemistry facilities desiring to be included on this mailing list of core facilities should provide their names to KRW.

VI. REFERENCE

1. Williams, K.R., Niece, R.L., Atherton, D., Fowler, A.V., Kutny, R., Smith, A.J. (In press) FASEB J.

11

PROTEIN SEQUENCING SAMPLE ADVISOR: AN EXPERT SYSTEM TO ASSESS SAMPLE SUITABILITY FOR MICROSEQUENCING

Joseph L. Modelevsky
Digital Equipment Corporation
Elk Grove Village, IL 60007-1018

Zafar I. Randawa
Otsuka Pharmaceutical Co.
9900 Medical Center Drive
Rockville, MD 20850

I. INTRODUCTION

Protein micro-sequencing technology is used to analyze the purity and sequence of protein preparations. Micro-sequencing can confirm the sequence of a purified protein or the major peptide in a sample, whether samples contain contaminating peptides, and identify any contaminants. Protein sequencing can provide an indication of the effectiveness of cumulative purification processes that generate a sample. Frequently, centralized protein microsequencing resources provide such analytical support to individuals and organizations.

In the pharmaceutical industry, regulatory agencies (e.g., the FDA) require sequencing of the first fifteen amino-terminal residues of a protein to positively identify a protein (1). For quantitative runs, the central laboratory must sequence a quantitative standard to optimize instrument sensitivity, before sequencing the experimental sample. Occasionally, however, the laboratory must determine 50 to 60 sequence residues of a sample or establish the entire sequence of a protein, which requires proportionally more labor.

Automated micro-sequencing is a very powerful and valuable tool, but there are significant costs associated with each sequencing run. Estimating cost of a run is difficult, but some of the costs follow. Minimal set-up cost for a sequencing run is approximately $125 to $200. Longer sequence runs cost approximately $20 to $30 per residue, plus expert labor for sample evaluation and interpretation of results. A typical amino-terminal identification performed to FDA specifications, including one day labor for sample evaluation and preparation, may cost from $650 to $1000, based upon materials and typical cost with overhead for a Research Scientist's/Laboratory Supervisor's labor. This estimate does not include the submitter's sample preparation costs, nor the cost of project delays. Every avoided failure could save at least $700 in material and labor costs. More importantly, it allows for more productive use of the microsequencing resource.

II. PROBLEM DEFINITION

A sequencing failure occurs when data generated in a sequencing run are not from the desired peptide, but are from contaminants, or when the target protein has undergone modifications that prevent sequencing. Sequencing failures waste submitted samples, preparation labor, sequencing time, materials, and expertise. Sample preparation is the major factor in a successful sequencing analysis: samples must be pure, unblocked, and of known quantity. Improper sample preparation, usually due to lack of understanding of chemical and technical requirements of micro-sequencing technology, causes more than half of all sequencing failures (i.e., up to 25% of all sequencing runs). Technical failures of the sequencer rarely occur. Improper sample preparation can modify a major protein to a form that cannot be sequenced or can result in isolation of co-eluting protein or peptide that is not the desired protein. Purity and suitability criteria can be complex.

Prior to a sequencing attempt, most microsequencing resources extensively assess sample purity. With each sample, submitters provide a brief, written description of the sample's preparation history. A microsequencing expert screens the preparation history of each submitted sample to assess sample suitability, prior to loading that sample onto the sequencer. Evaluation of sample suitability requires extensive protein sequencing experience. Oversight of any critical factors can induce failures.

In our experience, every sample submission has required extensive, additional consultation with the submitter (on average, one hour/sample) to diagnose and resolve problems in the sample's preparation history. A wide variety of preparation and handling constraints impact upon the quality and/or suitability of a sequencing sample, and all constraints must be checked for each sample. Some problems arise only under complex or unusual circumstances, and need not be explored for all sample submissions. Failure to investigate such problems when necessary can defeat a sequencing experiment. For these reasons, protein sequencing expertise is absolutely required to assess sample suitability prior to a sequencing run. Each sample can present unique sets of problems, which complicates generalized education of submitters.

An improved sample screening method could increase sequencing successes and provide specialized education to scientists who submit samples. A computer-based, self-explaining, sample evaluation tool (i.e., a sample evaluation expert system) could provide improved sample evaluation and preparation expertise, allow scientists to evaluate suitability of their own samples, train scientists in sample preparation practices and pitfalls specific to their samples and processes, facilitate optimal sample preparation and increase the likelihood of achieving desired sequencing results in the long term. Such a sample evaluation system could be readily expanded to serve as a general-purpose sample purity evaluation tool for other protein preparations. Such an expert system could provide significant returns by minimizing failures, and reducing risks and costs associated with sequencing.

III. RESULTS

We developed a prototype computerized

expert system (Sample Advisor) that can evaluate sample suitability and diagnose problems, before samples are submitted, to reduce or eliminate sequencing failures. Our objectives are to computerize sample evaluation expertise, a scarce and valuable commodity in high demand. In regular use, the Sample Advisor will allow scientists to rapidly evaluate their sequencing samples before submission (this will educate scientists, as well as evaluate samples), will reduce labor the microsequencing expert must apply to sample evaluation (1 hour per average sample), will reduce submission of improperly prepared samples (perhaps eliminate 50% of sequencing failures), to ultimately increase microsequencing productivity. Long term use of the Sample Advisor will educate, guide and help sample submitters to routinely generate samples suitable for sequencing; this will be a major benefit.

Protein microsequencing sample evaluation requires knowledge of protein chemistry, laboratory techniques, sequencing methodology, and the particular automated sequencer. A typical microsequencing expert can evaluate a sample's suitability in approximately one hour; the problem is tractable. We have codified, in the form of rules, knowledge required to evaluate sample suitability. Knowledge was extracted from a practicing microsequencing expert, primary references and knowledge resources (2). Our major objective is for the Sample Advisor to evaluate a sample's sequencing suitability based upon purity, quantity, volume, matrix, origin, solubility and likelihood of contamination. Suitability assessment must be equivalent to a human expert's assessment. The Sample Advisor must also be easy to use, understand and deliver to scientists. The Sample Advisor must explain its reasoning and its decision making processes. The Sample Advisor must generate a report of its rationale and findings, and ultimately must be capable of interaction with a program that can upload information for Sample Tracking. Consultation with the Sample Advisor should consume less than 15 minutes to make an evaluation.

The current system runs on any VAXstation(TM), with 4MB physical memory, 2000 blocks disk space, running MicroVMS(TM) Version 4.4 or later, VWS(TM) windowing software Version

3 or later, and NEXPERT/OBJECT(TM) Version 1.0 or later.

In the current phase of our work, we have engineered knowledge of the fundamental sample evaluation process, and developed a prototype expert system to execute this process. The Sample Advisor evaluates a sample's sequencing suitability based upon its current knowledge of purity, quantity, volume, matrix, origin, solubility and contamination criteria. The Sample Advisor also provides "red flags" for less than optimal preparation parameters. The Sample Advisor, currently delivered in the NEXPERT development environment, is easy to use and understand. NEXPERT's built in utilities provide rudimentary explanation and reporting functions. Because NEXPERT is a VAX layered product, the current system may be integrated with other applications software, including sample tracking.

IV. DISCUSSION

Many laboratories use protein microsequencing technology to determine amino acid sequence and purity of protein preparations. Improperly prepared samples cause most sequencing failures. Because most samples are prepared by individuals who are less familiar with the technical requirements of microsequencing chemistry, the sequencing expert spends considerable time evaluating the suitability of submitted samples prior to sequencing. We developed a rule-based expert system (the Sample Advisor) that evaluates sample history, diagnoses potential technical problems, and determines if the sample is ready for microsequencing before sample submission. The rule-base was engineered from a protein microsequencing expert's experience and from the documentation of Protein Chemistry Research Resource Facilities.

As protein microsequencing becomes routine, demands for sequencing expertise will increase. Using expert systems technology, we can deliver a dedicated, full-time, computerized expert sample advisor to scientists and save a human advisor for non-routine sample evaluation. This computerized advisor will also provide specialized education to scientists submitting samples which will provide long term improvements in sample quality. For example, a session with the advisor can identify and expose critical factors and remedies to a laboratory technician preparing a microsequencing sample for the first time.

Because the Sample Advisor was developed with an expert systems tool, knowledge can be added, modified, and deleted as necessary, at any time. Use of the Sample Advisor will reduce the time sequencing experts must invest in sample evaluation, and will reduce submission of improperly prepared samples and sequencing failures.

V. ACKNOWLEDGMENTS

The concept for this work is based upon an early prototype developed by the authors at Pitman-Moore, Inc. NEXPERT/OBJECT is a trademark of Neuron Data Corp. VAX, VAXstation, VWS and MicroVMS are trademarks of Digital Equipment Corporation.

VI. REFERENCES

1. Points to consider in the production and testing of new drugs and biologicals produced by recombinant DNA technology, April 10, 1985, Office of Biologics Research and Review, Center for Drugs and Biologics, FDA, Bethesda, MD, pp. 1-14.

2. Forms, forms, forms..., August 9, 1987, Research Resource Facilities Satellite Meeting, First Symposium of the Protein Society, San Diego, CA.

3. For further information on this expert system or the tools used for development, please contact the authors.

12

DETECTION OF PROTEIN STRUCTURAL FEATURES WITH PROFILE ANALYSIS[1]

Michael Gribskov[2]
David Eisenberg

Molecular Biology Institute
University of California – Los Angeles
Los Angeles, CA 90024–1570 USA

I. INTRODUCTION

The sequence of a protein is often determined in the hope that it will shed light on the structure and function of the protein. If a protein with a sufficiently similar sequence is found, it can be inferred that it and the newly sequenced protein are homologous, and that there are similarities in the structures and functions of the proteins. The search for similarity is usually performed by searching a protein sequence database using a hash–table based searching algorithm, finding individual sequences that are similar to the new sequence, and then making inferences about the structure and function of the newly sequenced protein based on the spectrum of sequences found.

We have approached the problem from a different direction. It is clear that certain three-dimensional structural *motifs* are shared by many proteins. These structural motifs have patterns in their amino acid sequences that permit them to be recognized from the sequences alone. Profile analysis (Gribskov *et al.* 1987, Gribskov *et al.* 1988)

[1]Supported by NSF grant CHE-85-09657 to D.E., and the following grants to M.G.: American Cancer Society post-doctoral fellowship grant PF-2649, NIH post-doctoral training grant CA-09056, and by a gift from Mrs. Lita Annenberg Hazen to M.G.

[2]Current address: NCI–Frederick Cancer Research Facility, BRI–Basic Research Program, Frederick, MD 21701–1013 USA.

allows us to create probes sensitive to structural patterns, or to the overall sequence characteristics of an entire protein family. When a sequence is found to be significantly similar to a profile that is specific for a three dimensional structure, one can immediately pose hypotheses about the structure of the newly sequenced protein. Profile analysis is also useful for determining whether a new sequence is related to one of the established protein families. In either case, because the comparison is made to a probe representing the overall sequence characteristics of a group of sequences, it is more sensitive to the overall similarity of the protein and the probe, and less dependent on finding a single sequence with relatively high similarity.

II. METHODS

A. Operating system and Availability

The profile analysis package now consists of the five programs shown in Table I. The programs are written in FORTRAN-77, using a limited number of the language enhancements available on VAX[3] computers, and the procedure library of the University of Wisconsin Genetics Computer Group (GCG) (Devereux *et al.* 1984). Versions independent of the GCG procedure library available. The programs and source code are available on request.

B. The Profile

The profile is a position-specific scoring table generated from a group of aligned sequences and a comparison table giving scores for the comparison of each pair of amino acid residues. Each row in the profile corresponds to one position in the aligned sequences. The columns of the profile contain the score for the alignment of each possible amino acid residue with each position in the aligned sequences. The scores are calculated as follows:
For a group of N aligned sequences of length L

[3]Digital Equipment Corp., Maynard, MA

TABLE I: Programs used in profile analysis

Program	Function
PROFILEMAKE	Generate a profile from a group of aligned sequences
PROFILEGAP	Align a sequence with a profile using dynamic programming algorithm
PROFILESEARCH	Compare a profile to each sequence in the PSQ database using dynamic programming algorithm
PROFILENORMAL	Normalize the results of **PROFILESEARCH** to remove systematic effects of sequence length
PROFILESCAN	Compare a sequence to each profile in the library of validated profiles and test for significant similarity

$$\begin{array}{ccccc} a_{1,1} & a_{1,2} & a_{1,3} & \cdots & a_{1,L} \\ a_{2,1} & a_{2,2} & a_{2,3} & \cdots & a_{2,L} \\ a_{3,1} & a_{3,2} & a_{3,3} & \cdots & a_{3,L} \\ & & \vdots & & \\ a_{N,1} & a_{N,2} & a_{N,3} & \cdots & a_{N,L} \end{array}$$

where $a_{i,j}$ is the amino acid residue in sequence i at position j. The profile will have a column for each character that occurs in the aligned sequences; for proteins there will be 20 columns. The scores in the profile at row r and column c are given by

$$\text{Profile}(r, c) = \sum_{d=1}^{20} W_d \text{Comp}(\text{residue}_d, \text{residue}_c)$$

where **Comp** is the value for the comparison of two amino acid residues; each possible residue (residue$_d$), and the amino acid residue represented by the column of the profile (residue$_c$). A weight (W$_d$) is applied that depends on the number of occurences of each kind of residue at each position, and the sum of the weighted number of sequences. We have used both linear weighting,

$$W_d = \frac{\sum_{i=1}^{N} w_i \delta_d}{\sum_{i=1}^{N} w_i} \qquad \delta_d = \begin{cases} 1 & \text{if} \quad a_{i,r} = \text{residue}_d \\ 0 & \text{if} \quad a_{i,r} \neq \text{residue}_d \end{cases}$$

and logarithmic weighting.

$$W_d = \frac{\ln\left\{1 - \sum_{i=1}^{N} w_i \delta_d / \left(1 + \sum_{i=1}^{N} w_i\right)\right\}}{\ln\left\{1 / \left(1 + \sum_{i=1}^{N} w_i\right)\right\}}$$

Using either kind of weighting $W_d = 0$ when a residue does not occur at a given position, and $W_d = 1$ when all residues at a position are identical. The weight on each sequence (w_i) is usually 1.0, but may be larger when it is desired to emphasize one of the aligned sequences more than the others. Logarithmic weighting results in a profile that is more specific for the probe sequences used to generate it; we normally use logarithmic weighting.

C. Dynamic Programming Algorithm

The implementation of the dynamic programming algorithm follows Smith and Waterman (Smith and Waterman, 1981). However, any version of the dynamic programming algorithm is easily modified for use with the profile system. Letting $S_{i,j}$ be the score for the alignment of the sequence and profile such that position i of the sequence is aligned with row j of the profile. The score $S_{i,j}$ for the alignment is defined recursively as

$$S_{i,j} = \text{Profile}(j, \text{column}_{a_i}) + \max \left\{ \begin{array}{l} S_{i-1,j-1}, \\ \max_{2 \le k \le j-1} \left(S_{i-1,j-k} - w_k\right), \\ \max_{2 \le l \le i-1} \left(S_{i-l,j-1} - w_l\right), \end{array} \right\}$$

where the penalties $\mathbf{w_k}$ and $\mathbf{w_l}$ are given by

$$w_k = p_{\text{open}}(j) + p_{\text{extend}}(j) \times (j - k + 1)$$

$$w_l = p_{\text{open}}(j) + p_{\text{extend}}(j) \times (i - l + 1)$$

with p_{open} and p_{extend} being the position-specific gap penalties in the profile. Note that the values of the penalties depend on the row in the profile.

D. Multiple Alignments

PROFILESCAN uses a dynamic programming algorithm modified to allow the production of all distinct alignments greater than a threshold score. During construction of the score matrix, a list of all scores greater than the threshold and their locations is saved. After completion of the score calculation, the list is sorted, and a traceback performed beginning at each position in the list beginning with the highest score $S_{i,j}$. As the traceback is carried out, the point at which the score falls below the threshold is stored in a second list, any other traceback that would contain this position is aborted and no alignment produced. This ensures that the alignments will be distinct, i.e. that they will not contain segments with identical alignments of the sequence and profile unless the alignments diverge before the threshold is reached.

E. Normalization for Length Effects

We have found empirically that, when there are no sequences related to the profile in the database, the distribution of comparison scores can be modelled by $S = Ce^{A \cdot L + B}$ where S is the score for comparison of a sequence of length L to a profile, and A and B are constants. The program **PROFILENORMAL** uses a non–linear curve fitting algorithm to fit the results of a database search to this equation and determine the constants A and B. The normalized results are generally easier to interpret, especially for when comparing the scores for sequences of different lengths.

III. THE PROFILE LIBRARY

We are assembling a library of profiles (Table II), each specific for a different structural or sequence motif. To be included in the library, the profile must be *validated*; the scores typical of protein sequences encoding the motif, and of those not encoding the motif, must be determined through a careful analysis of the comparison scores of the profile and each sequence in the PSQ[4] database. In the validation process, two threshold levels are determined. All sequences known to encode the motif should score above the **HIGH** level, and nearly all sequences thought to not encode the should motif score below the **INTERESTING** level (in practice the INTERESTING level is set at the 2.2σ or 3.0σ level for the comparison of the profile to se-

[4]Protein Identification Resource — Protein Sequence Database release 16.0, PSQ — New release 34.0, National Biomedical Research Foundation, Georgetown University Medical Center, Washington, D.C. 20007

TABLE II: The profile library

Profile	Specificity
Globin.prf	Globin fold
Ig-constant.prf	Immunoglobulin constant regions
Ig-variable.prf	Immunoglobulin variable regions
ZincI.prf	Type I zinc finger
ZincII.prf	Type II zinc finger
Ca–EFHand.prf	Calcium binding E–hand
Homeo.prf	Homeotic proteins
Cyclic.prf	cAMP and cGMP binding structure
Nuc.prf	Dinucleotide binding fold
HTH.prf	Helix–turn–helix DNA binding structure

quences not encoding the motif). We have previously described the derivation and validation of six profiles (Gribskov *et al.*, 1988) specific for the globin fold, immunoglobulin constant and variable regions, the Zinc finger motif, and the calcium binding EF–hand motif. Recently validated profiles are described below.

A. Homeotic Protein Profile (homeo.prf)

The sequences of the mouse m6 and Mo-10 proteins (PSQ entries WJMSM6 and WJMS10), african clawed frog AC1 and MM3 proteins (entries WJXLAC and WJXLMM), and fruit fly antennapedia, ultrabithorax and engrailed proteins (entries WJFFAP, WJFFBX, and WJFF4A) were aligned and used to generate a profile. All 26 homeotic proteins in the PSQ and New databases were clearly distinguished (scores from 97.25 to 34.61) from non–homeotic protein sequences (average: 12.33, standard deviation (σ): 2.04, highest: 21.82). The yeast mating hormone a1 protein was clearly related to the homeotic protein family with a score of 31.39 (9.33σ). The **HIGH** threshold has been set at 26.67 (8.0σ) and the **INTERESTING** threshold at 18.46 (3.0σ). The yeast mating hormone $\alpha 2$ falls in the interesting region with a score of 19.88 (3.70σ).

B. Cyclic Nucleotide Binding Profile (cyclic.prf)

This profile is specific for cAMP and cGMP binding structures of the type found in the cAMP receptor protein (CRP or CAP). The profile is based on the alignment of CRP, with the regulatory subunits of type I and type II cAMP dependent protein kinases, and the regulatory subunit of a cGMP dependent protein kinase (Weber *et al.*, 1987). The sequences used to generate the profile have comparison scores of 32.46 to 44.21, the highest score for an unrelated protein is 16.11, and the average score and standard deviation for unrelated sequences is 11.99 and 1.75. The **HIGH** threshold has been set at 29.44 (10.0σ for unrelated sequences) and the **INTERESTING** threshold at 15.83 (2.2σ).

C. Dinucleotide Binding Fold Profile (nuc.prf)

The sequences of mitochondrial malate dehydrogenase, cytoplasmic malate dehydrogenase, lactate dehydrogenase, liver alcohol dehydrogenase, glyceraldehyde-3-phosphate dehydrogenase, the FAD and NADP binding domains of glutathione reductase, lipoamide dehydrogenase, p-hydroxybenzoate hydroxylase, and dihydrofolate reductase, were aligned according to the structural alignment of Birktoft and Banaszak (1984). Only those parts of the sequences involved in the dinucleotide binding structure were included in the alignment. The regions of the sequences used correspond to the four β strands and one α helix that form the dinucleotide binding site.

Analysis of the results of the database search using this profile is difficult because the dinucleotide properties of many proteins are unknown or can only be discovered by a careful search of the literature. However, the proteins with the highest scores for comparison to this profile are mostly dehydrogenases and other oxidoreductases and therefore very likely to contain the dinucleotide motif. The **HIGH** threshold has been set by inspection at 8.65, the point where the first sequences unlikely to contain the dinucleotide fold motif are found. Among the sequences scoring above the **HIGH** threshold, in addition to those used to generate the profile, are NADH dehydrogenase, pyridine nucleotide transhydrogenase, D–amino acid oxidase, fumarate reductase, ketol–acid reductoisomerase, mercuric reductase, and 3–hydroxyacyl–CoA dehydrogenase. The **INTERESTING** level has been set arbitrarily to 7.20; at this level the majority of sequences seem to be of proteins that are unlikely to contain the dinucleotide fold motif.

D. Helix–Turn–Helix DNA Binding Profile (HTH.prf)

The 21 sequences identified by Pabo and Sauer (1984) as being likely to encode the DNA binding helix-turn–helix motif were used to generate a profile. The results of the database search with this profile are also more difficult than normal to interpret. The short length of the profile (20 residues) leads to a broad distribution of comparison scores. In addition most proteins thought to contain the helix–turn–helix motif have been identified based only on their observed binding to DNA, and the location of a sequence similar in some degree to that of the the proteins whose structure is known. This makes conclusive identification of sequences containing the helix–turn–helix motif difficult.

The **HIGH** threshold for this profile has been set at 4.11 (5.0σ and the **INTERESTING** threshold at 3.38 (3.0σ). In addition to the proteins expected to score above the **HIGH** threshold, the sequences of the T7 hypothetical gene 7.7 protein (Q7BP77), a *Neurospora crassa* intron encoded protein (B23431), and the T4 rIIB protein (Z2BPB4) were found. Some of the sequences used to generate the profile score in the **INTERESTING** region. These include the lac repressor (RPECL), and the arabinose operon repressor-activator protein (RGEBAT, RGECA, and A25027). In addition, most homeotic protein sequences score in the interesting region. Many of the sequences unlikely to encode the helix–turn–helix motif that score in **INTERESTING** region are those of proteins with high α helical contents. The sequences of many regulatory proteins score far below the **INTERESTING** threshold: the *Escherichia coli* lysR (RGECK), tyrR (RGECAY), ilvY (B26287), metJ (RGECMJ), phoB (RGECFB), phoM (RGECFM), and ompR (RGECOR) proteins, yeast GAL4 (A05022), and GAL80 (RGBYG8) proteins, phage P22 mnt (RGBPM2) and arc (RGBPA2) proteins, and plasmid encoded mercuric reductase operon regulatory proteins from *Pseudomonas aeruginosa* (RGPSHA) and *Shigella flexneri* (RGEBHD) are among these.

IV. DISCUSSION

A new sequence can be quickly compared to the profile library to check for the presence of structural features. For profiles based on known three–dimensional structures, the sequence similarity can immediately be translated into information about the identity and function of structurally important residues. This hypothetical structural information can then be directly checked by biochemical or genetic experiments.

Our goal is to provide validated profiles that will detect all known occurences of a given motif. However, this level of assurance is not critical. If a structural motif is not detected based on the analysis of its sequence, very little time or effort has been expended. However, if one or more structural motifs is detected, experiments can immediately be designed to test for the suggested functions, avoiding extensive screening for possible activities.

The profile analysis programs are available without charge. Any suggestions of well characterized motifs that would be suitable for inclusion in the profile library would be welcomed. For further information contact M.G. (electronic mail to GRIBSKOV@NCIFCRF.GOV if possible).

V. REFERENCES

Birktoft, J.J. and Banaszak, L.J. (1984) Structure–function relationships among nicotinamide–adenine dinucleotide dependent oxidoreductases. Peptide and Protein Rev., **4**, 1-46.

Dayhoff, M.O. and Schwartz, R.M. (1979) Matrices for detecting distant relationships. In Atlas of Protein Sequence and Structure, Dayhoff, M.O. (ed.) Vol. 5, Supp. 3, pp. 353-358.

Devereux, J., Haeberli, P, and Smithies, O. (1984) A comprehensive set of sequence analysis programs for the VAX. Nucl. Acids Res. **12**, 387-395.

Gribskov, M., McLachlan, A.D., and Eisenberg, D. (1987) Profile analysis: detection of distantly related proteins. Proc. Natl. Acad. Sci *USA* **84**, 4355-4358.

Gribskov, M., Homyak, M., Edenfield, J., and Eisenberg, D. (1988) Profile scanning for three-dimensional structural patterns in protein sequences. *CABIOS* **4**, 61-66.

Gribskov, M. and Burgess, R.R. (1986) Sigma factors from *E. coli, B. subtilis*,, Phage SPO1, and phage T4 are homologous proteins. Nucl. Acids Res. **14**, 6745-6763.

Pabo, C.O. and Sauer, R.T. (1984) Protein DNA recognition. Ann. Rev. Biochem. **53**, 293-323.

Smith, T.F., and Waterman, M.S. (1981) Comparison of biosequences. Adv. Appl. Math. **2**, 482-489.

Weber, I.T., Steitz, T.A., Bubis, J., and Taylor, S.S. (1987) Predicted structures of cAMP dependent protein kinase A and B domains. Biochemistry **26**, 343-351.

Section II

Introduction

Peptide and Protein Characterization by Mass Spectrometry

A.L. Burlingame

Department of Pharmaceutical Chemistry
Mass Spectrometry Facility
and The Liver Center
University of California
San Francisco, California 94143-0446

OVERVIEW

The classical methods of mass spectrometry have been successfully used in the qualitative and quantitative analysis of amino acids and small peptides for the last three decades (1). This work has been carried on under the express constraint that suitable chemical derivatization of the polar functional groups was required to effect sample vaporization without its pyrolytic degradation. Two of the more successful strategies have involved either acetylation and permethylation (2,3) of the labile hydrogens and chemical reduction of the amide linkages to the corresponding polyamino alcohols followed by C-terminal trimethylsilylation (4). These derivatives have then been analyzed by direct probe sublimation or gas chromatography inlet to electron impact (EI) ion sources respectively. However, two major problems remained, namely sample losses in the derivatization procedures which limit the overall sensitivity readily achievable to the low nanomole range, and volatility constraints which limit the tractable molecular sizes to 1000 - 1500 daltons.

Examples of early successes of the permethylation strategy with EIMS involved the structure and sequence determination of fortuitine (5), a peptidolipid isolated from Mycobacterium fortuitum, the characterization of the enkephalin neuropeptides (6), γ-carboxy glutamic acid (7), and the sequence determination of proteolytic peptides from several proteins (8). The reductive GC/MS methods culminated in sequence studies of the transmembrane protein bacteriorhodopsin (4b) and the partial sequencing of aminoacyl tRNA synthetases for the purpose of verification of the accuracy of cDNA sequences (9).

In summary, these rather heroic methods were tedious and cumbersome, and were additionally constrained to the rather low mass range accessible without thermal degradation (< 1000 daltons), viz. 5-10 amino acid residue oligomers. Since these methods do require tens of nanomoles of protein, clearly both the necessary level of technical and instrumental sophistication and the considerable capital cost have precluded general use of such methodology by protein biochemists.

However, this situation has changed dramatically over the last few years, triggered primarily by the discovery that free polar, thermally labile substances could be "trampolined" from the surfaces of viscous liquid solutions INTACT as desolvated protonated molecules by having been ejected ("sputtered") from an acidic medium, or as their deprotonated forms from an alkaline medium. Ejection of surface active substances, such as peptides of sufficient hydrophobicity (10), from viscous liquid solutions can be effected by irradiation of the surface by both atom and ion beams possessing 6 - 35 Kev kinetic energy, and are referred to as fast atom bombardment (FAB) (11) and cesium ion liquid secondary ion (12) mass spectrometry (LSIMS), respectively. A crucial point is that irrespective of whether the dissolved peptide or small protein is ejected from the liquid surface (the upper 50-100 Angstroms) as a protonated or unprotonated ionic species, it possesses sufficiently low vibronic energy content that little spontaneous unimolecular dissociation is usually observed. This situation is ideal for the sensitive determination of accurate[1] molecular weight of a peptide or small protein as $M+H^+$ or $M-H^-$ (13). However, the lack of an abundant and complete fragmentation pattern in LSIMS and FAB mass spectra has presented a major problem in the context of amino acid sequence determination. It is now well established that these low vibronic energy molecular species give rise to an unpredictable extent of fragmentation. This produces mass spectra showing an incomplete suite of product ions of relatively low abundance which are also confounded by liquid matrix associated chemical noise, particularly in the lower mass range. In the case of polypeptides, the occurrence of some sequence ions may provide some useful sequence related information, but rarely the complete unambiguous sequence of a peptide (14), even using nanomolar quantities of a given peptide. This undesirable situation has catalyzed the design and commercial fabrication of new mass spectrometer systems which are virtually ideally suited to sequence characterization of peptides, such as those obtained in fractions of HPLC eluants obtained from proteolytic digests of proteins (15,16). This is to say that mass selection of any specific molecular ion can be effected concomitant with its vibronic energy enhancement to produce a virtually complete fragmentation pattern free of

[1] The accuracy of molecular weight determination depends upon the type of mass analyzer employed as well as the molecular size range in question. This depends on three factors, namely the inherent mass resolution and precision of the ion optics of the instrument type in question, the stable isotope profile of the molecular ionic species and the increased probability of chemical effects which appear responsible for broadening the theoretical stable isotopic molecular profile as one goes to higher mass (> 3,000 - 4,000 daltons).

matrix associated chemical noise. This fragmentation pattern (mass spectrum) can be interpreted to yield the unambiguous sequence of the peptide component in question (14,15). This is achieved by tandem mass spectrometry (MS-MS) which is represented by diverse ion optical designs having widely different operational and analytical capabilities (15,16).

The practitioners must pay attention to the detailed technical considerations which determine the actual present capability of mass spectrometric methods for peptide and protein analysis and sequencing. These issues are necessarily complex even taken individually, let alone from the viewpoint of deriving an optimized strategy for dealing with peptide and protein primary structural analyses (16). This is due primarily to the fact that mass spectrometric methods rely on the purposeful manipulation of protein digestion and separation strategies and the chemistry of the sample. The major components of this rapidly evolving methodology include:

1. sample introduction: at the moment, batch introduction of single components or simple mixtures of similar hydrophobicities is preferred to minimize suppression of more hydrophilic components.
2. the surface active properties of the sample, viz., peptide in the liquid matrix being employed; a surface excess concentration of the peptide to be analyzed is required since the primary particle beam only sputters the surface layer of the liquid (16,17).
3. type of ion source and the temporal optimization of the molecular ion yield from a given limited quantity of peptide being introduced (18); Kambara pointed out that optimal temporal utilization of sample requires that only that portion of the surface which illuminates the source exit slit should be sputtered by the primary beam (19). This is easier to achieve with cesium ion LSIMS due to the ability to focus and control primary beam spot size (12).
4. the mass accuracy of isobaric molecular ion selection in MS-1 of a tandem instrument; selection of only the ^{12}C isobar of the stable isotope profile provides collision-induced dissociation (CID) spectra which are more easily interpreted (15,16).
5. the method and efficiency of vibronic energy enhancement of this mass selected monoisotopic molecular isobar (12-C isotopic component only).
   At the present writing only two routinely practical methods exist in more than one laboratory, namely low kinetic energy collisional activation of the selected molecular ion species by MS-1 (< 50 ev) and high kinetic energy collisional activation (> keV), a process in which either "low" or "high" translation energy molecular ionic species collide with helium atoms and in the collision lose a small portion of their translational energy and gain it back as vibronic energy. Thereby, the rate constants of the incident molecular ions' unimolecular dissociation reactions (fragmentation) are enhanced. Presently, it appears that the low energy collisions limit peptide molecular size to the order of 1,000 daltons, whereas the high

energy collisions extend the mass limit to the order of 3,000 to 4,000 daltons. In addition, there may be qualitative difference in the nature of the types of fragmentation processes observable with the two collision energies (15,16).

6. the method and efficiency of recording the collision-induced mass spectrum originating from the mass selected molecular ion. Since the process of internal energy enhancement of MH^+ by collision with helium results in very significant losses in sensitivity, the recent use of an electro-optical multi channel array detection system has been shown to increase the sensitivity of recording four sector CID spectra by approximately a factor of 100. This breakthrough in detector methodology permits recording of full CID spectra at the 10 picomole level (20).

In conclusion, the papers in this volume represent an unusually comprehensive treatment of the essential issues, the evolving micromethodologies and current results of the most advanced mass spectrometric methods for protein structural characterization and sequencing.

The contributors from the University of California, San Francisco and the University of Virginia illustrate picomole level peptide sequencing by both 4-sector tandem-array CID and tandem quadrupole FTMS photodissociation methods using cesium ion LSIMS sources. These methods are competitive in sample quantity with the micro-Edman method, but are capable of comparative high accuracy and speed of analysis, as well as delineation of post-translational modifications which may be encountered. Such MS-MS sequencing can be used to rapidly obtain sufficient sequence information on an unknown protein for construction of cDNA probes.

Further contributions describe methods for location of disulfide linkages in proteins and structural determination of N-linked glycosylation.

ACKNOWLEDGEMENT

Financial support was provided by NIH Division of Research Resources and the NSF Biological Instrumentation Program.

REFERENCES

1. Biemann, K. (1972). In "Biochemical Applications of Mass Spectrometry" (G.R. Waller, ed.), pp. 405-425. John Wiley and Sons, Inc., New York, ; and (1980). In "Biochemical Applications of Mass Spectrometry", (G.R. Waller and O.C. Dermer, eds.) First Supplementary Volume, pp. 469-525. John Wiley and Sons, Inc., New York.

2. Das, B. C., Gero, S. D., and Lederer, E. (1967). Biochem. Biophys. Res. Commun., 29, 211.
3. Morris, H. R. (1972). FEBS Lett., 22, 257.
4. a. Carr, S. A., et al. (1981). Biomed. Mass Spectrom., 8, 51-61.
   b. Herlihy, W. C., et al. (1981). Biomed. Mass Spectrom., 8, 62-69.
   c. Herlihy, W. C., et al. (1981). Biomed. Mass Spectrom., 8, 70-77.
5. Barber, M., Jolles, P., Vilkas, E., and Lederer, E. (1965). Biochem. Biophys. Res. Commun., 18, 469.
6. Hughes, J., Smith, T. W., Kosterlitz, H. W., Fothergill, L. A., Morgan, B. A., and Morris, H. R. (1975). Nature, 258, 577-579.
7. a. Magnusson, S., Sottrup-Jensen, L., Petersen, T. E., Morris, H. R., and Dell, A. (1974). FEBS Lett., 44, 189-193.
   b. Thorgersen, H. C., Petersen, T. E., Sottrup-Jensen, I., Magnusson, S., and Morris, H. R. (1978). Biochem. J., 175, 613-617.
8. Morris, H. R. (1979). Phil. Trans. Roy. Soc., A293, 39-51.
9. Gibson, B. W., and Biemann, K. (1984). Proc. Natl. Acad. Sci. USA, 81, 1956-1960.
10. For discussion of this phenomenon, see Burlingame, A. L., et al. (1988). Anal. Chem., 60, 307R-310R.
11. Barber, M., et al. (1981). In "Soft Ionization Mass Spectrometry" (H. R. Morris, ed.), pp. 137-152. Heyden, London.
12. Aberth, W., Straub, K. M., and Burlingame, A. L. (1982). Anal. Chem., 54, 2029-2034.
13. Aberth, W., and Burlingame, A. L. (1988). Anal. Chem., 60, 1426-1428.
14. Biemann, K., Gibson, B. W., Mathews, W. R., and Pang, H. (1985). In "Mass Spectrometry in the Health and Life Sciences" (A. L. Burlingame and N. Castagnoli, Jr., eds.), pp. 239-263. Elsevier, Amsterdam.
15. a. Biemann, K., and Scobel, H. A. (1987). Science, 237, 992-998.
    b. Biemann, K., and Martin, S. A. (1987). Mass Spectrom. Revs., 6, 1-77.
    c. Carr, S. A., Hemling, M. E., and Roberts, G. D. (1988). In "Macromolecular Sequencing and Synthesis. Selected Methods and Applications" (D. H. Schlesinger, ed.), pp. 83-99. A. R. Liss, Inc.
16. Burlingame, A. L., Maltby, D., Russell, D. H., and Holland, P. T. (1988). Anal. Chem., 60, 294R-342R.
17. a. Naylor, S., Findeis, A. F., Gibson, B. W., and Williams, D. H. (1986). J. Am. Chem. Soc., 108, 6359-6363.
    b. Naylor, S., Findeis, A. F., Gibson, B. W., Jiang, K., Walls, F. C. (1987). In "Proceedings of the 35th Annual Conference on Mass Spectrometry and Allied Topics" pp. 318-319.
    c. Ligon, W. V., Jr., Dorn, S. B. (1986). Int. J. Mass Spectrom. Ion Proc., 72, 317-319.
18. a. Falick, A. M., Wang, G. H., and Walls, F. C. (1986). Anal. Chem., 58, 1308-1311.
    b. Falick, A. M., Walls, F. C., and Laine, R. A. (1986). Anal. Biochem., 159, 132-137.

19. Kambara, H. (1985). In "Mass Spectrometry in the Health and Life Sciences" (A. L. Burlingame and N. Castagnoli, Jr., eds.), pp. 65-84. Elsevier, Amsterdam.
20. Burlingame, A.L. (1988). In "Proceeding of the 36th Annual Conference on Mass Spectrometry and Allied Topics", pp. 727-728.

13

STRUCTURAL ANALYSIS OF PROTEIN CARBOXYL METHYLTRANSFERASE UTILIZING TANDEM MASS SPECTROMETRY

William J. Henzel
Dana W. Aswad‡
John T. Stults

Department of Developmental Biology
Genentech, Inc.
460 Pt. San Bruno Blvd.
South San Francisco, CA 94080

Department of Psychobiology‡
University of California, Irvine CA 92717

I. INTRODUCTION

Protein carboxyl methyltransferase (PCMT) is an enzyme which selectively modifies L-isoaspartyl sites in peptides and proteins (1). Isoaspartyl bonds are presumed to arise largely from spontaneous deamidation of labile asparagine residues. The primary structure of PCMT is being determined by tandem mass spectrometry and automated Edman degradation.

Tandem mass spectrometry utilizing four-sector instruments represents one of the best means of obtaining sequence data for peptides by mass spectrometry. The mass range and resolving power permit reliable data to be obtained for a variety of peptides of up to about 25 residues in length. Complete data acquisition time for a peptide is typically only a few minutes. Newer techniques are pushing the mass range upward and reducing the amount of sample required. Typically, 0.5-1.0 nmol of peptide is necessary for obtaining collision spectra with reasonable ion signal levels.

Generally, MS1 is set to transmit only the ^{12}C isotope of the precursor mass. This selection separates the peptide of interest from other peptides and matrix ions, and provides monoisotopic peaks in the daughter spectrum. The precursor ions undergo collisions with helium atoms; the vibronic energy from the collision causes the precursor to undergo fragmentation, predominantly along the peptide backbone. The product ions are analyzed by MS2 to yield the daughter spectrum. For compounds with mass <2000 Da, available in >0.5 nmol amounts, these conditions usually provide sequence-rich spectra.

To facilitate sequencing, peptides were generated by digestion of 1 nmole of PCMT with Asp-N, trypsin, or cyanogen bromide followed with separation by HPLC. A 5% aliquot of each peptide was analyzed by fast atom bombardment mass spectrometry to determine molecular weight. Large peptides ($MH^+ > 2000$ AMU) or those which produced small MH^+ peaks were analyzed by conventional automated Edman degradation. The remaining proteolytic peptides were analyzed by tandem mass spectrometry.

The integration of mass spectrometry and automated microsequencing is shown to increase the efficiency and accuracy of protein sequencing. Tandem mass spectrometry produces sequence directly and quickly, and can give sequence data on individual components of peptide mixtures. When the mass range is exceeded for tandem mass spectrometry or the sample quantity is low, molecular weight determination in combination with automated Edman degradation can be used. Sample amounts required for mass measurement and automated microsequencing are comparable, on the order of 10-50 pmole. Knowledge of the molecular weight shows whether the peptide was sequenced to the C-terminus, improves confidence of residue assignments for low-level signals, and identifies the presence of post-translational modifications when they occur.

II. EXPERIMENTAL

Spectra were obtained with a JEOL JMS-HX110HF/HX110HF Tandem Double-Focusing Mass Spectrometer (E_1B_1-E_2B_2 configuration), equipped with a JEOL DA-5000 data system. Samples were dissolved in 0.1% TFA and 0.5 µl of each mixed 1:1 on the probed tip with 0.5 µl glycerol:DTT:DTE (6:5:1). Ionization was by fast atom bombardment using a JEOL FAB gun, 6 keV Xe atoms, 10 mA emission current. Precursor ions were selected with MS1 and fragmented by CID in the grounded collision cell-2 (10 keV energy, 75% attenuation of the precursor beam with helium). The fragments were analyzed with MS2. Data interpretation was aided by use of the program PLTSEQ (2) provided by Prof. K. Biemann of M.I.T.

Automated Edman degradation was performed with an Applied Biosystems model 470A protein sequencer and a 120A PTH analyzer. Sequence determination was performed on a VAX 11/785 as described (3).

III. SEQUENCE ANALYSIS BY TANDEM MASS SPECTROMETRY AND AUTOMATED EDMAN DEGRADATION

Trypsin and Asp-N protease were used to generate peptides for MS/MS analysis (Figures 1 and 2). Table 1 lists the major MH^+ ions observed when a 5% aliquot of each peak from the Asp-N digestion was analyzed.

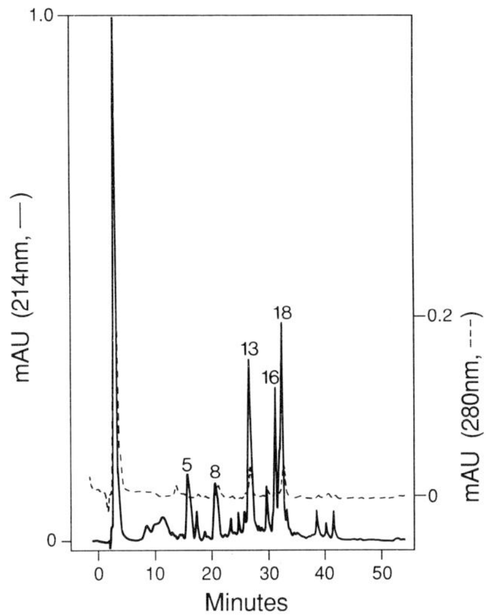

Fig. 1. Tryptic digestion of PCMT. Peptides were eluted with a linear gradient of 0.1% TFA to 50% acetonitrile in 50 min at a flow rate of 200 μl/min on a Synchron C4 4000Å column.

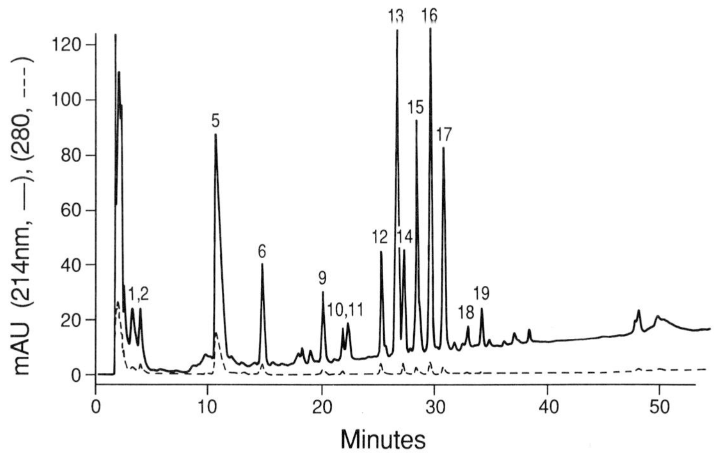

Fig. 2. Asp-N digestion of PCMT. HPLC conditions were the same as described in Figure 1.

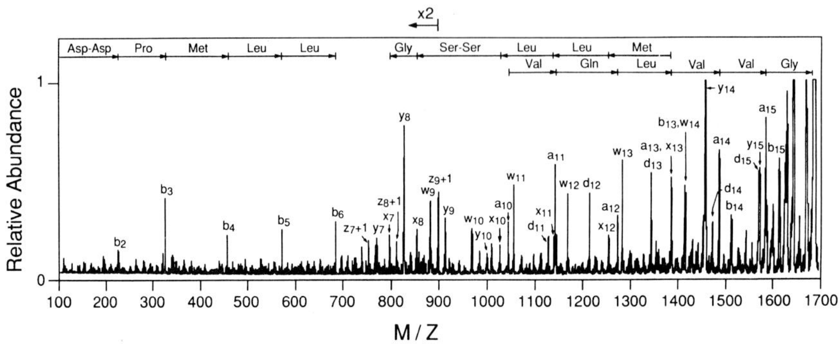

Asp-N 15

	D	D	P	M	L	L	S	S	G	R	V	Q	L	V	V	G
pmol	62.8	21.1	–	11.7	7.8	6.9	2.4	2.1	6.2	4.2	2.7	2.2	1.0	1.1	0.9	ND

Fig. 3. MS/MS daughter spectrum and automated Edman analysis of Asp-N peptide D14.

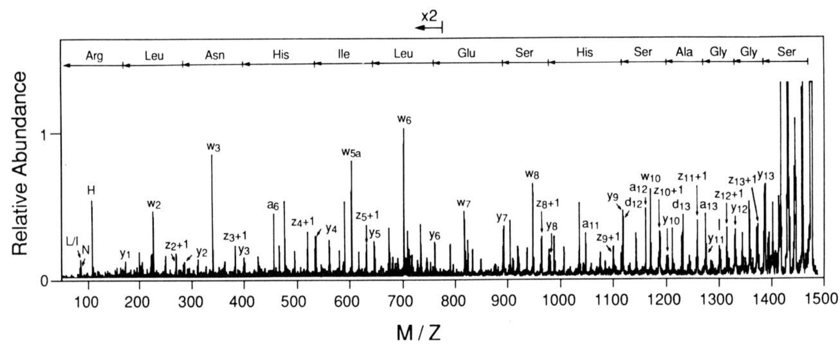

Tryptic T5

	S	G	G	A	S	H	S	E	L	I	H	N	L	R
pmol	16.9	16.0	18.8	18.0	3.6	14.3	2.5	5.3	4.0	6.6	3.2	4.2	2.9	2.7

Fig. 4. MS/MS daughter spectrum and automated Edman analysis of tryptic peptide T5.

TABLE 1. Major MH^+ ions of Asp-N digestion of PCMT

Peak No.	MH^+	Peak No.	MH^+
5	521.0[a]	13	1643.0[b]
6	1300.6[a]		1685.8[a]
	1390.7[a]	14	1641.8[a]
9	779.5[a]		1685.8[b]
10	779.5[a]		2660.8[b]
11	1152.5[a]	15	2343.6[b]
12	2661.8[b]	16	2653.0[b]
		17	1949.2[a]

[a]represents monoisotopic mass
[b]represents average mass.

The MS/MS daughter spectrum of Asp-N peptide D14 (Figure 3) shows a b-ion (4) series (b_2-b_6) which identifies the six residues at the peptide N-terminus. The anion series (a_{10}-a_{16}) gives the six residues at the peptide C-terminus. An x-ion series (x_7-x_{13}) identifies the remaining residues in the middle of the sequence and provides some overlap to the N-terminal residues.

The MS/MS daughter spectrum of tryptic peptide T5 (Figure 4) shows a complete y-ion series to give the whole sequence. Initial interpretation assigned Ser-Asn as the peptide amino-terminus rather than Ser-Gly-Gly because Asn and Gly-Gly have the same residue masses. Automated microsequencing confirmed the Gly-Gly identity. The presence of the w-ion series (4,5), often observed for tryptic peptides, permits distinction between Leu and Ile. However, the intense w-ions make data interpretation more difficult because they seldom give a long contiguous series. In order to confirm the MS/MS sequence interpretation, the peptides were removed from the probe and subjected to automated Edman degradation. Since the peptide can be recovered from the probe after MS/MS analysis, incomplete sequence assignment or suspected post-translational by modified residues can be confirmed by automated Edman degradation.

IV. CONFIRMATION OF SEQUENCE ACCURACY

With peptides of greater than 25 residues in length or with less than 0.5 nmol, tandem mass spectrometry may not be possible. However, molecular ion measurement can still provide confirmation of the accuracy of a sequence obtained by automated microsequencing. Three questions can be answered:

1. Were the correct residues assigned for each cycle?
2. Was the C-terminus of the peptide reached?
3. Were there any post-translational modifications?

1				5					10					15			
GLY	VAL	ILE	TYR	VAL	PRO	LEU	THR	ASP	LYS	GLU	LYS	GLN	TRP	SER	ARG	TRP	LYS
321	248	180	215	152	105	113	42.7	94.6	104	73.9	93.3	67.8	13.9	42.5	30.5	4.6	12.3

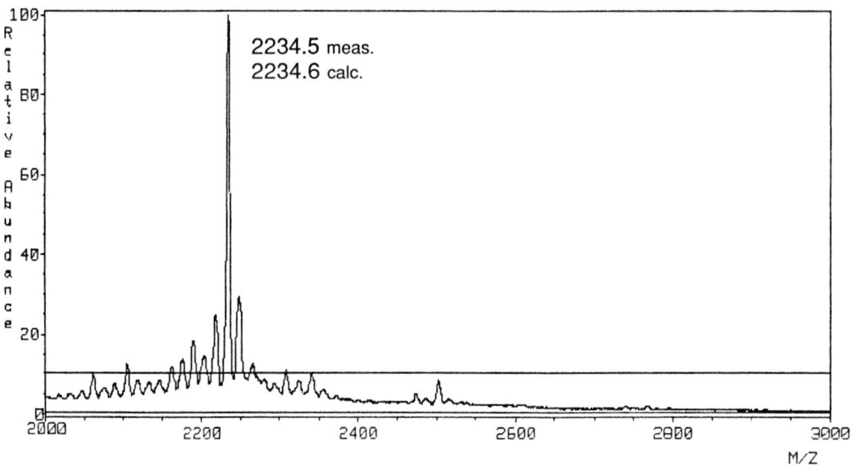

Fig. 5. FAB mass spectrum showing the molecular ion region of CNBr.

If a difference exists between the measured mass and the mass calculated from the sequence, the magnitude of the difference often points to the correction or modification.

Shown in Fig. 5 are the sequence data and the molecular ion region of the spectrum of a CNBr peptide of PCMT. The sequence data give the quantitation (pmol) for each residue and show diminishing levels near the C-terminus. The measured mass confirms that the sequence is correct and the C-terminus of the peptide was reached. The latter point is important since it shows that the C-terminus of the protein was identified because homoserine would otherwise be expected as the C-terminal residue of a CNBr fragment.

V. IMPROVED RELIABILITY OF MIXTURE SEQUENCING

Mixtures of 2, 3, or 4 peptides can be sequenced by automated microsequencing. When the quantities of the components are similar or levels are low, however, considerable ambiguity can exist in the result. Molecular weights for each of the components can be compared with the proposed sequence to sort out the assignments.

Shown in Figure 6 are the sequence data and mass spectrum of the components of a single HPLC peak #17 from an Asp-N digest of PCMT. Although the peak appeared as a sharp symmetrical peak, automated Edman degradation revealed a mixture of two sequences with similar quantitation. The FAB mass spectrum yielded two molecular weights, one of which was in complete agreement with the bottom sequence ($MH^+ = 2653.1$); however, only partial sequence was obtained from the top sequence (residues not identified are double underscored).

DGSVKMKPLMGVIYVPLT ($MH^+ = 1949.4$)

DQLKPGGRLILPVGPAGGNQMLEQY ($MH^+ = 2653.1$)

Utilizing this strategy mixture sequences can be sorted providing sequence information that may be useful for oligonucleotide probe design. We are developing a computer program that will utilize molecular ion and Edman degradation data to provide automatic interpretation of mixture sequences.

VI. CONCLUSION

Approximately 90% of the sequence of PCMT has been obtained, based on SDS-PAGE (M_r ~20 kDa). A contiguous sequence of 128 residues, including the C-terminus, has been determined. Further sequencing is in progress. The Asp-N digestion produces many peptides that match well the mass range requirements of tandem mass spectrometry. The lack of a basic residue at either end of the peptide, however, tends to promote several fragment ion series and make sequence interpretation more difficult.

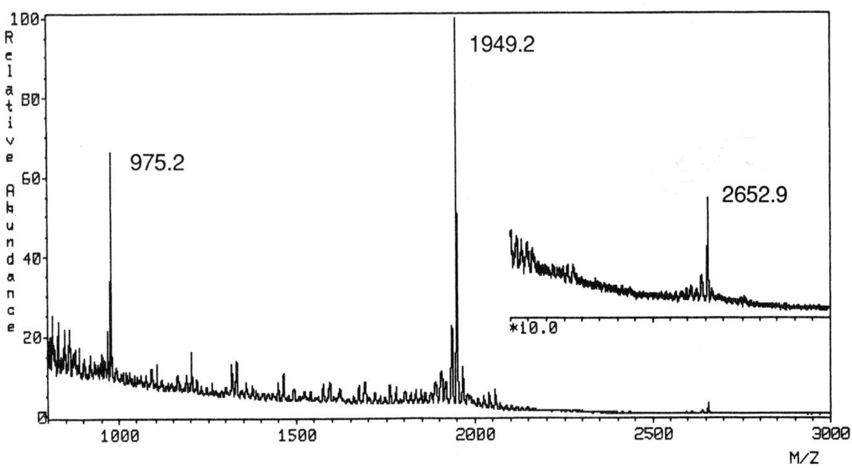

Fig. 6. FAB mass spectrum of PCMT Asp-N digest peak #17. The peak at M/Z 975.2 is an MH_2^+ ion.

Tandem mass spectrometry is shown to provide a rapid means of generating peptide sequence information for probe design, but limitations in sample amount prevent its routine use. Improvements that should permit sequence determination at or below the 10 pmol level would allow MS/MS to be competitive with current sensitivity of automated Edman degradation. Recent development of microchannel array detectors for sector instruments (6,7), and a tandem quadruple Fourier transform mass spectrometer (8) are allowing these sensitivity levels to be reached. Further improvements are likely to reduce the sample requirements even more.

REFERENCES

1. Aswad, D.W. and Johnson, B.A. *Trends Biochem. Sci.* **12**,155-158 (1987).

2. Scoble, H.A., Biller, J.E., and Biemann, K. Frezenius Z. *Anal. Chem.* **327**, 239-245 (1987).

3. Henzel, W.J., Rodriguez, H. and Watanabe, C. *J. Chromatogr.* **404**,41-52 (1987).

4. Johnson, R.S., Martin, S.A., Biemann, K., Stults, J.T., and Watson, J.T. *Anal. Chem.* **59**, 2621-2625 (1987).

5. Stults, J.T., and Watson, J.T. *Biomed. Environ. Mass Spectrom.* **14**, 583-586 (1987).

6. Hill, J.A., Martin, S.A., and Biemann, K. 36th ASMS Conf., San Francisco, 1988.

7. Elliot, G., Gallagher, R., Evans, S., Ashcroft, A., Gooch, G., Wright, A.D., Falick, A.M., Walls, F., and Burlingame, A.L. 36th ASMS Conf., San Francisco, 1988.

8. Hunt, D.F., Shabanowitz, J., and Yates, J.R. III, *J. Chem. Soc., Chem. Commun.* 548-550 (1987).

14

SEQUENCE DETERMINATION OF N-TERMINALLY BLOCKED AND INTERNAL PEPTIDES AT THE PICOMOLE LEVEL BY TANDEM MASS SPECTROMETRY WITH ARRAY DETECTION

Bradford W. Gibson, Zhonghua Yu, Beth Gillece-Castro, William Aberth, F.C. Walls and A.L. Burlingame*

The Mass Spectrometry Facility,
Department of Pharmaceutical Chemistry
and the Liver Center*
University of California
San Francisco, California 94143-0446

I. INTRODUCTION

Mass spectrometric methods for the analysis of peptides and proteins have undergone a number of revolutionary improvements since the introduction of liquid secondary ion mass spectrometry (LSIMS) [see reviews 1,2]. The initial application of LSIMS (or FABMS) to the structure determination of proteins has been used in combination with the Edman degradation to both sequence proteins directly, as in the case of two anti-tumor proteins, macromomycin and neocarzinostatin [3,4], and to determine the correctness of DNA-deduced protein sequences as shown for a series of amino acyl-tRNA synthetases [5]. Critical to success of these and similar efforts is the ease of identifying the relatively abundant molecular ions that are characteristic of LSIMS spectra to determine the precise molecular weights of underivatized peptides. This is true whether the peptide is present as a single component or in a complex mixture. This procedure is sometimes referred to as "peptide-mapping". An important aspect of this methodology is the ability to quickly identify numerous types of post-translational modifications [6,7], such as phosphorylation [8,9], glycosylation [7,10,11], and N-terminal blocking groups [12,13,14], as well as verify the fidelity in translation of recombinant proteins [7].

A wide variety of needs exist in our laboratory for the sequence and molecular weight analysis of peptides and covalently modified peptides derived from larger proteins and post-translationally modified proteins. While the original protocols described a few years ago required 10-20 nmoles [5], it is clear from the current problems in structural biology that these sample requirements need to be lowered by at least a factor of 100 or more, well into the low picomole range (e.g. 10-20 picomoles). To accomplish this, mass spectrometric methodology had to be improved in several areas, including sample preparation, ionization efficiency, optimization of the temporal use of a given sample quantity, instrument design and better ion detection systems. Indeed, significant improvements have been made in all of these areas over the last few years, particularly in the further development and application of new instrument designs such as high mass, four-sector tandem MS, FT-ICR and TOF mass spectrometers [see ref. 2].

While LSIMS methods alone have extended the mass range accessible for peptide detection such that the vast majority of components obtained by proteolytic digestion of a protein may be analyzed, the lack of abundant sequence ions in LSIMS spectra make unambiguous amino acid sequencing difficult or even impossible. This problem is compounded by the presence of significant chemical noise arising from the sputtering of the liquid matrix [2].

Thus, there was obviously a need to vibronically activate molecular ions throughout the mass range (< 3,500 daltons) to increase the abundance of fragment ions and at the same time eliminate or reduce the chemical noise. These needs quickly led to the design and commercial availability of several tandem double focusing mass spectrometers. These instruments are able to operate with high transmission and carry out high kinetic energy collision induction of sequence ions from any selected peptide [2], which increases the abundance of fragment ions with the concomitant elimination of chemical noise. Recently, these tandem MS techniques using high collision energies have been applied to various structure problems such as in the direct sequencing of E. coli thioredoxin [15] and the sequencing of the N-terminally blocked regions of rat and human fatty acid binding proteins [13,14].

In this paper, some aspects of our ongoing research efforts into peptide and protein analysis will be described, particularly those that relate to the isolation, purification and LSIMS analysis of peptides at the picomole level, such as those contained in the spots on 2-d gel electrophoretograms. Specific emphasis will be placed upon the use of a prototype Wien mass spectrometer equipped with a position sensitive channel electron multiplier array (CEMA) detector for molecular weight profiling at the picomole level [16], and the new Kratos tandem double focusing instrument system (model Concept II HH) which embodies the first electro-optical multi channel diode array detector [17] .

II. THE ANALYSIS OF LARGE PEPTIDES ON THE WIEN MASS SPECTROMETER

The prototype Wien instrument has been recently used to analyze high mass peptides and small proteins at the sub-nanomole level, including EGF [18], biotinylated-ß-endorphins [19], and tryptic peptides isolated from rat heart fatty acid binding protein (FABP) [13], and a genetically engineered mutant of rat trypsinogen [20]. For these samples, the Wien MS operated at its maximum magnetic field (**B**) and the electric field (**E**) was adjusted to center in the mass of interest, producing in each case a mass spectrum where ions were collected

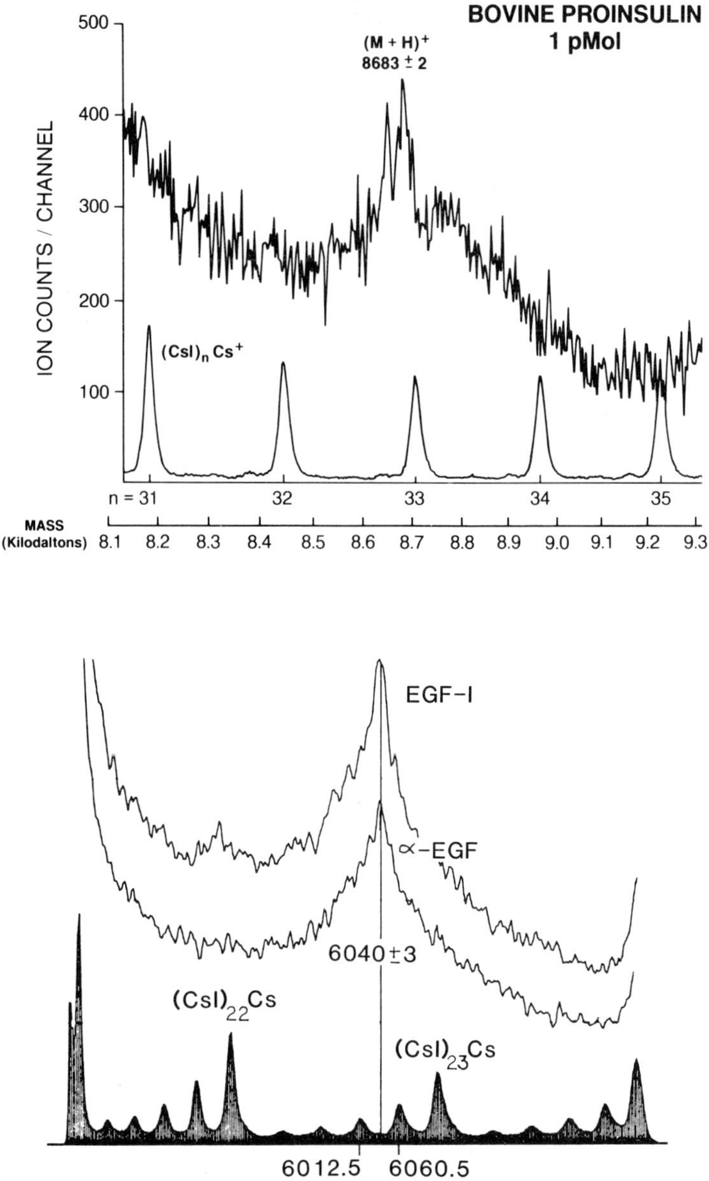

Fig. 1. Wien mass spectra of the molecular ion regions of bovine proinsulin (MH^+ 8,682) using one picomole of sample (top) and EGF and EGF-I using 20-40 picomoles (bottom). In each case, approximately 12% of the mass range was covered in a single 30-60 sec. acquisition.

simultaneously on the CEMA detector over a extended but limited mass range (approximately 12% at m/z 6000). Under these conditions, the highest sensitivity of detection is obtained with a working resolution of between m/z 500-1200, including the loss of resolution due to the CEMA detector itself. For example, Fig. 1 shows the spectrum of 1 picomole of bovine proinsulin (MH^+ 8,682) and 40 picomole of α-EGF and a novel deamidated form (EGF-I, MH^+ 6044). The advantages of using a Wien instrument for these types of analysis is not just its sensitivity, but its high mass capability. Unlike conventional magnetic sector instruments, there is no theoretical upper limit on the mass range at full accelerating voltage (40 keV). The major limitations in the present hardware configuration are the relatively small maximum magnetic field capacity of this prototype Wien MS (0.85 Tesla) and low 'practical' counting rate (30-50,000 counts/sec) of the CEMA detector using the resistive array read-out methods. However, both of these design and hardware limitations can be easily and dramatically improved with a new Wien MS design [H. Matsuda and H. Wollnick, personal communication] and by incorporating electro-optically coupled diode array detection [17].

III. ZOOM-IN/ZOOM-OUT PEPTIDE ANALYSIS

One of the major advantages of the Wien instrument for peptide analysis is its ability to expand ("zoom-out") and contract ("zoom-in") the mass range that can be simultaneously observed. The mass range that is focused on the detector is dependent on both the **E** and **B** fields. At maximum **B**, the **E** field determines the masses that will pass through the Wien and strike the CEMA detector,

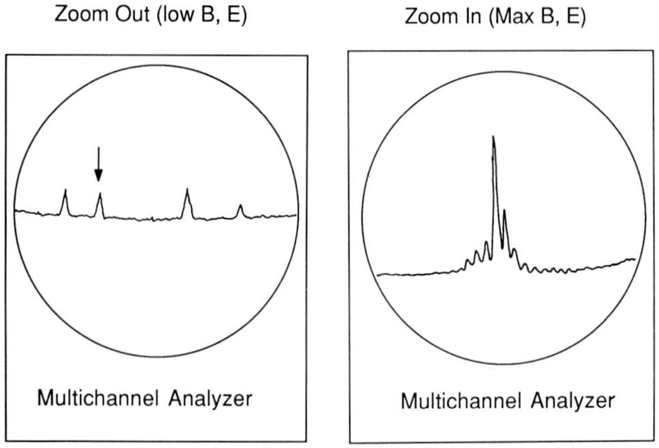

Fig. 2. Schematic representation of "zoom-out"/"zoom-in" mass acquisition capabilities of the Wien mass spectrometer.

producing the smallest mass window possible, but yielding the maximum resolution. By lowering the magnetic field **B**, and again adjusting **E** for a particular mass, a larger mass range can be simultaneously focused on the CEMA detector. This, in essence, allows one to "zoom-out" for simultaneous detection over a very large mass range (e.g. m/z 500-3000) and then "zoom-in" on a smaller mass range of particular interest (e.g. m/z 1000-1100) as shown schematically in Fig. 2.

In order to apply this approach and, at the same time, evaluate its sensitivity for the analysis of tryptic peptides isolated at the sub-nanomole level, 100 picomoles of rat heart FABP was digested with trypsin. The resulting mixture of tryptic peptides was then separated into 16 fractions by microbore HPLC (see Fig. 3). Under these conditions, each peptide fraction had an elution volume of approximately 50 μL. In analytical HPLC, typical flow rates and elution volumes are 10-20 times larger with an equivalent loss in UV

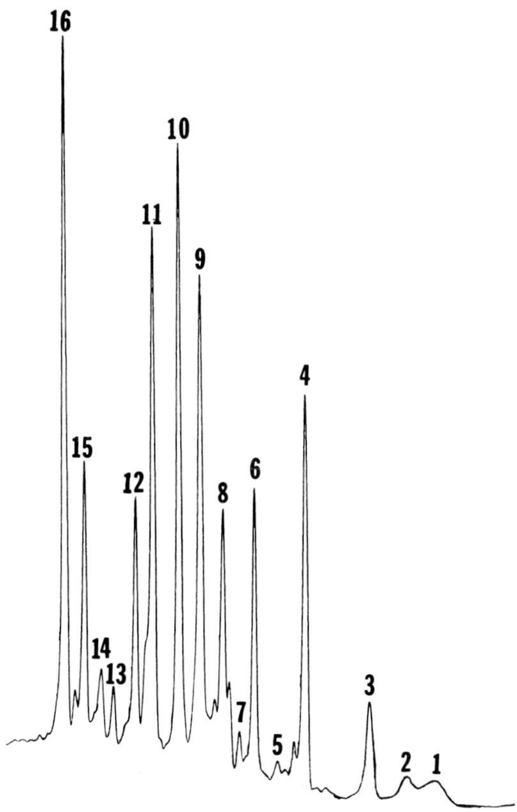

Fig. 3. HPLC separation of the tryptic peptides of rat heart FABP (100 picomoles) using a microbore HPLC column (Brownlee Aquapore-300, 25 cm x 1 mm)

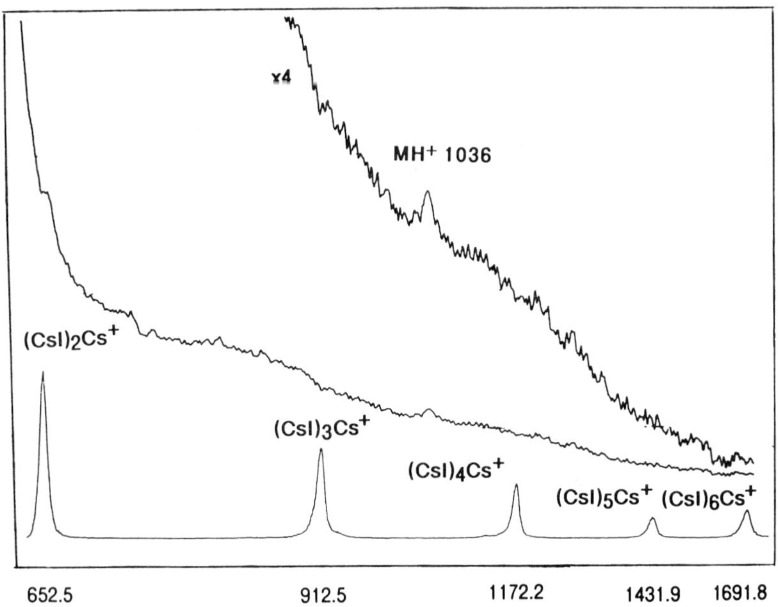

Fig. 4. "Zoom-out" mass acquisition of HPLC Fraction #16 over the mass range covering m/z 650-1700.

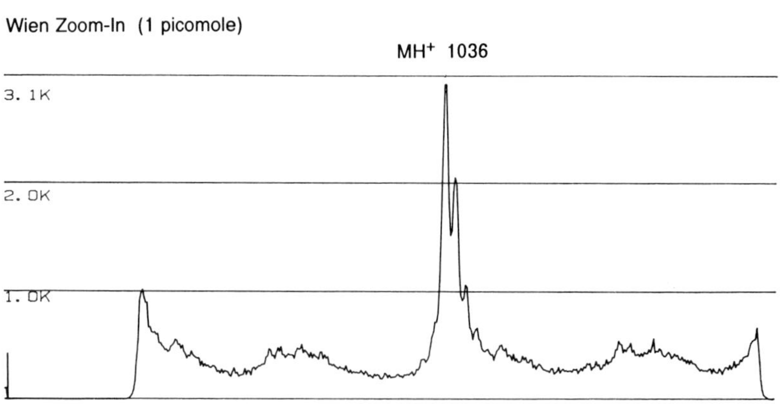

Fig. 5. "Zoom-in" mass acquisition of the region centered around m/z 1030 for Fraction #16 showing the resolved peak with MH^+ 1036 using a 1% aliquot (1 picomole).

detector sensitivity. In the first experiment, a 2 pmole equivalent (1µL) of HPLC Fraction 16 was directly analyzed over a mass range of approx. m/z 650-1,700 (see Fig. 4). Under these conditions, the primary Cs^+ current had to be reduced by a factor of about 100 (10^{-9}A) so that the CEMA detector would not be overloaded by an excessive count rate resulting from the increased signal contained in this larger mass window. However, even when using this extremely low Cs^+ flux, a spectrum was still obtained over this entire mass range which showed the presence of a significant ion cluster centered at around m/z 1030. Now, by adjusting **B** to its maximum and readjusting **E** to center this ion cluster, a 'zoom-in' mass acquisition can now be obtained showing a resolved, more abundant molecular ion specie (see Fig. 5). Thus, in screening a series of peptides of unknown mass, a "zoom-out"/"zoom-in" strategy is an extremely effective means of rapidly and accurately identifying the peptide molecular ions. As previously stated, "zoom-out" acquisitions often required that the Cs^+ beam be operated at a very small flux. While this had the effect of reducing the overall sensitivity, it had the advantage of conserving sample, since very little sample is consumed under these low Cs^+ flux conditions. Thus, by using an initial large mass range accumulation (or several large mass window segments) to detect the presence of peptide ions with a weak Cs^+ beam, the remaining sample still on the probe tip can be used more efficiently in the "zoom-in" mode to obtain resolved molecular ions or to select these identified molecular ions for structural analysis under MS/MS conditions. Ideally, the subsequent tandem experiments would be best carried out using the same Wien instrument as the first stage of a Matsuda-Wollnick MS/MS instrument with an inherent extended focal plane and multi channel array system in MS-II.

IV. ZOOM-IN PICOMOLE ANALYSES

The highest sensitivity (i.e. highest secondary ion beam intensity) of peptide analysis is clearly at maximum **B**, where a higher resolution and larger secondary ion beam can be tolerated on the resistive array type CEMA detector. In order to investigate the absolute sensitivity of combined microbore-HPLC and Wien MS analysis, a series of 100 picomole peptide fractions from the previous tryptic digest of FABP were directly analyzed on the Wien MS. In a series of experiments, a 10%, 1% and 0.1% aliquot of the 100 picomole peptide fraction was applied to the probe tip, which represents a maximum of 10, 1 and 0.1 picomoles of each peptide, respectively. Figure 6 shows the relative molecular ion abundances (MH^+) for fraction #15 which contained a single peptide with an MH^+ at m/z 1523. In this and similar cases, only weak signals were obtained at the 0.1 picomole level (less than 1000 counts at the peak maximum) while at the 1 and 10 picomole levels, the signals were clearly defined and well above the chemical background.

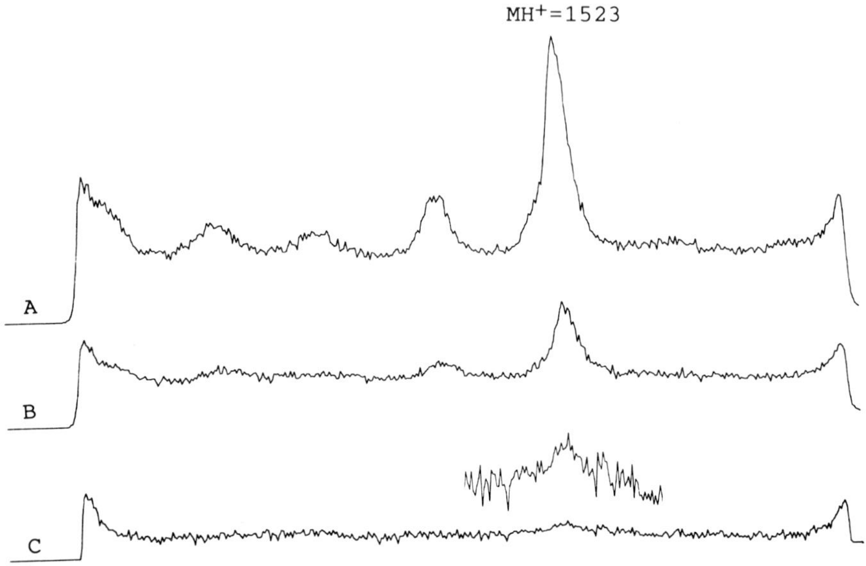

Fig. 6. "Zoom-in" mass acquisition of Fraction # 15 from the tryptic digest of FABP taken with (A) 10, (B) 1 and (C) 0.1 picomoles of sample showing the molecular ion at m/z 1523.

V. SAMPLE HANDLING TECHNIQUES

While unambiguous detection was obtained for many of the peptide samples at levels of 1 picomole and less, the question remains as to what level of combined HPLC isolation and mass spectrometric detection is realistically possible. Can a one pmole sample of peptide be isolated and analyzed directly? The answer to this question involves the more difficult question of sample handling procedures. In order to perform such an analysis properly, the peptide needs to be isolated at this level, a task that is extremely difficult or impossible to do with conventional HPLC combined with UV detection. However, in some cases peptides can be detected at this low level by immunochemical or bioassay techniques, but these methods are not generally applicable. Newer techniques such as capillary zone electrophoresis may enable this to be done more routinely in the future. The currently existing methodology is still not adequately suited for this task. Nonetheless, it may be possible in some limited cases to do this, or it can be done blindly by analyzing all fractions. However, a difficulty remains in how to effectively introduce this sample into the source of the mass spectrometer. Concentrating peptide samples to a more workable volume at this low level may lead to significant losses, as was found to be the case for the majority of the tryptic peptides from FABP.

In these studies, a comparison was made between the quality of the LSIMS spectra obtained by directly applying a 1 picomole aliquot (0.5 μL) to the probe tip from larger (100 pmole) HPLC peptide-containing fractions and loading a 1 picomole aliquot of a peptide sample that had first been dried and then re-dissolved in 0.1% TFA/H_2O. In both cases, an equivalent amount of matrix (0.5 μl thioglycerol/glycerol) was first added to the probe tip. This simple drying procedure had a dramatic effect on lowering the MH^+ abundances observed for most of the peptides tested, reducing their ion abundances between 3-10 fold. Figure 7 shows the effects of this drying step on two tryptic peptides from FABP. The number of counts obtained for the MH^+ peak maximums decreased dramatically from 12,000 to 3,500 counts for the MH^+ peak maximums in fraction #9 (MH^+ 861 and 875) In the case of the N-terminally blocked peptide (MH^+ 1036) the counts dropped from 3,000 to 1,300 counts (data not shown) and this pattern was repeated in a number of other tryptic peptides from FABP subjected to an identical drying step. Clearly, it is essential that the smallest possible volume be obtained in the chromatographic steps (microbore vs. analytical), since at these low levels, peptide losses are significant at steps that require the transfer or concentration of the sample.

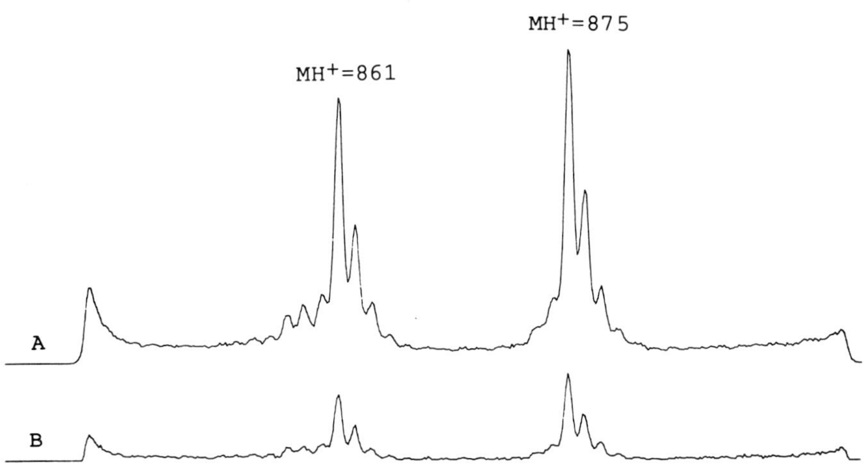

Fig. 7. Wien mass spectra showing a 1 picomole aliquot of HPLC Fraction #9 (A) before and (B) after drying the sample.

VI. TIME COURSE/SURFACE ACTIVITY

Another important advantage of using an array detector with a large mass range ion signal accumulation for peptide analysis is associated with the problems of radically different surface activities of peptides and the effect that this has on their surface excess concentration and hence their observed molecular ion abundance. In a previous paper, a hydrophobicity scale based on the indices of Bull and Breese was used to correlate the suppression of molecular ions of certain peptides in mixtures [21] This crude scale was surprisingly effective in predicting the peptides that were likely to be suppressed as being ones with a positive ΔG of transfer to the surface based on adding the individual experimentally determined ΔG values for each amino acid. It was also noted that these suppressed peptides are hydrophilic and generally have their highest ion abundances near the end of the life-time of the sample in the mass spectrometer, 5-10 minutes after initial bombardment. Having analyzed numerous peptides of differing types, it has been our experience that peptides follow distinctly different time courses in their relative ion abundances. This is presumably due to differences in their surface activity that confer at the extremes, a substantial surface deficit to small hydrophilic peptides like TRH (pyroGlu-His-Pro-NH_2) or a substantial surface excess to amphipathic, alpha-helical forming peptides like melittin and the magainins (or PGS) [22].

The practical consequence of this can be seen in Fig. 8, where a peptide fraction from the amphibian skin secretions of _Rana utricularia_ was analyzed in three separate scans on a conventional two sector magnetic instrument from m/z 2500 to 250 at 300 sec/decade. Thus, approximately six minutes separate each scan, of which only a small portion is shown in Fig. 8. Aside from sodium adducts (MNa^+) and peaks corresponding to deamination (MH^+-NH_2), there are clearly three MH^+ ions present at m/z 1248, 1296 and 1388. However, only MH^+ 1248 is observed in the first scan while onlyMH^+ 1296 and 1388 appear in the second and third scan only. In addition, the relative ion abundances are dramatically different in each scan with MH^+ 1388 clearly emerging as the dominant ion in the last scan. If the MH^+ 1388 ion had been at a higher mass (e.g. MH^+ 2000), it is unlikely that a second scan would have been taken over this entire mass range since a large peak was observed at MH^+ 1248 in the first scan. This would have resulted in the second higher mass peak having gone undetected. However, in a 'zoom-out' mass acquisition, all molecular ions have a much better chance of being detected as the accumulation continues over the entire mass range of interest. Even at low Cs^+ fluxes, the evaporation of the matrix will cause a buildup in concentration of the non-surface active peptides and eventually force them to the surface where they will be sputtered. Array acquisitions of a mass spectrum also has the advantage of offering a better "averaged" spectrum since it is, in effect, simultaneously integrating all ions over a large mass range during this time interval.

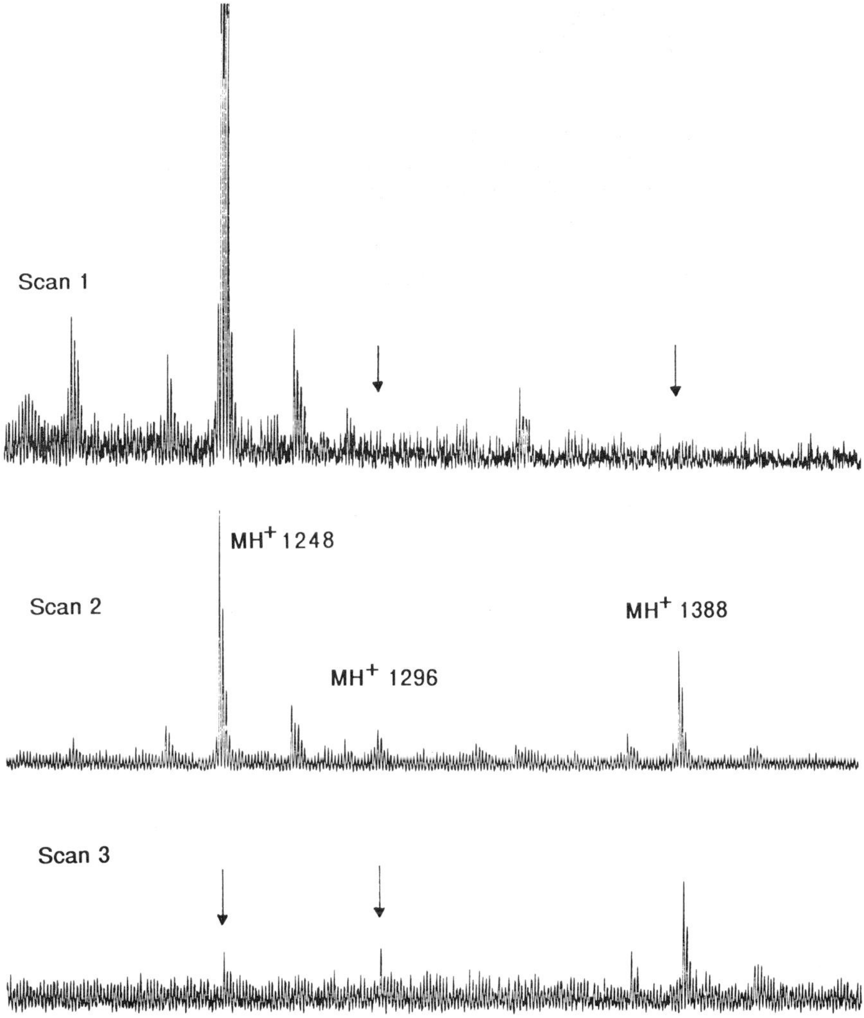

Fig. 8. Three consecutive scans of a peptide fraction from Rana utricularia taken approx. 5 minutes apart on the Kratos MS 50S mass spectrometer showing the changing relative abundances of three molecular ion species.

VII. ELECTRO-OPTICAL MULTI CHANNEL DIODE ARRAY DETECTORS AND TANDEM MS

As previously mentioned, one of the major drawbacks in the use of a CEMA detector on the Wien MS is its low counting rate. More recently, a 1,000 channel electro-optical diode array detector has become commercially available [17]. In its initial use on a double focusing mass spectrometer (Kratos MS 50RF), it has been shown to increase high mass detection efficiency by approximately a factor of 100 while maintaining a resolution of 3,500 daltons [17]. Our own experience using this instrument has allowed us to successfully analyze a series of high mass protein fragments from cytochrome b_5 using 100-300 picomoles of sample as well as a series of N-linked oligosaccharides from the acetylcholine receptor [11]. Figure 9 shows the spectra of rabbit trypsinized cytochrome b_5 (MH^+ 10145), in which the precise molecular weight information was used to determine the presence of the N-terminal blocking group (N-acetyl-glutamine) as well as the exact sequence domain of this trypsinized fragment. Clearly, such an electro-optical diode array detector would also be advantageous on the present Wien MS in order to eliminate the dynamic range limitation of the resistive readout. Indeed, the low abundances of daughter ions in tandem MS often require that the mass spectrometer operate at a lower resolution than is really ideal (500-1000 daltons) as well as limiting the useable mass range to peptides under 2,500 Da [1].

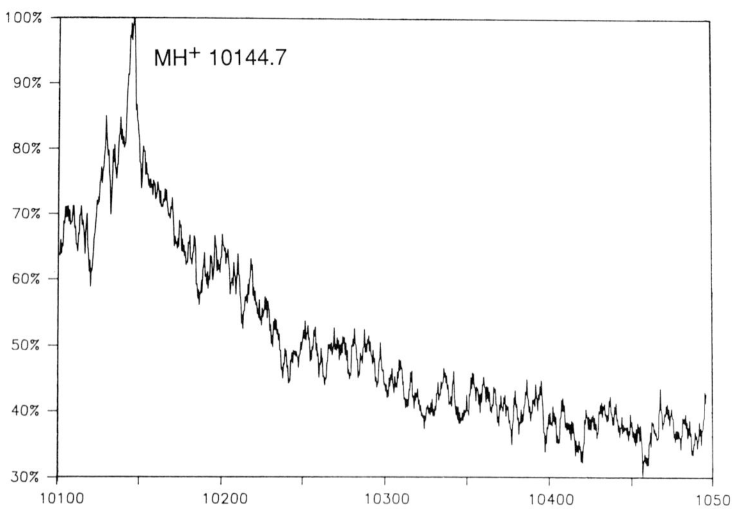

Fig. 9. Molecular ion region of trypsinized rabbit cytochrome b_5 taken with an MS 50RF mass spectrometer equipped with a diode array detector.

This laboratory has recently obtained a tandem four sector high mass magnetic instrument (Kratos, model Concept II HH, Manchester, England) with the first multi channel diode array detection system mounted on MS-II [25]. Even though ions of only one mass can be exactly in double focus at the normal collector slit location, there is a limited range of nearby masses that will fall on an approximate plane of double focus and be detected simultaneously. Initial experiments have shown that a mass range of 4% in MS-II can be simultaneously focused at a resolution of 3,000 with this instrument. The sensitivity improvement by simultaneous detection should equal the ratio of the array detector width to the normal collector slit width (approximately two orders of magnitude larger). In order to accumulate an entire collision induced dissociation spectrum (CID), the mass range is covered in sequential 4% mass windows under computer control. In this configuration, it is now possible to obtain complete MS/MS spectra of peptides at 10-100 picomole levels (and possibly lower) with an improved signal to noise ratio for the first time. This level of improvement in overall tandem instrument sensitivity is a pioneering breakthrough which has made these mass spectrometric based strategies competitive with the micro-Edman methods, but in addition brings a new realm of experimental flexibility, speed and precision of characterization to the field of protein biochemistry. In order to illustrate these points, the CID spectra of several peptides from rat heart fatty acid binding protein are shown which were obtained at the 100 to 10 picomole levels with this new tandem instrument.

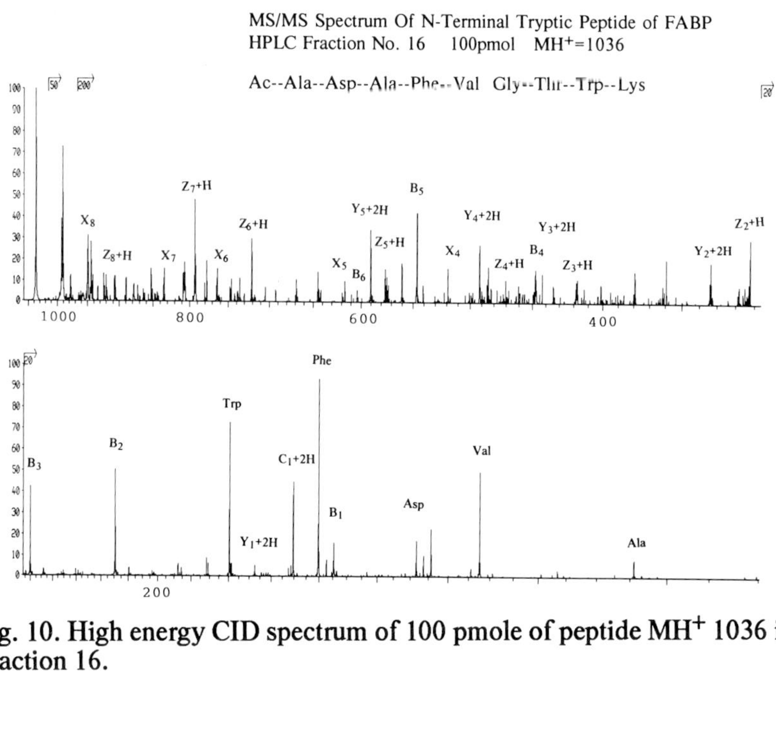

Fig. 10. High energy CID spectrum of 100 pmole of peptide MH$^+$ 1036 in Fraction 16.

As an example of determination of the sequence of an N-terminally blocked peptide, we have determined the high collision energy CID spectrum of tryptic Fraction 16 (see Fig. 3). The normal cesium ion LSIMS spectrum of Fraction 16 is shown in Fig. 5 as the characteristic protonated molecular ion with little interpretable fragmentation. Loading a total of 100 picomoles of Fraction 16 on a cooled direct inlet probe, sequential segments of the CID spectrum were recorded by stepping the magnetic field of MS-II through 4% mass windows under computer-control and accumulating for one second per step. The CID spectrum obtained is shown in Fig. 10. Partial series of X, Y+2 and Z+1 sequence ions with C-terminal charge retention are observed together with B sequence ions with N-terminal charge retention. In addition, some single immonium ions are observed for individual amino acids. This spectrum clearly illustrates that the N-terminal blocking group is N-acetyl and the sequence is: Ac-Ala-Asp-Ala-Phe-Val-Gly-Thr-Trp-Lys.

As an example of a peptide from Fraction 10, we recorded the CID spectrum of one of the two components (MH^+ 861 and 875) present in this fraction, namely that of MH^+ 875, using a total of only 10 picomole of this fraction. The CID spectrum in Fig. 11 shows a rich C-terminal sequence series, including W, X, Y+2 and Z+1 as well as the N-terminal series A, B, C+2 and D, yielding a sequence: Ser-Val-Val-Thr-Leu-Asp-Gly-Gly-Lys. Finally, the CID spectrum of substance P (Fig. 12) was also recorded using a total of 10 picomoles of this peptide.

VIII. CONCLUSIONS

As the data in this paper shows, a concerted effort in sample preparation, molecular ion detection, and the optimization of sensitivity in tandem MS/MS instrumentation is required for successful determination of the molecular weights and sequences of peptides and proteins at low picomole levels. As further improvements are made in the methodology for mass spectrometric protein and peptide sequencing, it is almost certain that mass spectrometry will become an essential part of the "classical' strategy now being used by most protein chemists.

IX. ACKNOWLEDGEMENTS

We would like to thank Syd Evans, Gerry Elliott and Geoff Gooch at Kratos (Manchester, England) for their help in acquiring data on the the tandem Concept-array mass spectrometer. We also thank Dr. N.M. Bass for the FABP samples. This research has been supported by grants from the National Institutes of Health (RR01614 to A.L.B), The National Science Foundation, Biological Instrumentation Program (DIR 87 00766), Liver Center (AM 24643 to A.L.B.) and the Cancer Research Coordinating Committee (B.W.G.).

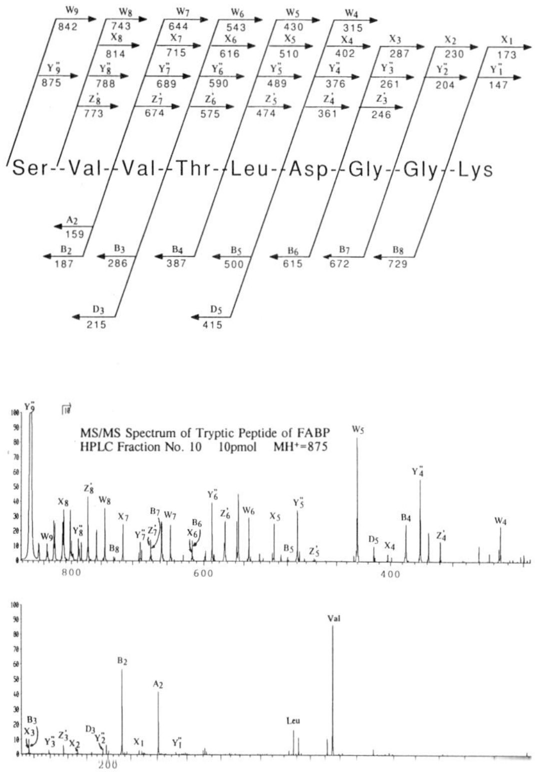

Fig. 11. High energy CID spectrum of 10 pmole of peptide MH$^+$ 875 selected from co-eluting mixture with MH$^+$ 861 from Fraction 10.

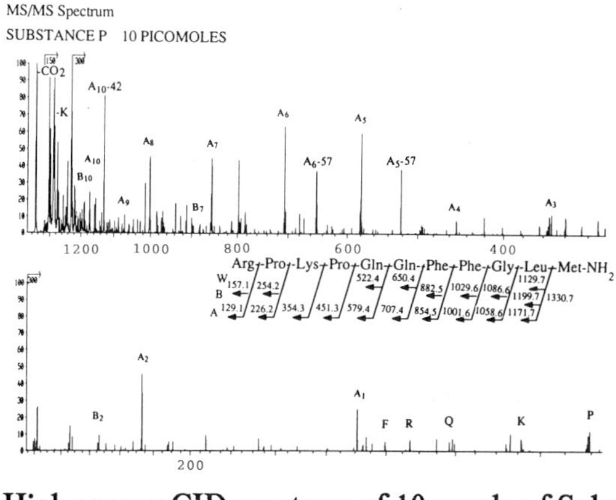

Fig. 12. High energy CID spectrum of 10 pmole of Substance P.

X. REFERENCES

1. Biemann, K., and Martin, S.A. (1987). Mass Spectrom. Revs., 6, 1-76.
2. Burlingame, A.L., Maltby, D., Russell, D.H. and Holland, P.T. (1988). Anal. Chem., 60, 294R-342R.
3. Samy, T.S.A., Hahm, K.S., Modest, E.J., Lampman, G.W., Keutmann, H.T., Umezawa, H., Herlihy, W.C., Gibson, B.W., Carr, S.A., and Biemann, K. (1983). J. Biol. Chem., 258, 183-191.
4. Gibson, B.W., Herlihy, W.C., Samy, T.S.A., Hahm, K.S., Meienhoffer, J., and Biemann, K. (1983). J. Biol. Chem., 258, 10637-10641.
5. Gibson, B.W., and Biemann, K. (1984). Proc. Natl. Acad. Sci. USA, 81, 1956-1960.
6. Carr, S.A., and Biemann, K. (1984). Meth. Enzym., 106, 29-57.
7. Carr, S.A., Hemling, M.E., and Roberts, G.D. (1988). In "Macromolecular Sequencing and Synthesis" (D.H. Schlessinger, ed.), pp. 83-99, A.R. Liss.
8. Gibson, B.W., Falick, A.M., Burlingame, A.L., Kenyon, G., Poulter, L., Williams, D.H., and Cohen, P. (1987). In "Methods in Protein Sequence Analysis" (K. Walsh, ed.), pp. 463-478, Humana Press.
9. Poulter, L., Ang, S.-G., Gibson, B.W., Williams, D.H., Holmes, C.F.B., and Cohen, P. (1988). Eur. J. Biochem., 175, 497-510.
10. Webb, J.W., Jiang, K., Gillece-Castro, B.L., Tarentino, A.L., Plummer, T.H., Byrd, J.E., Fisher, S.J., and Burlingame, A.L. (1988). Anal. Biochem, 169, 337-349.
11. Poulter, L., Ernest, J.P., Stroud, R.M., and Burlingame, A.L. (1988). Biomed. Mass Spectrom., in press.
12. Gibson, B.W., Daley, D.D., and Williams, D.H. (1988). Anal. Biochem., 169, 217-226.
13. Gibson, B.W., Yu, Z., Aberth, W., Burlingame, A.L., and Bass, N.M. (1988). J. Biol. Chem., 263, 4182-4185.
14. Offner, G.D., Brecher, P.B., Sawlivich, W.B., Costello, C.E., and Troxler, R.F. (1988). Biochem. J., 252, 191-198.
15. Johnson, R. and Biemann, K. (1987). Biochemistry, 26, 1209-1214.
16. a. Aberth, W. (1981). Int. J. Mass Spectrom. Ion Phys., 37, 379-382.
    b. Aberth, W. (1986). In "Advances in Mass Spectrometry" (J.F.J. Todd, ed.), Vol. 10, pp. 1469-1470, John Wiley & Sons, New York.
17. Cottrell, J., and Evans, S. (1987). Anal. Chem., 59, 1990-1995.
18. DiAugustine, R., Gibson, B.W., Aberth, W., and Walker, M. (1987). Anal. Biochem., 165, 420-429.
19. Hochhaus, G., Gibson, B.W., and Sadee, W. (1988). J. Biol. Chem., 263, 92-97.
20. Higaki, J., Gibson, B.W., and Craik, C.S. (1988). In "Evolution of Catalytic Function", pp. 615-621, Cold Springs Harbor Press.
21. Naylor, S., Findeis, A.F., Gibson, B.W., and Williams, D.H. (1986). J. Am. Chem. Soc., 108, 6359-6363.

22. Terry, A.S., Poulter, L., Williams, D.H., Nutkins, J.C., Giovannini, M.G., Moore, C.H., and Gibson, B.W. (1988). J. Biol. Chem., 263, 5745-5751.
23. Gibson, B.W., Walls, F.A., Falick, A.M., Lipka, J.J., and Waskell, L.A. (1987). In "Proceedings of the 35th Annual Conference on Mass Spectrometry and Allied Topics", pp. 66-67.
24. Burlingame, A.L. (1988). In "Proceedings of the 36th Annual Conference on Mass Spectrometry and Allied Topics", pp. 727-728.

15

A NEW STRATEGY FOR MAPPING THE TOPOGRAPHY OF A TRANSMEMBRANE PROTEIN USING MASS SPECTROMETRY[1]

Arnold M. Falick[2], Stephanie F. Mel[3], Robert M. Stroud[4], and A. L. Burlingame[2]

University of California, San Francisco, California 94143.

I. INTRODUCTION

In order to completely understand the mechanism of signal transduction across cellular membranes, one must first define the structure of the signal transducing molecule. To this end, we have undertaken to map the transmembrane topography of a bacterial ion channel, colicin Ia, using a combination of proteolytic digestion, HPLC, and mass spectrometric techniques. In this paper, we aim to demonstrate the power of mass spectrometry (MS), in particular tandem mass spectrometry (MS/MS), as applied to a specific problem in structural biology.

Colicin Ia is a 626 amino acid protein which forms an ion channel in the inner membrane of an *E. coli* and fatally depletes that cell of its electrical gradient (Konisky, 1978). Colicin is an unusual protein, leading a dual existence as both a water soluble and a transmembrane protein, a property that is required for its mode of action. It is made and released as a soluble protein during the SOS response in *E. coli.* It then binds to an iron receptor on the outer membrane of a sensitive strain, and subsequently inserts a channel forming region into the inner plasma membrane (Bowles, et al., 1983). Members of the colicin-producing strain of bacteria are immune to this killing

---

[1]This work was supported by the National Institutes of Health (Grant RR01614 to ALB and Grant GM24485 to RMS), the National Science Foundation Biological Instrumentation Program and the University of California, San Francisco (Claypole Fellowship and Graduate Division Mentorship to SFM).

[2] Mass Spectrometry Facility, Department of Pharmaceutical Chemistry

[3]Department of Pathology

[4]Department of Biochemistry and Biophysics

mechanism, as they have an immunity protein that resides in the inner plasma membrane and protects the cell (Weaver, et al., 1981).

Our overall strategy is to proteolytically digest colicin Ia in its soluble form, to separate the peptide fragments using HPLC, and to identify the individual fragments using liquid SIMS (Secondary Ion Mass Spectrometry) and MS/MS. Subsequently, the protein will be chemically modified in its membrane bound conformation. Analysis by MS/MS will allow us to determine the exact location of modified residues, which in turn will give precise structural information about the transmembrane conformation of the protein.

II. MATERIALS AND METHODS

Colicin Ia was purified from *E. coli* strain JK365, according to the method of Mel, et al. (1988).

Each proteolytic digestion was carried out on 2 nmoles of colicin Ia, with Endoproteinase Glu-C (EC-3.4.21.19) from Boehringer Mannheim (referred to below as *S. aureus* V8 protease) . Digestions were done at 37°C for 2 hours in 0.05M phosphate buffer, pH 7.8, and either 2.4M urea or 0.1% SDS at a substrate to enzyme ratio of 5:1.

For sequencing, protein was digested for 2 hours under the above conditions, and run on 16% polyacrylamide SDS gels. The bands were cut out of the gels and electroeluted for 21 hours at 4°C, at 200V. Edman sequencing was performed on an Applied Biosystems 470A Protein Sequencer.

HPLC separations were done on a Vydac C-18 reverse phase column (4.6 mm ID x 25 cm) using 0.1% trifluoroacetic acid in water (A) and 0.08% trifluoroacetic acid in acetonitrile (B). A linear gradient was used, 0-35% B in 105 min at a flow rate of 1 mL/min. The eluent absorbance was monitored at 215 nm.

Mass spectrometry was carried out on a Kratos MS-50S double focusing mass spectrometer equipped with a high-field magnet (mass range 3000 daltons at 8kV), a cesium ion liquid SIMS ion source (Aberth, et al., 1982; Falick, et al., 1986a), and a coolable sample introduction probe (Falick, et al., 1986b). Tandem mass spectrometry experiments were performed on a Kratos "Concept" 4-sector instrument fitted with a cesium liquid SIMS ion source, coolable probe and an electro-optical array detector (Cottrell and Evans, 1987; Elliott, et al., 1988).

III. RESULTS AND DISCUSSION

Our first goal was to proteolytically digest colicin Ia into component peptides that were of a convenient size to be identified by mass spectrometry. Figure 1 shows the time course of a 2 hour digestion of colicin Ia with *S. aureus* V8 protease in the presence of urea. Within 2 minutes, native colicin

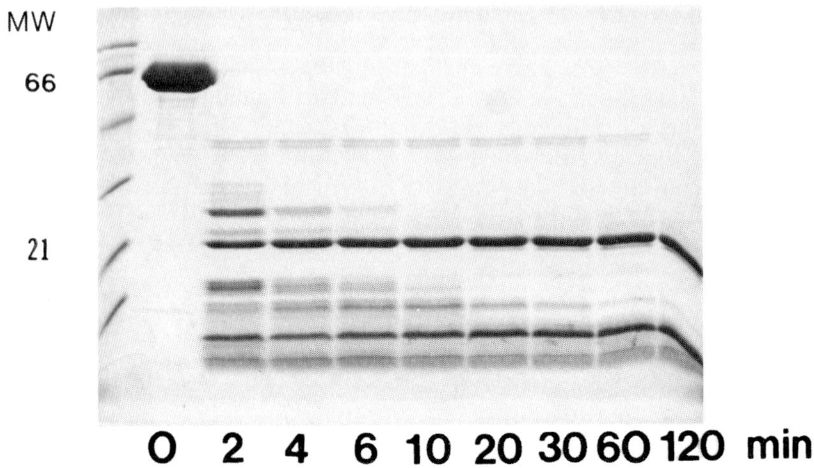

Fig. 1: Polyacrylamide gel (16%) of a colicin digestion. Lane 1 has low molecular weight standards, with 21kD and 66kD marked. Remaining lanes are colicin before the addition of *S. aureus* V8 protease (lane marked 0), and after 2, 4, 6,10,20,30,60, and 120 minutes of digestion. Note that colicin (MW 69kD), runs slightly aberrantly on a 16% gel.

completely disappeared from the gel, and a number of bands representing proteolytic fragments became apparent. At 2 hours, two major fragments that are more resistant to the action of the protease remain, one at approximately 20kD and the other at less than 14kD. After 19 hours at 37°, the strong 20kD band was still present on a gel (data not shown).

Following digestion, the fragments were separated on HPLC, as shown in Figure 2. Following separation by HPLC, fractions were analyzed by liquid SIMS. Figure 3 shows a portion of a typical mass spectrum of a single HPLC fraction. Multiple peptides are often present in a single fraction, but this generally presents no difficulty when using mass spectrometry, as the presence of each individual peptide is revealed by a peak in the mass spectrum at its specific mass. (For a recent comprehensive review of current developments in mass spectrometry, see Burlingame, et al., 1988). In the case of a known primary sequence, most of the peptides predicted from specific proteolytic enzyme cleavages have unique masses and are therefore easily identified. It does occur, however, that a small number of peptide pairs will, by chance, have very similar or identical masses and will therefore not be distinguishable on the basis of molecular mass measurements. Such cases occur with moderate frequency in relatively small proteins such as colicin, and with correspondingly greater frequency in digests of larger proteins. Some examples of such "isobaric peptides" from colicin Ia are shown in Table I. Any of these pairs is easily distinguishable by MS/MS (see below), which gives sequence information directly.

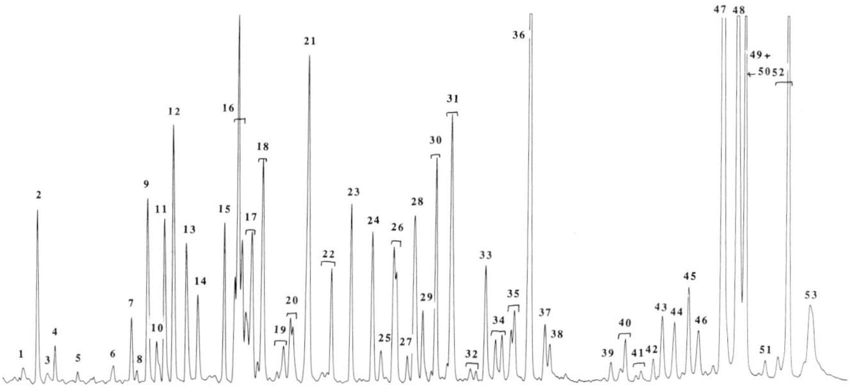

Fig. 2. HPLC chromatogram of *S. aureus* V8 digest of colicin Ia.

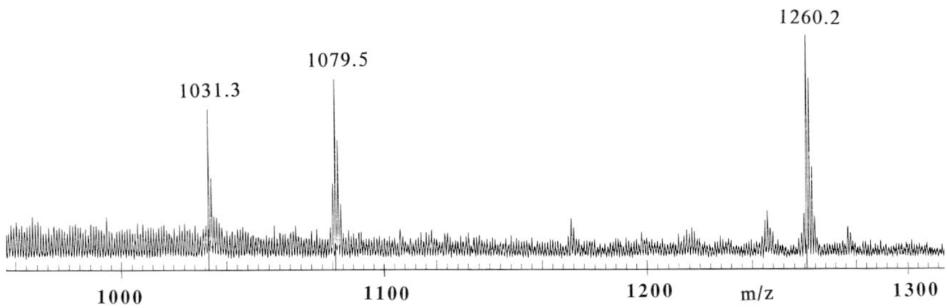

Fig. 3. A portion of the mass spectrum of Fraction 11. Three peptides were present in this fraction: AQNADKKAAD, m/z (calc. for MH^+) = 1031.5; SLGYDSDGHE, m/z (calc.) = 1079.4; and VVKPATMSNNAE, m/z (calc.) = 1260.6. The experimentally measured masses are noted above the peaks. The first two of these are predicted colicin peptides, while the third is an *S. aureus* V8 autodigestion product.

Tandem mass spectrometry was carried out on several representative peptides from the digest. In a tandem instrument, ions representing a variety of molecular species present in the sample are generated in the ion source of the first mass spectrometer (MS-1). A particular molecular species of interest is selected from this spectrum by means of the mass analyzer of MS-1, while ions representing other species are rejected (i.e., one selects only a single peak from the spectrum shown in Figure 3, for example). The selected ions are then focused into a collision cell for the purpose of generating structurally informative fragment ions via collision-induced unimolecular decomposition. Ions collected from the collision cell are refocused and mass-analyzed in a second mass spectrometer (MS-2). One thus obtains a clean parent plus daughter ion spectrum of one particular "pure" chemical species, with the secure knowledge that observed peaks are in fact logically related to the selected ions, with no interference from various background ions or other species present in the sample.

TABLE I. Examples of isobaric peptides in *S. aureus* digest of colicin Ia.

Sequence	Position/Length	Mass (MH^+)
WITE	551/4	548.2720
IMAVD	24/5	548.2754
KRQKD	188/5	674.3949
KRKQD	440/5	674.3949
SRMLFAD	162/7	839.4085
ALRLHTE	155/7	839.4739
FLKSVSEKYGAKAE	459/14	1556.8324
AGKRLSAAIAAREKD	98/15	1556.8872
RRKGILDTRLSELE	209/14	1685.9662
AGKRLSAAIAAREKDE	98/16	1685.9298

Figures 4 and 5 are MS/MS spectra of two peptides from the *S. aureus* V8 digest of colicin Ia. It has been established that high-energy gas collisions generate reproducible and interpretable fragmentation patterns from which the primary sequence of a peptide can be deduced (Biemann and Scoble, 1987; Johnson and Biemann, 1987; Johnson, et al., 1987; Crabb, et al., 1986; Anderegg, et al., 1988). In general, ion series are present that begin from both the C- and N-termini. Nomenclature for the various ion series observed has been summarized by Anderegg, et al. (1988). Some of the fragment ion mass differences that indicate the specific sequences present in Figures 4 and 5 are shown on the figures. Other ion series (not labeled in these figures) are also present, giving confirmatory evidence. It is clear from these spectra that covalent modification of any residue is easily detectable and that those residues that have been modified can be directly identified from the MS/MS spectrum. Furthermore, an MS/MS spectrum is rapidly obtainable on sub-nanomole quantities of material, even in the presence of other digestion fragments or other contaminants, and without regard to the presence of N-terminal blocking groups or other post-translational modifications of the peptide.

Figure 6 is a linear diagram of colicin Ia, showing those portions of the molecule that have been successfully mapped. Thus far, we have mapped approximately 67% of colicin Ia, using a combination of *S. aureus* V8 proteolytic digestion, HPLC separation, liquid SIMS, and MS/MS. Much of the remaining one third of the molecule that has not been mapped lies in the carboxy-terminal, channel-forming region. One explanation for the difficulty in mapping this region is that there are fewer appropriate cleavage sites in the C-terminal portion of the molecule. Alternatively, the results may indicate something of biological significance about colicin Ia. An analysis of the gel in Figure 1 shows several strong bands remaining after a 2-hour digestion in the

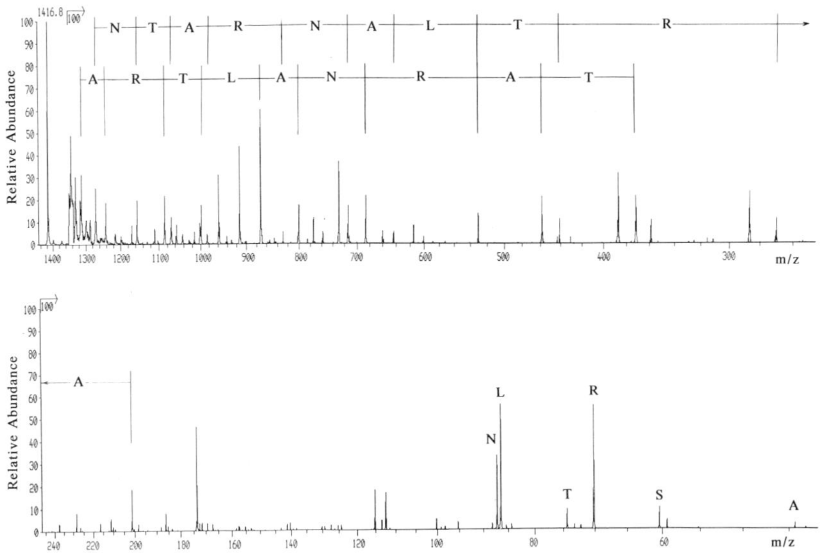

Fig. 4. MS/MS spectrum of the peptide SLNTARNALTRAE from Fraction 22, m/z (MH$^+$) = 1416.8. The spectrum contains several series of peaks from which the sequence can be confirmed or deduced. The series indicated in this figure are the "w" (upper) and "d" series (lower). The labeled low mass peaks provide confirmatory evidence of the presence of the residues noted.

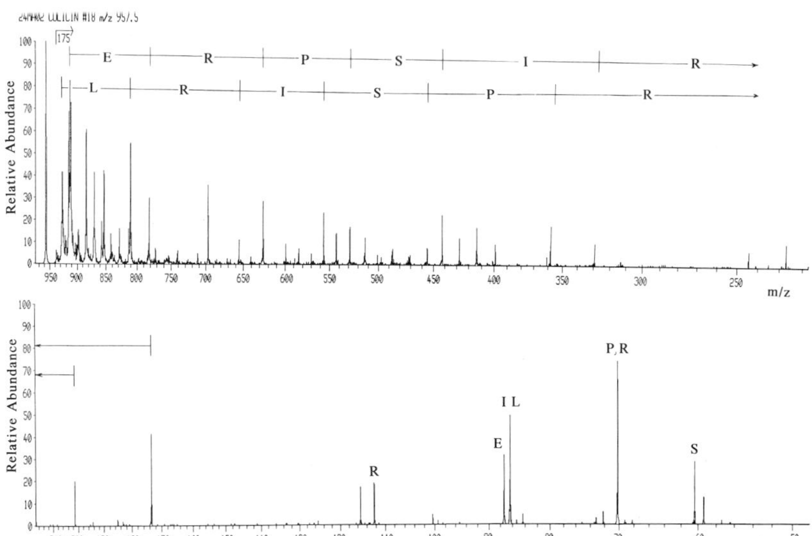

Fig. 5. MS/MS spectrum of the peptide SLRISPRE from Fraction 18, m/z (MH$^+$) = 957.5. The fragment ion series shown are the "a" series (upper) and the "w" series(lower).

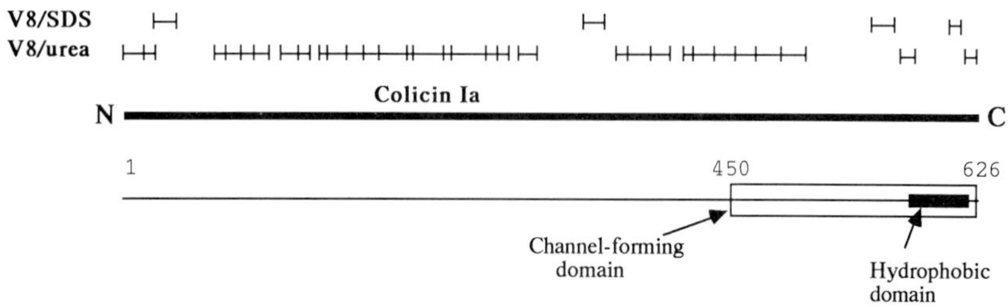

Fig. 6. Mapped portions of colicin Ia using enzymatic digestion and mass spectrometry. The *S. aureus* V8 digest with SDS produced all of the fragments that were found with *S. aureus* V8 protease in urea. The fragments labeled V8/SDS above represent additional pieces not found in the V8/urea digest.

presence of urea. Indeed, sequence analysis of the major band (20kD) reveals that, in fact, it corresponds to the channel-forming fragment of colicin Ia (Ghosh and Stroud, 1988). When SDS is used as a denaturant in the digestion, the 20 kD band is nearly absent from the gel, and addtional fragments are detected by HPLC and mass spectrometry. The resistance to digestion of this band suggests that this portion of the protein is folded very tightly in a domain-like structure.

Further support for this idea comes from electron microscopic studies of a 3-dimensional crystal lattice of colicin Ia (Choe and Stroud, 1988). An analysis of this crystal lattice has led these authors to propose a model for the structure of colicin in which three clearly visible sections of the molecule, as seen by EM, correspond to three functional domains (receptor-binding, translocation, and channel-forming), as described by biochemical studies (Konisky, 1982). Our present results support the idea that colicin Ia does have three functionally and structurally separate domains.

In this paper, we describe a strategy by which one can specifically define stuctural features of a molecule using standard biochemical analysis coupled with mass spectrometric techniques. We have proteolytically digested a bacterial ion channel, colicin Ia, and have identified the peptide fragments using liquid SIMS and MS/MS. Thus far, we have precisely mapped approximately two thirds of the molecule. Chemical modification of colicin Ia in its membrane-bound conformation, followed by MS/MS analysis to locate the modified residues will give us precise information about the structure of this ion channel. Mass spectrometry coupled with biochemistry promises to be an extremely powerful and important combination of techniques, allowing new insights into problems in structural biology.

IV. ACKNOWLEDGEMENT

The authors are indebted to F. C. Walls and B. L. Gillece-Castro for running the MS/MS spectra shown in Figures 4 and 5.

V. REFERENCES

Aberth, W., Straub, K.M. and Burlingame, A.L. (1982) *Anal. Chem.* **54**, 2029-2034.

Anderegg, R. J., Carr, S.A., Huang, I. Y., Hiipakka, R.A., Chang, C. and Liao, S. (1988) *Biochem.* **27**, 4214-4221.

Biemann, K. and Scoble, H.A. (1987) *Science* **237**, 992-998.

Bowles, L.K., Miguel, A.G., and Konisky, J. (1983) *J. Biol. Chem.* **258**, 1215-1220.

Burlingame, A.L., Maltby, D., Russell, D.H., and Holland, P.T. (1988) *Anal. Chem.* **60**, 294R-342R.

Choe, S.Y. and Stroud, R. (1988) manuscript in preparation.

Cottrell, J.S. and Evans, S. (1987) *Anal. Chem.* **59**, 1990-1995.

Crabb, J.W., Armes, L.G., Carr, S.A., Johnson, C.M., Roberts, G.D. Bordoli,R.S.,and McKeehan, W.L. (1986) *Biochem.* **25**, 4988-4993.

Elliott, G., Gallagher, R., Evans, S., Ashcroft, A., Gooch, G., Wright, A.D., Walls, F. and Burlingame, A.L. (1988) 36th ASMS Conf. on Mass Spectrom. & Allied Topics, June 1988, San Francisco, CA, paper #WPB 38.

Falick, A.M., Wang, G.H. and Walls, F.C. (1986a) *Anal. Chem.* **58**, 1308-1311.

Falick, A.M., Walls, F.C., and Laine, R.A. (1986b) *Anal. Biochem.* **159**, 132-137.

Ghosh, P. and Stroud, R.M. (1988) manuscript in preparation.

Johnson, R.S. and Biemann, K. (1987) *Biochem.* **26**, 1209-1214.

Johnson, R.S., Martin, S.A., Biemann, K., Stults, J. T., and Watson, J.T. (1987) *Anal. Chem.* **59**, 2621-2625.

Konisky, J. (1978) in *The Bacteria* (Ornston, L., and Sokatch, J.R., Eds.) Vol. 6 Academic Press, New York.

Konisky, J. (1982) *Ann. Rev. Microbiol.* **36**, 125-144.

Mel, S., Wolberger, C., and Stroud, R. (1988) manuscript in preparation.

Weaver, C.A., Redborg, A.H., and Konisky, J. (1981). *J. Bact.* **148**, 817-828.

16

LASER PHOTODISSOCIATION FOURIER TRANSFORM MASS SPECTROMETRY: NEW METHODOLOGY FOR SEQUENCE ANALYSIS OF OLIGOPEPTIDES AND LOCATION OF DISULFIDE BONDS

Patrick R. Griffin
Jeffrey Shabanowitz
John R. Yates, III
Nian Zhou Zhu
Donald F. Hunt

Department of Chemistry
University of Virginia
Charlottesville, Virginia 22901

I. INTRODUCTION

Protein sequence analysis by tandem mass spectrometry (1) involves digestion of the sample by site specific reagents such as cyanogen bromide or proteolytic enzymes, partial fractionation of the resulting mixture of oligopeptides by microbore, reverse-phase, high-performance, liquid chromatography (HPLC) and direct analysis of peptides in each HPLC fraction by particle-bombardment, collision activated dissociation (2) or laser photodissociation (3) mass spectrometry on a multianalyzer instrument. Here we describe amino acid sequence analysis by laser photodissociation on a newly developed tandem quadrupole Fourier transform mass spectrometer (TQ-FTMS) (4). This instrument is shown schematically in Fig. 1 and consists of a standard ion source, two sets of quadrupole rods, an ion cyclotron resonance (ICR) cell housed within a 7 Tesla superconducting magnet, and an argon fluoride excimer laser for photodissociation of trapped ions. Methodology for the location of disulfide bonds in proteins by mass spectrometry is also presented.

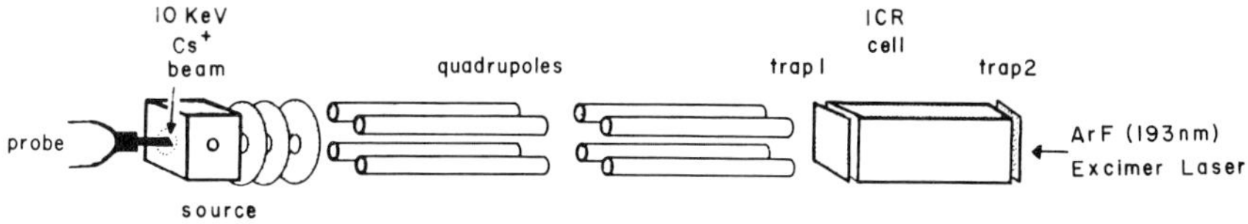

Fig. 1. Tandem quadrupole Fourier transform mass spectrometer.

II. EXPERIMENTAL

Sample analysis on the TQ-FTMS is performed by adding 0.5 ul of 0.1% trifluoroacetic acid solution containing peptides at the 5-30 pmol level to 0.5 ul of a matrix of glycerol: monothio-glycerol (1:3) on a gold-coated, stainless-steel, probe-tip that is inserted into the ion source of the instrument. Mass spectra are acquired by using a series of pulses controlled by the data system. The first, or beam pulse, lowers the voltage on the extractor lens of the Cs+ ion gun and allows 10 keV Cs+ ions to impact on the sample matrix for 4 ms. The same pulse places rf potential on the quadrupole rods and lowers the dc potential on trapping plate 1 from +3 to 0 V. This process allows efficient transport of the sputtered sample ions through the fringing fields of the magnet and into the ICR cell. The ions become trapped when the voltage on trapping plate 1 is raised back to a value of +3 V. After a 10 ms delay, the data system can also trigger a 10 ns pulse of light from the ArF excimer laser to photodissociate trapped ions. In the next step, 10 ms later, a rf pulse of 35 V, peak-to-peak, is applied to the transmitter plates (top) for 2.5 ms. This pulse contains a range of frequencies that accelerates all ions of all masses trapped in the cell into large orbits and causes them to move coherently at their characteristic cyclotron frequencies, values that are inversely proportional to the mass of the ion. Ion image currents induced by these orbiting ions on the ICR receiver (side) plates are monitored for 8 to 64 ms, amplified, and then digitized. Data from 30 to 200 such experiments are acquired, summed together and the resulting time-averaged signal is then converted to the desired mass spectrum by using Fourier transform analysis. The total experiment time to produce the required data is seconds, and the total sample ionization or consumption time is less than 1 s.

III. RESULTS AND DISCUSSION

A. Molecular Weight Analysis.

Although the above instrument is still under development, impressive results have already been obtained on a number of oligopeptide samples. Shown in Fig. 2a is the mass spectrum produced from 100 fmol of Substance P. A low resolution spectrum recorded on 10 pmol of porcine insulin, (M+H)+ = 5777.6, appears in Fig. 2b. In general, molecular weight information on samples in the mass range between 2 and 6 kDa can now be obtained routinely at the 5-20 pmol level. Detection limits for many peptides below mass 2,000 are in the high fmol range.

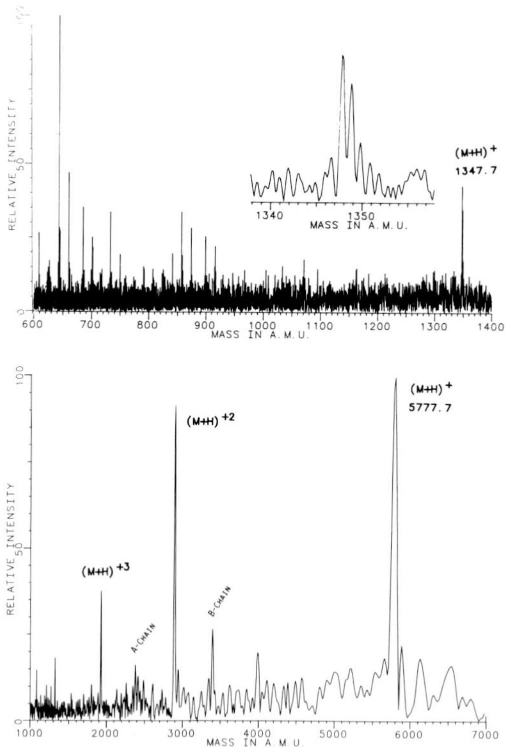

Fig. 2. (a) Mass spectrum recorded on 100 fmol of Substance P, RPKPQQFFGLM, (M+H)+ = 1347.7; (b) Low resolution spectrum of porcine insulin, (M+H)+ = 5777.6 recorded on 10 pmol of sample.

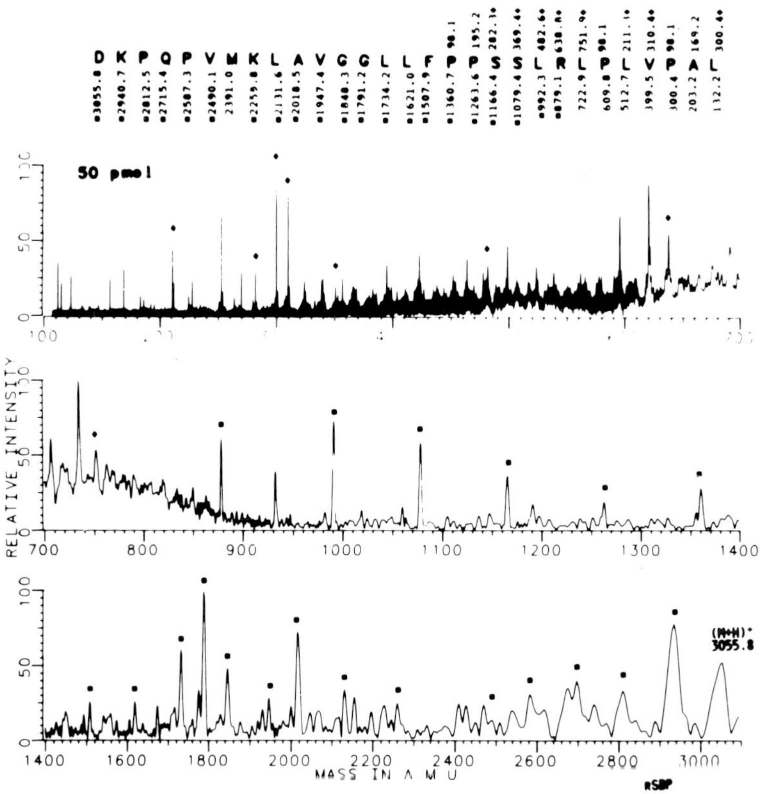

Fig. 3. Laser photodissociation mass spectrum recorded on 50 pmol of a 29 residue peptide from the rabbit sex steroid binding protein.

B. Amino Acid Sequence Analysis by Laser Photodissociation.

Structural characterization of peptides on the TQ-FTMS is accomplished by photodissociation of (M+H)+ ions with light from an ArF excimer laser. Results obtained by this approach on a tryptic peptide (10 pmol) from purple acid phosphatase (3) have already been published. Presented here (Fig. 3) is the photodissociation mass spectrum of a 29 residue peptide from rabbit sex steroid binding protein. Sample at the 50 pmol level was bombarded with 10 keV Cs+ ions for 4 ms and the resulting (M+H)+ ions were then stored in the ICR cell and exposed to a single 10 ns pulse of radiation at 193 nm from the ArF excimer laser. Light of this wavelength is absorbed efficiently by amide bonds in the peptide backbone. Absorption of a photon places 6.42 eV of additional energy in the (M+H)+ ion which then dissociates, primarily at the amide linkages, to produce the fragments required for structural characterization. Data from 50 such experiments on the same 50 pmol sample were acquired and summed in less than 20 s to produce the spectrum shown in Fig. 3. The spectrum contains a large number of fragment ions of type Y" (indicated by a solid box, ▪) derived from the parent molecule by loss of one or more amino acids from the N-terminus as well as fragments produced by in-chain cleavage at Pro residues (indicated by a triangle). Data from this single experiment allow placement of all but three amino acids in this large oligopeptide.

Laser photodissociation mass spectrometry can also be employed to sequence oligopeptides containing phosphorylated residues. Shown in Figure 4 is a laser photodissociation mass spectrum recorded on 30 pmol of an N-blocked, phosphorylated tryptic peptide derived form the N-terminus of photosystem II protein in spinach chloroplasts (5). The complete sequence of amino acids in this sample is obtained from a series of fragment ions of type Y". shown below the structure in Fig. 4. Assignment of the N-terminal residue as N-acetylphosphothreonine was made from the mass difference of 223.1 Da observed between the (M+H)+ ion and the fragment ion of type Y" formed by loss of a single residue from the N-terminus of the peptide. Note that signals due to Y_5" and Y_7" both occur at m/z values 101.0 Da above Y_4" and Y_6", respectively. If the threonine residue at either position 6 or 8 from the N-terminus was phosphorylated, the observed mass increment would be expected to be 181 Da (83 Da if loss of phosphoric acid accompanied formation of the above Y"-type ions). That this is not the case supports the assignment of the N-terminal threonine as the site of phosphorylation. Additional support for this assignment comes from the fact that all ions of type B observed in the spectrum lose phosphoric acid and therefore occur 98.0 Da lower than expected.

C. Assignment of Disulfide Linkages.

Location of disulfide bonds in proteins is greatly facilitated by mass spectrometry. A typical experiment involves digestion of the native protein with trypsin for 24 h in 100 mM Tris-HCl at pH 7.5 in the presence of 5 mM iodoacetic acid, fractionation of the resulting mixture by HPLC, and use of the TQ-FTMS to record the masses of all

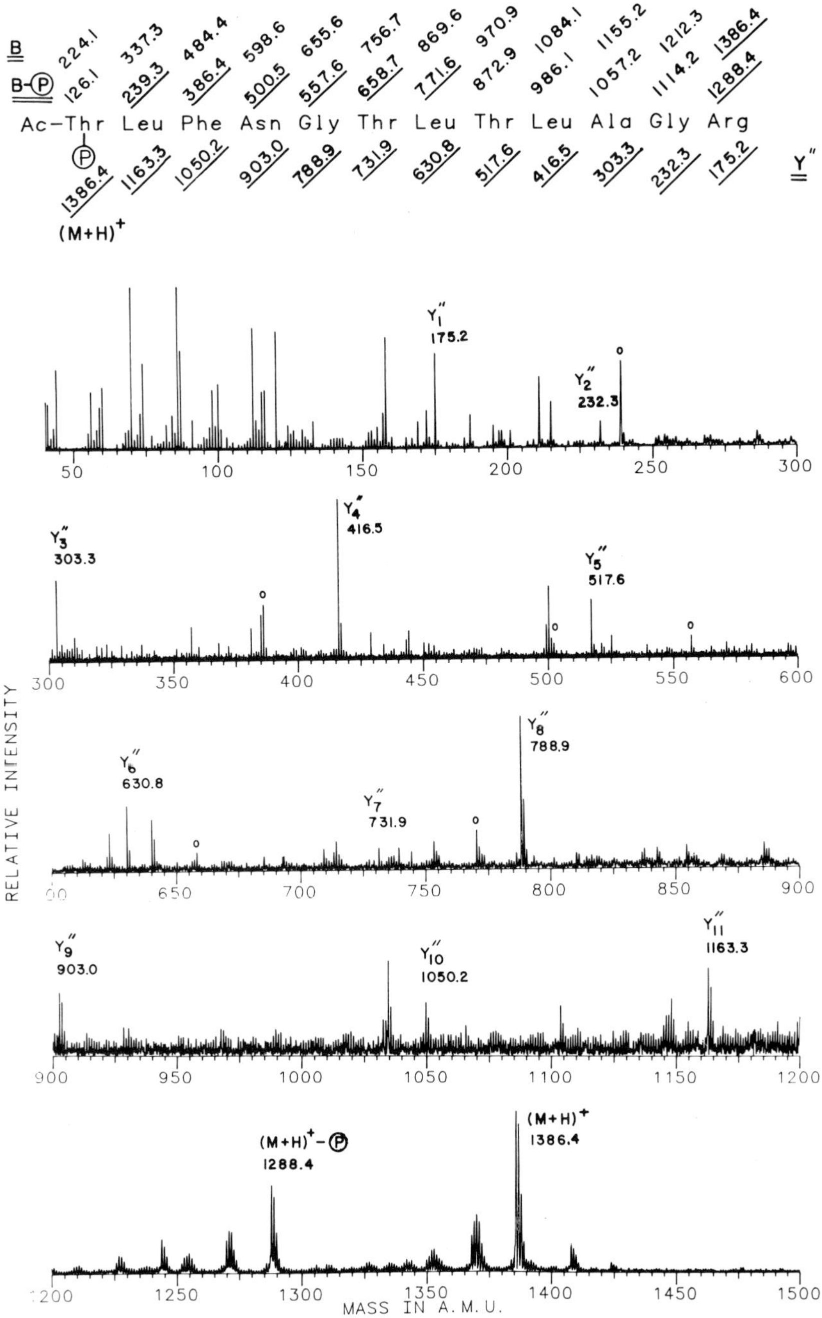

Fig. 4. Laser photodissociation mass spectrum recorded at the 30 pmol level on a phosphorylated tryptic peptide.

peptide (M+H)+ ions at the 5-20 pmol level in a matrix of 1:1 glycerol/thioglycerol. After each mass spectrum is recorded, the sample is withdrawn from the instrument and 0.5 ul of ammonium hydroxide is added directly to the sample probe. In the presence of excess base and thioglycerol, reduction of all disulfide linkages occurs in a few seconds. Following acidification, sample is then inserted back into the mass spectrometer and additional spectra are recorded. Newly produced (M+H)+ ions, those containing the reduced form of Cys, whose masses sum to a value three mass units higher than that of an (M+H)+ ion in the first mass spectrum are identified as being part of a disulfide linkage in the native protein. Peptide fragments containing two or more disulfide bonds are subjected to a second enzyme digest to cleave the molecule into fragments containing a single disulfide linkage. Alternately, collision activated dissociation on triple quadrupole instrument or laser photodissociation on the TQ-FTMS can employed to deduce the arrangement of disulfide bonds in peptides containing multiple disulfide linkages.

Results obtained from experiments to deduce the location of the three disulfide linkages in a protein toxin from the tarantula spider, Brachypelma smithi, are presented in Table I. The amino acid sequence of this 39 residue toxin was determined by a combination of classical Edman and tandem mass

Table 1. Oligopeptide fragments employed for the location of disulfide linkages in a tarantula spider protein toxin.

<table>
<tr><th>No</th><th>Digest</th><th>Sub-Digest</th><th>(M+H)+</th><th>Res. No.</th><th>NH_4OH (M+H)+</th><th>Sequence</th></tr>
<tr><td>1</td><td>Trypsin</td><td></td><td>2314</td><td>26-30</td><td></td><td>CSGGWK</td></tr>
<tr><td></td><td></td><td></td><td></td><td>+</td><td></td><td>|</td></tr>
<tr><td></td><td></td><td></td><td></td><td>1-12</td><td></td><td>IFECVFSCDIEK</td></tr>
<tr><td></td><td></td><td></td><td></td><td>+</td><td></td><td>|</td></tr>
<tr><td></td><td></td><td></td><td></td><td>31-32</td><td></td><td>CK</td></tr>
<tr><td>1a</td><td>Chymo.</td><td></td><td>1391</td><td>26-30</td><td>637</td><td>CSGGWK</td></tr>
<tr><td></td><td></td><td></td><td></td><td>+</td><td></td><td>|</td></tr>
<tr><td></td><td></td><td></td><td></td><td>1-6</td><td>757</td><td>IFECVF</td></tr>
<tr><td>2</td><td>Trypsin</td><td></td><td>1472</td><td>13-20</td><td>886</td><td>EGKPCKPK</td></tr>
<tr><td></td><td></td><td></td><td></td><td>+</td><td></td><td>|</td></tr>
<tr><td></td><td></td><td></td><td></td><td>35-39</td><td>589</td><td>LCLKL</td></tr>
</table>

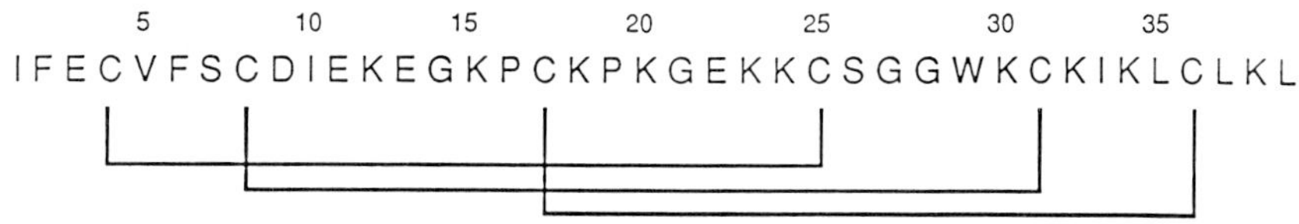

Fig. 5. Amino acid sequence of 39 residue protein toxin from the tarantula spider, Brachypelma smithi.

spectrometric methods and is shown in Figure 5. Digestion of native toxin with trypsin for 24 hr generated two peptides (#1 and #2 in Table I), the masses of which sum to that of the intact toxin. Treatment of peptide 2 (20 pmol) on the sample probe of the mass spectrometer with dilute ammonium hydroxide destroyed the (M+H)+ signal at m/z 1472 and produced new (M+H)+ ions at m/z 886 and 589. Since these are the expected m/z values for two of the predicted tryptic peptides from the reduced oligopeptide, residues 13-20 and 35-39, we conclude that the latter two peptides are connected by a disulfide linkage in the intact toxin.

Treatment of peptide #1 with chymotrypsin afforded a fragment whose (M+H)+ ion appeared at m/z 1391. Addition of ammonium hydroxide to the sample in thioglycerol destroyed this signal and produced new signals at the m/z values (637 and 757) expected for the trypsin/chymotrypsin peptides containing residues 26-30 and 1-6, respectively. This result confirms that the Cys residues at positions 4 and 26 are also connected by a disulfide bond. The only remaining location for the third disulfide bond involves Cys residues at positions 7 and 31.

IV. REFERENCES

1. Hunt D.F., Bone, W. M., Shabanowitz, J., Rhodes, J., and Ballard, J. M. (1981) Anal. Chem. **54**, 1704-1706.
2. Hunt, D. F., Yates, J. R., III, Shabanowitz, J, Winston, S., and Hauer, C. R. (1986) Proc. Natl. Acad. Sci., USA **83**, 6233-6237.
3. Hunt, D.F., Shabanowitz, J., and Yates, J. R., III, (1987) J. Chem. Soc., Chem. Commun., 548-550.
4. Hunt, D. F., Shabanowitz, J., Yates, J. R., III, Zhu, N. Z., Russell, D. H., and Castro, M. E. (1987) Proc. Natl. Acad. Sci., USA **84**, 620-623.
5. Michel H., Hunt, D. F., Shabanowitz, J., and Bennett, J. (1988) J. Biol. Chem. **263**, 1123-1130.

17

SEQUENCE ANALYSIS OF PROTEIN C-TERMINAL PROTEOLYTIC FRAGMENTS, PROTEIN ISOLATED FROM 2D-GELS, AND MURINE CALBINDIN: NEW METHODOLOGY

John R. Yates, III
Jeffrey Shabanowitz
Patrick R. Griffin
Nian Zhou Zhu
Donald F. Hunt

Department of Chemistry
University of Virginia
Charlottesville, Virginia 22901

I. INTRODUCTION

Protein sequence analysis by tandem mass spectrometry (1) involves digestion of the sample by site specific reagents such as cyanogen bromide or proteolytic enzymes, partial fractionation of the resulting mixture of oligopeptides by microbore, reverse-phase, high-performance, liquid chromatography (HPLC) and direct analysis of peptides in each HPLC fraction by particle-bombardment, collision activated dissociation (2) or laser photodissociation (3) mass spectrometry on a multianalyzer instrument. Here we describe use of the newly developed tandem quadrupole Fourier transform mass spectrometer (TQ-FTMS) (4) for the sequence analysis of proteins isolated from 2-dimensional gel electrophoresis and for the location and sequence analysis of protein C-terminal proteolytic fragments. Sequence analysis of a murine calcium binding protein by tandem mass spectrometry is also presented.

II. EXPERIMENTAL

Sample preparation and operation of the tandem quadrupole Fourier transform mass spectrometer in the laser dissociation mode for protein sequence analysis is described in an earlier paper in this volume.

III. RESULTS AND DISCUSSION

A. Sequencing of Proteins Isolated by 2D Gel Electrophoresis.

Two dimensional gel electrophoresis is a powerful analytical tool for the analysis of complex mixtures of proteins. Samples containing as many as 2,000 proteins can be resolved by this technique. Here we present preliminary results of a study to determine the feasibiity of employing TQ-FTMS for the sequence analysis of amino acids in proteins separated by the 2D-gel technique.

A sample of cerebrospinal fluid (20 ul) containing ~ 2,000 proteins was subjected to analysis by the 2D- electrophoresis. Apolipoprotein A-1, a 28 kDa protein of known sequence estimated from staining techniques to be present at the 40 pmol level on the gel, was electroblotted onto nitrocellulose, eluted from the nitrocellulose with 0.1% trifluoroacetic acid/ acetonitrile and digested with trypsin. Microbore HPLC was employed to fractionate the resulting peptides and 1/4 of each fraction (10 pmol) was then analyzed on the tandem quadrupole Fourier transform mass spectrometer to determine the mol mass of each tryptic peptide. To obtain sequence information laser photodissociation mass spectra were also recorded on 1/4 of each HPLC fraction (10 pmol) either before or after the samples had been converted to the corresponding methyl esters. Shown in Figure 1 is the laser photodissociation mass spectrum recorded on 10 pmol of the methyl ester of a tryptic peptide having a molecular mass 2002. Fragment ions of type Y" (labeled on the spectrum and under the structure by a solid ■) are observed for 13 of the 16 residues in this peptide. Note that the above information would be sufficient to construct an oligonucleotide probe based on the deduced sequence, QEFWDN. Sequence information on twelve of 37 tryptic fragments was obtained from this single proteolytic digest on apolipoprotein A-1.

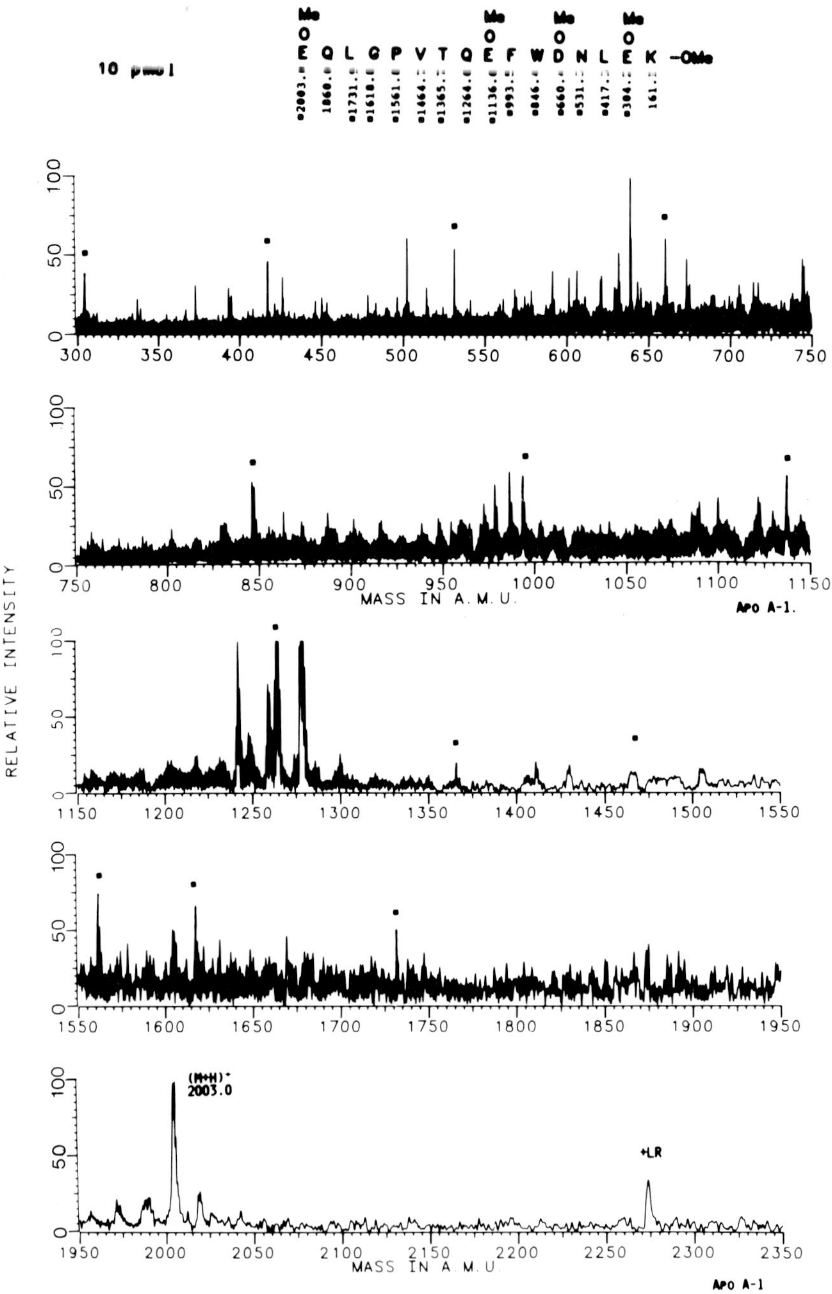

Fig. 1. Laser photodissociation mass spectrum recorded on the methyl ester of a tryptic peptide of molecular mass 2002 from apolipoprotein A-1

B. Identification and Analysis of Protein C-Terminal Peptides.

The importance of being able to locate and perform routine sequence analysis at the C-terminus of proteins stems from the need to; (a) verify the termination point of a protein sequence read from the corresponding cDNA, (b) detect ragged ends of proteins produced by recombinant DNA technology, and (3) provide sequence information for the construction of optimal oligonucleotide probes for mRNA. Here we describe a convenient procedure for the location and sequence analysis of protein C-terminal peptides based on mass spectrometry. Application of the procedure to ubiquitin is shown in Fig. 2.

Denatured protein in 8 M guanidine hydrochloride is treated with glycine methyl ester in the presence of a water soluble carbodiimide derivative. Carboxyl groups in the side chains and at the C-terminus of the protein are blocked as a result of this reaction. In the next several steps, protein

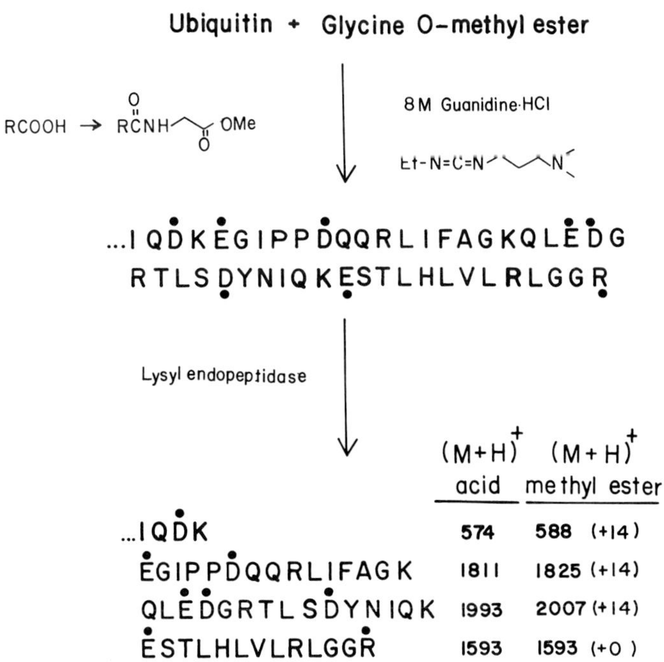

Fig. 2. Identification and analysis of protein C-terminal peptide by mass spectrometry (■ = glycine methyl ester group).

is digested with an enzyme such as endo Lys-C, the resulting sample is fractionated by HPLC, and mass spectra are recorded on all peptides before and after they are converted to the corresponding methyl esters with 2 N methanolic HCl. Following enzymatic digestion of the derivatized protein, only the C-terminal peptide is devoid of a free carboxylic acid moiety. Accordingly, (M+H)+ ions for all other peptides generated in the digest shift to higher mass by 14 mass units following treatment of the sample with methanolic HCl. The C-terminal peptide(s) already exists in the form of a methyl ester prior to this step. The (M+H)+ ions for this peptide fails to shift to higher mass and is thus easily pinpointed. Once the C-terminal peptide is located, sequence analysis is performed by either collision activated dissociation on our triple quadrupole or laser photodissociation on the TQ-FTMS.

C. Sequence Analysis of a Murine Calcium Binding Protein by Tandem Mass Spectrometry.

Described in this section is the sequence analysis of the major isoform of a small calcium binding protein from mouse intestine. Methods employed to attack this problem are those being used routinely to solve the primary structures of more than 20 proteins presently under study in the our laboratory. Calbindin-D_{9k}, a 78- residue, vitamin-D dependent, protein involved in calcium transport, was isolated from mouse intestine by a combination G-100 column chromatography, preparative slab gel electrophoresis, and reverse-phase HPLC chromatography. Exposure of the purified protein to several cycles of Edman degradation on a gas phase sequenator failed to remove a phenylthiohydantoin derivative and suggested that the protein contained a blocked N-terminus.

The next set of experiments involved digestion of the protein at the 100 pmol level in separate experiments with trypsin, elastase, endo-Glu-C, and dilute acid. Progress of each cleavage reaction was monitored by microbore HPLC and the molecular weight distribution of the resulting oligopeptide fragments in the various fractions was then examined on the TQ-FTMS instrument. Since treatment of the protein with dilute acid produced large fragments (#1-6 in Table I) that could be easily separated by HPLC, this procedure was repeated on a 5 nmol scale to generate sufficient material for analysis of subdigests on the triple quadrupole mass spectrometer.

Subdigestion of fragment #1, (M+H)+ = 2256, with endopeptidase Glu-C afforded two smaller oligopeptides, #1a and #1b, which in turn exhibited (M+H)+ ions at m/z 773 and 1501, respectively. Conversion of peptide #1a to the

Table I. Oligopeptide fragments from the major isoform of mouse calbindin-D_{9k} characterized by tandem mass spectrometry.

No	Digest	Sub-Digest	(M+H)+[a]	Res. No.	Sequence[b]
1	Acid		FT-2256	1-20	[Ac-AEKSPAEMKSXFQKYAAKEG]
1a		Glu-C	773	1-7	Ac-AEKSPAE
1b		Glu-C	1501	8-20	MKSXFQKYAAKEG
2	Acid		FT-2787[c]	24-48	[QXSKEEXKXXXQSEFPSXXKASSTX]
3	Acid		1164	38-48	FPSXXKASSTX
4	Acid		1177	50-59	NXFKEXDKNG
5	Acid		FT-2039	61-77	[GEVSYEEFEAFFKKXSQ]
5a	Trypsin		1582	61-73	GEVSYEEFEAFFK
6	Acid		1088	61-69	GEVSYEEFE
7	Elastase		1033	1-9	Ac-AEKSPAEMK
8	Elastase		622	10-14	SXFQK
9	Elastase		FT-2035	15-32	[YAAKEGDPDQXSKEEXKX]
10	Elastase		1207	15-25	YAAKEGDPDQX
11	Elastase		846	26-32	SKEEXKX
12	Elastase		1147	33-42	XXQSEFPSXX
13	Elastase		FT-2938	45-70	[SSTXDNXFKEXDKNGDGEVSYEEFEA]
13a		Glu-C	1154	45-54	SSTXDNXFKE
13b		Glu-C	847	55-62	LDKNGDGE
14	Glu-C		1444	8-19	MKSXFQKYAAKE
15	Glu-C		FT-2836	30-54	[XKXXXQSEFPSXXKASSTXDNXFKE]
16	Glu-C		943	30-37	XKXXXQSE
17	Glu-C		FT-2186	48-66	[XDNXFKEXDKNGDGEVSYE]
18	Glu-C		969	70-77	AFFKKXSQ
19	CNBr		916[d]	1-8	Ac-AEKSPAE

[a]Mass values labeled FT, were recorded on the TQ-FTMS instrument and represent the average mass rather than the monoisotopic mass. [b]Sequences displayed in brackets were not determined directly by CAD experiments (see text for discussion). [c]The observed (M+H)+ ion corresponds to residues 24-48 having pGlu at the N-terminus. [d]The m/z value recorded for the (M+H)+ is that for the corresponding methyl ester.

corresponding methyl ester shifted the observed (M+H)+ ion to higher mass by 42 units and thus confirmed the presence of two acidic residues within the fragment. The sequence of amino acids in the methyl ester was deduced as Ac-AEKSPAE from the collision activated dissociation mass spectrum. Assignments for the N- terminal residue as Ac-Ala rather than Ile or Leu and for residue three as Lys rather than Gln were made by recording the collision activated dissociation mass spectrum of the (M+H)+ ion at m/z 857 for the resulting monoacetylated derivative. Fragment ions that contain residue three all shift to higher mass by 42 mass units in the CAD spectrum of the (M+H)+ ion derived from the monoacetylated derivative. Residue three is therefore assigned as Lys rather than Gln. Failure to introduce a second acetyl group a the N-terminus of the above peptide methyl ester under the above reaction conditions can only be explained if the N-terminal residue is

```
                     10                  20                  30
Ac-A E K S P A E M K S X F Q K Y A A K E G D P D Q X S K E E X
\--------------FT-A-2256-----------------/        \--FT-A-2787-
\---A-Ec-773---/ \------A-Ec-1501---------/            \--E-846--
\---CBr-916------/             \------E-1207--------/
                            \----------FT-E-2035-----------

                     40                  50                  60
   K X X X Q S E F P S X X K A S S T X D N X F K E X D K N G D
   -------------------FT-A-2787-----/    \----A-1177--------/
   ---/ \--E-1147--------/        \-----Ec--1154-----/
   ---Ec-943----/ \---A-1164----------/                \--Ec-847---
   --/                            \----------FT-E-2938-----------

                    70
  G E V S Y E E F E A F F K K X S Q
  \------------FT-A-2039----------/
  \-------A-T-1582---------/
  ---/                \----Ec-969---/
  -----------------/
```

Fig. 3. Amino acid sequence data obtained on the major isoform of mouse calbindin-D_{9k} protein by tandem mass spectrometry. Oligopeptide fragments sequenced by CAD on the triple quadrupole instrument are labeled by enzyme and mass of the $(M+H)^+$ ion. Enzymes are designated as follows: Ec = endo-Glu-C; E = elastase; T = trypsin. Fragments resulting from mild acid hydrolysis are designated by A-. Oligopeptides labeled FT- were characterized by enzyme specificity and mass of the $(M+H)^+$ ion observed with the TQ-FTMS instrument. Mass measurements on methyl ester $(M+H)^+$ ions were used to support assignments made by this latter technique. The predicted incremental mass shift of 14 daltons/ -COOH was observed in all cases. An X in the above sequence indicates that the residue was identified by mass spectrometry as either Leu or Ile, both of which have the same molecular mass. [a]Measured mass corresponds to peptide with pGlu at the N-terminus.

already blocked. Accordingly, the N- terminal residue is defined as Ac-Ala rather then Ile or Leu. Assignments for amino acid residues 8-20 in the protein were obtained from the CAD spectrum of the (M+H)+ ion (m/z 1501) for fragment #1b.

Oligopeptide #5 from the acid hydrolysis reaction (Table I) is placed at the C-terminus of the calbindin isoform as a result of an endo-Glu-C digest of the intact protein. The latter experiment produced five major fragments, only one of which, peptide #18, failed to contain Glu at the C-terminus. Since endo-Glu-C shows high specificity for cleavage on the C-terminal side of Glu residues, fragment #18 is assigned to the C-terminus of the intact protein. Appearance of four identical residues at the N-terminus of peptide #18 and at the C-terminus of the tryptic peptide, #5a, allowed the sequence to be extended back an additional 9 residues from the C-terminal end of the protein. Since the predicted mass for the (M+H)+ ion of the resulting 17 residue peptide matched that observed for oligopeptide #5 from acid hydrolysis, the latter fragment was assigned to the C-terminus of protein.

Sequence analysis of two additional endo-Glu-C fragments, peptides #13a and #13b, provided the overlap information to extend the sequence back to residue 45 and to establish the connectivity of the acid hydrolysis fragments as 1-2-3-4-5 (Table I). Overlapping sequence information derived from peptides #3, #10, #11, #12, and #16 (Table I) allowed the remaining primary structure of the protein to be deduced (Fig. 3). Note that the (M+H)+ ion for acid hydrolysis fragmont #2 occurs at a m/z value corresponding to residues 24-48 having pGlu at the N-terminus. Elimination of ammonia from the peptides #3, #10, #11, #12, and #16 (Table 2), allowed the N-terminal Gln to form pGlu is not unexpected under the conditions employed in the acid hydrolysis. Sequence analysis of oligopeptide #10 obtained in the elastase digest confirmed the identity of residue 24 as Gln in the intact protein.

IV. REFERENCES

1. Hunt D.F., Bone, W. M., Shabanowitz, J., Rhodes, J., and Ballard, J. M. (1981) Anal. Chem. **54**, 1704-1706.
2. Hunt, D. F., Yates, J. R., III, Shabanowitz, J, Winston, S., and Hauer, C. R. (1986) Proc. Natl. Acad. Sci., USA **83**, 6233-6237.
3. Hunt, D.F., Shabanowitz, J., and Yates, J. R., III, (1987) J. Chem. Soc., Chem. Commun., 548-550.
4. Hunt, D. F., Shabanowitz, J., Yates, J. R., III, Zhu, N. Z., Russell, D. H., and Castro, M. E. (1987) Proc. Natl.Acad. Sci., USA **84**, 620-623.

18

LOCATION OF DISULFIDE BONDS IN PROTEINS BY PARTIAL ACID HYDROLYSIS AND MASS SPECTROMETRY[1]

Yiping Sun
Zhongrui Zhou
David L. Smith

Department of Medicinal Chemistry and Pharmacognosy
Purdue University
West Lafayette, Indiana

I. INTRODUCTION

Disulfide bonds, which are essential to the structure and function of many proteins, have traditionally been located by identifying peptides formed after the protein has been cleaved between the half-cystinyl residues. Highly specific chemical or enzymatic reactions have been preferred for the fragmentation step because they facilitate identification of the disulfide-bonded peptides. This approach frequently fails to give peptides diagnostic for the disulfide linkages because the amino acids required for the specific cleavage reactions may not be present in protein segments joining half-cystinyl residues. The problem of producing structurally diagnostic fragments is further confounded by the fact that even when the residues required for a specific cleavage reaction are present, the enzymes often have very low efficiencies because of steric hinderance. As a result, experience gained with a protein that has been reduced and carboxymethylated may not be relevant to cleavage of the same protein when the disulfide bonds are intact.

---

[1] *Supported by a grant from the National Institutes of Health GM 40384.*

Partial acid hydrolysis is an alternative means of producing peptides which are diagnostic for the location of disulfide bonds in proteins. This approach is attractive because there is a possibility that any amide bond may be hydrolyzed. In addition, hydrolysis is expected to proceed despite steric hinderance, although perhaps at a reduced rate. The use of partial acid hydrolysis to produce disulfide-bonded peptides in insulin (1) and lysozyme (2) is well established. Despite these early successes, partial acid hydrolysis has not reached the mainstream of methods for fragmenting proteins because it usually results in the formation of a complex mixture of peptides, which is difficult to analyze. The low specificity of partial acid hydrolysis is attractive because it makes the method, in principle, applicable to all proteins. However, this lack of specificity results in the production of so many different peptides that their isolation, purification, and identification by amino acid composition or sequence is difficult. Partial acid hydrolysis is also attractive because the disulfide bonds are unlikely to undergo rearrangement reactions in dilute acid.

Fast atom bombardment mass spectrometry (FABMS) (3-5) is an excellent method for identifying peptides because it is fast, and because it is effective for the analysis of peptide mixtures. In addition to its use for identifying disulfide-bonded peptides (6-8), FABMS has been especially useful for the investigation of peptides and proteins which have been posttranslationally modified (9). The purpose of this investigation is to determine the practicality of FABMS for identifying disulfide-bonded peptides in partial acid hydrolysates of proteins.

II. RESULTS AND DISCUSSION

Bovine insulin has been used as a model protein because it has three disulfide bonds, and, with only 51 residues, is relatively small. The structure of insulin is given in Fig. 1. The first step in this investigation was to identify hydrolysis conditions that give a high yield of peptides with molecular weights between 1000 and 3000. This molecular weight range is desirable because peaks corresponding to smaller peptides are often hidden by matrix peaks which are always present in FAB mass spectra, while the signals for peptides with molecular weights greater than 3000 are frequently weak. Although early investigations of partial acid hydrolysis used concentrated acid to fragment

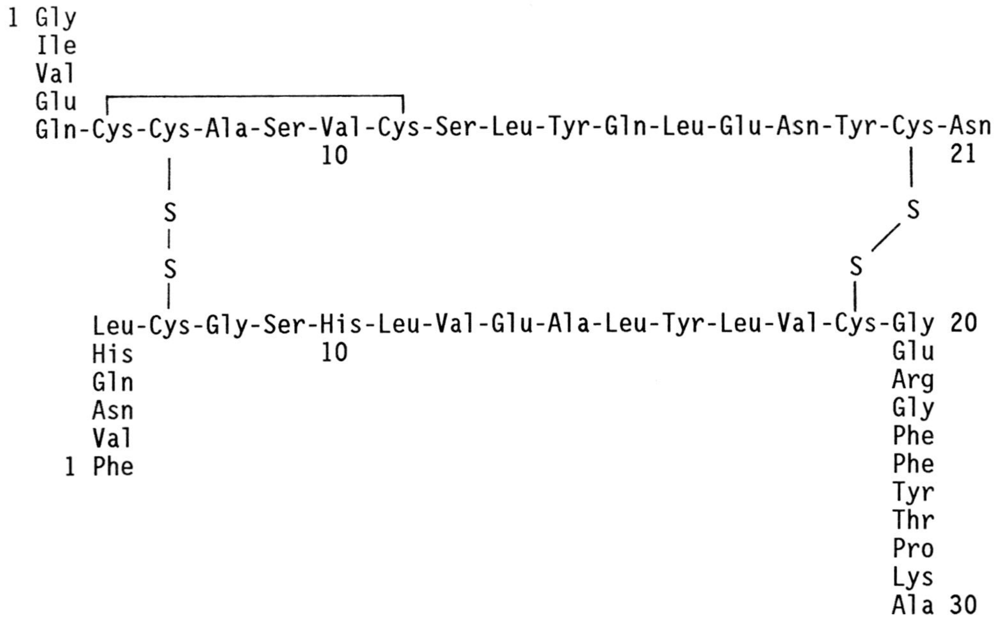

Figure 1. Structure of bovine insulin illustrating the amino acid sequences of the A (upper) and B (lower) chains, as well as the two disulfide bonds joining the chains.

the protein (1,2), there is now evidence that a higher degree of selectivity is achieved with dilute acid (10,11).

The HPLC chromatograms of insulin hydrolyzed in 0.03N HCl for 2, 8, and 16 hrs are given in Figure 2. From these data it is apparent that the degree of fragmentation, (i.e., the number of peptides) varies with the hydrolysis time. After two hours (Figure 2a), only one broad peak is found, indicating that very little fragmentation has occurred. After 16 hrs (Figure 2c), many peaks are present, indicating that many different peptides have been formed. Analysis of fractions collected from Figure 2c by FABMS shows that nearly all of these peptides have molecular weights less than 1000.

The 8 hr hydrolysate, Figure 2b, proved to be the most useful for analysis of disulfide-bonded peptides in insulin. A portion of the FAB mass spectrum of the fraction indicated by the arrow in Fig 2b is given in Fig 3a. The FAB mass spectrum of the same fraction after treatment with performic acid is given in Fig 3b. Performic acid is used here to indicate the presence of a disulfide-bonded peptide by converting the disulfide moiety to two cysteic acid residues (12). Since the peak at m/z 2782 is not present after oxidation, we concluded that it is most likely due to a disulfide-bonded peptide. The m/z of the peptide at 2500 increased by 48 mu, indicating that this peptide has one sulfhydryl group.

Although FABMS has been used extensively to identify peptides formed by specific cleavage reactions, it has not been used to identify peptides formed by non-specific fragmentation of proteins. One of the first questions to be addressed, is "How many different segments of the protein could have the same molecular weight, if it is assumed that any amide could be hydrolyzed?" We have investigated this question by developing a computer program to list all segments of the protein that have a specified molecular weight. For example, a search of the B chain of insulin shows that there are nine segments which have molecular weights of 2500 $\pm$ 50. The coordinates and monoisotopic masses of the neutral peptides are given in Table I. Since the peptides are identified by molecular weight, it is important to know the accuracy with which the molecular weight is determined. The mass spectra given in Figure 3 were acquired with an oscillographic recorder. Since the m/z of ions in this mass range can be determined with an accuracy of $\pm$ 2 mu by this method, it is apparent from the results in Table I that there is only one segment of insulin

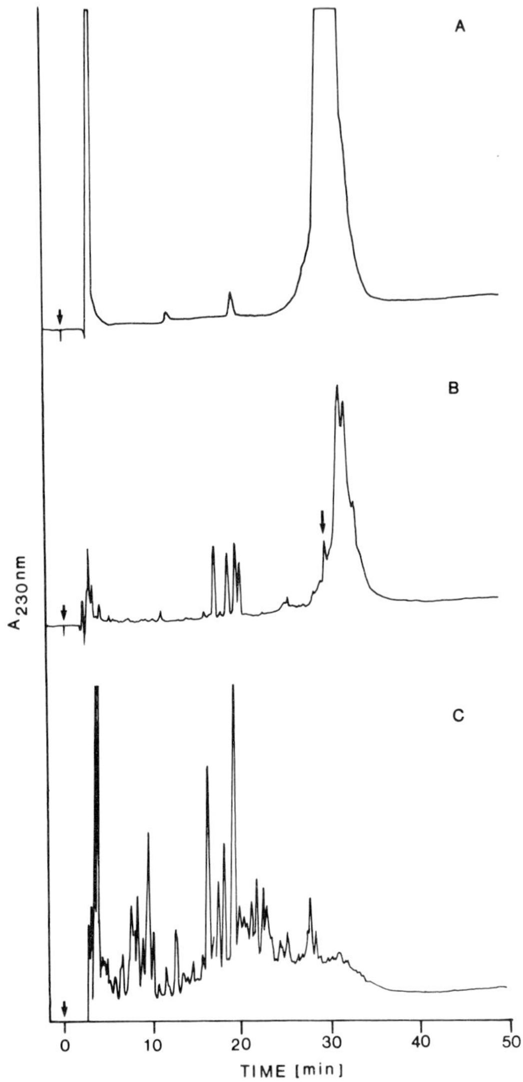

Figure 2. HPLC chromatograms of hydrolysates of insulin. (a) 2 hrs, (b) 8 hrs, (c) 16 hrs. Insulin was dissolved in 0.03N HCl, sealed in glass tubes under vacuum, and hydrolyzed at 110°C. The hydrolysate was fractionated by reversed phase HPLC using a Synchropak RP-8 column, and eluted with a water/acetonitrile/TFA mobile phase (linear gradient from 10% to 60% acetonitrile in 0.1% TFA). Peptides were detected by their absorbance at 230 nm.

(B9-30), that could give a peak at m/z 2500 ± 2. The molecular weight of this peptide is expected to increase by 48 mu when the hydrolysate is treated with performic acid because the peptide contains one cysteinyl residue, which will be converted to a cysteic acid residue upon oxidation. This spectrum is illustrated in Figure 3b. It should be noted that the m/z of ions in this range can be determined with an accuracy of ± 0.2 mu with some computerized data acquisition systems, and with higher accuracy by the "peak-match" method.

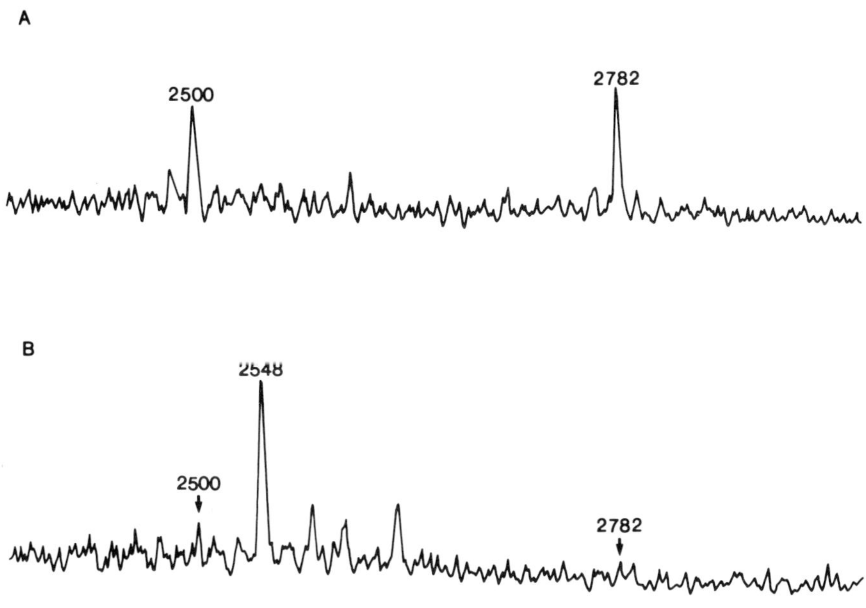

Figure 3. FAB mass spectra of the HPLC chromatographic fraction of the 8 hr hydrolysate of insulin indicated by the arrow in Figure 2b. Spectra were taken before (a) and after (b) treatment of the fraction with performic acid. Results obtained with a Kratos MS-50 high resolution mass spectrometer and a 50/50 (V/V) mixture of glycerol and 1-thioglycerol as the matrix. The spectra were acquired with an oscillographic recorder and a mass resolution of 1000.

Table I. Monoisotopic molecular weights[a] and coordinates of all peptides with molecular weight of 2500 ± 50 that could be derived from the B chain of bovine insulin.[b,c]

Coordinates	Molecular Weight
7-28	2460.1
6-27	2476.2
4-25	2477.2
8-29	2485.2
1-22	2486.2
9-30	2499.2
5-26	2512.2
2-24	2543.2
1-23	2543.2

[a] *See reference 13 for a discussion of terminology used to describe molecular weight in FABMS.*

[b] *Since Asn is rapidly converted to Asp during acid hydrolysis, Asn-3 has been changed to Asp in the program.*

[c] *Molecular ions (MH^+) for these peptides would be found at m/z one mass unit higher in the mass spectrum because they have an additional proton.*

Since the peak at m/z 2782 was not present in the FAB mass spectrum of the oxidized chromatographic fraction (Figure 3b), it was concluded that this peptide likely consists of a pair of peptides joined by a disulfide bond. If this is a disulfide-bonded peptide, treatment with performic acid will give two peptides, each with a cysteic acid residue (11). Another computer program has been developed to list all possible peptide pairs which might be joined by a disulfide bond to give a peptide of a specified molecular weight, and which may be derived from a particular protein. When the A and B chains of insulin are searched for such peptide pairs which would give a protonated molecule ion (MH^+) at m/z 2782 ± 2, five possibilities were found. Two of the disulfided-bonded peptides have a molecular ion at m/z 2782, and are illustrated in Table II.

Table II. Peptide pairs which could be derived from bovine insulin to give disulfide-bonded peptides with MH^+ at m/z 2782.

Chain	Coordinates	Sequence	Peptide MW[a,b]
A	19-20	YC	284.1
B	9-30	SHLVEALYLVCGERGFFYTPKA	2499.2
A	15-20	QLENYC	768.3
B	10-26	HLVEALYLVCGERGFFY	2015.0

[a] *Monoisotopic mass of the neutral constituent peptide.*
[b] *The mass of Asp has been used in place of Asn because Asn undergoes rapid deamidation during hydrolysis.*

Yazdanparast et al. (7) have reported that disulfide-bonded peptides undergo reduction during analysis by FABMS. As a result, FAB mass spectra of such peptides usually have peaks corresponding to the molecular ions of the disulfide-bonded peptide as well as both of the constituent peptides. As a preliminary screen, this observation can be used to make the correct assignment of the peak at m/z 2782 ± 2. Although the data are not presented here, the remainder of the FAB mass spectrum shown in Figure 3a has no other significant peaks above m/z 1000. Below m/z 1000 there are many peaks, which are likely due to low molecular weight peptides, or matrix cluster ions. Of the possible assignments of the peak at m/z 2782 ± 2, only the peptide pair comprised of A19-20/B9-30 has a constituent peptide with MH^+ at m/z 2500. Hence the peak at m/z 2782 is uniquely assigned to the A19-20/B9-30 disulfide-bonded peptide. Finding this peptide in the partial acid hydrolysate of insulin clearly places a disulfide bond between Cys-20 of the A chain, and Cys-19 of the B chain. In the event that the peptide molecular weights derived from the FAB mass spectrum cannot be uniquely related to a specific portion of the parent protein, verification by auxillary means may be necessary. For example, a one-step Edman degradation reaction may be performed to effect a predictable and highly specific change in molecular weight.

Although hydrolysis conditions have not been optimized to give peptides diagnostic for the locations of the other disulfide bonds in insulin, this preliminary investigation has demonstrated that a combination of partial acid hydrolysis and FABMS is the basis of a powerful method for locating disulfide bonds in proteins. Of particular importance, this study has demonstrated the specificity with which peptide molecular weight can be related to unique segments of a small protein. Based on our experience with insulin and hen egg-white lysozyme (results to be presented elsewhere), it appears that partial acid hydrolysis, coupled with FABMS, is a useful complement to other methods, which rely on highly specific cleavage reactions.

III. REFERENCES

1. A.P. Ryle, F. Sanger, L.F. Smith, and R. Kitai, The Disulphide Bonds of Insulin (1955) Biochem. J. 60:541-556.
2. A. Thompson, Amino Acid Sequence in Lysozyme (1955) Biochem. J. 60:507-515.
3. S.A. Carr and K. Biemann, Identification of Posttranslationally Modified Amino Acids in Proteins by Mass Spectrometry (1984) Methods in Enzymology 106:29-58.
4. T. Takao, M. Yoshida, Y. Hong, S. Aimoto, and Y. Shimonishi, Fast Bombardment (FAB) Mass Spectra of Protein Digests: Hen and Duck Egg-white Lysozymes (1984) Biomed. Mass Spectrom. 11:549-556.
5. D.F. Hunt, J.R. Yates III, J. Shabanowitz, S. Winston, and C.R. Hauer, Protein Sequencing by Tandem Mass Spectrometry (1986) Proc. Natl. Acad. Sci USA 83:6233-6237.po
6. H.R. Morris and P. Pucci, A New Method for Rapid Assignment of S-S Bridges in Proteins (1985) Biochem. Biophys. Res. Com. 126:1122-1128.
7. R. Yazdanparast, P. Andrews, D.L. Smith, and J.E. Dixon, A New Approach for Detection and Assignment of Disulfide Bonds in Peptides (1986) Anal. Biochem. 153:348-353.
8. R. Yazdanparast, P.C. Andrews, D.L. Smith, and J. E. Dixon, Assignment of Disulfide Bonds in Proteins by Fast Atom Bombardment Mass Spectrometry (1987) J. Biol. Chem. 262:2507-2513.
9. P.C. Andrews and J.E. Dixon, Application of Fast Atom Bombardment Mass Spectrometry to Posttranslational

Modifications of Neuropeptides (1988) Methods in Enzymology: Hamme Action Pact L: Neuroendocrine Peptides 168.

10. A. Light, Partial Acid Hydrolysis (1967) Methods in Enzymology XI:417-420.

11. J. Schultz, Cleavage at Aspartic Acid (1967) Methods in Enzymology XI:255-263.

12. Y. Sun and D.L. Smith, Identification of Disulfide-Containing Peptides by Performic Acid Oxidation and Mass Spectrometry (1988) Anal. Biochem. 172:130-138.

13. J. Yergey, D. Heller, G. Manson, R.J. Cotter, and C. Fenselau, Isotopic Distributions in Mass Spectra of Large Molecules (1983) Anal. Chem. 55:353-356.

19

Evaluation of Methods for the Analysis of Disulfide Containing Peptides by Fast Atom Bombardment Mass Spectrometry

Henry Rodriguez
Byron Nevins
John Chakel

Department of Protein Chemistry
Genentech Inc.
South San Francisco, California

I. INTRODUCTION

The assignment of disulfide bridges is an essential element in the complete determination of a protein's primary structure. The advent of recombinant DNA methods has provided us with the ability to obtain the complete primary structure of a protein through cloning. However, the disulfide bridge structure cannot be deduced in this way and may be essential for biological activity. Even if a protein is cloned, the production of that protein in an expression vector with the correct disulfide orientation is not routine. The increasing number of proteins being cloned requires an efficient method for the rapid determination or verification of disulfide bridges in the native or recombinant protein, respectively. Verification that a disulfide bridge exists between two peptides can be quickly and conclusively determined by Fast Atom Bombardment Mass Spectrometry (FABMS) if the molecular ion of the disulfide linked peptide can be dissociated into its constituent peptides.

A classical approach to the identification of disulfide linked peptides is to chemically or enzymatically fragment the protein and compare the resulting chromatographic profile under reducing and non-reducing conditions. Peaks which change their retention time under reducing conditions are identified as disulfide linked peptides. Typically, the separated peptides can be identified by protein sequence or amino acid analysis. This method has been successfully employed to determine the disulfide bridges in proteins such as interferon (1), T4 lysozyme (2), and human growth hormone (3). However, this approach is time consuming and impractical for larger polypeptides. The speed and accuracy of peptide identification can be greatly enhanced over conventional methods by FABMS analysis. Recently, FABMS has been utilized for the determination of disulfide bonds (4,5,6).

There are two principal methods for cleaving a disulfide bridge: 1) the reduction of cystine to cysteine in an alkaline solution in the presence of a reducing agent such as dithiothreitol (7), 2) and oxidation of cystine to cysteic acid by performic acid (8). Although these methods are very effective at dissociating the disulfide bridge FABMS requires that the sample be free of salts, detergents and other excipients which may affect ionization of the peptide. Additionally, the FABMS sample probe has a volume restriction of 1-2 microliters. These requirements have prompted us to evaluate reduction and oxidation for the analysis of disulfide linked peptides by on-the-probe (OTP) FABMS analysis.

II. METHODS

<u>A. HGH Disulfide Peptides</u> Human growth hormone was digested with trypsin and the disulfide linked peptides were isolated by high performance liquid chromatography (HPLC) using a non-volatile buffer as previously described (9). The disulfide peptides were desalted using a volatile buffer system consisting of 0.1% trifluoroacetic acid and acetonitrile before being divided into 200 picomole aliquots. All experiments were performed using the same digest.

<u>B. FABMS Analysis</u> All FABMS analysis were performed using a JEOL JMS HX110HF/HX110HF tandem MS in a two sector mode. Peptides were dissolved in 1-2 ul of 10% acetic acid. One microliter of the sample was mixed with 0.6 ul of the matrix (glycerol) on the sample probe. Typically, spectra were obtained with an instrument resolution of 1000.

<u>C. OTP Analysis</u> Reduction was performed with either 0.2M N-ethylmorpholine(NEM), pH8.5/1M dithiothreitol (DTT) mixed 1:1 (v/v) or ammonium hydroxide/1-thioglycerol (1:1, v/v). The sample was dried OTP (in the vacuum inlet) and 1 ul of reductant was added, mixed, and allowed to react for 10 minutes at room temperature. The sample was dried again and 0.5 ul of 10% oxalic acid and 0.7 ul of thioglycerol were added and the sample analyzed. Alternatively, after FABMS analysis of the unreduced disulfide peptide in a thioglycerol matrix, the same sample could be reduced OTP. In this procedure 1ul of the ammonium hydroxide/thioglycerol reductant was added to the sample, allowed to react 10min., excess ammonia removed in the vacuum inlet, acidified

with 0.5 ul of 10% oxalic acid, and analyzed.

OTP oxidation was accomplished by addition of 1ul of 3% performic acid to the sample dissolved in the glycerol matrix. Ten minutes was allowed for oxidation to occur prior to analysis. Alternatively, after FABMS analysis of the unoxidized disulfide peptide the same sample could be oxidized OTP.

III. RESULTS

The disulfide containing tryptic peptides of recombinant human growth hormone (3,9) were used to evaluate the reductive and oxidative cleavage methods described above. The protein sequence of the HGH disulfide linked peptides as well as their molecular ions under reducing, non-reducing, and oxidizing conditions are shown in Table 1.

HGH Disulfide Designation	Peptide Sequence	Molecular ions (MH+)		
		Oxidized	Reduced	Non-Reduced
T6	YSFLQNPQTSLCFSESIPTPSNR	2666	2617	3764
T16	NYGLLYCFR	1197	1149	
T20	IVQCR	666	618	1401
T21	SVEGSCGF	833	785	

Table 1 Sequence and molecular ions for the reduced, non-reduced, and oxidized disulfide containing tryptic peptides of recombinant hGH.

A. OTP Reduction The disulfide peptides T6-16 and T20-21 were reduced on the MS probe by the addition of ammonia in thioglycerol or by the addition of NEM/DTT. The spectra of both buffers used for OTP reduction gave virtually identical spectra with very little background. The spectra generated with NEM/DTT is shown in Figure 1 demonstrating the generation of the reduced peptides.

With OTP reduction a molecular ion can be obtained on a peptide and that same peptide can then be reduced OTP and reanalyzed for reduction products. The utility of this method is demonstrated on an HGH disulfide peptide T6-16 (Fig. 2). After a molecular ion was obtained on the disulfide peptide (m/z 3764, Fig. 2A) the same sample remaining on the probe was reduced by the addition of ammonia (Fig. 2B). This allowed identification of the reduced peptides T6 and T16 at m/z 2617 and 1149, respectively. Some reduction of the disulfide linked peptide occured OTP (in situ) under non-reducing conditions as can be seen by the presence of the T6 and T16 peptides at m/z of 2617 and 1149, respectively (Fig. 2A). Additionally, a small amount of the doubly charged parent ion was

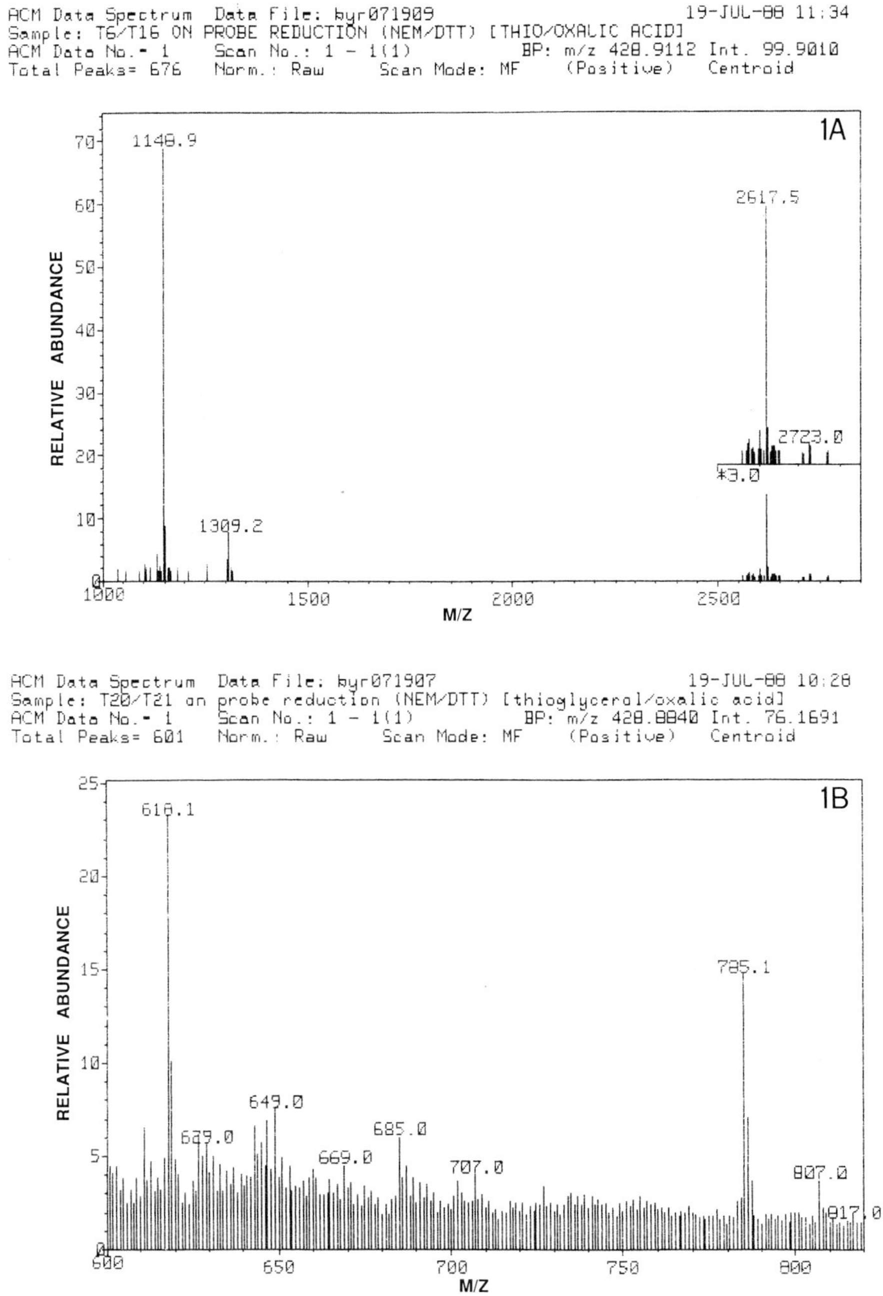

Figure 1 Spectra of peptide T6-16 (1A) and T20-21 (1B) reduced on the probe with NEM/DTT.

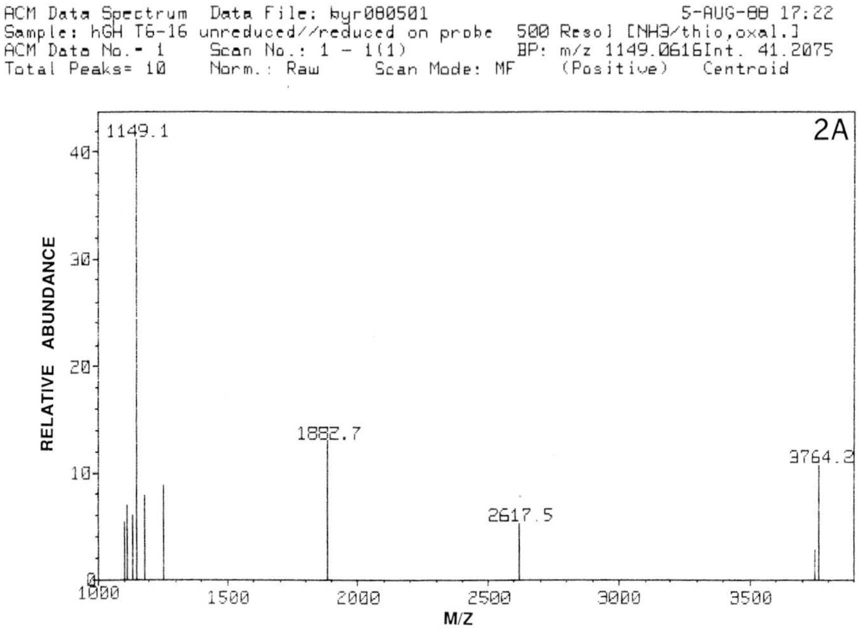

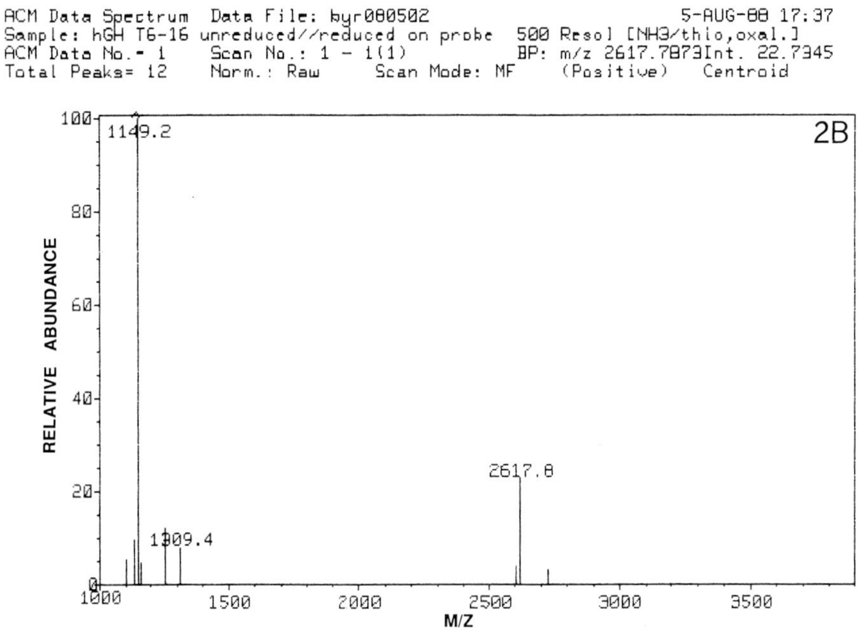

Figure 2 Molecular ions of unreduced tryptic T6-16 (2A) followed by OTP reduction by the addition of ammonia to the thioglycerol matrix (2B).

observed at m/z 1882. The energy emitted by the FAB gun (6kV, 10mA) is sufficient to dissociate a small percentage of the disulfide bonds under non-reducing conditions (5). The speed and simplicity of this method makes it possible to evaluate an entire tryptic map in a relatively short period of time. This method gave a high relative abundance of ions and a negligable amount of adduct ions allowing unambiguous peptide identification.

B. OTP Oxidation - Performic acid is an extremely powerful oxidant. It oxidizes both cysteine and cystine to cysteic acid and methionine to methionine sulfone. Tryptophan is degraded to several products (10). Destruction of tryptophan is a major drawback to the use of performic acid for cleaving disulfide bonds. However, if the degradation of tryptophan occurs in a predictable manner this method may be useful in disulfide assignment. Tyrosine, serine, and threonine are also oxidized, but very slowly and only under drastic conditions.

Several synthetic model peptides (Table 2) were analyzed before and after performic acid oxidation to establish the predictablity in forming the oxidation products expected. Peptides containing methionine, tryptophan, and cysteine which are residues known to oxidize, under these conditions, were investigated. Methionine and cysteine were quantitatively converted to their oxidation products methionine sulfone

Peptides*	Molecular Ions (MH+) Non-Oxidized	Oxidized	Mass Shift	Time (hr)
1. Ac-ELRWAG	773	789 805 821	16 32 48	1
2. Ac-ELRMAG	718	750	32	4
Ac-ELRMAG	718	750	32	2.5
Ac-ELRMAG	718	750	32	1
Ac-ELRMAG	718	750	32	0.5
Ac-ELRMAG	718	750	32	10min.
Ac-ELRMAG	718	734 750	16 32	5min.
Ac-ELRMAG	718	734 750	16 32	1min.
3. Ac-ELRHAG	725	725	0	1
4. Ac-ELRSAG	674	674	0	1
5. AEDEDNQQGQGEGLC	1592	1640	48	1
6 RRLIEDAEYAARG	1519	1519	0	1

Table 2 Molecular ions of unoxidized and performic acid oxidized model peptides showing the mass shift upon oxidation. * (Ac designates acetyl)

(+32 amu) and cysteic acid (+48 amu), respectively (Table 2). While, the peptide which contained tryptophan showed the addition of one (+16 amu), two (+32 amu), and three (+48 amu) oxygen atoms. When peptide 2 (Table 2) was treated with performic acid for less than 10 minutes an ion at m/z 734 corresponding to the conversion of methionine to methionine sulfoxide (+16) was observed. Beyond that time complete conversion of methionine was observed. Under the conditions used no other amino acid was found to react with performic acid.

OTP oxidation of peptide T20-21 (Fig. 3) gave excellent results compared to off-the-probe oxidation of the same HGH peptide (data not shown) which showed a lower ion abundance and signal to noise ratio. This lower signal magnified the Na^+ and K^+ adducts and obscurred peptide identification with off-the-probe analysis. However, when the model peptides (Table 2) were treated off-the-probe with performic acid their oxidation products yielded mass spectra of high relative abundance and the peptides were clearly identifiable. The main difference between off-the-probe performic acid oxidation of the HGH peptide which gave poor results and the test peptides which gave good results was the amount of material used. Approximately 100-200 pmoles of the HGH peptides was oxidized of which 50-100 picomoles was analyzed compared to 1 nanomole for the test peptides of which 500 picomoles was analyzed. When OTP oxidation was compared with off-the-probe oxidation on 100 picomoles of model peptide 2 similar results were observed as with the HGH peptide. The additional handling of the sample for off-the-probe analysis most likely accounts for the decrease in sensitivity.

IV. DISCUSSION

OTP performic acid oxidation and OTP reduction proved to be effective and desirable methods for the cleavage of disulfide linked peptides. OTP analysis generated higher quality spectra and provided higher sensitivity when compared to the same disulfide cleavage methods performed off-the-probe. However, when OTP reduction was compared with off-the-probe reduction, results were similar. But, because off-the-probe reduction methods involves additional steps and handling OTP analysis is preferred. Of the two OTP disulfide cleavge methods described reduction may be preferred because it is less harsh than the oxidation procedure. Additionally, oxidation of disulfide peptides containing multiple methionine and tryptophan residues may complicate identification. However the mass change afforded by performic acid oxidation may allow a more definitive verification that a disulfide exists. The choice of which of the two OTP methods to use depends on the user but the data presented clearly demonstrates that OTP analysis is preferred over other methods. The simplicity of OTP analysis can be further appreciated when disulfide bond assignment is being determined on small quantities of a large protein. The minimal handling of the sample employing OTP reduction or

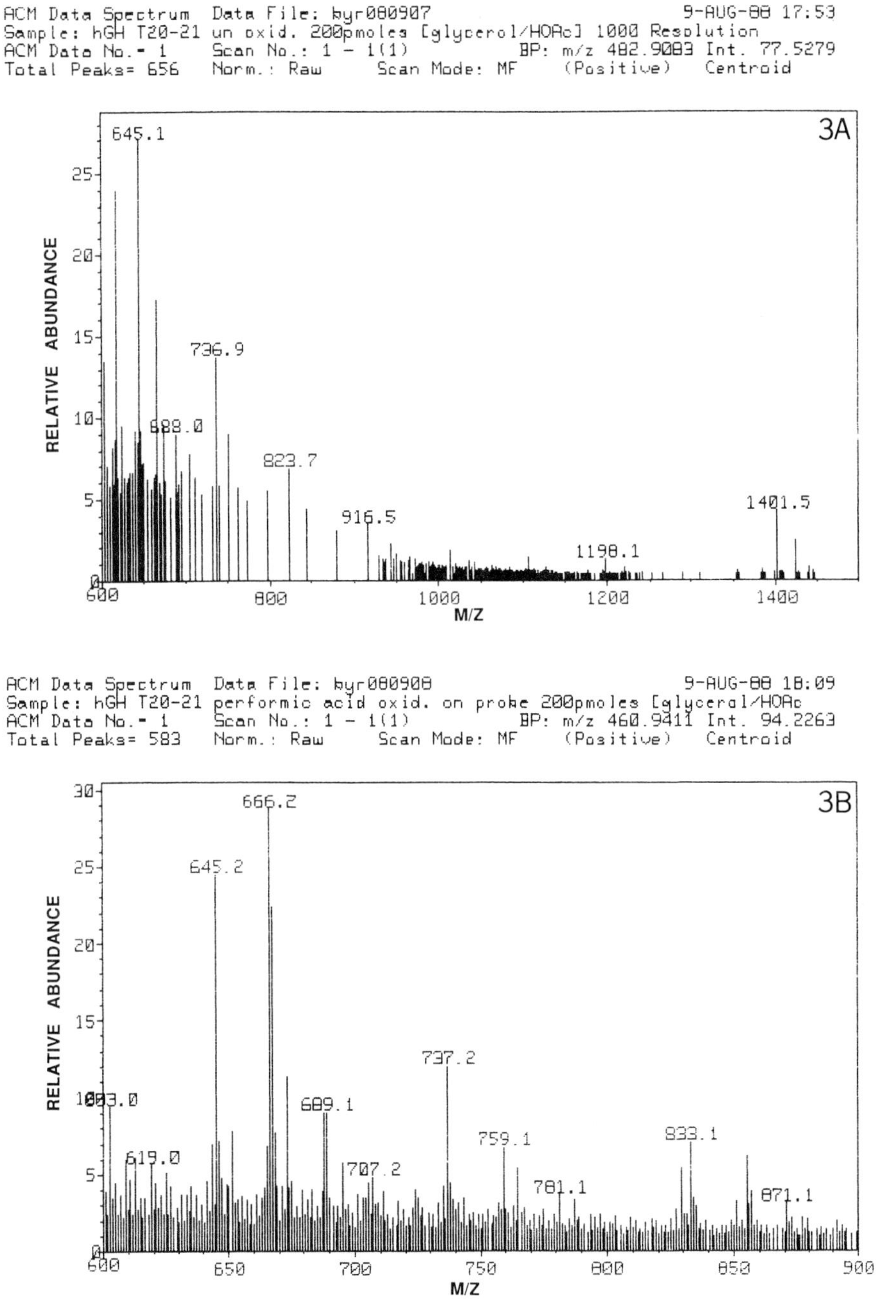

Figure 3 Spectra of unoxidized HGH tryptic T20-21 (3A) followed by OTP oxidation (3B).

oxidation allows for the greatest degree of sample recovery.

OTP analysis has significant advantages over off-the-probe methods. It requires the least amount of sample manipulation making it more suitable for sub-nanomole analysis. Additionally, it allows the direct analysis of the same peptide sample before and after cleavage allowing quick confirmation of the presence of a disulfide.

Acknowledgments

The authors wish to acknowledge Victor Ling for his helpful suggestions in establishing the OTP methods. The authors also wish to acknowledge Dr. James Marsters for kindly providing the synthetic peptides and Dr. John Stults and James Bourell for their assistance in obtaining the mass spectral data.

V. REFERENCES

1. Wetzel, R. (1981). Nature, 289,606.
2. Perry, J.L. and Wetzel, R. (1984). Science, 226, 555.
3. Kohr, W. J., Keck, R., and Harkins, R. N. (1982). Anal. Biochem., 122, 348.
4. Sun, Y. and Smith, D. L., (1988). Anal. Biochem., 172, 130.
5. Yazdanparast, R., Andrews, P. C., Smith, D. L., and Dixon, J. E. (1987). J. Biol. Chem., 262, 2507.
6. Morris, H. R., and Pucci, P., (1985). Biochem. Biophys. Res. Commun., 126, 1122.
7. Cleland, W. W., (1964). Biochem., 3, 480.
8. Sanger, F., (1949). Biochem. J., 44, 126.
9. Hancock, W. S., Canova-Davis, E., Chloupek, R. C., Wu, S-L., Baldonado, I. P., Battersby, J. E., Spellman, M. E., Basa, L. J., and Chakel, J. A., (1988). in Banbury Report 29:Therapeutic Peptides and Proteins, Cold Spring Harbor Laboratory.
10. Hirs, C. H. W., (1967) in Methods in Enzymology (Hirs, C. H. W., and Timasheff, S. N. Eds.), Vol. 6, 197, Academic Press, New York

20

EXAMINATION OF THE POSTTRANSLATIONAL MODIFICATIONS OF CARCINOEMBRYONIC ANTIGEN[1]

Stanley A. Hefta, Raymond J. Paxton, Laura J. F. Hefta, Terry D. Lee, and John E. Shively

Division of Immunology
Beckman Research Institute of the City of Hope
Daurte, CA 91010.

I. INTRODUCTION

Carcinoembryonic antigen (CEA) is a 180 kD glycoprotein found in increased amounts on the surface of colorectal tumor cells. The chemistry, biochemistry, clinical aspects, and tissue distribution of CEA have been reviewed (1). The complete amino acid sequence of CEA has been elucidated by a combination of protein (2) and cDNA sequencing techniques (3). The cDNA data predicts a structure having an 108 amino acid N-terminal domain, followed by three domains that share extensive sequence homology with immunoglobulin molecules, and a short hydrophobic domain suggested to be the membrane-anchoring region. The Ig-like domains each contain two disulfide loops and have a total of 26 N-linked glycosylation sites. Recent studies have shown that CEA is processed posttranslationally to remove the predicted hydrophobic C-terminal domain with the subsequent

[1]Supported by NIH grant CA37808 and NCI grant 33572.

addition of an ethanolamine-containing phosphatidylinositol glycan (PI-G tailed) moiety (4). This paper will describe the techniques used to study the posttranslational modifications of CEA.

II. METHODS

A. Preparation of CEA

CEA was isolated from liver metastases of colon tumors and purified as described (2). Deglycosylation was accomplished with anhydrous hydrofluoric acid (HF) (5) or with trifluoromethanesulfonic acid:anisole (TFMSA) (6). Following reduction and alkylation (7), the samples were digested with the appropriate enzymes, and the peptides purified by reverse-phase HPLC.

B. Structural analysis of CEA

Sequence analysis was performed on polyvinylidene difluoride membranes (Millipore) treated with polybrene using gas-phase sequencers built at the City of Hope (8) and equipped with a continuous flow reactor (9). For fast atom bombardment mass spectrometry (FAB-MS), the samples were concentrated to dryness in polypropylene microcentrifuge tubes, redissolved in 2 µl of 5% (vol/vol) aqueous acetic acid and added to 2 µl of the liquid matrix on a 1.5 x 6-mm stainless-steel sample stage. For analysis of larger peptides (>800 amu) the liquid matrix consisted of dithiothreitol/dithioerythritol (1/5, wt/wt) containing 6 mM camphorsulfonic acid; for analysis of smaller peptides (<800 amu) the liquid matrix consisted of glycerol or thioglycerol. Positive ion FAB mass spectra were obtained using a JEOL HX-100HF high-resolution, double-focusing, magnetic-sector mass spectrometer operating at 5 kV accelerating potential and a nominal resolution of 3000 or 5000. Sample ionization was accomplished using a 6 keV Xe atom beam. For normal spectra, the electric sector of the mass spectrometer was set to transmit all ions at source potential (5 kV), and the magnetic sector was scanned over a given mass range. For B/E-linked scans (10), the electric-sector voltage and magnetic-sector field strength were linked such that the ratio was constant.

III. RESULTS AND DISCUSSION

A. Identification of Glycosylation Sites

Proteins glycosylated to the extent of CEA are generally resistant to enzymatic digestion (2). Deglycosylation with TFMSA or HF leaves the terminal N-acetylglucosamine(s) at the site of glycosylation (5, 6). The site of glycosylation can be identified directly during sequence analysis as the PTH-Asn(GlcNAc) derivative (2). Figure 1A shows the first three cycles of a peptide (NVTRNDTASY) which contains two N-linked glycosylation sites. The PTH-Asn(GlcNAc) in cycle 1 is indicated. The identity of this peak was confirmed by comparing the retention times of a synthetic PTH-Asn(GlcNAc) derivative (2) with the natural product. Although the presence of N-acetylglucosamine does not interfere with repetitive yields, the yield of PTH-Asn(GlcNAc) at any given cycle is generally less than that of other residues. This reduction in yield is probably due to a lower solubility of the aminothiazolinone derivative of Asn(GlcNAc) in the extracting solvent, butyl chloride, relative to other amino acid aminothiazolinone derivatives (2).

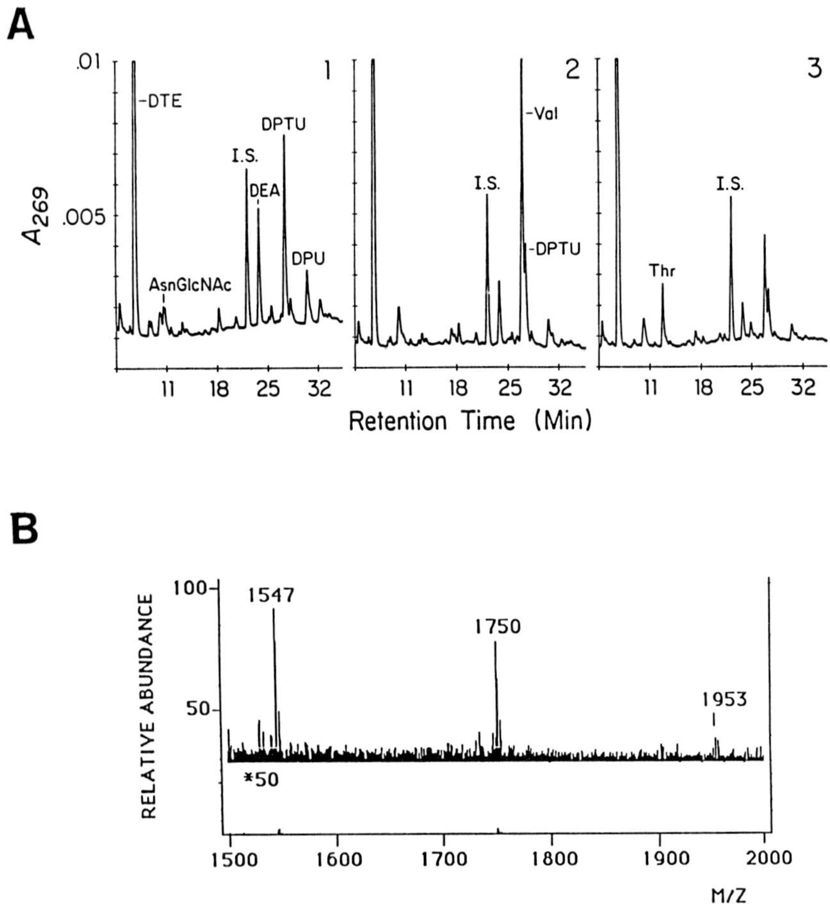

Figure 1. Identification of CEA glycosylation sites. (A) Reverse-phase HPLC chromatograms for sequencer cycles 1-3 of a chymotryptic peptide from TFMSA-treated CEA. DTE, dithioerythritol; I.S., internal standard; DEA, diethylamine; DPTU, diphenylthiourea; DPU, diphenylurea.
(B) Positive ion FAB-MS spectrum of the molecular ion region of the intact peptide using the DTT/DTE/camphor-sulfonic acid matrix described previously.

Positive ion FAB-MS analysis of the molecular ion region for the above peptide is shown in Figure 1B. The molecular ion with a mass-to-charge ratio (m/z) of 1547 corresponds to the predicted weight of this peptide plus two N-linked N-acetylglucosamine moieties (1140 amu for the peptidyl portion and 203 amu for each acetylglucosamine). The molecular ions at m/z 1750 and 1953 represent forms of this peptide containing 3 and 4 N-acetylglucosamine moieties, respectively. Although we can not rule out the possibility that the lower molecular weight ions are fragments of a relatively unstable m/z 1953 parent ion, we believe that these ions represent separate glycosylated forms of this peptide which coeluted during the reverse-phase purification. The reason for not observing PTH-derivatives containing 2 N-acetylglucosamine moieties during the sequence analysis of this peptide (Fig. 1A), is probably due to the insolubility of such derivatives in butyl chloride. Using this approach, a total of 28 N-linked sites were identified in CEA: 26 in the three Ig-like domains and 2 in the N-terminal domain (2).

B. Mapping of Oligosaccharides

Analysis of oligosaccharide structures in the picomolar or low nanomolar range presents several problems to the researcher. Purification schemes are complicated by the difficulty of separating and detecting the highly polar, nonchromophoric molecules. Several studies have shown the utility of derivatizing the molecules with chromophores such as ethyl *p*-aminobenzoate (11,12), or direct detection by amperometric detection (13). Both of these methods have been shown to allow detection at the low picomolar range. The use of compounds such as ethyl *p*-aminobenzoate has the added benefit of incorporating a protonation site for subsequent FAB-MS analysis (11,12). For the purpose of mapping oligosaccharide structures in CEA, we have centered on the isolation and structural analysis of glycopeptides. Presence of the peptide bonds provides a means for detection, and the peptidyl portion of the molecule provides a protonation center for positive ion FAB-MS analysis. The general approach employed in these studies involves using amperometric detection for determination of the oligosaccharide composition, exoglycosidases for sequential analysis, and FAB-MS for structural analysis.

Although heavily glycosylated proteins are generally resistant to proteolysis, we have been able to obtain several glycopeptides from tryptic and chymotryptic digestions of native CEA. Figure 2 shows the results of positive ion FAB-MS analysis on 100-200 pmoles of one such glycopeptide (LQLSNGNR). The spectra shows a prominent molecular ion at m/z 2118 and several fragment ions of lesser intensity. The peptidyl portion of this glycopeptide accounts for 901 amu of the observed mass.

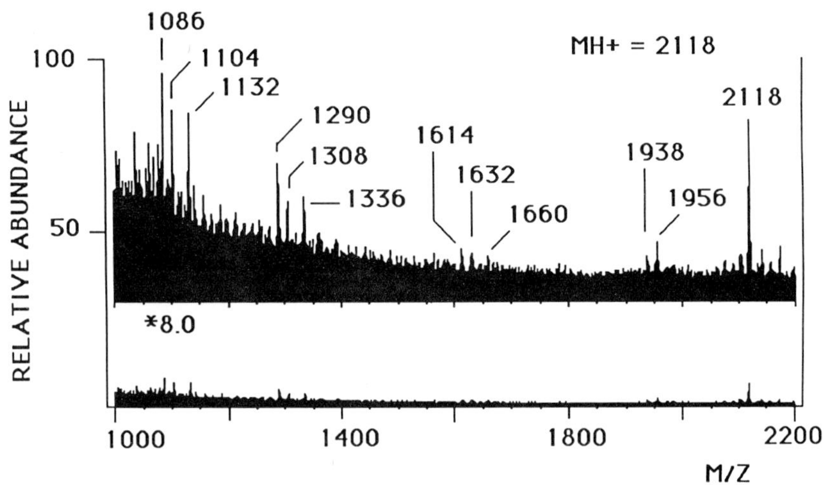

Figure 2. Positive ion FAB-mass spectrum of 100-200 pmoles of a glycopeptide (LQLSNGNR) isolated from a tryptic digest of native CEA. The DTT/DTE/champhor-sulfonic acid matrix described previously was used for this analysis.

Examination of the fragmentation pattern of m/z 2118 provides sufficient information to define the oligosaccharide structure on this glycopeptide. The proposed structure and predicted fragmentation series are shown in Fig. 3. As indicated in Fig. 3B, cleavages of

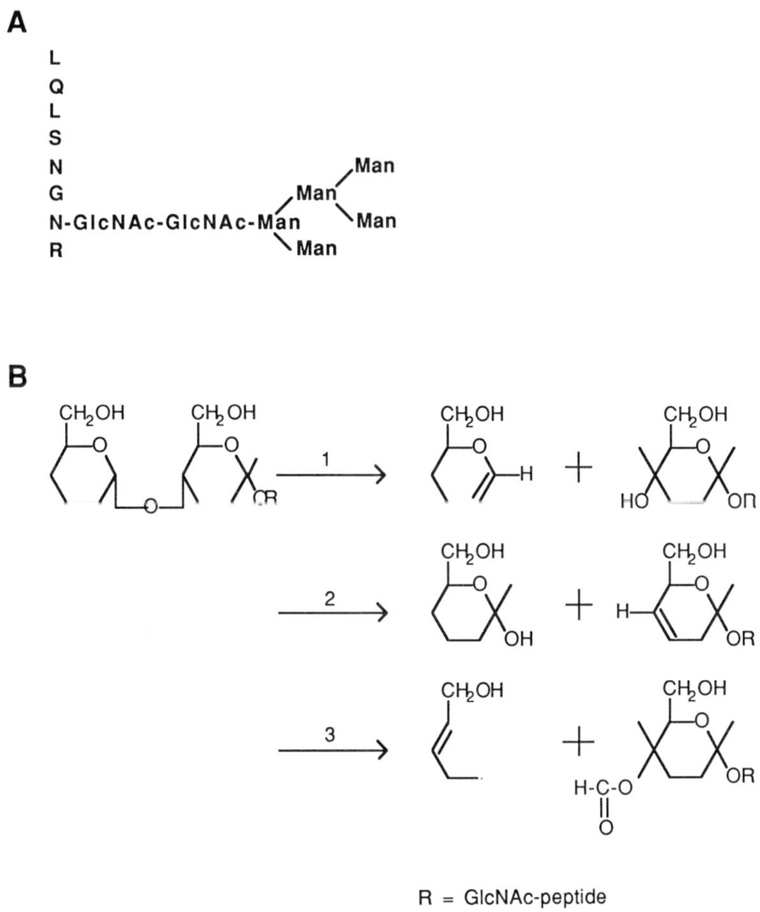

Figure 3. (A) Proposed structure of the oligosaccharide present on the glycopeptide shown in Fig. 2. (B) Predicted FAB-MS oligosaccharide fragmentation pathways.

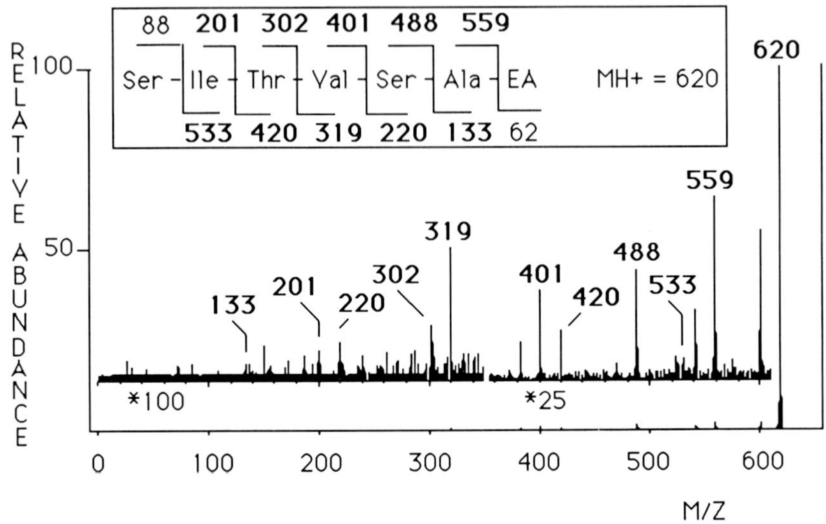

Figure 4. B/E-linked scan FAB-MS spectrum for a tryptic peptide, T12, (SITVSA) from HF-treated CEA using a glycerol matrix. The insert defines the N-terminal (B series) and C-terminal (Y" series) fragmentation series of m/z 620. EA, ethanolamine.

IV. CONCLUSIONS

This paper has described the uses of FAB-MS for the characterization of posttranslational modifications in carcinoembryonic antigen (CEA). FAB-MS has proven invaluable during sequence analysis for confirming peptide sequences and for identifying glycosylation sites. Recent studies on CEA have established the importance of FAB-MS for analyzing the structures of the N-linked oligosaccharides, and for detecting the presence of ethanolamine at the C-terminus of the mature protein. This later feature is consistent with CEA being attached to the plasma membrane through an ethanolamine-

O-glycosidic bonds result in a reduction of either 162 (series 1), 180 (series 2), or 134 (series 3) mass units depending on the type of fragmentation series (11).

The fragments ions resulting from the loss of one (m/z 1956), three (m/z 1632), and five (m/z 1308) anhydromannose (series 1) are clearly evident in Fig. 2. Associated with these ions are fragment ions from series 2 and 3 type fragmentations. The loss of one, three, and five mannoses can occur from single cleavages of the branched structure (Fig. 3A), and therefore should appear in significant abundance. The losses of two and four mannoses, however, requires multiple, simultaneous cleavages and are expected to be in lesser abundance, and therefore not observed above the background. This observed fragmentation pattern is characteristic of a branched oligosaccharide relative to a linear structure which fragments with sequential losses of hexose units (11). The remainder of the oligosaccharide structure (Fig. 3A) is defined by the ion at m/z 1104 which corresponds to the loss of $Man_5GlcNAc$ leaving the peptidyl portion plus the N-linked N-acetylglucosamine. These results are consistent with the structures of high mannose type oligosaccharides previously published for CEA (14).

C. Posttranslational modifications of the C-terminus in CEA

The cDNA sequence data (3) predicts the presence of a 26 residue hydrophobic peptide at the C-terminus of CEA. Owing to the length and hydrophobic nature of this peptide, it was proposed that this region constituted the membrane binding domain in the mature protein (3). Recently, a mechanism for attachment of proteins to membranes has been described that involves a phosphatidylinositol-glycan complex (so called PI-G tail) covalently linked to the C-terminal residue of the mature protein through an ethanolamine moiety (for review, see ref. 15). Attachment of the PI-G tail occurs posttranslationally and replaces a short (generally less than 30 residues) hydrophobic peptide at the C-terminus of the protein. Because protein-sequence analysis of CEA failed to provide evidence for the hydrophobic peptide predicted by the cDNA data, we concluded that this domain was missing in the mature protein.

From a tryptic digest of deglycosylated CEA we isolated and sequenced a hexapeptide, T12. The sequence determined for T12 (SITVSA) corresponded to the terminal six residues (residues 637-642) of the third Ig-like domain in the mature protein. This region is immediately adjacent to the proposed membrane-binding C-terminal domain (3). Results from FAB-MS analysis of T12 indicated a protonated ion 43 amu higher than that calculated for the sequence. This deviation in mass suggested the presence of an ethanolamine moiety coupled covalently to the carboxyl group of the terminal alanine.

To establish the presence of ethanolamine, we analyzed T12 by FAB-MS using B/E-linked scan analysis. In B/E-linked scan analysis, the electric and magnetic sectors are scanned simultaneously to permit observation of daughter ions generated by the fragmentation of a preselected parent ion in the first field-free region of the mass spectrometer. Figure 4 shows the results of the B/E-linked scan on 126 pmol of T12 (m/z 620). The insert is provided to aid in identification of the N-terminal and C-terminal fragment ion series. Ions corresponding to these series are labeled in the figure. Of particular interest to this analysis is the ion at m/z 559 which arose from the loss of ethanolamine from the peptidyl portion of T12. As illustrated, this type of analysis provided enough information to define the entire peptide structure. Confirmation of CEA being anchored to the plasma membrane through a PI-G tail structure was obtained by indirect immunofluorescence and flow cytometry after treatment of a CEA-expressing cell line with phosphatidylinositol-specific phospholipase C (4).

phosphotidylinositol-glycan complex. Future studies will focus on further elucidation of posttranslational modifications in CEA and related proteins.

V. REFERENCES

1. Shively, J. E. & Beatty, J. D. (1985) *CRC Crit. Rev. Oncol./Hematol.* **2,** 355-399.
2. Paxton, R. J., Mooser, G., Pande, H., Lee, T. D. & Shively, J. E. (1987) *Proc. Natl. Acad. Sci. USA* **84,** 920-924.
3. Oikawa, S., Nakazota, H. & Kosaki, G. (1987) *Biochem. Biophys. Res. Comm.* **142**, 511-518.
4. Hefta, S. A., Hefta, L. J. F., Lee, T. D., Paxton, T. D. & Shively, J. E. (1988) *Proc. Natl. Acad. Sci. USA* **85,** 4648-4652.
5. Mort, A. J. & Lamport, C. T. A. (1977) *Anal. Biochem.* **82,** 289-309.
6. Edge, A. S., Faltynek, C. R., Hof, L., Reichert, Jr., L. E. & Weber, P. (1981) *Anal. Biochem.* **118,** 131-137.
7. Waxdal, M. J., Koningsberg, W. H., Henley, W. L. & Edelman, G. M. (1968) *Biochemistry* **9,** 1959-1966.
8. Hawke, D. H., Harris, D. C. & Shively, J. E. (1985) *Anal. Biochem.* **147,** 315-330.
9. Shively, J. E., Miller, P. & Ronk, M. (1987) *Anal. Biochem.* **163,** 517-529.
10. Wetson, A. F., Jennings, K. R., Evans, S. & Elliot, R. M. (1976) *Int. J. Mass Spectrom. Ion Phys.* **20,** 317-327.
11. Webb, J., W., Jiang, K., Gillece-Castro, B. L., Tarentino, A., L., Plummer, T. H., Byrd, J. C., Fisher, S. J. & Burlingame, A. L. (1988) *Anal. Biochem.* **169,** 337-349.
12. Wang, W. T., LeDonne, Jr., N., C., Ackerman, B. & Sweeley, C. C. (1984) *Anal. Biochem.* **141,** 366-381.

13. Hardy, M. R., Townsend, R. R. & Lee, C., Y. (1988) *Anal. Biochem.* **170,** 54-62.
14. Yamashita, K., Totani, K., Kuroki, M., Matsuoka, Y., Udea, I. & Kobata, A. (1987) *Cancer Research* **47,** 3451-3459.
15. Low, M. G. & Saltiel, A. R. (1988) *Science* **239,** 268-275.

21

THE ASSIGNMENT OF PROTON RESONANCES IN 2D NMR SPECTRA OF PROTEINS[1]

S. Walter Englander

Department of Biochemistry and Biophysics
University of Pennsylvania
Philadelphia, Pennsylvania 19104

A. Joshua Wand and Sarah J. Nelson

Institute for Cancer Research
Fox Chase Cancer Center
Fox Chase, Pennsylvania 19111

I. INTRODUCTION

Proton NMR spectra of proteins can provide, in principle, many hundreds of probe points for the study of structure, dynamics and function. First, however, it is necessary to assign the various NMR resonances to their parent protons. With the advent of two dimensional NMR methods (1-5), this daunting problem has now been solved for a number of small proteins. As larger proteins are examined, the assignment problem becomes rapidly more difficult. So far only a few proteins larger than 100 residues have been solved, namely lysozyme at 129 residues (6), thioredoxin at 108 residues (7), and equine cytochrome c at 104 residues, both in its reduced (8) and oxidized (9) forms. Here we briefly introduce the problems, the data, and some of the strategies used in these efforts.

Assignments achieved so far have been accomplished *by hand*. This article describes an approach, based on the main chain protons of proteins, that is algorithmic in nature and promises to reduce the assignment problem to a computer exercise.

[1]Supported by NIH research grants GM 31847 and DK 11295 (SWE), and GM 35940 (AJW).

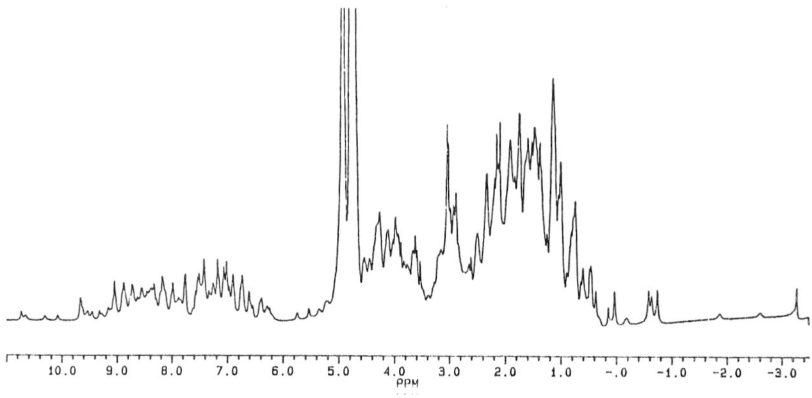

Fig. 1. Proton NMR spectrum of reduced cytochrome c at 500 MHz.

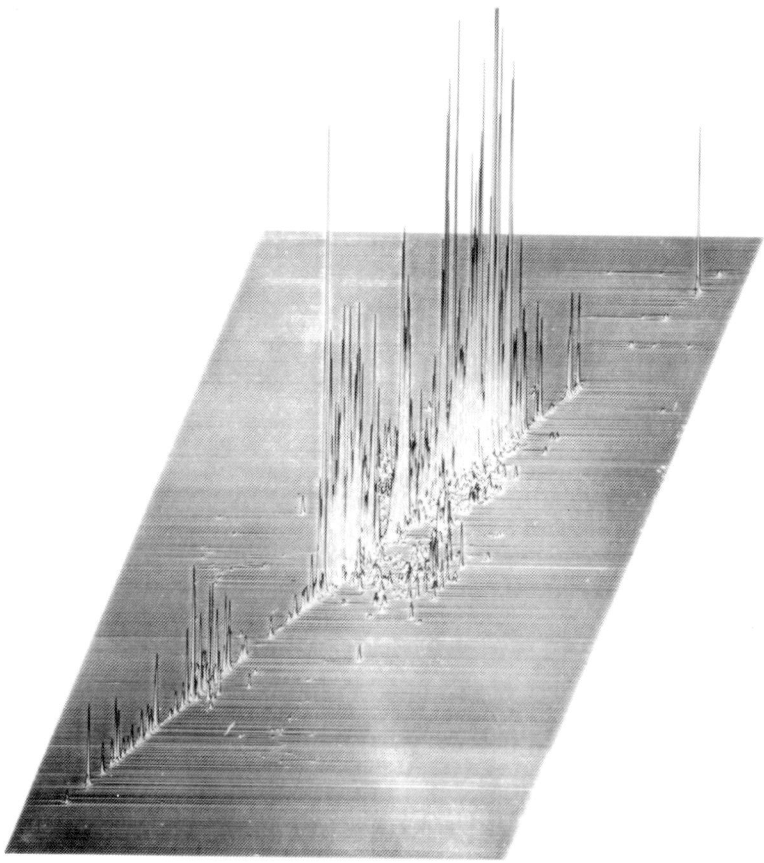

Fig. 2. Stacked COSY plot of 2D ^{1}H NMR spectrum (reduced cyt c).

II. TWO DIMENSIONAL NMR SPECTRA

Fig. 1 shows a standard 1D spectrum of reduced cytochrome c. Each of the 800 protons in the molecule produces a resonance peak of unit area, and the protons in the various amino acid residues spread out over the chemical shift axis with the amide NH protons in general farthest downfield, neighbored by the $C_{\alpha}H$ (with the aromatic ring CH interspersed), then the $C_{\beta}H$, and farthest upfield the protons of the aliphatic side chains. The proton resonances are badly overlapped, and the spectrum provides little information that could allow the individual resonances to be assigned.

Fig. 2 is a 2D spectrum, presented as a stacked plot. The 1D spectrum appears along the diagonal, and in addition off diagonal cross peaks are easily discernable although they are relatively small. These are best studied in the contour representation of Fig. 3. In the 2D plot, both axes are in terms of chemical shift, so that the two coordinates of each on-diagonal peak are identical and equal to the chemical shift in the 1D spectrum. An off diagonal cross peak has two different chemical

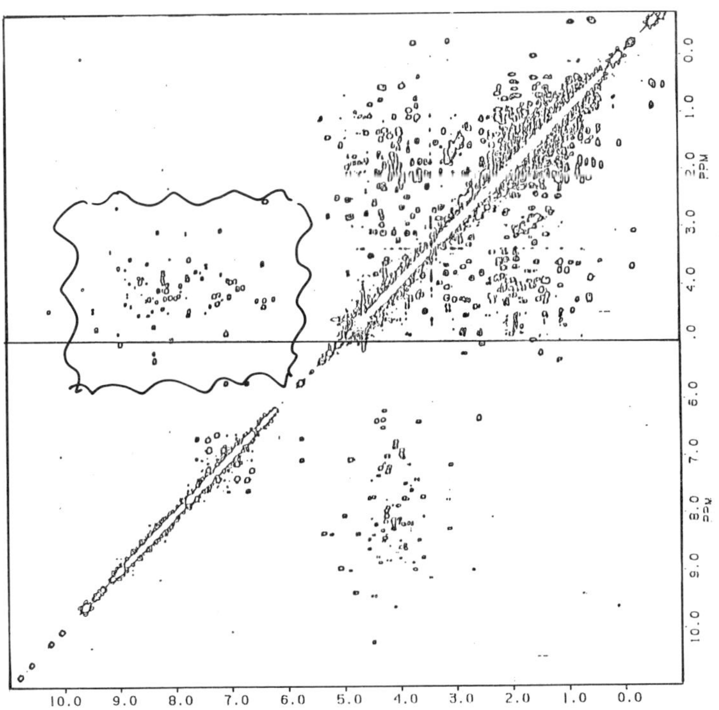

Fig. 3. Contour plot of the 2D COSY spectrum of Figure 2.

shift coordinates, which represent the chemical shift of two different protons that produce the cross peak by virtue of the particular proton-proton spin interaction selected in the NMR experiment. In Fig. 3 the paired protons are correlated by a through-bond interaction that is exploited to produce the J-coupled or COSY (correlated spectroscopy) spectrum. For example, each cross peak in the boxed, "fingerprint" region (and its symmetrical counterpart below the diagonal) represents a J-coupled NH-C_{α}H pair within a given amino acid residue, and each cross peak can be pictured to appear at the intersection of horizontal and vertical lines drawn from the resonance positions of the two individual protons on the diagonal. Similarly much of the next clump of cross peaks to the right along the diagonal represents C_{α}H-C_{β}H COSY cross peaks.

More information appears in the COSY spectrum than is apparent from the long range viewpoint of Fig. 3. The detail present can be better seen in Fig. 4, an expansion of the fingerprint region, and Fig. 5 which shows a section farther upfield. Both Figs. 4 and 5 are displayed in the phase sensitive mode, which develops the antiphase substructure of the cross peaks.

Additional information can be obtained by exploiting related kinds of coherent magnetization transfer. A relayed coherence transfer (RCT or relayed COSY) experiment can develop cross peaks that represent

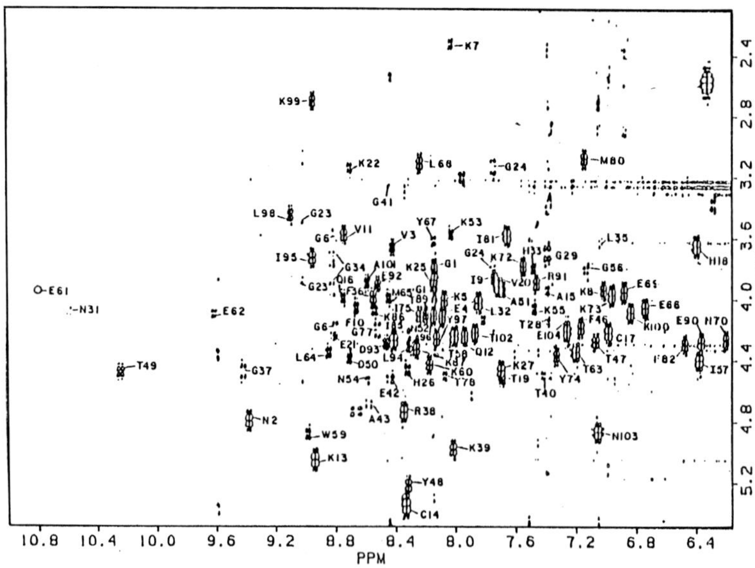

Fig. 4. Expansion of the COSY fingerprint region, boxed in Fig. 3, shown in the phase sensitive mode.

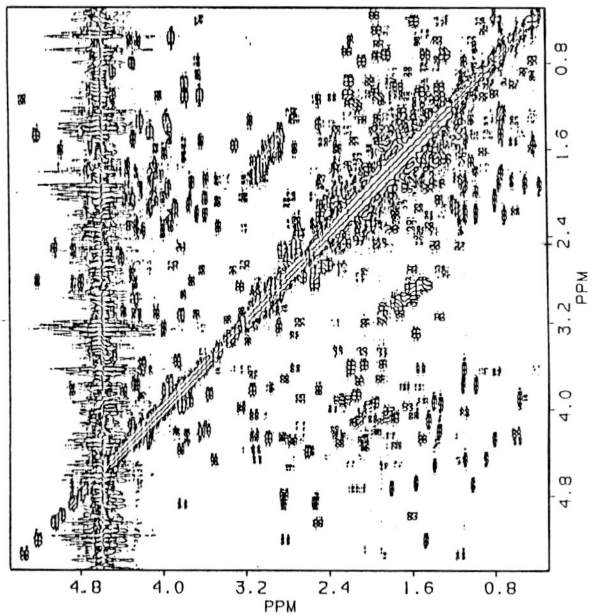

Fig. 5. (Top) Expansion of the far upfield region of the COSY spectrum in Fig. 3, containing aliphatic side chain cross peaks (phase sensitive).

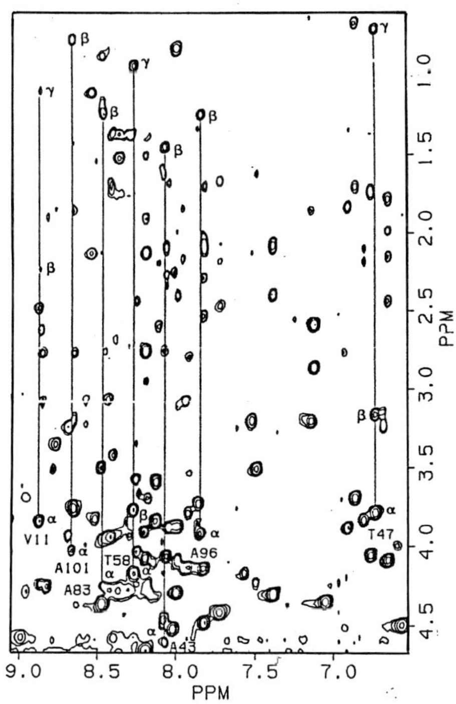

Fig. 6. TOCSY spectrum. Cross peaks are due to magnetization exchange between each amide NH and all the protons in its amino acid.

interactions between second nearest neighbor cross peaks, for example between amide NH and their $C_\beta H$. These cross peaks then appear in the previously unpopulated NH-$C_\beta H$ area. The TOCSY experiment (total correlation spectroscopy) carries this motif further by transferring magnetization through the entire amino acid spin system, and can display all the J-coupled resonances in an amino acid residue in a single line of cross peaks (e.g. Fig. 6).

The J-correlated spectra shown so far are limited to interactions within the same amino acid residue, since amino acids in a polypeptide are separated by a carbonyl group which interrupts cross-residue coherence transfer. A different kind of 2D experiment displays spatial information in the form of cross peaks due to the nuclear Overhauser effect between protons closer than about 4 A or so (see NOESY below). The NOESY spectrum duplicates many of the intraresidue COSY cross peaks, but provides in addition cross peaks between protons in different amino acids that are near in space. Information obtained from the NOESY spectrum is crucial for solving the assignment problem, as we shall see. In addition, it is the spatial information provided by the NOESY that has opened the possibility for NMR-based determinations of the structure of proteins in solution.

Two dimensional NMR spectra afford far better resolution of individual proton signals than can be obtained in 1D, and they provide a vast base of information that can help to assign the individual resonances. However, even for a rather small and well behaved protein like cytochrome c, the cross peak information is dense, often ambiguous, and incomplete. The various spectra present an enormous puzzle, the solution of which is by no means straightforward. In pursuing this kind of challenge, one can adopt different strategies. We consider here some ways in which this information has been handled, initially describing the sequential assignment strategy and then a different main chain directed (MCD) approach.

III. SEQUENTIAL ASSIGNMENT

The sequential assignment strategy was developed in the laboratory of Kurt Wüthrich in work directed at assigning a number of small proteins, in the range of 50 to 80 residues (10-12). In this approach one attacks the assignment puzzle by initially attempting to define amino acid side chain spin systems in coherence transfer spectra (COSY, relay, TOCSY). Amino acids are then knit together using NOESY information.

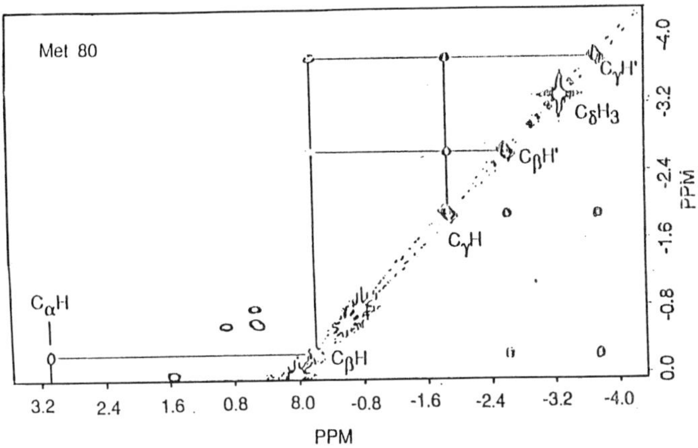

Fig. 7. The Met 80 side chain subspin system in the far upfield region of the reduced cytochrome c COSY spectrum. See also Figs. 1 and 2.

A. SIDE CHAIN SPIN SYSTEMS

As an atypically clean example, the COSY spectrum in Fig. 7 shows the connectivities that define the side chain spin system of Met 80 in reduced cytochrome c. Fig. 7 clearly displays a network of 4 J-coupled cross peaks connecting $C_\gamma H$ with $C_\gamma H'$, $C_\beta H$ with $C_\beta H'$, and each of the two $C_\gamma H$ protons with one of the two $C_\beta H$ protons. Even in this crystal clear case, the other two potential $C_\gamma H$-$C_\beta H$ crosspeaks are not seen (low coupling constant based on unfavorable dihedral angle), and $C_\delta H_3$ to $C_\gamma H$ cross peaks do not appear (low coupling constant due to 4 bond separation). The continuation of the side chain subspin system to the $C_\alpha H$ is also indicated in Fig. 7, and a further cross peak in the fingerprint region, not shown, leads to the amide NH resonance.

The Met 80 resonances in cytochrome c are exceptionally well resolved and can be seen as individual protons even in the 1D spectrum in the far upfield region (Fig. 1) and similarly as separate peaks in the diagonal of the 2D spectrum. The sulfur atom of Met 80 ligands the heme iron, so that these protons are directly apposed to the heme group and experience an extreme upfield ring current shift that places their resonances in an unpopulated spectral region.

The sequential assignment method assumes that most side chain spin systems can be defined in this way and extended downfield to discover their $C_\alpha H$ and amide NH resonances. Note that the definition of spin systems in this way does not yet indicate which spin system corresponds to which amino acid in the protein sequence or even necessarily to which amino acid type. J-correlated spectra can at best display unique spin-spin coupling patterns for 9 of the 20 amino acid

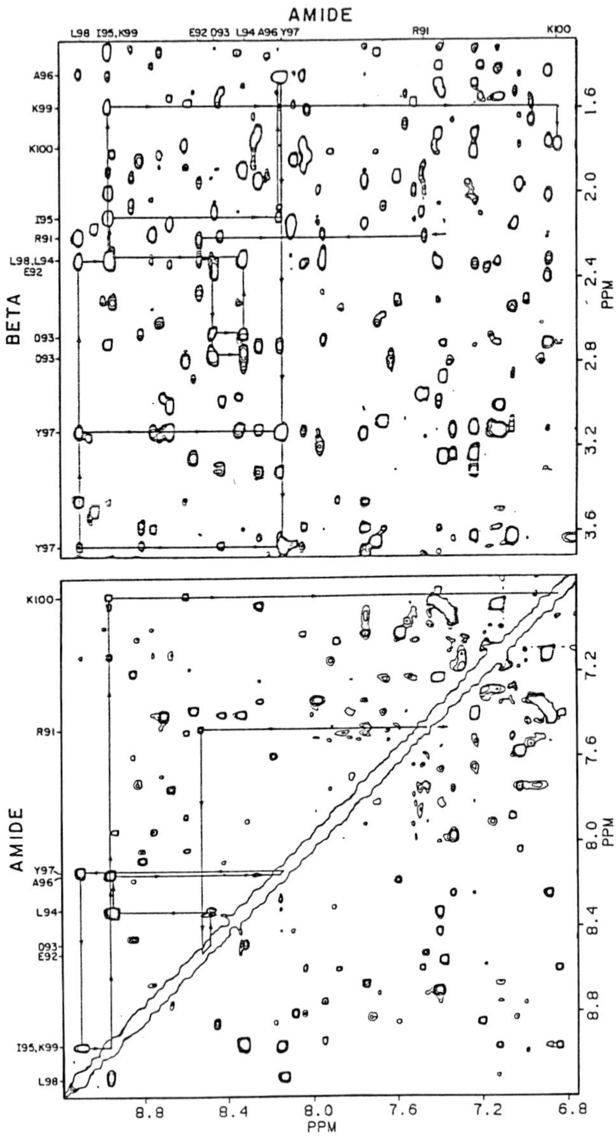

Fig. 8. Sections of a NOESY spectrum for reduced cytochrome c with connectivities that define the 90s helix.

types, can distinguish Glx (Glu or Gln), and can recognize the remaining 9 amino acid types (including the aromatics) as an AMX_2, NH-C_αH-$C_\beta H_2$ fragment. In working out spin systems, one often depends upon matching their chemical shift patterns with those found in free amino acids (13) and small peptides (14).

B. NOESY CONNECTIVITIES

In order to identify the spin systems found in J-correlated spectra, one relies on the NOESY spectrum, which can indicate through-space connectivities between protons in different residues that are brought within ~4 A of each other in secondary structural elements or in tertiary folding. As an example, the bottom panel of Fig. 8 shows the NOESY cross peaks that connect the sequential amide NHs in the 90s helix of reduced cytochrome c. The top panel shows confirming cross peaks that connect each C_βH with its own and the next residue NH and each NH with its own and the prior C_βH. The definition and confirmation of NOESY cross peaks between defined spin systems can lead to the identity of the parent amino acids. For example, the demonstration that several known spin system types are sequence neighbors is often enough to specify their identity upon comparison with the known amino acid sequence.

C. SOME PROBLEMS

The sequential assignment method has been used successfully to solve a number of small proteins, but the problems mount as larger proteins are attempted. One can begin to appreciate the problems by looking at the far upfield aliphatic side chain region shown in the COSY section in Fig. 5. Could the Met 80 cross peak network (Fig. 7) have been found if it appeared at +1 ppm rather than -1 ppm? The side chain COSY region (Fig. 5) is far more congested than the more downfield fingerprint region (Fig. 4) or even the far downfield NH-NH region in the NOESY (Fig. 8 bottom), even though the NOESY spectrum has overall about 3 times as many cross peaks as the COSY. The problem is compounded by the fact that side chain spin system networks are often not very well behaved. As an example, Fig. 9 compares the standard COSY cross peak network defined for glutamic acid in a simple oligopeptide (panel a) with some cross peak networks found for glutamic acid residues in several proteins (panels b-e), which show chemical shift inversions and missing cross peaks. In sizeable proteins, the density of side chain cross peaks in J-correlated spectra, the high degree of chemical shift degeneracy, and the plasticity of the patterns sought can effectively block a sequential assignment effort at the first step. Problems with degeneracy can also become difficult in attempting to trace out sequential connectivities in the NOESY spectrum (see e.g. Fig. 8).

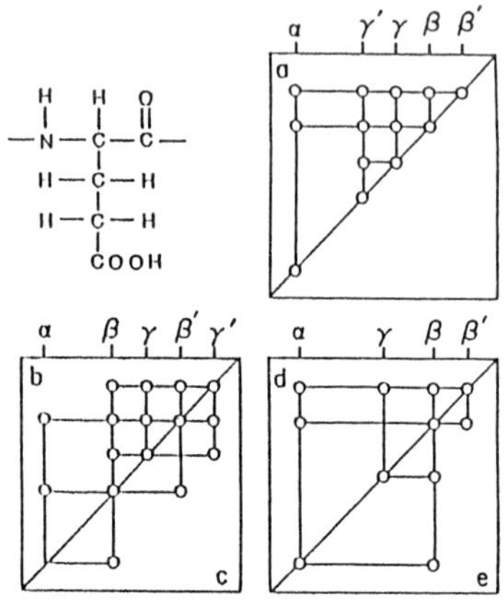

Fig. 9. Variable COSY cross peak patterns found for glutamic acid residues in an unstructured oligopeptide (a) and several proteins(b-e).

IV. THE MAIN CHAIN DIRECTED (MCD) ALGORITHM

For the reasons just outlined, we were unable to make reasonable progress with the sequential assignment of reduced cytochrome c (104 residues), and were driven to develop an alternative strategy. We put aside the search for side chains, and focussed on determining NH, $C_{\alpha}H$, and $C_{\beta}H$ resonances. These form a simple, linear subspin system that is common to all residues, occurs in the less congested downfield spectral region, and can be determined with high reliability. This turned out to provide the first step in an algorithmic main-chain-directed (MCD) approach to the assignment problem. The second step consists of a series of searches of the NOESY spectrum, with each search designed to find some defined NOE connectivity pattern that involves only the NH-$C_{\alpha}H$-$C_{\beta}H$ (NAB) protons. Each NAB set has 3 (or 4) protons and these provide a large target that leads to multiple inter-NAB connectivities. The particular patterns searched for involve linked connectivities that form cyclic, therefore self checking, patterns. The results of these searches organize the NAB sets into groupings based on secondary structural elements. The recognition of a few of the easiest side chain spin systems then places each linked NAB grouping within the protein amino acid sequence. This also determines the amino acid identity of all the rest of the NAB sets in that group. With foreknowledge of each amino acid type, one can then finally attack the side chain assignment problem with some confidence. The following discussion briefly displays some salient points of the MCD strategy.

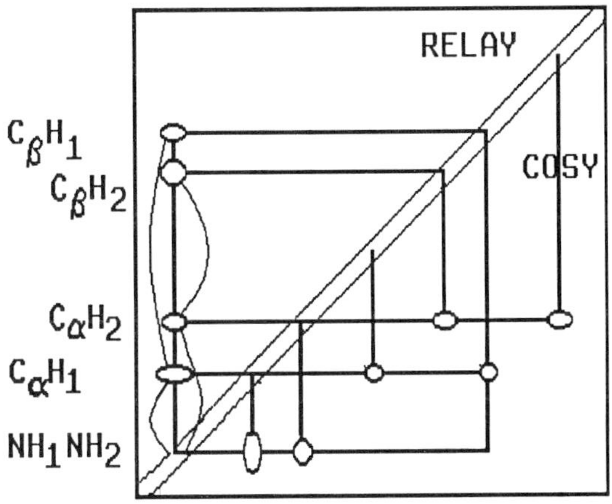

Fig. 10. Sections of hypothetical COSY and RCT spectra joined at their diagonals. Joint examination of the two spectra unscrambles the ambiguities due to a two-fold NH degeneracy and multiple $C_\alpha H$ degeneracies.

A. NAB SETS

NAB sets can be reliably found, for example by joint use of the COSY and relayed COSY spectra. The diagram in Fig. 10 joins sections from hypothetical COSY and RCT spectra at their diagonal and connects the pertinent cross peaks. In this example, two NH protons have degenerate chemical shift, so that two aligned NH-$C_\alpha H$ cross peaks appear in the COSY, and each $C_\alpha H$ appears to J-couple with two (or more) potential $C_\beta H$. The RCT spectrum, which aligns authentic $C_\alpha H$ and $C_\beta H$ resonances at the chemical shift of their NH resonance, is similarly ambiguous in respect to deciphering the true NAB sets. However examination of the two spectra resolves the ambiguities and selects out the correct sets (left as an exercise for the reader; the correct answer is at given at the left of Fig. 10). This approach can in principle pick out the correct NAB sets in the face of any degree of degeneracy.

When cross peak density becomes disruptive, as for larger proteins, the spectra can be simplified, and made to resemble the situation for smaller proteins, by taking advantage of the ability to make invisible some of the spin systems while retaining others. Since we are dealing with NH-linked spin systems, one can resort to a kinetic labeling approach (15) that uses H-D exchange to erase any desired number of amide NH resonances and their associated cross peaks. Here exchange of the protein in H_2O and D_2O is used to set a window such that only amide NH sites with H-exchange rates within any selected time scale are protonated and therefore NMR visible.

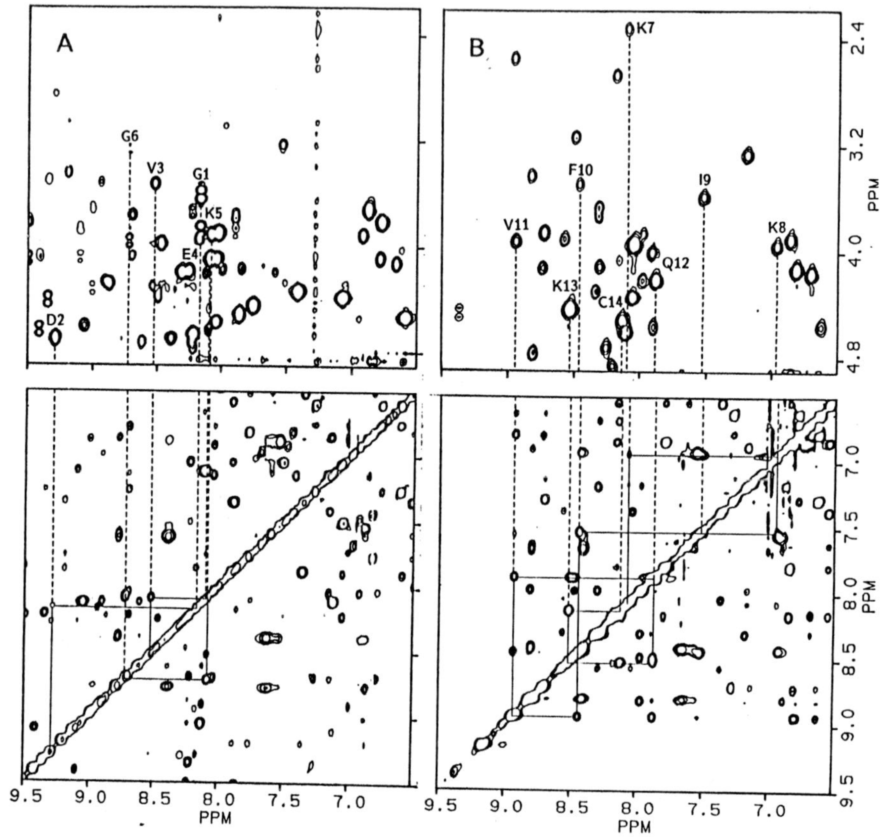

Fig. 11. Aligned sections of NOESY (lower panels) and COSY (upper panels) spectra for oxidized cytochrome c. Identical regions of the spectra are shown for samples A (left) and B (right). Different protons appear because A and B were selectively H-D exchange labeled. The fast exchanging NH and their respective cross peaks are present in A, while the slower H-bonded NH appear in B.

The value of this approach is suggested in Fig. 11, which aligns sections of NOESY and COSY spectra for samples of oxidized cytochrome c treated in this way. Differential H-D exchange was used to make visible in Fig. 11A those residues with fast exchanging NH (the N-terminal residues are noted) and in Fig. 11B only slow exchanging residues (H-bonded residues in the first helical segment are noted).

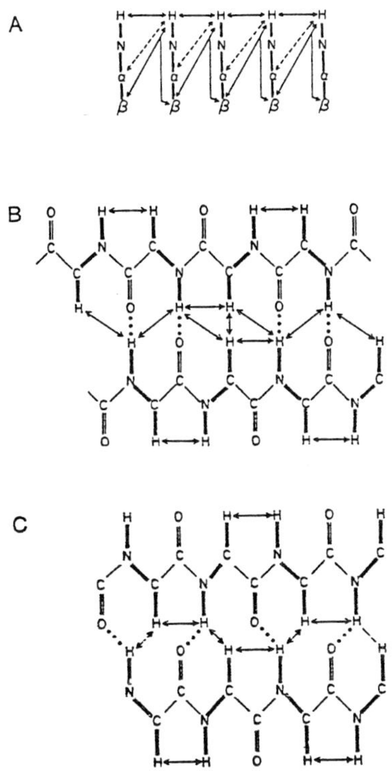

Fig. 12. MCD-defined closed loop patterns for helix (A), antiparallel sheet (B), and parallel sheet (C). J-coupled connectivities are shown as heavy lines and NOEs as arrows.

B. MCD PATTERNS

The second stage of the MCD algorithm attempts to organize NAB sets into groups that correspond mainly to particular secondary structural elements. Fig. 12 diagrams NOE patterns expected for residues in helix, antiparallel sheet, and parallel sheet. The heavy lines in Fig. 12 connect J-coupled members of an NAB set and the arrows represent NOE connectivities. The reliability (fidelity) and relative success ratio (frequency) of the various possible patterns were tested in a large scale statistical survey of known high resolution structures. As an example, suppose that the mixing time in a 2D experiment is set so that NOE cross peaks will appear for protons up to 4.0 A apart. The survey showed that a search of the 4 A cutoff NOESY spectrum for the "full loop" antiparallel sheet pattern (consisting of a joined "inner loop" - the trapezoid-shaped cycle in Fig. 12, and an "outer loop" - the roughly pentagonal cycle) will find 73% of all true "full" groups in the structure and the fidelity will be 100%, i.e. <u>all</u> the apparent patterns found will in fact represent protons in the structure arranged in this way. A search for the "double loop" helical pattern at a 4.0 A NOE cutoff will find 90% of all true double loop groups and have a 93% fidelity score. In each search, the cyclic nature of the patterns being sought helps to guarantee against error. For example, if one suspects that two NH-NH cross peaks represent a sequential helical segment (NH_i-NH_{i+1}-NH_{i+2}), then the helix double loop criterion requires that 5 other NOESY cross peaks with predefined coordinates must also be present ($C_\beta H_i$ to NH_i and NH_{i+1}, $C_\beta H_{i+1}$ to NH_{i+1} and NH_{i+2}, $C_\beta H_{i+2}$ to NH_{i+2}). Otherwise the hypothesis can be rejected.

The MCD-defined NOEs that identified the 90s helix in reduced cytochrome c are shown in Fig. 8. The NOE map of Fig 13 shows these and other NOE connectivities found to join the NAB protons in the 90s helix. The additional NOEs (e.g. $C_\alpha H_i$-NH_{i+3}, etc.), not utilized in the initial MCD search, provide an additional level of confirmation.

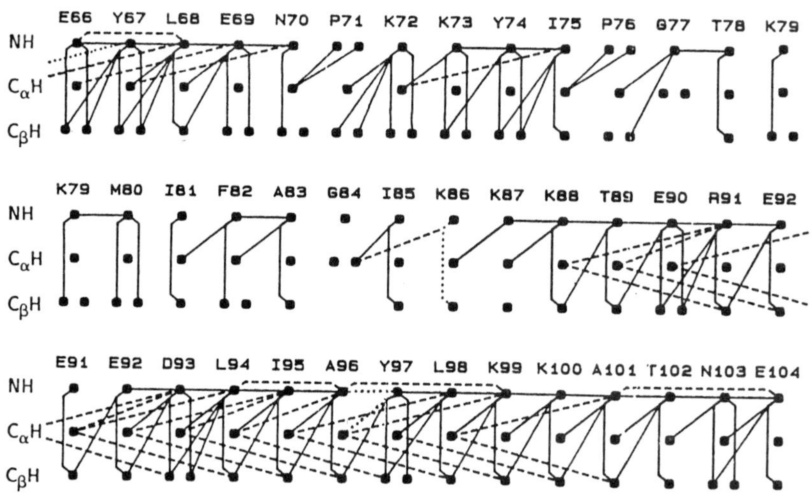

Fig. 13. NOE map for some of the NH, $C_\alpha H$ and $C_\beta H$ protons in reduced cytochrome c. Solid lines indicate NOEs within residues or between neighbors, dashed lines are medium range NOEs within secondary structural elements, dotted lines indicate ambiguities.

C. MCD SEARCHES

In the MCD algorithm, the different high fidelity-high frequency NOE patterns are sought one at a time in a prescribed order. Helices are sought first, while all other potential patterns are ignored. After the helix search, one removes from further consideration all confirmed helical cross peaks, and next searches for antiparallel sheet patterns. In turn these cross peaks are expunged, and a search is made for parallel sheet. These searches find also type II and type I reverse turns, respectively. Finally in the NOESY spectrum much simplified by the peak subtraction procedure, the difficult extended strand pattern, which has no cyclic, self-checking property (NH-C_αH only), is searched for.

At this point one has grouped the known NAB sets into self consistent assemblies, but has as yet no information about amino acid identity or type. To place these groupings in the polypeptide sequence, it is now necessary to move into the side chain region. Here one need only find a few of the easiest amino acid types, such as Gly, Ala, Thr and Ser, which can be done using the J-correlated spectra (e.g. Fig. 6). The identification of the amino acid types of several NAB sets within any sequential grouping is sufficient to place it in the amino acid sequence and thus specify the identity of all the residues in the sequence.

Finally one can attempt to assign side chain protons. The exercise just outlined predefines the amino acid associated with each NAB set. The TOCSY spectrum in Fig. 14 was able to identify some of the most difficult Lys and Arg spin systems in cytochrome c when the residue type of each was known in advance.

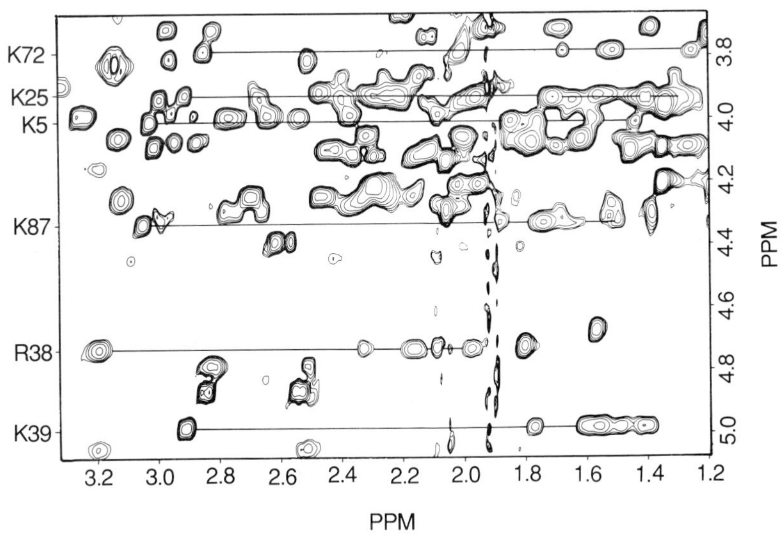

Fig. 14. TOCSY spectrum showing C_αH to side chain connectivities.

V. SUMMARY

The development of two dimensional NMR techniques has overcome the proton resolution problem for small proteins and provides information sufficent for assigning the great majority of the resonances to their parent protons. The effort to solve the assignment problem has so far been limited by the human problem of digesting and integrating the vast quantity of information provided by a set of 2D spectra. So far most successful assignment efforts have used the sequential assignment method, often together with a range of opportunistic approaches, with the operator depending on his/her own ingenuity at each step along the way. Ideally informational problems of this magnitude should be handled by computers. The main chain directed method outlined here of course uses the very same data utilized by all workers before, and it recommends defined steps that have undoubtedly also been used before along with innumerable other steps that the MCD method does not formally incorporate. The MCD method avoids a dependence on the difficult side chain protons and bases the analysis of NOE information on cyclic connectivity patterns involving the amide NH, $C_\alpha H$, and $C_\beta H$ protons. The MCD method is methodical, basically algorithmic in nature. Attempts are under way to computerize the exercise (16).

VI. REFERENCES

1. Jeener, F. Ampere Summer School, Basko Polje, Yugoslavia (1971).
2. Aue, W. P., Bartholdi, E, & Ernst, R. R. (1976) J. Chem. Phys. 64:2229-46.
3. Sorensen, O. W., Eich, G. W., Levitt, M. H., Bodenhausen, G. & Ernst, R. R. (1983) Prog. Nucl. Magn. Reson. Spectrosc. 16:163-192
4. Braunschwieler, L., Bodenhausen, G. & Ernst, R. R. (1983) Mol. Phys. 48:535-560.
5. Wider, G., Macura, S., Kumar, A., Ernst, R. R. & Wüthrich, K. (1984) J. Magn. Res. 56:207-234.
6. Redfield, C. & Dobson, C. M. (1988) Biochemistry 27:122-36.
7. LeMaster, D. M. & Richards, F. M. (1988) Biochemistry 27:142-50.
8. Wand, A. J., DiStefano, D. L., Feng, Y., Roder, H. & Englander, S. W. (1988) Biochemistry, in press.
9. Feng, Y., Roder, H., Englander, S. W., Wand, A. J. & DiStefano, D. L. (1988) Biochemistry, in press.
10. Wagner, G. & Wüthrich, K. (1982) J. Magn. Resn. 33:675-80.
11. Billeter, M., Braun, W. & Wüthrich, K. (1982) J. Mol. Biol. 155:321-346.
12. Wüthrich, K. (1983) Biopolymers 22:131-138.
13. McDonald, C. C. & Philips, W. D. (1973) Biochemistry 12:3170-86.
14. Bundi, A. & Wüthrich, K. (1979) Biopolymers 18:285-298.
15. Calhoun, D.B. & Englander,S. W. (1985) Biochemistry 24:2095-2100.
16. Wand, A. J. & Nelson, S. J., in progress.

22

SOLUTION STRUCTURE OF TGFα BY 2D ^{1}H NMR AND MOLECULAR MODELLING

Bradley A. Katz*, Marian Seto*, Rick Harkins*, James C. Jenson*, and Brian D. Sykes#

* Triton Biosciences Inc.
1501 Harbor Bay Parkway
Alameda, California 94501

NMR Group in Protein Structure & Function and Department of Biochemistry, University of Alberta, Edmonton, Alberta T6G 2H7 Canada

I. INTRODUCTION

Recent advances for solving three dimensional structures of small proteins by 2D NMR techniques are now providing guidance for molecular design in the biotechnology industry. Since the pioneering research started in 1977 by Wüthrich to develop strategies for determining the NMR solution structures of small proteins, NMR structures of macromolecules began appearing in the literature about 4 years ago. Recently there has been an explosion of NMR structures of proteins, peptides, carbohydrates, and oligonucleotides, along with improved techniques for obtaining and refining them.

In this article we describe a preliminary model of the solution structure of a small protein, human transforming growth factor alpha (TGFα) (1a, b, c), along with the methods used to determine and refine it: molecular modelling based on homology, high resolution proton 2 dimensional nuclear magnetic resonance spectroscopy (^{1}H 2D-NMR), computer graphics, molecular mechanics, and automated analysis of high resolution x-ray protein crystal structures. As the techniques in this field are quickly advancing, it is expected that by the time this article is published, some of the methods and programs described or referred to here will be somewhat outdated.

TGFα, a 50 amino acid mitogenic protein, is produced extracellularly by tumor cells, and also by virally or chemically transformed cells in culture (2). TGFα exerts its effects (or at least some of them) through binding to the epidermal growth factor receptor (EGFR) (for reviews of EGFR see references 3 and 4).

The molecular mechanisms underlying most cancer developments are at present poorly understood. However, it has been observed that both TGF*α* (1a) and EGFR (5-7) are overexpressed in tumours and tumour cell lines. Thus it is hoped that an understanding of the 3 dimensional structure of TGF*α* will aid in protein engineering efforts to develop both diagnostics and therapeutics for some cancers.

Gene Synthesis, Cloning, Folding, and Bioassays

The human TGF*α* gene was synthesized according to the DNA sequence reported by Derynck et al. (8), and cloned as a *trp* LE fusion protein in *E. coli*, containing an additional 15 N-terminal amino acids derived from the E gene of the tryptophan operon. A 3 residue segment, Glu-Phe-Met, was inserted between the *trp* E fragment and the TGF*α* coding sequence. The fusion protein was expressed in *E. coli* and purified to homogeneity by reverse phase HPLC following cell lysis and extraction. The fusion protein was cleaved with cyanogen bromide in 0.1 N HCl and mature TGF*α* was separated by HPLC. Recombinant TGF*α* purified in this manner is not fully biologically active because of mispairing of its 6 cysteines in improper disulfides. In order to generate properly folded TGF*α*, scrambled TGF*α* was fully reduced with DTT, and then allowed to refold by air oxidation. Under these conditions, properly folded TGF*α* is readily separable from inactive species by HPLC. The correct disulfide pairing of purified TGF*α* was confirmed by peptide mapping of the disulfide containing peptides by HPLC following digestion with thermolysin. The properly folded TGF*α* was shown to be fully active when tested for its ability to displace ^{125}I labelled murine EGF from the EGFR present on A431 cells, a human epidermal carcinoma cell line (9).

II. MODELLING BASED ON HOMOLOGY

There are several descriptions of, and strategies for modelling the structures of proteins based on homologous x-ray crystal structures (10-12). For TGF*α* we first modelled the structure based on careful examination of a published stereo diagram of the NMR structure of the homologous EGF molecule (13), and based on structural information on EGF contained in references 14 and 15.

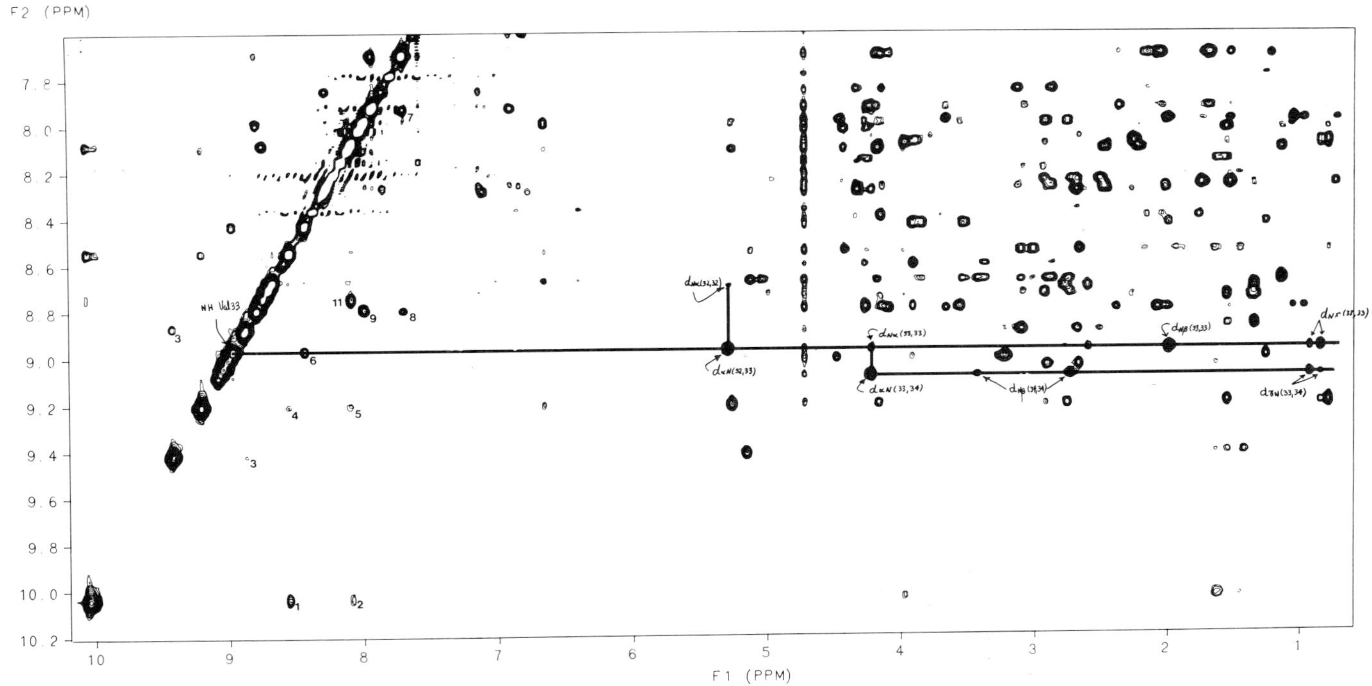

Figure 1.

Portion of the NOESY 2D Spectrum of TGFα in H_2O

III. MEASUREMENT OF 2D ^{1}H NMR STRUCTURAL INFORMATION

Six 2D-NMR data sets were collected at 500 MHz on a Varian VXR500 NMR spectrometer for a 3 mM solution of TGFα dissolved in 10 mM potassium phosphate buffer at pH 6.40 and 30 deg C. Three of these spectra were collected for the protein in a D_2O solution, and three were in an H_2O solution. Whereas in D_2O only the non-exchangeable CH protons are observed, in H_2O the exchangeable NH protons can also be observed. The three spectra collected are described by the acronyms DQF-COSY (double-quantum filtered COSY: an experiment which correlates nuclei which are coupled to one another through chemical bonds), RELAY (an experiment which relays the COSY correlations to additional nuclei which are also coupled to the reference ones through chemical bonds), and NOESY (an experiment which correlates nuclei that are close to one another in space). The above procedures are described in great detail in reference 16.

Of particular interest are the cross-peaks which correlate the resonance frequencies of the backbone NH and αCH protons on each amino acid. Under the experimental conditions used we observed 37 out of a possible 49 NH protons, demonstrating that about two thirds of the backbone NH protons are protected from exchange by the structure of the protein, ie they either make hydrogen bonds or are buried.

The second stage of the assignment relies on the internuclear contacts observed in the NOESY spectra to determine the sequential conductivities along the polypeptide chain. Fig. 1 shows the region of the NOESY 2D-spectrum in H_2O, correlating the NH resonances (7.6 - 10.2 ppm) with the other NH resonances and the αCH, $\beta\beta'CH_2$, CH_3 and other resonances (0.6 - 5.6 ppm).

As an example, the NOE contacts which correspond to the sequence Cys 32, Val 33, Cys 34 are highlighted in Fig. 1 and depicted in Fig. 2. The sequential NOE cross peaks observed for this section of the polypeptide chain include $d_{N\alpha}$[32,33], $d_{N\alpha}$[33,33], & $d_{N\alpha}$[34,34] (each of which has a corresponding peak in the DQF-COSY spectrum), and $d_{\alpha N}$[32,33] & $d_{\alpha N}$[33,34]. These are indicated by the thick double headed arrows in Fig. 2. Other peaks include $d_{N\beta}$[33,33] & $d_{N\gamma}$[33,33] for Val 33, $d_{\gamma N}$[33,34] between Val 33 & Cys 34, and $d_{N\beta}$[34,34] for Cys 34.

Once the NOESY cross-peaks have been correlated with the amino acid sequence, other cross-peaks such as those representing the NH, NH contacts (some of which are labelled in Fig. 1), and the αCH, αCH contacts (not shown), help to define the secondary structure of TGFα.

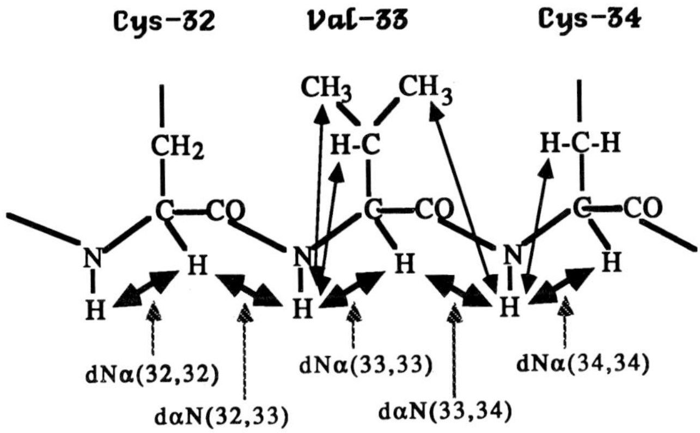

Figure 2.

Depiction of the NOE contacts corresponding to the sequence Cys 32, Val 33, Cys 34.

IV. MODEL BUILDING VIA INTERACTIVE COMPUTER GRAPHICS AND MOLECULAR MECHANICS

An initial model of TGF*α* was built (by referring to the EGF structure) with BIOGRAF software (17) on an Evans & Sutherland color computer graphics system (PS 330). As the model was built its potential energy was minimized using the united AMBER force field (18). Energy minimization removes local strain and restrains bond lengths, bond angles, dihedral angles, hydrogen bonds, non-bonded and hydrophobic interactions to 'ideal' or standard values.

Energy minimization is a non linear, iterative process that provides a local minimum in the 'conformational space' of the protein, but rarely the global minimum which is sought (19-21). Thus the converged model after several steps of energy minimization is similar in structure (within about 1 Å) to the initial model. Since some native active 3 dimensional protein structures may be determined in part kinetically during folding, determination of the global minimum in potential energy may not always be equivalent to determination of the native active conformation of a protein (21). Indeed, there are protein structures that exhibit regions of local strain presumably to aid in their function (22).

Local minima can be surmounted by simulating the motions and dynamics of the atoms in the protein molecule. The collection of atoms is assigned velocities with a Boltzmann distribution so that a temperature is defined. Newton's third law of motion is integrated numerically, atomic positions and kinetic and potential energies are calculated, and the structure of a molecule that was trapped in a local minimum can evolve with time to one of lower potential energy. The molecular dynamics calculations are time consuming and usually only short simulations (about 100 picoseconds) are practical.

V. INCORPORATION INTO THE MODEL OF NMR RESTRAINTS AND REFINEMENT OF THE TGFα STRUCTURE

The initial model of TGFα was refined by cycles of energy minimization, molecular dynamics, and model building. As NMR distance and dihedral restraints were determined, they were incorporated into the model as pseudopotential energy terms (23). Thus the structure is determined in part theoretically and in part experimentally. An explicit all atom force field that includes hydrogens (24) was used when the NOE restraints were incorporated. Non-bonded parameters were taken from reference 25. A program was written to compare observed and calculated distance (or dihedral) restraints, to calculate R-factors (or rms deviations from restrained dihedrals) and to analyze distance restraints as a function of their proximity to the proton pair of a flagged restraint. Such an analysis proved useful in identifying misassigned restraints: the spectra were reanalyzed in the light of the R-factor analysis, and a few restraints were reassigned and reincorporated into the model.

We found that the 'tertiary' restraints (eg Phe 5 close to Val 25) had to be carefully 'manually' modelled before running the molecular mechanics calculations or the restraints would force transitions of peptide bonds from trans to cis as well as amino acids from L to D. Indeed, the process of solving a protein structure by 2D NMR is by no means automatic, and, perhaps not surprisingly, the molecular mechanics programs can not be treated as black boxes. The capability to interactively observe the structural changes during molecular mechanics calculations and other computer graphics manipulations was found to be critical.

VI. DESCRIPTION OF THE 3 DIMENSIONAL STRUCTURE OF TGFα

The compact structure of TGFα, homologous to that of EGF, is highly constrained by 3 disulfide crosslinks. TGFα contains 3 domains: an N-terminal domain, a C-terminal domain, and a domain comprising residues 8 through 16 (the top region of the molecule as it is depicted in Fig. 3. The N-terminal domain comprises a stretch of antiparallel β-sheet. TGFα differs from EGF in that Pro 30 in TGFα (Tyr in EGF) inturrupts the β-sheet structure. In the present model it forms a β-bulge, and after the bulge (toward the left hand part of the diagram) there could be an additional hydrogen bond between Leu 24 NH and O=C Asp 28. The portion between the 2 strands of sheet, Val 25 through Asp 28, appears to comprise an Ω loop (26) structure. In the model there is a salt bridge between Arg 22 and Asp 28. The first 4 residues of the N-terminus appear to be flexible and disordered.

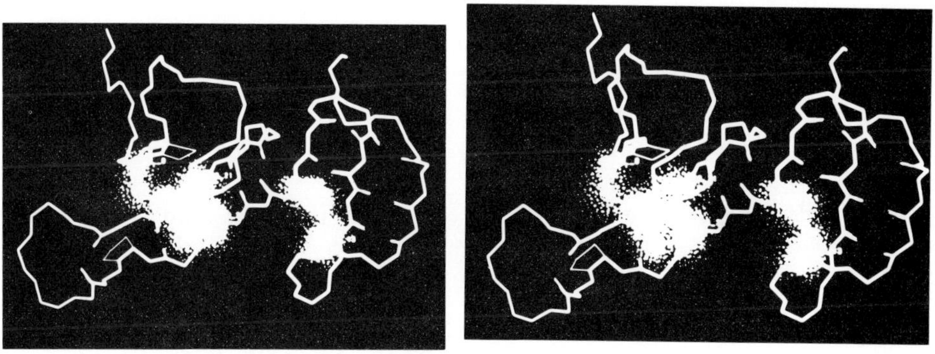

Figure 3.

Model of backbone of TGFα showing regions of β-sheet and disulfide bonding pattern

The C-terminal portion of TGFα is made up of an antiparallel triple stranded β-sheet structure. In the present model there are hydrogen bonds between His 35 NH and O=C Tyr 38, His 35 C=O and HN Tyr 38, Val 39 C=O and HN His 45, Val 39 NH and O=C His 45, and Asp 47 NH and O=C Gly 37. There is a salt bridge between Arg 42 and Glu 44. The C-terminal residues 48 through 50 appear to be flexible and disordered.

VII. MODELLING REGIONS OF TGFα BASED ON AUTOMATED BROOKHAVEN DATA BASE ANALYSIS

A complemenatary approach that we are using to determine and/or confirm the structure of regions of TGFα is to try to correlate amino acid sequences with structural elements based on the Brookhaven x-ray crystallographic data base (27). For example it was found that the sequence SerGly is found much more often in a type II β-turn than in a type I β-turn. Based on this information the Ser36Gly37 turn in TGFα (Fig. 4) would be more likely to comprise a type II β-turn (as it does in the EGF structure). Other turns in TGFα such as Ala41Arg42 are currently under analysis (Fig. 4)

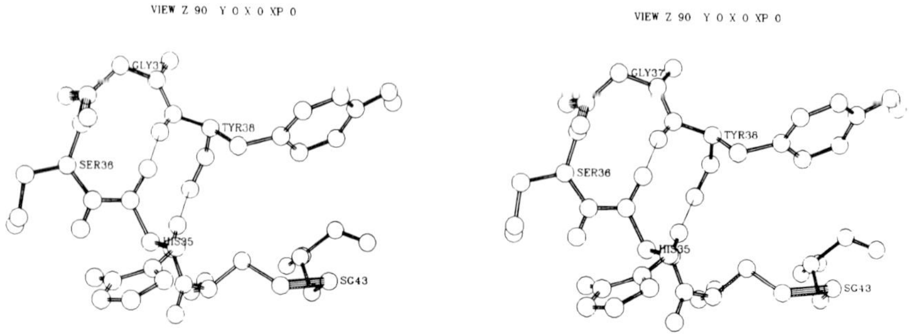

Figure 4.

Portion of the C-terminal region of the molecule showing type II β-turn accommodated by Ser36 and Gly37

VIII. METHODS UNDER DEVELOPMENT FOR FURTHER REFINEMENT AND ANALYSIS OF THE TGFα STRUCTURE

New methodology for determining solution structures of proteins from NMR data is appearing in the literature (28-33). Further automated Brookhaven data base analysis will yield more insight into the structure of TGFα and into protein structure in general. It is hoped that improvements in the software for correctly implementing NOE restraints involving mobile protons (such as those in rapidly flipping phenyl rings or in rotating methyl groups) may help with structural refinement.

Clearly, molecular mechanics simulations on a protein model should be more instructive if the surrounding solvent is included in the calculations. One major advantage is an appreciation of potential locations of ordered water molecules, which in part determine the structure. Most published NMR structures are really a family of structures, each of which is low energy and satisfies equally well the NOE restraints. The ambiguity may arise because there is flexibility in the protein or because the NMR data is insufficient to unamiguously define some regions (or both). We are using a distance geometry program DISGEO (16, 13-38) to determine a number of structures of TGFα each one equally compatible with the current data. The near future in the field of NMR structural determination in conjunction with molecular modelling promises to be both quickly evolving and exciting. Molecular modelling is fast becoming less of an art and more of a science.

IX. ACKNOWLEDGEMENTS

We are grateful to John Brandis and Hsiao-Lai Liu for the gene synthesis, cloning and expression of TGF-*α*.

X. REFERENCES

1a. R. Derynck J. Cell. Biochem. **32**, 293-304 (1986).
1b. G.L. Todaro, J.E. De Larco, and S. Cohen, Nature, **264**, 26-31, 1976.
1c. J.P. Tam, M.A. Maqsood, D.S. Soloman and L. Ossowski, *Proc. Natl. Acad. Sci. USA*, **83**, 8082-8086 (1986).
2. J.E. De Larco and G.J. Todaro, *Proc. Natl. Acad. Sci. USA*, **75**, 4001-4005 (1978).
3. Y. Yarden and A. Ullrich, BIOCHEM., **27**, 3113-3118 (1988).
4. J. Schlessinger, BIOCHEM., **27**, 3119-3123 (1988).
5. R. Jungsil, S.M. North, G.E. Gallick, G.N. Hortobagyi, J.U. Gutterman, and M. Blick, CANCER RESEARCH, **48**, 161-164 (1988).
6. M. Haeder, M. Rotsch, G. Bepler, C. Henning, K. Havemann, B. Heimann, and K. Moelling, CANCER RESEARCH **48**, 1132-1136 (1988).
7. C.R. Carlin, D. Simon, J. Mattison and B.B. Knowles, Mol. Cell. Biol., **8**, 25-34 (1988).
8. R. Derynck, A.B. Roberts, M.E. Winkler, E.Y. Chen, D.V. Goedell, Cell, **38**, 287-297 (1984).
9. J. Massague, J. Biol. Chem., **258**, 13614-13620 (1983).
10. T.L. Blundell, B.L. Sibanda, M.J.E. Sternberg and J.M. Thornton, Nature, **326**, 347-352 (1987).
11. T.A. Jones and S. Thirup, EMBO, **5**, 819-822 (1986).
12. N.L. Summers, W.D. Carlson and M. Karplus, *J. Mol. Biol.* **196**, 175-198 (1987).
13. R.M. Cooke, A.J. Wilkinson, M. Baron, A. Pastore, M.J. Tappin, I.D. Campbell, H. Gregory and B. Sheard, NATURE, **327**, 339-341 (1987).
14. G.T. Montelione, K. Wuthrich, E.C. Nice, A.W. Burgess and H.A. Scheraga, *Proc. Natl. Acad. Sci. USA*, **83**, 8594-8598 (1986).
15. K. Makino, M. Morimoto, M. Nishi, S. Sakamoto, A. Tamura, H. Inooka, and K. Akaska, *Proc. Natl. Acad. Sci. USA*, **84**, 7841-7845 (1987).
16. K. Wuthrich in *'NMR of Proteins and Nucleic Acids'*, John Wiley & Sons, NY (1986).
17. BIOGRAF is the name of a program available from BIODESIGN Inc, 199 South Los Robles Avenue, Suite 270, Pasadena, CA 91101.
18. S.J. Weiner, P.A. Kollman, D. Case, U.C. Singh, C. Gio, G. Alagona, S. Profeta, Jr. and P. Weiner, J. Amer. Chem. Soc. **106**, 765-784 (1984).

19. P. Kollman and W.F. van Gunsteren, in *'Methods in Enzymology'*, R. Wu and L. Grossman, Editors, Vol **154**, pages 430-449 (1987), Academic Press, Inc.
20. W.F. van Gunsteren, Protein Engineering, **2**, 5-13 (1988).
21. G.M. Crippen and P.K. Ponnuswamy, J. Comp. Chem., **8**, 972-981 (1987).
22. P.A. Karplus and G.E. Schulz, *J. Mol. Biol.* **195**, 701-729 (1987).
23. M. Nilges, A.M. Gronenborn, A.T. Brunger and G.M. Clore, Protein Engineering, **2**, 27-38 (1988).
24. S.J. Weiner, P.A. Kollman, D.T. Nguyen and D.A. Case, J. Comp. Chem., **7**, 230-252 (1986).
25. W.L. Jorgensen and J. Tirado-Rives, J. Amer. Chem. Soc. **110**, 1657-1666 (1988)
26. J. F. Leszczynski and G. D. Rose, Science, 234, 849-855 (1986).
27. F.C. Bernstein, T.F. Koetzle, G.J.B. Williams, E.F. Meyer, M.D. Brice, J.R. Rodgers, O. Kennard, T. Shimanouchi and M. Tasumi, *J. Mol. Biol.* **112**, 535-542 (1977).
28. O. Lichtarge, C.W. Cornelius, B.G. Buchanan and O. Jardetzky, PROTEINS **2**, 340-358 (1987).
29. P.J. Kraulis and T.A. Jones, PROTEINS **2**, 188-201 (1987).
30. M. Nilges, G.M. Clore, and A.M. Gronenborn, FEBS LETT, **229**, 317-324 (1988).
31. A.G. Anderson and J. Hermans, PROTEINS, **3**, 262-265 (1988).
32. H. Oschkinat, C. Griesinger, P.J. Kraulis, O.W. Sorensen, R.R. Ernst, A.M. Gronenborn, and G.M. Clore, Nature, **332**, 374-376 (1988).
33. C. Griesinger, O.W. Sorensen, and R.R. Ernst, J. Amer. Chem Soc., **109**, 7227-7228 (1987).
34. G.M. Clore and A.M. Gronenborn, Protein Engineering, **1**, 275-288 (1987).
35. G. Wagner, W. Braun, T.F. Havel, T. Schaumann, N. Go, and K. Wuthrich, *J. Mol. Biol.* **196**, 611-639 (1987)
36. G.M. Clore, A.T. Brunger, M. Karplus, and A.M. Gronenborn, *J. Mol. Biol.* **191**, 523-551 (1986).
37. J. de Vlieg, R.M. Scheek, W.F. van Gunsteren, H.J.C. Berendsen, R. Kaptein, and J. Thomason, PROTEINS **3**, 209-218 (1988).
38. G.M. Clore, A.M. Gronenborn, M. Nilges, and C.A. Ryan, Biochem., **26**, 8012-8023 (1987).

23

STUDY OF A PHOSPHORYLATED PROTEIN BY TWO-DIMENSIONAL NMR SPECTROSCOPY[1]

Michael G. Wittekind and Rachel E. Klevit, Department of Biochemistry, University of Washington, Seattle, WA 98195

E. Bruce Waygood, Department of Biochemistry, University of Saskatchewan, Saskatoon, Saskatchewan S7N 0W0

I. INTRODUCTION

The application of two-dimensional ^{1}H NMR (2DNMR) spectroscopy to the study of proteins has been shown to be capable of yielding high resolution structures of proteins in solution (for a recent review, see Kaptein *et al.*, 1988). In addition to using 2DNMR methods to obtain solution structures of small, soluble proteins, a particularly exciting application of these powerful new techniques involves the study of species that are not long-lived enough for crystals to grow. In one example of such possibilities, we have studied the phosphorylated form of a phosphotransfer protein, HPr, by 2DNMR.

HPr is one of the protein components in the phosphoenolpyruvate:sugar phosphotransferase system in bacteria. The mechanism by which a phosphoryl group is transferred from phosphoenolpyruvate to hexoses involves a series of phosphoprotein intermediates, each of which is phosphorylated on a histidine residue. *E. coli* HPr is phosphorylated at one of its two histidine residues, His-15. One dimensional ^{1}H NMR (1DNMR) measurements showed that the pK_a of His-15 changes from an unusually low value of 5.6 in the unphosphorylated protein to an unusually high value of 7.8 in the phosphorylated protein (Dooijewaard *et al.*, 1979). These pK_as seem optimally suited to allow HPr to perform as both a phosphoryl group acceptor and a phosphoryl group donor. At neutral pH, unphosphorylated His-15 will be deprotonated due to its low pK_a, making it a good nucleophile, while phosphorylated His-15 will be protonated at pH 7, making the phosphoryl substituent a better leaving group. Since

[1]Supported by National Institutes of Health Grant RO1 DK35187 and an operating grant from the Medical Research Council of Canada

phosphorylation of an imidazole ring as such does not result in a change in pK_a (Jencks & Gilchrist, 1965), the environment of the phosphorylation site in HPr must be responsible for this unusual property. 2DNMR studies of HPr in both the unphosphorylated and phosphorylated forms have been conducted in order to obtain detailed structural information.

HPr from *E. coli* has 85 residues and is soluble at millimolar concentrations, making it well-suited for study by 2DNMR techniques. Complete sequence-specific assignments for the spectrum of the unphosphorylated form have been reported and a model for the three-dimensional structure of the protein has been built, based on the observed nuclear Overhauser effects (NOEs) (Klevit *et al.*, 1986; Klevit & Waygood, 1986). The topology of the protein consists of a four-stranded anti-parallel β-sheet with three α-helices crossing over one side of the sheet. Determination of the structure using the distance geometry algorithm is currently in progress and the preliminary structures all possess the same topology as the original model based on a qualitative analysis of the observed NOEs (R. Klevit, J. Herriott, and D. Hare, unpublished observations).

II. RESULTS AND DISCUSSION

The phosphorylation site has His-15 in close proximity to an arginine two residues away in the primary structure. These two residues are on the surface of the protein. Two glutamate residues, Glu-83 and Glu-85 (the C-terminal residue of HPr) give NOEs to His-15 and Arg-17 and lie below the phosphorylation site in the model. The presence of one or more of these carboxylate groups is thought to be responsible for the observed pH dependence of the rate of phosphohydrolysis (Waygood *et al.*, 1985). As shown in Figure 1, the rate decreases with an apparent pK_a of 3.5, suggestive of the involvement of a carboxylate group. At pH values above neutral, the rate decreases with increasing pH, with an apparent pK_a of 7.8, the pK_a of 1-P-His-15 (Dooijewaard *et al.*, 1979). While this property is an extremely interesting aspect of HPr, it is also what makes performing structural studies on P-HPr so difficult.

The 2DNMR spectra used to determine the structure of unphosphorylated HPr were obtained at pH 6.5, 30°C. In general, pH values below neutral are used in 2DNMR studies of proteins in order to slow amide proton exchange rates. As shown in Figure 1, the rate of hydrolysis of 1-P-His-15-HPr (P-HPr) is maximal at pH 6.5, with a half-life of only 5 minutes at 37°C--just barely enough to obtain a simple 1DNMR spectrum. The shortest 2DNMR experiment we wished to perform was a magnitude COSY, which can be collected in about 10-11 hours for a sample in D_2O. Furthermore, although a COSY spectrum gives useful information regarding chemical shift perturbations due to the phosphorylation, the information necessary to assess conformational changes is contained in a NOESY spectrum, which requires at least 16

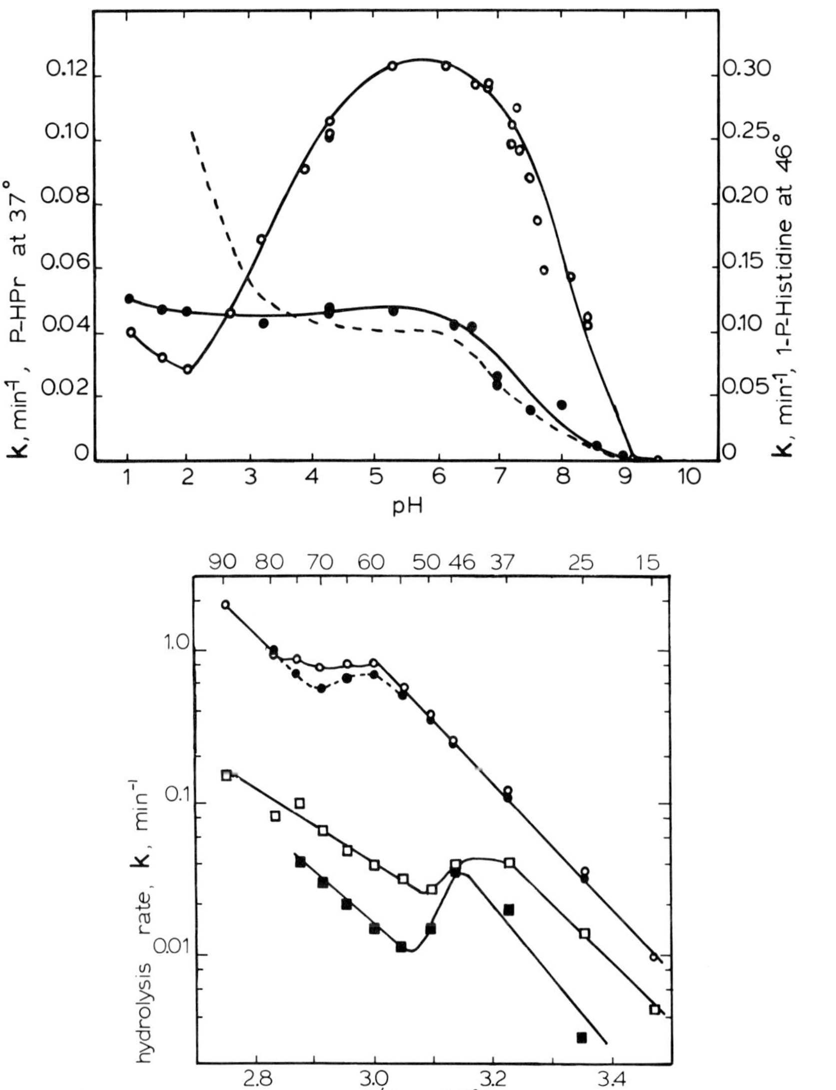

Fig. 1. (a) pH dependence of the rate of hydrolysis of PHPr. Open circles, native HPr; Closed circles, denatured HPr; Dashed line, 1-phosphohistidine. (b) Temperature dependence of the hydrolysis of 1-P-histidnyl residue in phospho-HPr and phospho-HPr-1 (frozen). Hydrolysis rates of the 1-P-histidinyl residue were determined at pH 6.3 (circles) and pH 8.4 (squares) for both P-HPr (open symbols) and P-HPr-1 (closed symbols) for the temperatures over the range 15°C - 90°C. The lower T is °K, the upper scale is °C. Taken from Waygood et al, 1985. Also see this reference for a description of Phospho-HPr-1.

hours to acquire. Thus, it was necessary to find conditions under which P-HPr could be maintained in the NMR tube for the required time and would still give a high quality NMR spectrum.

We reasoned that a half-life approximately equal to the duration of the experiment to be performed should be sufficient. Under such conditions, the sample would still be 50% phosphorylated at the end of the data collection. Since peak intensities decrease due to T_1 relaxation as the 2DNMR experiment progresses, the reappearance of unphosphorylated protein during the last half of the experiment should not result in significant contribution to the spectrum. The pH-dependence of phosphohydrolysis indicated that working at pH's higher than pH 6.5 would prolong the half-life of the P-HPr in solution. Since it had been reported that P-HPr undergoes irreversible denaturation at high pH (Dooijewaard *et al.*, 1979), we were wary of working at pH values greater than pH 8.5. Increasing the pH of the NMR sample from pH 6.5 to pH 8.3 would result in an increase in the half-life of P-HPr from 5 minutes to aproximately 14 minutes--a factor of 3, but still not yet approaching a useful value.

The temperature dependence of the rate of phosphohydrolysis was measured and was found to decrease with decreasing temperature (Figure 1). However, the linewidths of the NMR spectrum of HPr increase with decreasing temperature, compromising the quality of the 2D spectra. Thus, using temperatures near freezing was not a viable alternative. Since a decrease in temperature from 37°C to 15°C results in a 10-fold decrease in the rate of hydrolysis while not significantly increasing the NMR linewidths, the 2DNMR experiments were performed at 15°C. Based on the pH- and temperature-dependence studies, we estimated that P-HPr at pH 8.3, 15°C would have a half-life greater than 2 hours.

In an early 1DNMR study, Dooijewaard *et al.* (1979) had shown that the presence of a catalytic amount of the phosphorylating enzyme, Enzyme I, in the NMR sample was undetectable in the resulting spectrum. Furthermore, they showed that the presence of 5mM pyruvate (the product of the phosphorylation reaction) does not inhibit the regeneration of P-HPr. Therefore, a catalytic amount (1U) of purified Enzyme I and a 5-10 molar excess of phosphoenolpyruvate were included in the NMR tube as a regeneration system. As in the earlier 1DNMR work, no lines from the high molecular weight Enzyme I were detected. The two resonances due to phosphoenol-pyruvate did not interfere with the protein spectrum and served as a useful monitor of the status of the sample during the experiment. The phosphoenolpyruvate was usually completely converted to pyruvate during the course of a 2DNMR experiment. Since the conditions required to study P-HPr were different from those used in our original studies, control spectra were taken of the unphosphorylated form of HPr under identical conditions. In practice, spectra were obtained consecutively. Since the phosphorylation reaction is slower at 15°C, complete phosphorylation was obtained within minutes after adding Enzyme I to the NMR tube at 37°C. The sample was then cooled to 15°C and a 2DNMR spectrum was collected. Immediately after, the sample was

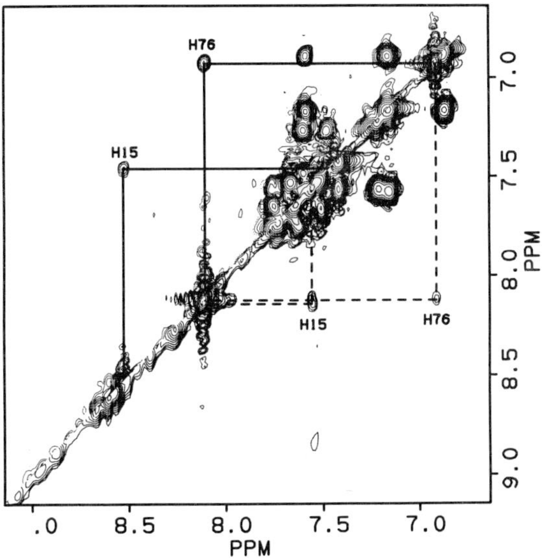

Fig. 2. The region of the COSY (HPr, lower right) and RELAY (PHPr, upper left) spectra showing the aromatic resonances. The resonance of the C2 proton of His-15 moves downfield upon phosphorylation at the N1 position of the imidazole ring.

allowed to completely dephosphorylate (by warming to 37°C for several minutes) and the control 2DNMR spectrum was obtained on the identical sample. In this way, spectra were directly comparable. The 2DNMR spectrum of HPr at pH 6.5 and pH 8.3 were nearly identical and the sequence-specific assignments were directly transferrable to the control spectrum.

COSY spectra were analyzed first to ascertain the degree of chemical shift perturbation present in the P-HPr spectrum. Although a significant number of cross peaks show a reproducible chemical shift difference, the majority of cross peaks are superimposable in the two spectra, signifying that phosphorylation does not result in a global change in the protein. Indeed, when the resonances that are shifted in the P-HPr spectrum are plotted on the three-dimensional model, they all cluster locally around the phosphorylation site. Figure 2 shows a small region of the two coherence transfer spectra, illustrating both shifted and non-shifted cross peaks. Since the majority of peaks did not shift, it was not necessary to reassign the P-HPr spectrum; even the shifted resonances could be assigned by inspection.

Once the resonance positions in the P-HPr spectrum had been assigned, the NOESY spectra could be analyzed. As with the COSY spectra, the majority of cross peaks are identical, confirming the conclusion that the overall structure of HPr is not changed upon phosphorylation. New

NOE cross peaks, indicative of new close through-space interactions have been identified for protons in the phosphorylation site, most notably involving protons of His-15, the residue that is phosphorylated. On the basis of these observations, we conclude that phosphorylation of His-15 causes a local change in conformation at the phosphorylation site. Complete analysis of the NOESY spectra obtained with different mixing times will allow a detailed description of the nature of this change.

We feel that the approach described in this paper has general applications to other systems involving species with limited half-lives. If the system can be manipulated through use of a regeneration system to extend the half-life of the species of interest to about 10 hours, 2DNMR spectra can be acquired that will allow for the determination of the structure of such species in solution.

III. REFERENCES

Dooijewaard, G., Roosien, F. F., and Robilliard, G. T. (1979) Biochemistry 18, 2996.

Jencks, W. P. & Gilchrist, M. (1965) J. Am. Chem. Soc. 87, 3199.

Kaptein, R., Boelens, R., Scheek, R. M., and van Gunsteren, W. F. (1988) Biochemistry 27, 5389.

Klevit, R. E. and Waygood, E. B. (1986) Biochemistry 25, 7774.

Klevit, R. E., Drobny, G. D., and Waygood, E. B. (1986) Biochemistry 25, 7760.

Waygood, E. B., Erickson, E., El-Kabbani, O. A. L., and Delbaere, L. T. J. (1985) Biochemistry 24, 6938.

24

ANALYTICAL ASPECTS OF 2D ^{1}H NMR: PURITY VERSUS CONFORMATIONAL HETEROGENEITY IN CALBINDIN D_{9k}[1]

Johan Kördel, Sture Forsén and Walter J. Chazin[2]

Department of Molecular Biology
Res. Inst. of Scripps Clinic
La Jolla, CA 92037, USA

and

Department of Physical Chemistry 2
Chemical Centre, University of Lund
S-221 00 Lund, Sweden

I. INTRODUCTION

In the past ten years Nuclear Magnetic Resonance (NMR) has become an increasingly important tool for the study of proteins in solution. The development of two-dimensional (2D) ^{1}H NMR techniques has had the greatest impact because these experiments provide a means for completely assigning the complex ^{1}H NMR spectra and for determining the three-dimensional structure of small proteins (<15 kDa, to date) in solution (1).

In the developmental stages of applying 2D ^{1}H NMR to the study of proteins much of the attention has been focussed on working towards the three-dimensional structure of small proteins. There are however a number of important and very

---

[1]This work was supported by an operating grant (to W.J.C.) from the National Institutes of Health and by a graduate fellowship (to J.K.) and operating grants (to S.F.) from the Swedish National Science Research Council.

[2]Address correspondence to this author at the Research Institute of Scripps Clinic.

useful applications of 1D and 2D NMR methods to problems in analytical aspects of protein chemistry. The most obvious application is the determination of purity. ^{1}H NMR can readily provide information at the level of a few percent on the homogeneity of protein preparations, the presence of free peptides produced in the course of purification, the occlusion of small molecules in proteins produced by recombinant DNA technology, and on the amount of residual buffer components. In the course of detailed analysis, ^{1}H NMR can also provide a check on the chemical determination of amino acid composition and protein sequence (1,2).

^{1}H NMR also offers a very rapid and powerful method for comparative analysis of different protein preparations, chemically modified derivatives, and homologous proteins. Even without specific resonance assignments, straightforward comparison of 1D and 2D spectra can be utilized to establish the relative similarity of different protein solutions. With sequence-specific assignments, it is possible to obtain direct information on the nature of the differences (eg., Ref. 3). This has vast potential as a screening method for studies utilizing protein mutagenesis. Complete sequence-specific ^{1}H NMR assignments also provide sufficient information for the identification and characterization of multiple conformations of pure proteins. In the following we briefly describe ^{1}H NMR studies of calbindin D_{9k} which demonstrate that this protein exhibits two native conformations in solution (W.J. Chazin, J. Kördel, T. Drakenberg, E. Thulin, P. Brodin, T. Grundström & S. Forsén, submitted). By obtaining complete sequence-specific assignments we were able to demonstrate that this effect arises from *cis-trans* isomerism of Pro 43.

Calbindin D_{9k} is a member of the troponin C superfamily of calcium regulatory proteins that have common helix-loop-helix calcium binding sites known as EF-hands (4). It is an attractive model system for exploring the relationship between structure, dynamics and function in the EF-hand family, because it is the smallest protein known (M_r 8500, 75 amino acids) with a pair of EF hands and its structure resembles closely the globular domains of the homologous proteins calmodulin, parvalbumin and skeletal muscle troponin C. The protein used in our studies is produced by a synthetic bovine calbindin gene expressed in <u>E. coli</u> (r-calbindin D_{9k})(5), or isolated from bovine intestines. These studies were originated as a first step towards our long term goal of determining the conformational basis for function of calcium regulatory proteins, by examining the changes that accompany binding of the Ca^{2+} ion. These studies will form the background for further analysis of the factors determining calcium affinity and specificity in this class of proteins, utilizing the power of site-directed mutagenesis.

II. MATERIAL AND METHODS

Recombinant calbindin D_{9k} was purified as described by Linse *et al.* (6). Protein from bovine intestine was purified according to Hitchman, Kerr & Harrison (7). Samples were made up 1 to 4 mM in protein at pH 6.0, with no added buffer. ^{1}H NMR experiments were performed at 300 and 330 K with Bruker AM-500 spectrometers equipped with Aspect 3000 computers and digital phase shifting hardware. 2D NMR spectra were recorded in the phase-sensitive mode with quadrature detection in the ω_1 dimension achieved using time-proportional phase incrementation (8). A total of 2048 quadrature pairs of data points were collected in the sequential acquisition mode for each t_1 value. In scalar correlated experiments that give rise to peaks with anti-phase multiplet-structure, baseline artefacts were suppressed through the use of strong sine bell weighting functions. In experiments that give rise to peaks with in-phase multiplet-structure, the data were treated with weak sine bell weighting functions followed by spline function baseline correction after the first Fourier transform (9). Data that were cosine modulated in the ω_1 dimension were routinely treated for suppression of t_1 ridges by dividing the first FID by 2 (10). The final data block was typically 1024 points in ω_1 and 2048 points in ω_2, giving a digital resolution of 4.6 Hz/point and 3.1 Hz/point, respectively. Further details are given in Ref. 11.

Standard pulse sequences and phase cycling were utilized to perform phase sensitive 2D scalar correlated spectroscopy (COSY) experiments (8) in ^1H_2O. Typically 40 scans per t_1 value and 950 t_1 values ($t_{1\,max}$=101 ms) were collected in 17 h. Pure absorption nuclear Overhauser effect spectroscopy (NOESY) spectra were obtained using the standard pulse sequence and phase cycling (12). All but the high temperature spectra were acquired with Hahn-echo observation (Dr. Mark Rance, unpublished). One experiment was acquired from ^1H_2O solution with a 200 ms mixing period, 96 scans per t_1 value and 512 t_1 values ($t_{1\,max}$= 43 ms) in 27 h. One experiment was acquired from ^2H_2O solution with a 200 ms mixing period, 64 scans per t_1 value and 450 t_1 values ($t_{1\,max}$= 49 ms) in 14 h. The exchange NOESY experiment (13) was carried out at 330 K with a 1.4 s mixing period, 128 scans per t_1 value and 350 t_1 values ($t_{1\,max}$= 70 ms) in 45 h. High digital resolution was imperative for this particular experiment, so the spectrum was folded in the ω_1 dimension, leading to a digital resolution was 2.4 Hz/point in ω_1 and 2.7 Hz/point in ω_2.

III. RESULTS AND DISCUSSION

Each protein preparation was shown to be purified to homogeneity by extensive biochemical characterization (including gel electrophoresis, iso-electric focussing, amino acid composition, amino acid sequence). However all ^1H NMR spectra show that approximately half of the proton resonances are resolved into major and minor signals in a ratio of 4:1. Fig. 1A shows a small region of the backbone fingerprint in a COSY spectrum of r-calbindin D_{9k}. The resolution of major and minor resonances is indicated and it is clear that the chemical shift difference between the two forms is variable.

The sequence-specific assignment of all resonances in the ^1H spectrum are needed to gain further insight into the origin of the two sets of resonances. This work will be reported in detail elsewhere (J. Kördel, S. Forsén & W. J. Chazin, manuscript in preparation). In summary, a strategy integrating

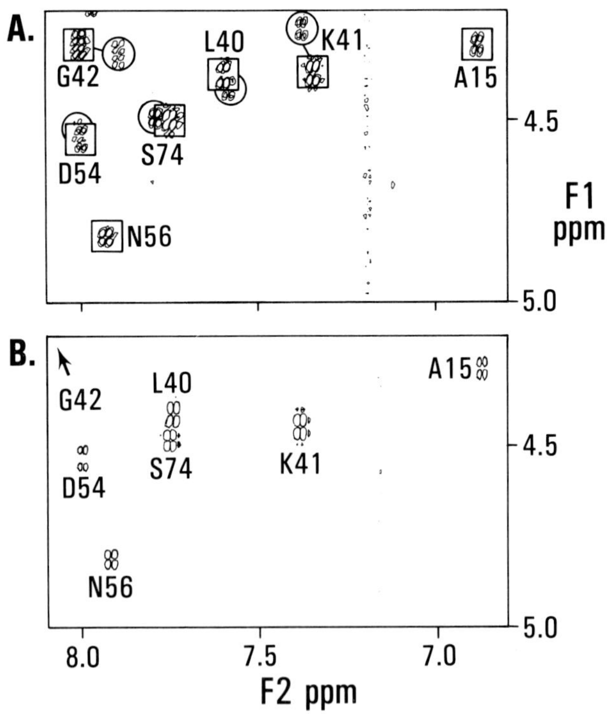

Fig. 1. Expanded COSY spectrum of A) native r-calbindin D_{9k} and B) r-calbindin D_{9k} Pro 43 → Gly mutant. Major resonances are boxed and, when distinguishable, minor resonances are circled.

relayed coherence transfer and multiple quantum techniques (11) was used for the identification of the spin systems of each amino acid. The sequential resonance assignment procedure (14) was followed to assign the amino acid spin systems to their appropriate location in the sequence. For a general review, see Ref. 1.

A region of a NOESY spectrum containing cross peaks indicating close proximity of amide protons is shown in fig. 2. This spectrum shows that it is possible to identify separate pathways of NOE connectivities for the major and minor forms of the protein and obtain sequence-specific assignments for both species.

Using NMR to analyze the short proton-proton distances along the backbone (15), we are able to establish that the peptide bond between Gly 42 and Pro 43 is different in the two

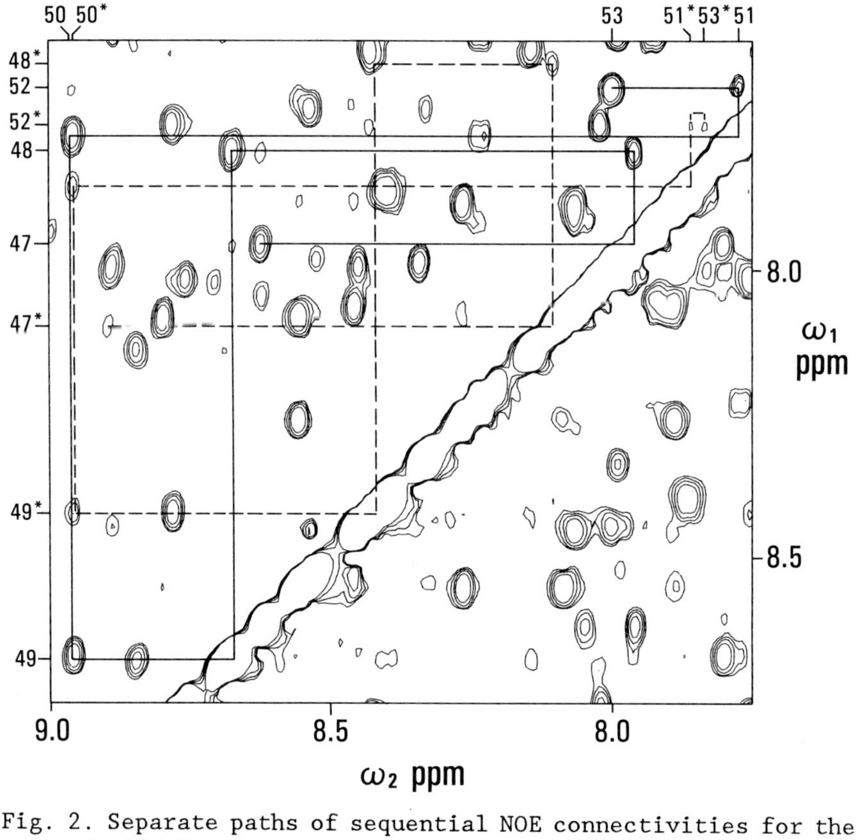

Fig. 2. Separate paths of sequential NOE connectivities for the major (solid line) and minor (dashed line) forms of r-calbindin D_{9k} in the Hahn-echo NOESY spectrum.

forms. For the major form the NOE connectivities between the C^{α} protons of Gly 42 and the C^{δ} protons of Pro 43 (fig. 3, boxed) indicate a *trans* peptide bond in extended conformation. The minor form shows NOE connectivities between the C^{α} protons of Gly 42 and the C^{α} of Pro 43 (fig. 3, circled) indicating a *cis* peptide bond between these two residues. While these data strongly implicate proline *cis-trans* isomerism, it remains necessary to prove that the minor form signals originates from the same molecular species as the major, and not from a protein impurity.

An indirect method involving site directed mutagenesis suggested that the source of the conformational heterogeneity was isomerization at Pro 43. A Pro 43 → Gly mutant was prepared by alteration of the synthesized gene (6). 2D ^{1}H NMR analysis of this mutant demonstrates, as can be seen in the COSY

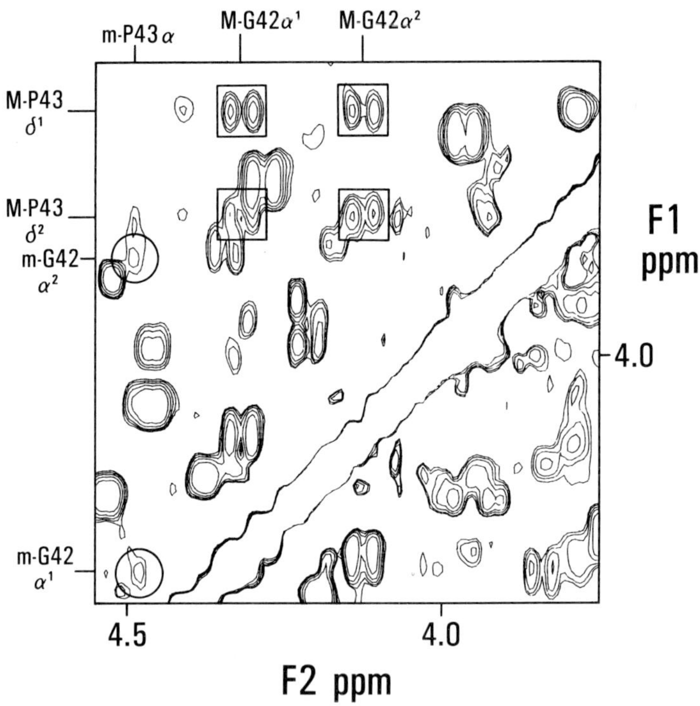

Fig. 3. Characteristic NOEs identifying trans Pro 43 in the major form (boxed cross peaks) and cis Pro 43 in the minor form (circled cross peaks). Region of a Hahn-echo NOESY spectrum acquired from $^{2}H_2O$ solution.

spectrum pictured in fig. 1B, that the global conformation of the mutant is the same as that of the two wild type forms, but there is no conformational heterogeneity. Interestingly, the local conformation of the region near Pro 43 in the mutant appears to be intermediate between the conformations of the native *trans* (major) and *cis* (minor) forms. However, a conclusive statement on this issue must await the results of detailed analysis of this mutant.

To demonstrate unambiguously that the two forms in solution are attributable to the same chemical species, 2D ¹H NMR exchange experiments were carried out. The experiments were performed at elevated temperatures, to promote the exchange process. To ensure that resonances could be resolved and clearly identified, a special preparation of r-calbindin D_{9k} with all amino acids deuterated except serine and valine (P. Brodin, T. Drakenberg, E. Thulin, S. Forsén & T. Grundström, manuscript in preparation) was utilized. Fig. 4 shows an expanded part of an exchange NOESY spectrum acquired from this sample. The boxed exchange cross peaks arise from a methyl group that resonates at the frequency of one form in the first dimension, undergoes chemical exchange during the mixing period of the experiment and resonates at the methyl group frequency of the other form in the second dimension. The rate of transfer from one form of the protein to the other is estimated (13) to be 10^{-1} - 10^{-2} s^{-1}, in the range expected for proline isomerism.

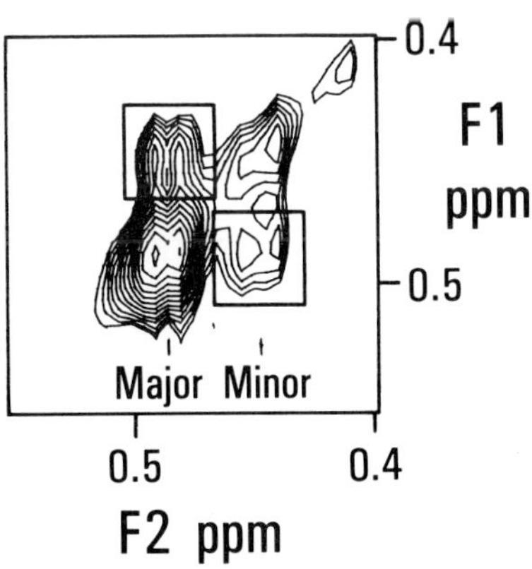

Fig. 4. Conclusive experimental evidence that the two species in solution are in equilibrium. An expanded region of an exchange NOESY spectrum acquired at 330K is shown.

IV. CONCLUSIONS

2D ^{1}H NMR can be a powerful analytical tool for the study of proteins in solution. Our results on calbindin D_{9k} provide convincing evidence that, whereas the three-dimensional structure of globular proteins are commonly viewed in terms of single native conformations, proteins may exhibit multiple native conformations in solution (see also Ref. 16 and 17). Detailed 2D ^{1}H NMR analysis shows that in this case the two conformations arise from *cis-trans* isomerism at Pro 43 (W.J. Chazin, J. Kördel, T. Drakenberg, E. Thulin, P. Brodin, T. Grundström & S. Forsén, submitted). We therefore demonstrate that 2D ^{1}H NMR is a unique method for examining protein conformational heterogeneity because it can provide the identity of the specific residue(s) in chemical equilibrium, as well as information on the extent of differences in the local and global conformations, an accurate measurement of relative concentrations of the various species, and valuable insights into their rate(s) of interconversion.

V. ACKNOWLEDGEMENTS

We acknowledge helpful discussions with Drs. Torbjörn Drakenberg, Theo Hofmann and Mark Rance.

V. REFERENCES

1. Wüthrich, K. (1986) "NMR of Proteins and Nucleic Acids", Wiley, New York.
2. Moore, J.M., Chazin, W.J., Powls, R. & Wright, P.E. Biochemistry, in press.
3. Chazin, W.J., Goldenberg, D.P., Creighton, T.E. & Wüthrich, K. (1985) Eur. J. Biochem. 152, 429-437, and references therein.
4. Kretsinger, R.H. & Nockolds, C.E. (1973) J. Biol. Chem. 248, 3313-3326.
5. Brodin, P., Grundström, T., Hofmann, T., Drakenberg, T., Thulin, E. & Forsén, S. (1986) Biochemistry 25, 5371-5377.
6. Linse, S., Brodin, P., Drakenberg, T., Thulin, E., Sellers, P., Elmdén, K.,Grundström, T. & Forsén, S. (1987) Biochemistry 26, 6723-6735.
7. Hitchman, A.J.W., Kern, M.K., & Harrison, J.E. (1973) Arch. Biochem. Biophys. 155, 221-222.

8. Marion, D. & Wüthrich, K. (1986) Biochem. Biophys. Res. Commun. 113, 967-974.
9. Zuiderweg, E.R.P., Boelens, R. & Kaptein, R. (1985) Biopolymers 24, 601-611.
10. Otting, G., Widmer, H., Wagner, G. & Wüthrich, K. (1986) J. Magn. Reson. 66, 187-193.
11. Chazin, W.J.C., Rance, M., Wright, P.E. (1988) J. Mol. Biol. 202, 603-622.
12. Bodenhausen, G., Kogler, H. & Ernst, R.R. (1984) J. Magn. Reson. 58, 370-388.
13. Jeener, J., Meier, B.H., Bachmann, P. & Ernst, R.R. (1979) J. Chem. Phys. 71, 4546-4554.
14. Billeter, M., Braun, W. & Wüthrich, K. (1982) J. Mol. Biol. 155, 321-346.
15. Wüthrich, K., Billeter, M. & Braun, W. (1984) J. Mol. Biol. 180, 715-740.
16. Fox, R.O., Evans, P.A. & Dobson, C.M. (1986) Nature 320, 192-194.
17. Evans, P.A., Dobson, C.M., Kautz, R.A., Hartfull, G. & Fox, R.O. (1987) Nature 329, 266-268.

Section III

Introduction

AMINO ACID ANALYSIS TECHNIQUES

John A. Smith

Departments of Molecular Biology and Pathology, Massachusetts General Hospital, and the Department of Pathology, Harvard Medical School, Boston, MA 02114

The purification and characterization of minute quantities of proteins continue as undiminished challenges to protein chemists, as well as molecular and cell biologists. Amino acid analysis occupies a central position among the armamentarium of protein chemistry [1]. Amino acid analysis is used routinely to estimate the amount and to determine the composition of a protein. Further, it is utilized to confirm the composition of peptides generated by enzymic digestion or chemical cleavage, which will also be analyzed by chemical microsequencing and/or by fast atom bombardment or plasma desorption mass spectrometry.

The evolution of contemporary amino acid analysis is inextricably related to the development of modern chromatographic methods by Martin and Synge (1,2). Moore and Stein were the first to recognize that the analysis of the composition of proteins and peptides by elemental analysis and chemical methods specific for each amino acid would be extremely laborious and time-consuming, and in the 1940's they began using starch columns with alcohol-acidic aqueous eluents and quantitating each separated amino acid from its visible absorbance following post-column derivatization with ninhydrin (1 μmol/amino acid)(3). In the 1950's, they began using 0.9 cm ion-exchange columns packed with sulfonated polystyrene (4,5) and in cooperation with Darrel Spackman automated the process of amino acid analysis (200 nanomole/amino acid)(6). Later, the commercial Beckman 120-series analyzers significantly improved on this sensitivity (greater than 10-fold). The introduction of high performance liquid chromatographic (HPLC) methods by Hare (7) and Hamilton (8) lead to still greater improvement in speed and sensitivity (< 1 nanomole/amino acid).

There are two methodological approaches used for amino acid analysis referred to as post- and pre-column derivatization. The post-column method developed by Stein

[1] The importance of amino acid analysis to protein chemistry has been hallmarked by the award of the 1972 Nobel Prize in Chemistry to William Stein and Stanford Moore and later by the establishment of the Stein/Moore Award by the Protein Society.

and Moore involves: (i) hydrolysis of proteins (or peptides) to their constituent amino acids; (ii) separation of the individual amino acids from one another by ion-exchange chromatography using buffers of increasing ionic strength and pH, as well as higher column temperature; and (iii) reaction of the resolved amino acids with ninhydrin and quantitation by visible detection at 570 nm (primary amino groups) and 440 nm (secondary amino groups). The pre-column methods require chemical derivatization of the constituent amino acids, following hydrolysis but prior to chromatographic separation, with a reagent that quantitatively attaches a chemical moiety to the amino groups of amino acids and that can be readily quantitated by fluorescence, ultraviolet or visible absorbance after the derivatized amino acids are resolved by reversed-phase HPLC. Various pre-column reagents have been introduced including: o-phthalaldehyde (OPA) (reviewed by Jones (9)), 9-fluorenylmethylchloroformate (FMOCl or FMOC)(see the chapters by Smith et al. and by Carlton and Morgan in this section), and dansyl chloride (10) detected by fluoresence; phenylisothiocyanate (PITC)(see the chapters by Atherton, by Dupont et al., and by West and Crabb in this section) detected by ultraviolet absorbance; and by dimethylaminoazobenzene sulfonyl chloride (DABS-Cl) (see chapter by Chang et al. in this section) detected by visible absorbance.

There are two methods for HCl hydrolysis: (a) proteins are dissolved in a 6 N HCl solution and hydrolyzed or (b) proteins are hydrolyzed in an HCl/water vapor. The later procedure results in considerably lower levels of background contamination and has become the method of choice for analyzing low picomole levels of amino acids. This and other sources of contamination are discussed in detail in the chapter by Atherton in this section. Further, it should be emphasized that hydrolysis of proteins under acidic conditions (i.e., 6 N HCl at 110 °C for 24 hr), regardless of method or temperature, will partially destroy certain amino acids (e.g., serine, threonine, and tyrosine) and completely destroy other amino acids (e.g., cysteine, cystine, tryptophan, asparagine, and glutamine), as well as most post-translational chemical modifications (e.g., γ-carboxylglutamic acid, N-glycosylation and phosphorylation). However, an accurate analysis of serine, threonine, and tyrosine can be determined by carrying out hydrolyses for 24, 48, and 72 hours and extrapolating to zero time. Cysteine/cystine can be identified as S-carboxymethyl-cysteine, if the protein is reduced and alkylated prior to hydrolysis (described in (9)) or as cysteic acid, if the protein is treated with performic acid (11). Tryptophan can be estimated if hydrolysis is carried out in methanesulfonic acid rather than HCl (12), although the high salt concentration that remains may affect pre-column derivatization of certain amino acids. Asparagine and glutamine can be quantitated, if an enzymatic hydrolysis with protease(s)(described in (9)) rather than an acid hydrolysis is carried out. Further, it should be emphasized that the rate of peptide bond hydrolysis is affected by amino acid sequence (e.g., Ile-Ile and Ile-Leu are not hydrolyzed completely using routine hydrolysis times), and this may lead to an underestimation of various amino acids. Post-translational modifications are best determined by a comparison of the data from amino acid analysis and mass spectrometry (e.g., see the chapters by Violand et al. and Hartmanis and Engstrom in this section).

The precision of amino acid analysis varies with method employed, the amount of protein being hydrolyzed, and the level of endogenous and exogenous contamination. Commercially available amino acid analyzers utilizing ion-exchange chromatography and post-column derivatization with ninhydrin are not discussed in this section but continue to be the method of choice for routine amino acid analysis at > 100 pmol/amino acid. State of the art amino acid analysis is capable of estimating low picomole levels of amino acids, and the first six chapters in this section deal with the strengths and weaknesses of three different amino acid analysis methods based on pre-column derivatizations with FMOCl, PITC, and DABS-Cl. The last two chapters of this section deal with the use of amino acid as an adjunct to mass spectrometry for identifying post-translational modifications in recombinant proteins.

There is an continuing need to improve the precision and sensitivity of amino acid analysis.

References

1. Martin, A.J.P. (1979), in 75 Years of Chromatography-A Historical Dialogue, Ettre, L.S., and Zlatkis, A., eds. , Elsevier, Amsterdam, pp. 285-296.

2. Synge, R.L.M. (1979), in 75 Years of Chromatography-A Historical Dialogue, Ettre, L.S., and Zlatkis, A., eds. , Elsevier, Amsterdam, pp. 447-452.

3. Stein, W.H., and Moore, S. (1949) J. Biol. Chem. 178, 79-91.

4. Moore, S., and Stein, W.H. (1951) J. Biol. Chem. 192, 663-681.

5. Moore, S., and Stein, W.H. (1954) J. Biol. Chem. 211, 893-906.

6. Spackman, D., Stein, W.H., and Moore, S. (1958) Anal. Chem. 30, 1190-1205.

7. Hare, E. (1966) Fed. Proc. 25, 709

8. Hamilton, P.B. (1967) Methods Enzymol. 11, 15-27.

9. Jones, B.N. (1986), in Methods of Protein Microcharacterization, Shively, J.E., ed., Humana Press, Clifton, NJ, pp. 121-151.

10. DeJong, C., Hughes, G.J., Van Wieringen, E., and Wilson, K.J. (1982) J. Chromatogr. 231, 345-359.

11. Hirs, C.H.W. (1967) Methods Enzymol. 11, 59-62.

12. Simpson, R.J., Neuberger, M.R., and Liu, T-Y. (1976) J. Biol. Chem. 251, 1936-1940.

25

AN EVALUATION OF AN AUTOMATED HIGH SENSITIVITY AMINO ACID ANALYZER BASED ON THE 9-FLUORENYLMETHYLCHLOROFORMATE (FMOCl) CHEMISTRY

Alan J. Smith and Jack M. Presley

Department of Biological Chemistry
University of California, Davis
California 95616

William McIntire

Department of Biochemistry-Biophysics
University of California, San Francisco
California 94143

I. INTRODUCTION

Amino acid analysis has been recognized as an essential component of protein structure analysis since its introduction in 1963. The availability of automated amino acid analyzers has made the handling of 15-20 analyses a day relatively routine. Until recently such analyzers were based almost exclusively on ion exchange chromatography and post-column derivatization with ninhydrin (1). This has proved to be a very versatile and stable chemistry although there is a significant loss in precision when quantitating amino acids below the 100 picomole level. Today the ability to sequence ever smaller amounts of protein has placed considerable pressure on automated amino acid analyzers to keep pace. As a result, other derivatization and separation techniques for amino acid analysis have evolved.

Post-column derivatization with orthophthalaldehyde (OPA) (2) offers increased sensitivity but poor quantitation of the secondary amino acids. The recent introduction of an automated analyzer employing pre-column derivatization with OPA and fluorenylmethylchloroformate (FMOCl), followed by reverse phase chromatography (RP-HPLC) may offer some improvements in this respect. A more frequently used alternative is the pre-column derivatization with phenylisothiocyanate and separation of the subsequent phenylthiocarbamyl (PTC) derivatives by RP-HPLC (3). This chemistry offers good sensitivity but can suffer from variable recoveries of lysine and aspartic and glutamic acids due to the presence of salts and/or metals in the samples. This chemistry is capable of providing excellent analyses on fractions obtained from RP-HPLC (4). However, commercially available analyzers utilizing this chemistry are not fully automated and require significant operator attention.

An automated amino acid analyzer based on pre-column derivatization with FMOCl and subsequent separation by RP-HPLC (5) offers several advantages. The reaction occurs rapidly at room temperature, under aqueous conditions to yield FMOC amino acid derivatives that are stable and have a high fluorescent yield. The major disadvantage is that the hydrolysis product, fluorenylmethyl alcohol (FMOH), elutes in the middle of most RP-HPLC separations systems. In addition, both the mono- and di-FMOC derivatives of lysine, and histidine and tyrosine can be formed. The automated FMOC analyzer incorporates an autosampler capable of performing extractive chemistry which significantly reduces the amount of FMOH present in the separation mixture. The derivatization reaction is buffered at pH 8.5 which produces the di-FMOC derivatives of lysine, histidine and tyrosine. This study provides a comparison of the automated FMOC analyzer with an automated ninhydrin analyzer and a semi-automated PTC analyzer on hydrolysates of standard proteins. In addition, some unique features of the FMOC chemistry that can be used to monitor the efficiency derivatization will be discussed.

II. MATERIALS AND METHODS

Three commercially available protein standards, bovine serum albumin, sperm whale apomyoglobin and oxidized β-chain of insulin were subjected to identical hydrolytic conditions of 6N HCl, 110°C for 24 hrs. Phenol at a level of 0.1% was included as an antioxidant. For normal sensitivity studies the amount of protein hydrolyzed was adjusted such that 20% of the hydrolysate would contain an average of 250 picomoles of the amino acids on-column. This translated into 50, 200 and 500 picomoles of BSA, apomyoglobin and β insulin respectively in each hydrolysate. For the high sensitivity studies ten-fold less protein was hydrolyzed in order to obtain on-column recoveries which averaged 25 picomoles of the amino acids. This corresponded to 5, 20 and 50 picomoles of the respective protein standards in each hydrolysate. For the high sensitivity analyses blank hydrolyses containing no protein were always included.

The hydrolysates were quantitated on a Beckman 6300 analyzer with ninhydrin detection, a PTC analysis system (6), and a Varian Amino Tag FMOC analyzer. The FMOC separation conditions were as follows:

Solvent A contained 15 mM sodium acetate and 10 mM tetramethylammonium chloride to pH 2.85 with phosphoric acid. Solvent B contained 90% solvent A to pH 4.50 with phosphoric acid and 10% methanol, Solvent C was 100 % acetonitrile. The following gradient conditions were used:

Time	% A	% B	% C
0.0	73	0	27
11.5	58	0	42
13.0	58	0	42
13.1	0	63	37
18.0	0	62	38
25.0	0	30	70
27.0	0	25	75
32.0	0	25	75
32.1	73	0	27

The sample dilution buffer was 0.1 M sodium bicarbonate, 0.1 M sodium borate pH 8.5 which contained 5 picomoles per microliter of β-thionyl alanine (βTA) as internal standard.

Lysine, histidine and tyrosine were quantitated as the di-FMOC derivatives. Low recoveries of these amino acids could indicate the presence of the mono-FMOC derivatives.

Molar ratios were calculated by dividing the amount of each amino acid by the theoretical number of residues for that amino acid and averaging these values for all the amino acids to obtain an integer value.

III. RESULTS

A typical separation of 15 picomoles of amino acid standards is shown in Fig. 1.

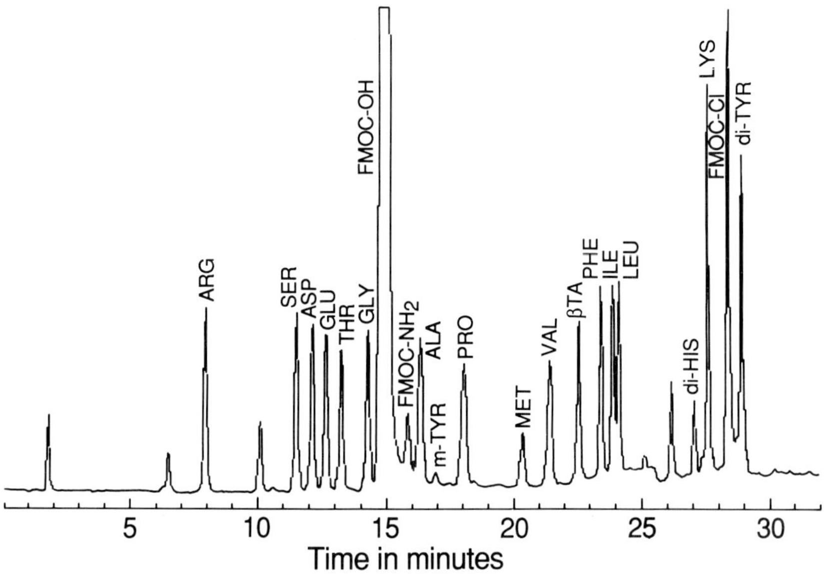

Figure 1. Separation profile of FMOC derivatives of 15 picomoles of standard amino acid mixture. FMOC-Cl, FMOC-OH and FMOC-NH_2 represent the chloroformate, alcohol and amine, respectively.

The 3 standards proteins were selected to cover a wide molecular weight range. The commercially obtained preparations were hydrolyzed without any pretreatment. Tables I, II and III compare the compositions obtained from these 3 proteins with the various analyzer methodologies. In all instances 20% of the total hydrolysate was analyzed in order to accommodate the sample loading requirements of the analyzers. Data represents an average of 5 separate hydrolysates. At this analysis level of approximately 250 picomoles of amino acid on-column, the FMOC analyzer was competitive with the other chemistries.

TABLE I. Amino Acid Composition of Bovine Serum Albumin.

		FMOC		NINHYDRIN	PTC
Amino Acid	Theoretical Molar Ratio	Molar Ratio	Pmoles	Molar Ratio	Molar Ratio
ASP	53	52	(306)	48	31
THR	34	31	(182)	34	35
SER	28	26	(154)	32	26
GLU	78	82	(480)	77	63
PRO	28	26	(152)	28	28
GLY	15	18	(105)	16	17
ALA	46	49	(285)	47	46
VAL	36	37	(220)	36	33
MET	4	0	0	0	4
ILE	14	14	(83)	13	13
LEU	61	63	(373)	57	59
TYR	19	18	(108)	18	21
PHE	26	25	(149)	25	28
LYS	59	56	(326)	60	51
HIS	17	12	(70)	15	17
ARG	23	21	(124)	23	25

Data derived from 20% of the total hydrolyzed sample which was 50 picomoles of protein

TABLE II. Amino Acid Composition of Apomyoglobin

		FMOC		NINHYDRIN	PTC
Amino Acid	Theoretical Molar Ratio	Molar Ratio	Pmoles	Molar Ratio	Molar Ratio
ASP	8	7.8	(199)	8.2	2.4
THR	5	5.5	(141)	4.9	4.5
SER	6	5.9	(150)	6.3	6.2
GLU	19	18.4	(470)	18.3	12.5
PRO	4	4.9	(125)	3.9	4.8
GLY	11	12.0	(306)	12.5	12.5
ALA	17	17.5	(455)	17.2	16.8
VAL	8	8.5	(218)	7.7	8.6
MET	2	0	0	0	0
ILE	9	9.6	(244)	7.6	8.6
LEU	18	19.3	(493)	17.8	17.9
TYR	3	1.6	(41)	1.8	1.9
PHE	6	6.2	(158)	5.5	6.1
LYS	19	17.8	(451)	18.3	17.2
HIS	12	6.6	(168)	9.4	6.9
ARG	4	4.4	(113)	4.5	3.9

Data derived from 20% of the total hydrolyzed sample which was 200 picomoles of protein.

TABLE III. Amino Acid Composition of oxidized β Chain of Bovine Insulin

		FMOC		NINHYDRIN	PTC
Amino Acid	Theoretical Molar Ratio	Molar Ratio	Pmoles	Molar Ratio	Molar Ratio
ASP	1	1.0	(83)	1.2	0.9
THR	1	0.9	(73)	0.9	1.0
SER	1	1.0	(79)	1.1	1.0
GLU	3	3.1	(255)	3.0	2.6
PRO	1	0.8	(66)	1.2	1.0
GLY	3	2.9	(241)	3.2	3.3
ALA	2	2.0	(164)	2.1	2.0
VAL	3	2.9	(238)	2.9	2.8
MET	0	0	0	0	0
ILE	0	0	0	0	0
LEU	4	3.9	(333)	3.9	4.0
TYR	2	1.8	(150)	1.6	2.0
PHE	3	2.7	(227)	2.6	2.7
LYS	1	0.8	(67)	1.1	0.9
HIS	2	1.4	(111)	1.8	1.9
ARG	1	1.0	(85)	1.2	1.2

Data derived from 20% of the total hydrolyzed sample which was 500 picomoles of protein.

When ten-fold less of the standards were hydrolyzed the compositional data showed greater scatter (Tables IV, V and VI). At this sensitivity level the ninhydrin data showed a wide variability and was not included in the comparison. Data represents an average of 5 hydrolysates from which 3 averaged blank hydrolyses have been subtracted. In general, the FMOC analyses compare favorable with the PTC data. However poor stoichiometry is seen for glycine and alanine due to interference by the FMOH peak.

Considerable improvement can be achieved by lowering the reagent concentration from 4 mM to 1 mM (data not shown). The poorest stoichiometry for any FMOC amino acid was for histidine. However, this was not due to incomplete formation of the di-FMOC derivative since no mono-FMOC histidine was observed. It is possible that fluorescent products from phenol, which tend to elute at the end of the chromatogram may compromise the quantitation of di-FMOC histidine.

TABLE IV. Amino Acid Composition of Bovine Serum Albumin

		FMOC		PTC	
Amino Acid	Theoretical Molar Ratio	Molar Ratio	Pmoles	Molar Ratio	Pmoles
ASP	53	51	(25.5)	26	(17.3)
THR	34	30	(14.9)	38	(25.3)
SER	28	36	(18.2)	29	(20.0)
GLU	78	80	(40.0)	56	(37.1)
PRO	28	26	(13.1)	28	(18.6)
GLY	15	20	(10.1)	23	(15.5)
ALA	46	42	(21.0)	45	(30.0)
VAL	36	37	(18.5)	30	(20.3)
MET	4	0	0	0	0
ILE	14	19	(9.7)	21	(13.9)
LEU	61	59	(29.3)	56	(37.5)
TYR	19	16	(8.2)	22	(14.9)
PHE	26	26	(13.1)	24	(15.7)
LYS	59	59	(29.6)	40	(26.6)
HIS	17	18	(9.2)	15	(9.8)
ARG	23	26	(13.2)	20	(13.5)

Data derived from 20% of the total hydrolyzed sample which was 5 picomoles of protein.

TABLE V. Amino Acid Composition of Apomyoglobin

Amino Acid	Theoretical Molar Ratio	FMOC Molar Ratio	FMOC Pmoles	PTC Molar Ratio	PTC Pmoles
ASP	8	8.2	(23.0)	5.9	(19.6)
THR	5	4.9	(13.7)	4.1	(13.6)
SER	6	6.9	(19.4)	8.5	(28.1)
GLU	19	17.5	(49.0)	14.1	(48.0)
PRO	4	4.5	(12.6)	5.4	(18.0)
GLY	11	16.8	(47.1)	18.0	(60.0)
ALA	17	14.9	(41.6)	15.4	(50.7)
VAL	8	7.7	(21.6)	8.4	(27.8)
MET	2	0	0	0	0
ILE	9	8.6	(24.0)	8.4	(27.7)
LEU	18	18.3	(51.6)	17.0	(56.0)
TYR	3	0	0	0	0
PHE	6	5.3	(14.9)	5.9	(19.5)
LYS	19	15.9	(44.6)	13.1	(43.0)
HIS	12	9.5	(26.6)	0	0
ARG	4	5.1	(14.3)	3.8	(12.5)

Data derived from 20% of the total hydrolyzed sample which was 20 picomoles of protein.

TABLE VI. Amino Acid Composition of oxidized β Chain of Bovine Insulin

Amino Acid	Theoretical Molar Ratio	FMOC Molar Ratio	FMOC Pmoles	PTC Molar Ratio	PTC Pmoles
ASP	1	1.0	(6.0)	0.7	(6.4)
THR	1	0.7	(4.6)	0.9	(8.0)
SER	1	1.0	(6.1)	1.1	(9.8)
GLU	3	3.1	(18.7)	2.4	(21.5)
PRO	1	1.0	(6.2)	1.2	(10.2)
GLY	3	3.9	(24.0)	3.9	(35.5)
ALA	2	1.5	(9.0)	2.0	(18.2)
VAL	3	3.3	(20.4)	3.7	(33.2)
MET	0	0	0	0	0
ILE	0	0	0	0	0
LEU	4	4.8	(29.5)	4.2	(37.4)
TYR	2	0	0	0	0
PHE	3	3.0	(18.3)	2.8	(25.0)
LYS	1	0.8	(4.6)	0.9	(8.3)
HIS	2	1.1	(6.6)	1.4	(12.1)
ARG	1	1.0	(6.0)	0.8	(7.0)

Data derived from 20% of the total hydrolyzed sample which was 50 picomoles of protein.

IV. CONCLUSIONS

The performance of the FMOC analyzer on standard protein hydrolysates compared favorably with ninhydrin and PTC-based systems at normal sensitivity levels. Since lysine, histidine and tyrosine were quantitated as the di-FMOC derivatives there appeared to be little contribution from the mono-FMOC derivatives.

At high sensitivity levels where ten-fold less standards were hydrolyzed the FMOC and PTC analyzers gave greater scatter. At this sensitivity level little usable data was generated by the ninhydrin analyzer and was not reported. When performing hydrolysis at these levels the presence of an excess of anti-oxidant was essential. That this was not the case in the present study is demonstrated by the low recoveries of methionine, tyrosine and histidine. In addition, it was imperative that control hydrolyses in the absence of protein be included in order to subtract background contamination. Given these caveats the FMOC analyzer compares favorable with the PTC analyzer with the exception of elevated levels of glycine which elutes adjacent to the FMOH peak.

A unique feature of the FMOC chemistry is the ability to monitor the efficiency of the derivatization reaction in situ. For optimal quantitation it is necessary to maintain at least a ten-fold excess of FMOCl over total amine in the sample. Since the sample dilution buffer contains a known amount of internal standard (βTA), a lower than expected recovery is an indication of insufficient reagent. The derivatization reaction is buffered at pH 8.5 which favors the formation of di-FMOC lysine, histidine and tyrosine. The presence of significant amounts of mono-FMOC tyrosine is a clear indication of sub-optimal pH of derivatization.

Where high sensitivity amino acid analyses are routinely required the concentration of FMOCl can be reduced from 4 mM to 1 mM with greatly improved quantitation of glycine and alanine. However, the recovery of internal standard must be carefully monitored to ensure an adequate excess of reagent.

The automated FMOC analyzer is a useful addition to the high sensitivity amino acid analysis field. However amino acid analysis of samples at levels less than 50 picomoles of amino acids is not a routine methodology. Considerable care must be exercised with sample preparation and hydrolysis conditions if useful analyses are to be obtained at this level with any of the existing derivatization chemistries.

REFERENCES

1. Spackman, D.H., Stein, W.H., and Moore, S. Anal. Chem. 30, 1190-1206 (1958).
2. Lee, K.S. and Drescher, D.G., Int. J. Biochem. 9, 457-467 (1978).
3. Heinrikson, R.L. and Meredith, S.C., Anal. Biochem. 136, 65-74 (1984).
4. Bergman, T., Carlquist, M. and Jornvall, H. in "Advanced Methods in Protein Sequence Analysis", B. Wittmann-Liebold, Ed. pp 45-55, Springer-Verlag (1986).
5. Cunico, R., Mayer, A.G., Wehr, C.T., and Sheehan, T., Bio Chromatography 1, 6-14, (1986).
6. McIntire, W., Singer, T.P., Smith, A.J. and Mathews,F.S., Biochem. 25, 5975-5981 (1986).

26

SIMPLE, ECONOMICAL AMINO ACID ANALYSIS BASED ON PRE-COLUMN DERIVATIVIZATION WITH 9-FLUORENYLMETHYL CHLOROFORMATE (FMOC)[1]

James E. Carlton and William T. Morgan

Bio-Technology Unit
and Department of Biochemistry and Molecular Biology
Louisiana State University Medical Center, New Orleans, LA 70112

I. INTRODUCTION

Amino acid analysis remains a challenging problem in biochemical analysis. Einarsson and co-workers (1-3) introduced derivativization of amino acids with 9-fluorenylmethyl chloroformate (FMOC) before analysis by reversed-phase HPLC. This methodology has recently been reviewed by Betner and Foldi (4) and Smith (5). Our goal was to determine whether the FMOC analysis approach can provide laboratories with modest needs for amino acid analysis a simple and economical method for composition analysis of proteins and peptides. The convenience of the FMOC technique, the intensity of the FMOC chromophore and the wide availability of HPLC systems with binary gradient capability and UV-filter detectors imply that uncomplicated procedures based on the use of FMOC have promise for routine amino acid analysis. We report one such method here.

II. MATERIALS AND METHODS

The solvent delivery system consisted of a Beckman HPLC system using two 112 pumps, a 340 injection block and a 421 gradient controller. UV detection was performed using an Isco UA-5 absorbance monitor with a type 6 optical unit containing a 254 nm filter and a 1.6 μl flow cell of 5 mm path length. Quantitation was achieved with a Hewlett-Packard 3390A

---

[1]This work was supported in part by the Bio-Technologies Unit, LSU Medical Center, New Orleans and NIH grants DK-27237 and HL-37570 to W.T.M.

integrator. A Rainin type 3 column heater and two reverse phase columns (Varian "Amino-Tag", 5 um ODS-80 TM, 150 mm x 4.6 mm and an Analytichem International (Harbor City, CA) C_{18}, 5 micron, 150 mm x 4.6 mm) were used in this study. In our experience a variety of comparable equipment should be able to be substituted for this arrangement with only minor changes in the method. References 4, 5 and 6 can be consulted for information useful for such modifications. For example, satisfactory results (not reported here) have also been obtained with the two columns noted above using a Varian 5020 HPLC with Varian UV detector.

The amino acid calibration standard was obtained from Pierce. Tetramethylammonium chloride was obtained from Fluka, 9-fluorenylmethyl chloroformate (FMOC) and homo-arginine from Sigma and the other reagents from J.T. Baker. All solvents were HPLC grade. Bovine serum albumin (BSA) obtained from Sigma and a peptide (NH_2-S-V-R-N-D-R-N-K-K-K-K-E-V-P-K-P-E-C-S-E-S-Y-T-COOH) synthesized in the Bio-Technologies Unit, LSU Medical Center, New Orleans were used as analytical samples. Protein and peptide samples were placed in 6 M HCl, de-aerated with N_2 and hydrolyzed at 110 ^{o}C under vacuum for 24 hours.

Pre-column derivativization with FMOC was carried out after taking the standard mix (1.5 nmol) or hydrolyzed samples to dryness under nitrogen or air. To derivatize the amino acids, 200 μl of 0.1 M boric acid, 0.1 M sodium bicarbonate, pH 8.9. The pH is adjusted with 1 M NaOH. FMOC (200 μl of 4 mM in dry acetone) was then added, and the solution was vortexed and incubated at ambient temperature for 10 minutes with occasional mixing. The derivativization mix was then extracted with 400 μl of pentane:ethyl acetate (80:20 v/v) 2 or 3 times to remove excess FMOC-OH. It is important that the standard and sample be extracted in exactly the same way since differential losses of FMOC-amino acids otherwise will occur. As an alternative to extraction excess FMOC can be reacted with 1-aminoadamantane (ADAM) before chromatography (7).

The solvent system used for analysis was:
Reservoir A: 0.05 M tetramethylammonium chloride, 0.05 M sodium citrate, pH 2.25 (set with phosphoric acid);
Reservoir B: (58:26:16 v/v/v) acetonitrile: reservoir A solution: methanol.

Chromatography was carried out with a flow rate of 2 ml/min at 33 ^{o}C, and the gradient employed was:

time	%B
0	30
1.0	39
6.0	55
23.0	70
39.5	100
48.0	30

Ten minutes re-equilibration is needed before the next analysis.

III. RESULTS AND DISCUSSION

Many laboratories require only a modest number of amino acid analyses per year but do not have convenient access to a dedicated amino acid analyzer. Our goal was to find a method to quantitate amino acids that was simple and adaptable to HPLC equipment commonly found in laboratories. The method reported here employs pre-column derivativization with FMOC, a binary gradient solvent system and UV detection.

As shown in Fig. 1, good resolution can be obtained in a run of less than 40 minutes. The sensitivity achieved (about 100-200 pmol) is sufficient for most applications, and higher sensitivity can be gained by employing a fluorescence detector. Moreover, the resolution obtained with a Varian "Amino-Tag" column (Fig. 1) was nearly matched by an Analytichem column (data not shown) without any attempt to optimize the chromatography conditions. Only leucine and phenylalanine were not separated.

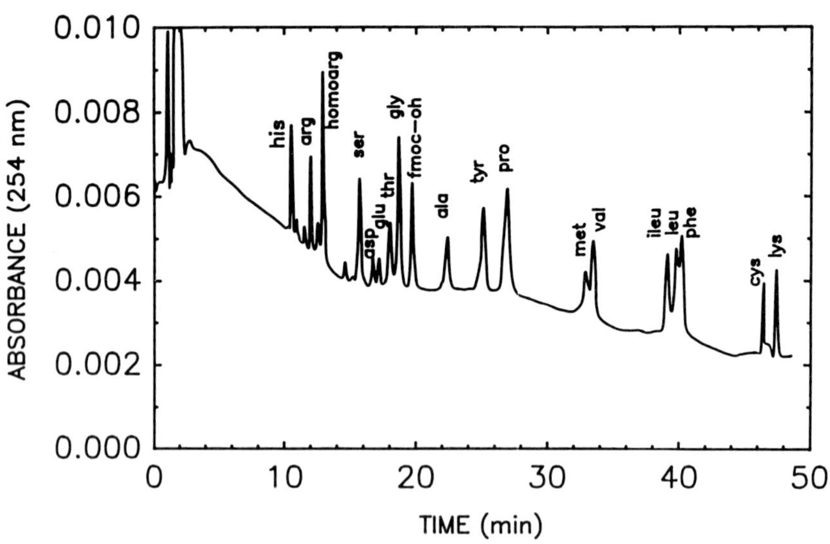

Fig. 1. Chromatography of FMOC-Amino Acid Standard Mixture.

The derivatized amino acid standard (180 pmol) was using the conditions presented in the Experimental section. The injected volume was 30 μl. The internal standard homo-arginine was present at 225 pmol.

The stability of the derivatized amino acids with the exception of methionine was found to extend over two weeks at-20 °C (Table I). FMOC-amino acids are also stable for 24 hours at room temperature (8), but solvent evaporation is more difficult to control.

TABLE I. Stability of FMOC-Amino Acids[a]

Amino Acid	Day 7	Day 14
	%	%
his	97	90
arg	100	92
ser	100	87
asp	96	84
glu	95	85
thr	105	105
gly	88	97
ala	85	107
tyr	68	75
pro	69	83
met	40	10
val	78	96
ile	75	96
leu	85	115
phe	77	103
lys	94	111

[a]To examine the stability of the FMOC-amino acids produced under these conditions, a standard mix (2.5 nmol total) was derivatized, and 200 pmol in 20 μl was analyzed on day 1. The remainder of the standard was stored at -20 °C, and subsequent analyses were made on day 7 and day 14. The results on day 1 were defined to be 100%.

The accuracy of the method is comparable to that of other methods for amino acid analysis. As shown in Fig. 2 and Table II, the results from a typical run on a sample of hydrolyzed synthetic peptide are in good agreement with the expected composition. Just as with other methods of analysis, the accuracy of the estimation is enhanced by carrying out multiple determinations on multiple samples (data not shown). Moreover, absolute sample recovery determined by using an internal standard (homo-arginine) was about 60%, falling in the range obtained with other methods. Similarly, the amino acid composition derived from an acid hydrolyzed sample of BSA (Fig. 3) agrees with the expected values.

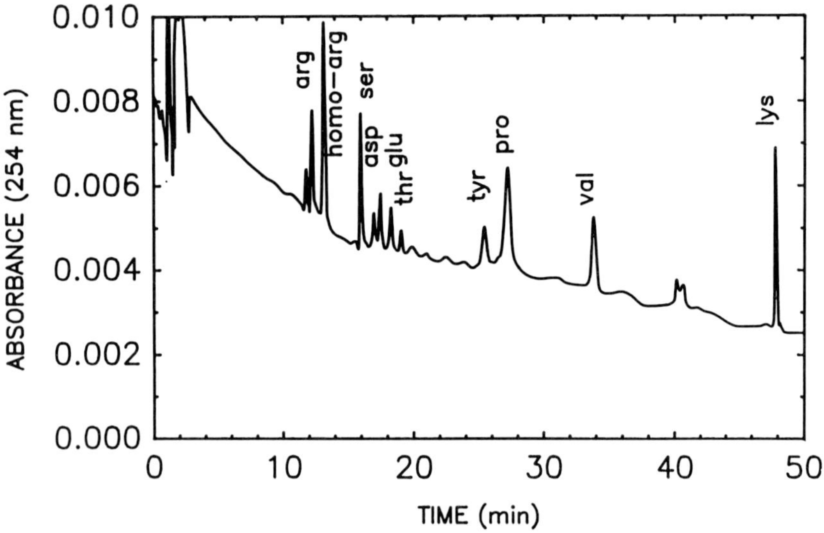

Fig. 2. FMOC-Amino Acid Analysis of a Synthetic Peptide.

Approximately 1.8 nmol of the synthetic peptide NH_2-S-V-R-N-D-R-N-K-K-K-K-E-V-P-K-P-E-C-S-E-S-Y-T-COOH was hydrolyzed, derivatized and chromatographed using the conditions presented in the Experimental section. The injected volume (25 μl) contained an estimated 180 pmol of hydrolyzed peptide and 400 pmol of the internal standard homo-arginine.

TABLE II. FMOC-Amino Acid Analysis of a Hydrolyzed Synthetic Peptide[a]

	Theoretical	Observed
arg	2.0	1.9
ser	3.0	2.5
asp	3.0	2.0
glu	3.0	3.3
thr	1.0	1.3
tyr	1.0	0.9
pro	2.0	1.8
val	2.0	1.9
lys	5.0	3.8

[a]The recovery of the amino acid residues was calculated from the recovery of the internal standard (homo-arginine) after hydrolysis, derivativization and analysis of the synthetic peptide, NH_2-S-V-R-N-D-R-N-K-K-K-K-E-V-P-K-P-E-C-S-E-S-Y-T-COOH. Approximately 57% recovery and good agreement with the expected values for all amino acids except aspartic acid and lysine was obtained in this run. Lysine loss may be due to a difference in extraction between sample and standard. Problems with glutamic acid by-products (9) were not encountered.

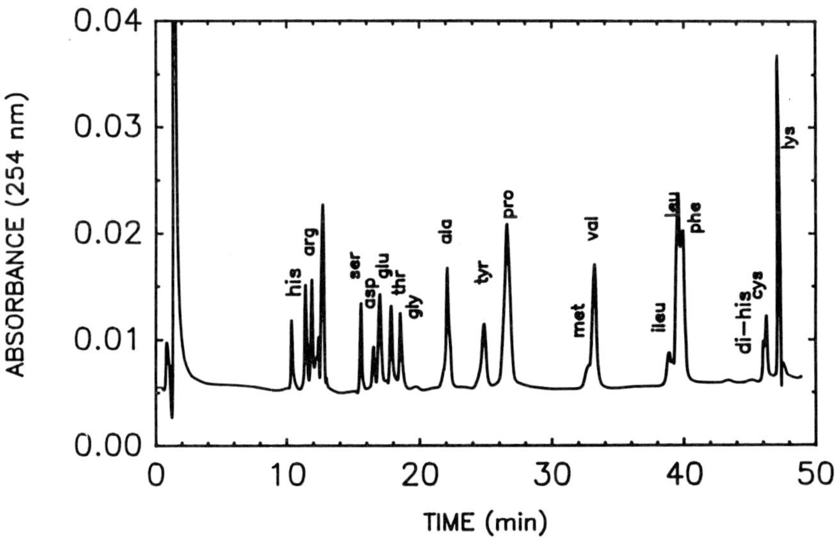

Fig. 3. FMOC-Amino Acid Analysis of Hydrolyzed BSA.

BSA (2.5 nmol) was hydrolyzed, derivatized and chromatographed as described. Ten μl containing ca. 100 pmol of hydrolyzed protein was injected. No homo-arginine was added, although a peak is evident at the same elution position.

Homo-arginine was evaluated as an internal standard here and shows promise as a standard employed in place of, or in addition to, other standards such as norleucine. Homo-arginine has the advantage of eluting earlier in this chromatography than other common internal standards aiding quantitation and method development. However, some protein samples (e.g. BSA, see Fig. 3) may contain a contaminant which elutes at the same position requiring caution to be taken in the use of this standard.

IV. CONCLUSION

Amino acid analysis based on pre-column derivativization with FMOC provides accurate and reliable amino acid analysis data. The technique is simple and economical since it can be adapted to a wide variety of columns and HPLC equipment with a minimum of developmental work. Therefore, virtually any laboratory with access to standard HPLC equipment can perform amino acid analyses on a variety of samples using the FMOC technique.

V. ACKNOWLEDGEMENTS

The authors thank Mr. Anton G. Mayer of Varian Instruments and Dr. Al Smith of UC Davis for helpful discussions.

VI. REFERENCES

1. S. Einarsson, B. Josefsson and S. Lagerkvist, J. Chromatogr. 282, 609-618 (1983).
2. S. Einarsson, J. Chromatogr. 348, 213-220 (1985).
3. S. Einarsson, S. Folestad, B. Josefsson and S. Lagerkvist, Anal. Chem. 58, 1638-1643 (1986).
4 I. Betner and P. Foldi, LC-GC 8, 832-840 (1988)
5. A. J. Smith, J.M. Presley and W. McIntire, this volume (1988)
6. G. Ogden, H. Hedstrom and P. Foldi, "Amino Acid Analysis, Theory & Laboratory Techniques", Pharmacia LKB Biotechnology AB, Bromma, Sweden (1987)
7. I. Betner and P. Foldi, Chromatographia 22, 381-387 (1986)
8. G. Ogden and P. Foldi, LC-CC 5, 28-40 (1987)
9. H.J. Keller, K. Quang Do, N. Zollinger, K.H. Winterhalter and M. Cuenod, Anal. Biochem. 166, 431-434 (1987)

27

SUCCESSFUL PTC AMINO ACID ANALYSIS AT THE PICOMOLE LEVEL[1]

Donna Atherton

The Rockefeller University
Protein Sequencing Facility
New York, New York

I. INTRODUCTION

The goal of amino acid analysis is to determine the quantity of each amino acid present in a protein. The most widely used method, developed by Moore and Stein (1), has been routinely used from the micromole to the nanomole level. There is, however, a continual need to perform such analyses at the lowest possible level. This demand has led to the development of new instruments and various pre- and post-column derivatization techniques to increase amino acid chromophore sensitivity.

Phenylisothiocyanate (PITC) derivatization of amino acids to form the phenylthiocarbamyl derivatives (PTC) has provided the most promising approach (2-4). PTC amino acid analysis requires acid hydrolysis of the polypeptide followed by separation by reverse phase HPLC and detection in the UV (ultraviolet) range. Unfortunately, not all of these aspects of PTC analysis have been studied sufficiently to provide accurate compositions routinely at the picomole level. For example, when 100 and 900 pmol of a test peptide, synthesized and distributed by the Research Resource Group (5) to multiple laboratories, was analyzed, only 45% of the 49 reported compositions correctly detected the proper amounts of at least 34 of the 40 amino acids in the peptide.

[1] supported in part by funds provided by the Office of Naval Research for the purchase of equipment.

In the Rockefeller University Protein Sequencing Facility we have focused on optimizing the PITC pre-derivatization technique and instituting routine controls for instrument performance, random contaminants and hydrolytic destruction, all of which can dramatically influence the results. Here we report a summary of the sources of contamination which we have identified and describe the practical limits of the overall procedure in our laboratory.

II. MATERIALS AND METHODS

PITC, standard amino acid mixture H and constant boiling 6N HCL (Sequanal grade) were purchased from Pierce in 1 ml vacuum sealed ampules. Triethylamine (TEA; Gold Label) was from Aldrich; redistilled phenol (Molecular Biology grade) from Boehringer Mannheim; ethanol (100%) from Industrial Chemical Co.; and acetonitrile (HPLC grade) from Burdick & Jackson. Sodium acetate (HPLC grade) was from Fisher and glacial acetic acid (aldehyde free) from J.T. Baker.

Water was distilled daily (distillation apparatus had no exposed teflon or polyethylene tubing, water was collected in glass only, glassware was dedicated and pyrolyzed), and filtered through Millipore Norganic cartridges (replaced every 15 days) and finally through a 0.22 μm Millipore filter GSWP 047. Prepurified Nitrogen, 99.998%, from Matheson Gas Products was filtered through a Millipore 0.22 μm Millistak-GS disposable filter (replaced every three months).

PYREX hydrolysis tubes were from Corning, glass storage vials from VWR Scientific, WISP glass and disposable limited volume inserts from Waters, disposable 1.5 ml microfuge tubes from Sarstedt and RT20 Clean-Pak pipette tips from Rainin.

Waters Pico-Tag Workstation and HPLC two pump gradient system was used with a Model 490 multi-wavelength detector, WISP, and Model 840 data station. Plate counts of each HPLC column were determined by the five sigma method (NOVA-PAK Column Care and Use Manual from Waters) and columns used only if they exceeded 8500. Because of instrument performance problems which manifested at the lower levels only, the instrument was periodically tested for band spreading (also the five sigma method, except that a connector replaces the column.)

Amino acid analysis: Hydrolysis of samples in acid vapor (6N HCl, 0.1% phenol, 20 hr, 110°C), derivatization of the free amino acids with PITC, and separation by HPLC was performed essentially as described by Bidlingmeyer et al. (4)

and as outlined in the Waters operator's manual for Pico-Tag Amino Acid Analysis System, but with three critical modifications: 1) after hydrolysis, samples are dried under vacuum below 50 millitorr for a minimum of 30 minutes, 2) after the "redrying procedure", they are dried again under vacuum for 30 minutes, and 3) following the coupling reaction, the drying time is extended to 90 minutes.

Hexosamine analysis: After vapor hydrolysis (4N HCl, 7 hrs, 100°C) and PITC derivatization (4) of the hexosamines, HPLC separation (6) was performed at 1 ml/min and 38°C using solvent A (0.14M sodium acetate, pH 5.3) and solvent B (60% acetonitrile). Starting conditions were 10% B, 10 minutes of curve 5 to 51% B, isocratic at 51% B for 2 minutes, a column wash for 2 minutes at 100% B, and 5 minutes equilibration at 10% B. Detection was at 254 nm.

III. RESULTS

PTC amino acid analysis includes hydrolysis of the protein, derivatization with PITC, HPLC separation of the PTC amino acids, UV detection of the chromophore, and data collection and analysis. The units used to designate the quantity of a peptide or protein (e.g., micrograms versus picomoles) have different relevance in each of the above five steps involved in analysis. The failure to realize this often leads to an inappropriate assessment of the limits of performance of each of the steps. The quantity of free amino acids obtained on hydrolysis of 10 pmol of a peptide with a molecular weight of 1000 daltons is about 90 pmol. Ten pmol of a protein with a molecular weight of 100,000 daltons, however, yields 9000 pmol of free amino acids. It is more useful to think in terms of micrograms of protein or peptide to compare results for hydrolysis and derivatization. In contrast, for HPLC separation, UV detection, and data integration and analysis, the quantity of free amino acids is most usefully discussed in picomoles.

The lack of a clear distinction of the quantity of protein hydrolyzed, quantity derivatized, and quantity injected for HPLC separation and UV detection has obfuscated the practical performance limits of amino acid analyses at low levels. For example, if one derivatizes 1 nm of standard amino acids, then injects only 10 pmol for HPLC separation, and obtains a clean, clear chromatogram, the performance of the HPLC system has been documented at 10 pmol, that of hydrolysis and derivatization, however, has been documented only at one nm, not 10 pmol. There is no guarantee that one can hydrolyze or derivatize only 10 pmol and obtain a readable HPLC chromatogram when injecting all 10 pmol, because the sample will include all reagent and sample contaminants. This, in fact, is often the case. The practical limits of each step in amino acid analysis need to be determined; the weakest step will determine the usable limit of the analysis.

A. Characterization and Reduction of Background Contaminants

In assessing the various problems with PTC analyses, the most obvious one was the large quantity of background contaminants and ammonia, which overwhelmed the adjacent amino acids. We also detected one contaminant that coeluted with methionine. We first characterized and reduced the contaminants arising from reagents and labware used during the derivatization with PITC, and then reduced those arising from hydrolysis.

To characterize the contaminants arising from the PTC chemistry, empty tubes were subjected to the conditions of PITC derivatization and the products were analyzed by HPLC. By simultaneous monitoring at both 269 nm and 254 nm, optical density 269/254 ratios of the peaks were determined and ranged from 0.13 to 0.76. The ratio for the species eluting at the methionine position was 0.52 indicating that the component was not an amino acid. The absorbance of most PTC amino acids at 269 nm is about 70% of their maximum at 254 nm; while that of the contaminants varies from 2% to 50% of their absorbance at 254 nm. Based on these results, we now monitor analyses at 269 nm because less interference is obtained from artifact peaks. The contaminants were reduced in amount significantly by increasing the drying times following hydrolysis and derivatization (see Materials and Methods) and by diluting the PITC 1:10 (vol/vol). With these modifications, a 20-fold decrease in the reagent peaks was obtained and no contaminating amino acids were observed at 0.005 AUFS.

To characterize the contaminants produced during acid hydrolysis, empty tubes were subjected to the conditions of hydrolysis as well as derivatization, and the products analyzed by HPLC. The major contaminants in this case were amino acids. The use of tubes pyrolyzed at 500°C for 8 hours eliminated the 0.05 to 0.3 μg of amino acids found in each non-pyrolyzed tube. Even after pyrolysis two to three pmol each of glutamate, serine and glycine remain, as reported previously (8,9). Because these amounts are variable, background subtraction has not proven useful and this contamination remains a limiting factor.

Other random contaminants ranging from picomole to nanomole quantities have been encountered. A summary of these contaminants is listed in Table I. As a control for random contaminant peaks, an empty tube is now subjected to the conditions of hydrolysis and derivatization with each group of samples in every reaction vial.

Of particular note in Table I is the fact that some nitrogen tanks or their fittings can be the source of nanomoles of amino acid contaminants; our tank is fitted with an in-line Millipore filter which is replaced every three months. In addition, some nitrogen has been contaminated with oxygen, causing reduced yields, sometimes drastic, sometimes subtle, during hydrolysis. We test each tank by using the nitrogen for hydrolysis of 100 pmol of amino acids and hexosamines.

Dust or other airborne contaminants have often been cited as the source of unexplained contaminants. To test this hypothesis, dust was allowed to collect in tubes placed at various locations in the laboratory, and the contents of each tube subjected to the complete analysis procedure. After one week, even the worst samples contained only 28 pmol of total amino acids. Dust has therefore not been a source of major contamination in our laboratory.

B. Analysis of Standard Amino Acids at Picomole Levels

The limits of PTC amino acid analysis are dependent on many factors, including the effect of acid hydrolysis, the coupling efficiency of PITC, the stability of PTC amino acids, and the ability of the HPLC system to separate, detect, and integrate low picomole levels. We began optimizing our system by focusing on the HPLC system first and working back through each step.

After HPLC chromatographic separation of all the amino acids was achieved, the range of linear response and the limit of sensitivity for each amino acid on our analysis system were determined. The standard deviations for retention times for one to 100 pmol of each amino acid remained consistently less than 1%. However, the standard deviations for areas increased from 5% (100 pmol of each amino acid injected) to 10% (20 pmol), and up to 20% (10 pmol injected). At 1 pmol, some amino acids were not detected or the data system failed to integrate the peaks; no quantitation was attempted. These poor results are probably due to elution problems from the column and/or inaccurate detection at these sensitivities.

The amount of PITC required for 100% coupling efficiency per μg protein was also critical. Standard amino acids at 10 pmol each (0.02 μg total), 100 pmol (0.2 μg), 1 nm (2 μg), 10

TABLE I. Sources of Contaminants

Solvents and Reagents: every new bottle is tested for contaminants by its use in the analysis of the contents of a blank tube after subjection to the conditions of hydrolysis and PITC derivatization then labelled for AAA ONLY.

HCL: contains amino acids; vapor hydrolysis is used.

Phenol: about 1/2 ml aliquots are stored at -20°C in pyrolyzed vials or vials cleaned with methanol, one vial opened weekly, contaminants develop with age or improper storage.

PITC: stored under nitrogen in vials described above, -20°C, replaced biweekly. PTC amino acid losses occur with age or improper storage.

TEA: stored under nitrogen, -20°C.

HPLC buffers: prepared with freshly prepared water; sodium acetate replaced every six months (two bottles have generated HPLC baseline problems); commercial buffers were of unreliable quality).

Water: prepared as described in Materials and Methods.

Millipore Norganic cartridges and filters: can be source of HPLC artifact peaks. As capacity of the cartridge is exceeded, it adds instead of deletes contaminants; replace biweekly; six month shelf life.

Distillation Apparatus: drained and rinsed with methanol every 2-3 months or as needed to eliminate high ammonia peaks from the chromatogram. Cleaned with dilute nitric acid once in the last four years.

Glassware: dedicated to AAA use; pyrolyzed at 500°C, 8 hr. Hydrolysis tubes pyrolyzed in lots of tens in large KIMAX culture tubes, then stored sealed in the same tubes.

Glass storage bottle for 100 ml water: replaced semiannually, refilled with fresh distilled water daily.

Reaction vials: developed leaks with continued use; tested by analysis for increased hydrolysis destruction of 100 pmol of hexosamines and amino acids.

WISP disposable limited volume inserts: one contaminated batch was found; each lot is tested before use. Glass inserts contained random contaminants.

Microfuge tubes: contaminated tubes have been obtained from other investigators. Preliminary studies showed that from an extract of 10 tubes, a total of 2.4 μg amino acids was found in tubes originally planned for the test peptide. After washes with 50% acetonitrile/0.05% TFA, less that 0.2 μg was recovered (7)

Savant SpeedVac: pump oil changed biweekly, centrifuge cleaned with methanol weekly, liquid nitrogen cold trap emptied and cleaned with acetone or methanol daily.

TABLE I. (continued)

Nitrogen: oxygen contamination of each tank determined by the hydrolysis and analysis of hexosamines and amino acids.

Gloves: powders are a probable contaminant.

HPLC system: pump seals, in-line filters, and WISP have generated contaminants; system washed with methanol weekly.

Refrigerator: all ammonia solutions are stored in separate refrigerator to prevent high ammonia contamination.

---

nm (20μg), and 50 nm (100 μg) were derivatized and analyzed by HPLC to determine yields. Based on the results, the following guidelines were developed: for samples with less than 1 μg protein to be derivatized, PITC is diluted 1:10 to minimize reagent peak interference; for samples of 1 to 20 μg, the PITC solution is used without dilution to avoid exhausting the reagent and generating poor yields of PTC derivatives.

The stability of picomole levels of standard PTC amino acids at 25°C and at -20°C did not differ significantly from those reported in the literature (3). However, real samples exhibited greater instability when stored at -20°C for several days. Samples are analyzed with 24 hours of derivatization and not allowed to remain at 25°C more than six hours.

The destruction of amino acids during vapor hydrolysis was tested by hydrolyzing and derivatizing 1 nm (2 μg total) and 100 pmol (0.2 μg total) of each amino acid of a mixture of 17 amino acids. Ten per cent (100 pmol) and 20% (20 pmol) respectively, were analyzed. Under optimum hydrolysis conditions, no significant destruction was observed. At 1 nm, serine (15%) and threonine (5%) destruction was as expected; however, at 100 pmol no apparent loss was observed, most likely because any destructive loss was compensated for by an increase in the proportion of free amino acid contaminants at this level. In general, significant destruction of amino acids was only observed when a hydrolysis vial leaked or oxygen contaminated the nitrogen supply, and is, of course, proportionately greater at lower picomole levels.

Based on the combined results, separate calibration tables and integration parameters were established for each sensitivity level; 1 nm, 100 pmol, and 20 pmol. Routine controls include: 1) analysis of an HPLC blank (injection of sample buffer), 2) analysis of a calibration standard of PTC amino acids at the appropriate concentration, 3) analysis of the contents of a blank tube after subjection to the conditions of hydrolysis and PITC derivatization, and 4) analysis of an hydrolyzed and derivatized standard mixture of amino acids.

C. Practical Limits of Picomole Analysis

Several proteins and peptides of known composition including ribonuclease, β-lactoglobulin, myoglobin, ACTH, cytochrome c, and synthetic peptides were hydrolyzed and analyzed. The analysis of several micrograms of these standard proteins is reasonably straightforward; analysis of 1 μg or below required the essential controls listed in the preceding paragraph. A graph of micrograms hydrolyzed of various standard proteins and peptides plotted against per cent correct of the known compositions is presented in Fig. 1. The per cent correct of total composition is defined as the number of amino acid residues correctly detected divided by the total known number of amino acids in the polypeptide. No corrections were made for serine, threonine, tryptophan or cysteine destruction, or for peptide bonds resistant to cleavage. For β-lactoglobulin with 162 amino acids, the correct assignment of 154 residues gives a 95% correct composition. This calculation allows us to compare the compositional data obtained from the analysis of known peptides and proteins of various sizes and quantities. It does not replace and is not to be confused with the classical quantitation method which requires several hydrolyses and analyses to correct for hydrolytic destruction and bonds resistant to cleavage prior to the calculation of standard deviations for the quantity of each amino acid present in a protein. In contrast, per cent correct of total composition is an estimate of the real composition based on a single hydrolysis and analysis. It lacks accuracy, but by performing such analyses on proteins of known composition we can determine the practical limits of our system and discern the level of confidence in data obtained from the analysis of samples of limited quantity which cannot be analyzed by the classical method.

In Fig. 1, 1 μg or more of hydrolyzed protein or peptide provides a composition that is over 90% correct. Above 20 μg the accuracy of the compositions decline, due probably to the exhaustion of reagents, a change in pH, or the presence of salts. Below 1 μg obtaining correct analyses is more difficult. During our initial studies, we had only a 50% chance of obtaining a reasonable composition below 1 μg of a known protein such as ribonuclease. Subsequent studies showed that solvent contaminants contributed significantly to these poor results. One ml of glass distilled water, lyophilyzed to dryness, hydrolyzed, derivatized, and analyzed yielded a total of almost 600 pmol amino acids. This represented a significant level of contamination, so water is now prepared

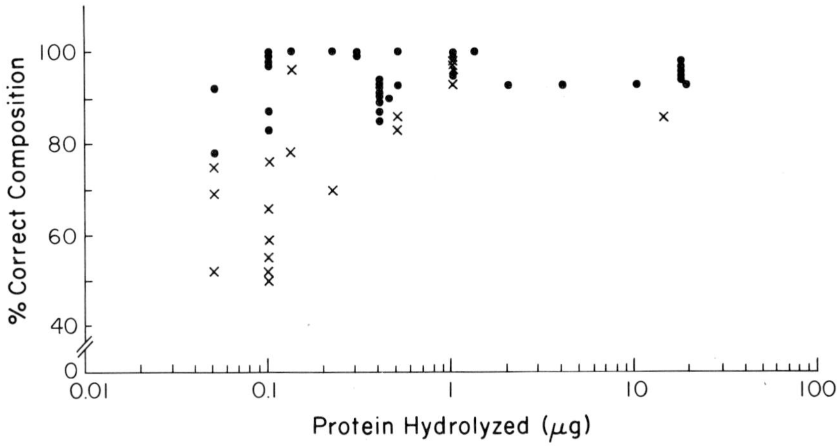

Figure 1. Effect of solvent contaminants on compositions of proteins/peptides. Samples (X) in contaminated solvents, and samples (•) in solvents free of contaminants were hydrolyzed, derivatized, and analyzed by HPLC.

as described in Materials and Methods and contains less than 170 pmol/ml.

When the solvents used to prepare the samples as well the entire system and reagents are known to be free of contaminants, amino acid analyses of peptides below 1 μg give correct compositions plus or minus 10%. At the same level, many protein standards, however, only yield values plus or minus 20% correct. The increased error may be due to free amino acids which adhere to charged protein molecules; free amino acids are often seen in the first cycle of protein sequence analysis.

As a service facility, we also have to consider sources of contamination derived from the user. Samples of solvents provided by users of the facility were tested; 100 μl of 0.1% trifluoracetic acid yielded 0.08 μg (700 pmol amino acids); 100 μl commercial HPLC grade water, 0.05 μg (322 pmol); 1 ml of NanoPure water, 0.03 μg (260 pmol); and 1 ml of water from a Milli-Q system, 0.05 μg (400 pmol). Additional sources of contamination are from HPLC columns which have been used previously to purify larger quantities of material, and from glass storage bottles previously used for media or other solvents. From these results it was clear that controls for sample associated contaminants were needed. Although a reagent blank for background subtraction is obtained from the investigator, it does not provide adequate correction for the presence of free amino acids in most cases, probably due to free amino acids associated with the polypeptide.

To correct for interference with the coupling reaction caused by trace amounts of contaminants such as buffer salts or SDS, we obtain in addition to the blank control tube, a second blank and add to this, standard amino acids at concentrations to match the sample; analysis of this blank provides the necessary corrections. The results have yielded data comparable to published compositions of related proteins.

IV. DISCUSSION

Successful picomole amino acid analysis cannot be considered a routine procedure. It is dependent on several factors including appropriate hydrolysis procedures, a well-defined chemistry, a reliable instrument, detection of the chromophore, and data acquisition, integration and interpretation, all of which must be demonstrated to be able to perform at picomole levels. For samples less than 1 μg, the hydrolysis of the sample in the absence of oxygen is crucial, and the PTC chemistry must be performed under optimum conditions if it is to provide accurate data. With the proper precautions, our instrument will provide usable quantitative data down to the level of 10 pmol per amino acid. Undoubtedly, improvements could be made to extend the limits of sensitivity; however, the major limitation of successful analyses at these sensitivities is the presence of contaminants.

Multiple sources of contaminants dictate that additional controls must be run with each sample. In spite of these controls, persistent and significant amino acid contaminants introduced during purification of real samples define the sensitivity limits of picomole amino acid analysis. We have correctly analyzed small peptides of known composition on as little as 70 nanograms; however, the practical limits based on empirical experience of analysis of real samples are that the average sample of one or more μg of peptides or protein yields a 90% or greater correct composition, 0.1 to 1 μg of proteins give results that are only better than 80% correct, whereas in this same range (0.1 to 1 μg) peptides continue to give greater than 90% correct composition.

Despite the inability to obtain precise results in some cases, PTC amino acid analysis provides useful, and much needed data which is not obtainable with classical ninhydrin analysis at the picomole level. Amino acid analysis data can be used not only to determine the quantity of each amino acid present, but also the relative amounts and number of each amino acid in a polypeptide, as well as the total quantity of protein present. For example, when protein samples give no sequence data, PTC amino acid analysis can confirm the presence or absence of protein with a probable blocked amino terminus. It can also identify a protein as a member of a specific family of proteins. In addition, quantitation of proteins prior to enzymatic digestion can best be determined by amino acid analysis.

These uses, however, do not preclude the need for precise acquisition of data at the submicrogram/picomole level. Further developments in HPLC analysis can lead to improved accuracy, but the results described here suggest that the focus must first be on solvent purity and the elimination of free amino acid contaminants in protein samples. Until methods are developed to reduce free amino acids from samples, instrument and chemistry improvements to increase sensitivity are of little practical utility.

V. ACKNOWLEDGEMENTS

The excellent technical assistance of both Farzin Gharahdaghi and Debra Lieblich is gratefully acknowledged.

VI. REFERENCES

1. S. Moore, D.H. Spackman, and W.H. Stein, Anal. Chem., 30, 1185 (1958).
2. D.R. Koop, E.T. Morgan, G.E. Tarr, and M.J. Coon, J. Biol. Chem., 257, 8472 (1982).
3. R.L. Heinrikson, and S.C. Meredith, Anal. Biochem., 136, 65 (1984).
4. B.A. Bidlingmeyer, S.A. Cohen, and T.L. Tarvin, J. Chromatogr., 336, 93 (1984).
5. R.L. Niece, K.R. Williams, C. Wadsworth, J. Elliot, K. Stone, W.J. McMurray, A. Fowler, D. Atherton, R. Kutny, and A. Smith, this volume.
6. S.A. Cohen, and D.J. Strydom, Anal. Biochem. (in press).
7. K.R. Williams, private communication.
8. K.L. Stone, and K.R. Williams, J. Chromatogr., 359, 203 (1986).
9. R. Knecht, and J.Y. Chang, Anal. Biochem., 58, 2375 (1986).

28

A COMPREHENSIVE APPROACH TO AMINO ACID ANALYSIS

David R. Dupont, Pamela S. Keim, Arthur H. Chui,
Rose Bello, MeriLisa Bozzini and Kenneth J. Wilson

Research and Development
Applied Biosystems, Inc.
Foster City, CA 94404

I. INTRODUCTION

There is more to amino acid analysis than derivatization and chromatography alone. Sample preparation and hydrolysis, however, have been largely ignored as recent development efforts have focused on derivatization chemistry. The goal has been a method combining sensitivity at the femtomole level with the reliability and reproducibility of an ion exchange / ninhydrin system at the nanomole level. Yet the accuracy of the final results is more dependent on the integrity of the sample than on the method of analysis. While affording higher sensitivities, precolumn chemistries are subject to interferences from the sample matrix, a consequence of derivatizing without prior ion exchange separation. Higher sensitivity allows smaller sample amounts to be hydrolyzed, increasing the possibility of hydrolytic losses. A derivatization method cannot differentiate the amino acids which compose a sample from those introduced by fingers, glassware or reagents.

The utility of an amino acid analysis method, then, cannot be determined by its ability to quantitate femtomoles of amino acid standard when a typical hydrolysis blank contains picomoles as background. A truly useful method must provide reproducible quantitation of unknown samples which have been isolated in limited amounts and hydrolyzed at the desired analysis level. Logically then, an integrated approach is demanded. It begins with purification procedures which minimize potential interferences in subsequent hydrolysis and derivatization steps. To plan these procedures successfully requires an understanding of the ways in which various buffer salts, detergents and common contaminants interact with the derivatization chemistry. The next step is to incorporate a hydrolysis scheme designed to give the lowest possible backgrounds and a high degree of reproducibility. This is a function of both the actual hydrolysis conditions and their compatibility with the derivatization chemistry. Finally, the derivatization chemistry itself must be simple, reliable and quantitative.

Described here is an approach which begins to address the entire process of amino acid analysis from sample preparation through derivatization. While far from a completed method, the information contained here should be useful in designing successful strategies for amino acid analysis. Although the specifics are geared towards optimizing the results of coupling with PITC to form PTC-amino acids, the general principles are applicable to any derivatization chemistry.

II. MATERIALS

Amino acid analyses were performed on the Applied Biosystems Model 420A Derivatizer / Analyzer, an automated PTC amino acid analysis system. The instrument consists of a Model 420 Derivatizer unit, the Model 130 Analyzer, and the Model 920 Data Analysis Module. Samples are applied to glass frit supports, which deliver samples via a turntable mechanism to a flow through reaction chamber. After derivatization, the PTC-amino acids are transferred on-line to the HPLC, and the frit and chamber are washed in preparation for the next sample.

The reagents used for derivatization, 5% phenylisothiocyanate in heptane, diisopropylethylamine and methanol, and the sodium acetate buffers and acetonitrile used for sample transfer and chromatography were supplied by Applied Biosystems Inc. Figure 1 shows a typical chromatogram obtained when 100 picomoles of amino acid standards were derivatized on the described system. Reagent and reaction by-product peaks are identified.

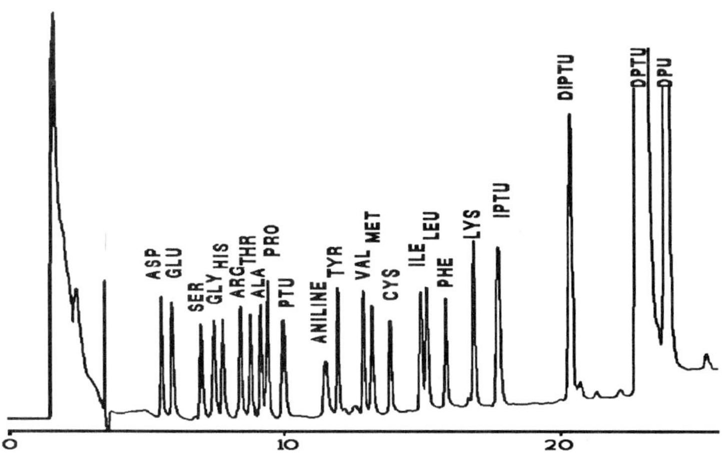

Fig. 1. Chromatogram of 100 picomoles of an amino acid standard

Amino acid standard, constant boiling HCl, septa and hydrolysis vials were purchased from Pierce Chemical Co. Protein standards were purchasd from Sigma Chemical Co. The peptide standard was synthesized on an Applied Biosystems Model 430 Peptide Synthesizer. EDTA, free acid, di- or tetrasodium and dipotassium, and the various hydrolysis additives were supplied by Aldrich Chemical Co., Inc., except for liquified phenol from ABI. Hydrolysis tubes were purchased from Corning Glass Works.

III. SAMPLE PREPARATION

Minimize sample handling. Every time a sample is manipulated it can be lost or contaminated.

There are a number of ways to reduce sample losses. Substitute polypropylene centrifuge tubes for glass tubes when appropriate to minimize losses due to surface adsorption. Problems with accurately pipetting small amounts of liquid into these tubes due to the lack of a hydrophilic surface for adsorption can be overcome by pipetting directly into a small volume of water previously introduced into the tube. Concentrate rather than dry samples to prevent losses due to insolubility upon reconstitution. Never completely dry a sample in a tube unless it will be hydrolyzed in that tube. Even after hydrolysis, samples should not be dried for longer than 10-20 minutes. Extended drying can cause significant losses of the charged amino acids, particularly the basics.

Chromatographic separations should be performed on short, narrow or microbore columns (2 or 1 mm ID by 15-30 mm long). Organize isolation protocols to minimize the number of drying, dialysis, or lyophilization steps necessary. This can be accomplished by arranging chromatographic steps appropriately: ion-exchange or hydrophobic interaction before reverse phase, reverse phase before SDS-PAGE, etc.

SDS-PAGE may serve as an alternative to chromatography for sample isolation when followed by either electroelution or electroblotting. In particular, samples which have been electroblotted onto PVDF membrane are suitable both for direct microsequencing and for amino acid analysis. The procedures for SDS-PAGE and for electroblotting onto PVDF membrane have been described (1, 2). For PTC-amino acid analysis, the addition of 0.1% Na_2EDTA to the electroblotting buffer improves the accuracy and reproducibility of aspartic acid, glutamic acid and lysine quantitation (see Table V).

Contamination of the sample is more difficult to control. Cleanliness of all surfaces the sample contacts is essential. A first step is to use dedicated or disposable tubes, pipettes, etc. to help prevent the introduction of foreign materials. Any equipment which contacts the sample should be rinsed before use and kept covered when not in use. A speck of dust will not only contribute free amino acid contamination to a sample, but since a high percentage of dust is proteinaceous, it will mask the true composition of the sample if hydrolyzed. Careful opening and transferring of the sample, careful preparation and storage of solutions and sample diluents, and the purification of reagents, if necessary in a laminar flow hood, also reduce the risk of contamination.

Traces of salts, metals or detergents whether intentionally included in the sample or picked up from contaminated glassware may alter subsequent results. Ideally, samples for precolumn amino acid analysis should be free of salts, buffers and detergents prior to hydrolysis. Realistically, it is impossible to perform purification protocols without introducing buffer salts, and the losses incurred when desalting small amounts of sample can be unacceptably high. The use of buffer systems which are compatible with PITC derivatization can effectively eliminate the need for desalting samples. The key is to know the specific effects of the salts on the chemistry used.

Table I summarizes the effects of some common buffer salts and detergents on the derivatization of an amino acid standard. The table shows that ammonium and tertiary ammonium buffer salts are quite compatible with PITC derivatization. At these levels, there is no chromatographic interference from the phenylthiourea peak generated by the reaction of PITC with ammonia. In contrast, the sodium analogs of the same salts interfere to varying degrees. Note the sodium phosphate effects in particular. Although this would indicate that all sodium salts should be avoided in PITC chemistry, some, e.g. sodium chloride and sodium borate, do not interfere. The commonly used detergents are compatible, affecting yields only slightly.

TABLE I. Effects of common buffer salts and detergents on derivatization[a]

Additive	Effects and % Recovery
Ammonium acetate	No negative effect
Sodium acetate	HIS, CYS ~ 50%; ILE, LEU, PHE, LYS ~ 80%
Triethylammonium acetate	HIS,THR and PHE ~ 90%
Ammonium bicarbonate	THR ~ 90%
Sodium bicarbonate	MET ~ 40%; HIS,TYR ~ 60%; ILE, LEU, PHE and LYS ~ 80%
Sodium borate	No negative effect on results
Sodium chloride	No negative effect on results
Sodium phosphate	Very low and variable yields, all < 50%
TEA phosphate	No negative effect on results
CAPS	Very large late eluting peak obscures PHE, LYS
HEPES	Gives artifact peak which co-elutes with MET
TRIS	HIS ~ 60%, artifact peak co-elutes with TYR
SDS	HIS and THR ~ 90%, CYS and LYS ~ 110%
Triton X-100	HIS and THR ~ 90%, CYS and LYS ~ 110%

[a]Data presented are the result of derivatizing 100 picomoles of hydrolysate amino acid standard in the presence of 1 micromole of salt or 20 μL of a 0.1% solution of detergent.

In order to assess the effects of hydrolyzing a sample with buffer salts present, a synthetic peptide containing one of each of the hydrolysate amino acids was used as a standard. Aliquots of 300 picomole each were hydrolyzed in the presence of 1 micromole of salt, and duplicate 100 picomole portions were analyzed. These results are presented in Table II. The data were standardized to the composition obtained from the peptide hydrolyzed alone and normalized to norleucine. Any differences in the results should be attributable to the effects of the salts.

The results improved significantly for several of the sodium salts when present during both hydrolysis and derivatization. Sodium acetate and sodium bicarbonate had only a slight effect on the results. This is presumably because the anions can be volatilized during hydrolysis and subsequent drying, leaving only sodium, in the form of sodium chloride. Results do not improve for samples containing mono- or dibasic sodium phosphate. These salts would be unchanged through the hydrolysis procedure, and so the results, as compared with those in Table I, should not change appreciably.

TABLE II. Effects of buffer salts and detergents on amino acid yields from hydrolysates

		Acetate			Bicarbonate		Borate	Phosphate		TRIS
AA	H_2O	NH_4	Na	TEA	NH_4	Na	Na_2	NaH	Na_2	
ASP	1.1	1.1	1.0	1.1	1.0	1.0	0.9	0.9	0.6	1.1
GLU	1.1	1.1	1.1	1.1	1.0	1.0	1.0	1.0	0.7	1.1
SER	1.1	1.0	1.0	1.0	1.0	0.9	0.9	0.8	0.5	1.0
GLY	1.1	0.9	0.9	0.9	0.9	0.8	0.8	0.7	0.5	0.8
HIS	0.9	1.0	1.0	1.0	1.0	0.9	0.9	0.9	0.5	1.2
ARG	1.1	1.1	1.0	1.1	1.1	0.9	0.9	0.9	0.6	1.1
THR	1.0	1.1	1.0	1.0	1.0	0.9	0.9	0.9	0.5	1.1
ALA	1.1	1.1	1.1	1.1	1.0	0.9	0.9	0.9	0.5	1.0
PRO	1.1	1.1	1.1	1.0	1.1	0.9	0.9	1.0	0.6	1.1
TYR	1.1	1.0	1.0	1.1	1.1	0.9	0.9	1.0	0.8	1.2
VAL	1.1	1.1	1.1	1.1	1.1	1.0	1.0	1.0	0.8	1.0
MET	1.1	1.1	1.1	1.1	1.2	1.0	1.0	0.8	0.6	0.8
ILE	1.0	1.0	1.0	1.0	1.0	0.9	0.9	1.0	1.1	0.9
LEU	1.0	1.0	1.0	1.0	1.0	0.9	0.9	1.0	1.1	1.0
PHE	1.0	1.0	1.1	1.1	1.0	1.0	1.0	1.1	0.9	1.0
LYS	1.1	1.0	1.1	1.1	1.1	1.0	1.0	1.0	0.5	1.0

Some sodium salts interfere in derivatization with PITC as shown in Table I. Salts of di- and trivalent metals, because of their ability to complex amino acids, are even more problematic. The effects of some common metal contaminants are described in Table III. By redissolving samples in a solution of EDTA in high-purity water, the effects of metal contamination can be reversed(see Table IV for comparative data).

TABLE III. Effects of trace metal contamination on derivatization yields[a]

	% RECOVERY								
AA	Al	Ca	Cu	Fe	Pb	Mg	Mn	Ni	Zn
ASP	12	60	42	90	86	90	15	6	17
GLU	12	81	64	64	105	89	32	9	27
SER	25	69	90	45	85	92	34	4	50
GLY	31	95	111	105	102	106	97	14	103
HIS	28	88	15	53	97	58	24	3	14
ARG	36	97	97	38	110	73	62	9	52
THR	33	84	101	63	96	110	70	77	93
ALA	30	98	100	100	100	91	97	9	100
PRO	56	95	126	106	101	116	99	69	110
TYR	27	91	78	88	97	60	63	5	89
VAL	36	95	85	110	98	97	98	11	108
MET	16	30	73	56	42	48	56	6	19
CYS	9	61	6	0	30	32	6	0	0
ILE	49	95	63	100	92	99	95	17	106
LEU	47	98	75	95	94	95	93	8	106
PHE	39	94	58	85	89	85	90	9	103
LYS	18	88	8	14	67	74	47	7	16

[a]Data presented are the result of derivatizing 100 picomoles of hydrolysate amino acid standard in the presence of 1 - 2 nanomoles of metal salt.

IV. HYDROLYSIS

Acid hydrolysis of samples may be accomplished using liquid-phase or vapor-phase methods. In liquid-phase hydrolysis, acid is added directly to the sample tube. This method is well-suited to hydrolyzing large amounts of sample, or complex substances such as grain and other agricultural products. It is most frequently used in conjunction with post-column ion-exchange methods where contaminating salts, etc. are chromatographically removed prior to derivatization.

When limited amounts of samples are available, a preferable method is to use a vapor-phase hydrolysis. That is, small tubes containing dried samples are sealed into a larger vessel containing the acid. Upon heating, only the acid vapor comes into contact with the sample, thus excluding non-volatile contaminants. A vapor phase protocol which is particularly suitable for the analysis of sub-microgram sample quantities by pre-column derivatization with PITC has recently been described (3).

Solutions of samples to be hydrolyzed should be introduced into the bottoms of 6 x 50 mm Pyrex tubes which have been pyrolyzed overnight at 500°C. Small volumes (< 5 μL) can be more accurately and reproducibly transferred when pipetted into ~ 20 μL of high purity deionized water with a disposable glass micropipet. Solid samples for analysis, including resin-bound synthetic peptides and electroblotted proteins, need further preparation. Lyophilized samples should be dissolved or suspended in a volatilo colvont co that thoy can bo driod ac a uniform film on tho bottom of the hydrolysis tube. PVDF bands should be rinsed twice with 200 μL water directly in the hydrolysis tube. A few microliters of water added to resin beads in the bottom of the hydrolysis tube before purging will prevent them from blowing out. Prior to hydrolysis, all samples should be dried under vacuum with centrifugation. Note that for resin beads and PVDF strips, a *very gentle* argon stream should be used for purging.

When preparing samples for hydrolysis, routinely include a hydrolysis blank, an empty tube which has been handled in the same way as those containing samples. This gives a measure of the background arising from the handling and hydrolysis procedures, and is necessary when dealing with very small sample amounts. Hydrolyzing a reagent blank, e.g. an aliquot of HPLC elution buffer or a band excised from a "non-lane" region of electroblotted PVDF membrane will also yield useful information to assist in data interpretation. Including a tube containing a standard protein or peptide provides a control in cases where the unknown samples give unusual or unexpected results.

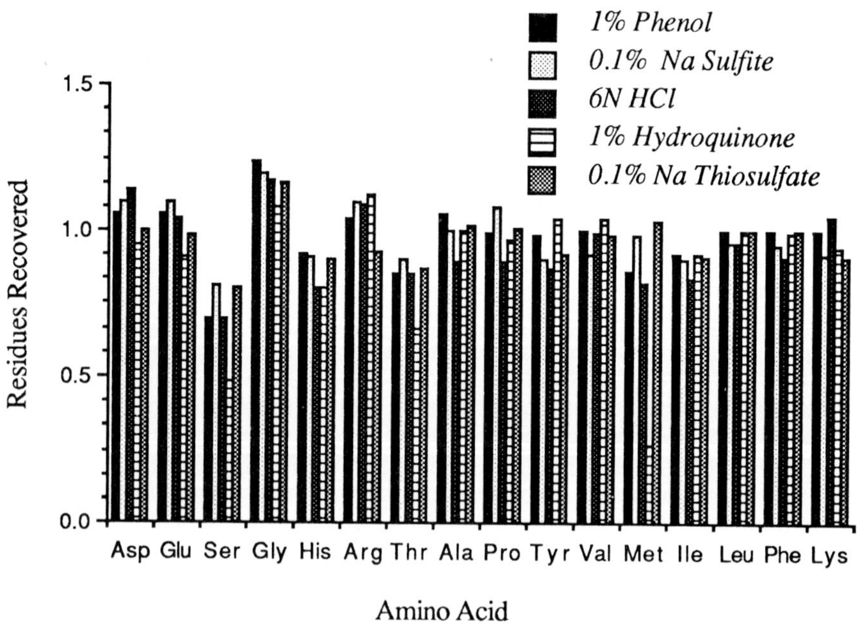

Fig. 2. The effects of several scavengers on yields of amino acids from hydrolyzed peptide. Amount hydrolyzed = 100-150 pmoles (~ 0.25 μg). Hydrolysis conditions: 6N HCl, vapor phase, argon blanket, 165°C, 60 min.

If an unknown sample contains a significant amount of a buffer salt, adding a similar amount of that salt to one of the standard samples and noting the effects may greatly assist in deciphering results.

The presence of appropriate anti-oxidants / scavengers during hydrolysis with hydrochloric acid can prevent some losses of labile residues. The uses of phenol for tyrosine, thioglycolic acid for tryptophan (4) and sodium sulfite for methionine and cystine (5) have been described for liquid phase hydrolysis protocols followed by analysis on a post-column ninhydrin system. Fig. 2 shows the effects of several scavengers when used in the vapor phase hydrolysis of a synthetic test peptide containing one of each of the common hydrolysate amino acids. The hydrolysis conditions and procedures were those described (3).

As Fig. 2 illustrates, each substance improved the yield of one or more sensitive residues. Sodium sulfite improved recoveries of methionine and also protected serine and threonine. Sodium thiosulfate had similar effects. Some additives preserved one residue at the expense of others. Hydroquinone, for example, increased yields of tyrosine, but significantly lowered serine, threonine and methionine recoveries. Phenol was equally good for tyrosine, without the deleterious effects on other residues.

The difficulties of recovering tryptophan and cystine at these hydrolysis levels were not addressed by the data presented in Fig. 2. While both sodium sulfite and sodium thiosulfate improved yields of unmodified cystine residues, recoveries were less than 50% when 100 picomoles of peptide was hydrolyzed. It is still necessary, therefore, to reduce and alkylate cystine for good recoveries. Pyridylethylation produces a cysteine derivative which is hydrolysis-stable and which is chromatographically well-resolved as a PTC-derivative. The hydrolysis protocols for tryptophan utilizing non-volatile acids did not work for this amount of sample. Numerous interfering artifacts were generated, precluding the interpretation of data. Several other potential scavengers, particularly mercaptans, produced PITC-reactive artifacts during hydrolysis which interfered with quantitation of some residues. Thioglycolic acid, for example, produced a byproduct which co-elutes with lysine in this method. Promising results for tryptophan were obtained with mercaptopropionic and thiodipropionic acids, but are still preliminary at this time.

To date, a combination of 1% phenol and 0.1% sodium sulfite has been the most promising scavenger formula, giving good recoveries of tyrosine, threonine, serine and methionine without lowering other recoveries or producing interfering artifacts.

Perhaps the best general recommendation regarding the hydrolysis of sub-microgram sample amounts is this: Evaluate the protocol which you intend to use by hydrolyzing both reagent blanks and model samples of known compositions at the desired levels before attempting the analysis of unknown samples.

V. SAMPLE RECONSTITUTION

When ready for analysis, hydrolysates should be dissolved in a solution of 250 ppm EDTA in high purity, deionized water. Including EDTA prevents the variable yields of aspartic and glutamic acids which have been a characteristic problem of PTC-amino acid analysis. The poor yields are most likely due to trace amounts of aluminum salts leached from the glass tube used for hydrolysis.

Table IV compares compositional results generated from protein hydrolysates redissolved in water only versus an EDTA solution. The comparisons are of separate hydrolyses of the same proteins, not of two aliquots taken from the same hydrolysate. Hydrolyses were done in 6N HCl with 1% phenol at 165°C for one hour. In both cases, the samples dissolved in the EDTA solution show significantly more accurate recoveries for aspartic acid, as well as slight improvements in glutamic acid, histidine and lysine yields. Solutions of up to 1000 ppm EDTA can be used without adverse effects. Of the salts readily available, dipotassium EDTA appears to be the best choice in terms of purity and ease of use.

TABLE IV. Comparison of compositional data from hydrolysate samples dissolved in water or 250 ppm EDTA solution[a]

Sample: **a-Lactalbumin** Amount hydrolyzed: 40 pmol				Sample: **Carbonic Anhydrase** Amount hydrolyzed: 15 pmol			
AA	Exp.	No EDTA	EDTA	AA	Exp.	No EDTA	EDTA
Asp	21	15.4	20.5	Asp	32	22.3	31.5
Glu	13	11.4	12.4	Glu	23	19.8	23.0
Ser	7	4.0	3.9	Ser	16	10.4	9.9
Gly	6	6.5	6.2	Gly	20	20.8	21.2
His	3	2.5	2.8	His	11	6.2	9.1
Arg	1	1.1	1.0	Arg	9	7.7	8.4
Thr	7	5.4	5.3	Thr	14	10.0	10.6
Ala	3	3	3	Ala	17	17	17
Pro	2	2.0	1.9	Pro	19	17.7	18.4
Tyr	4	3.6	3.9	Tyr	8	6.7	7.5
Val	6	6.2	5.7	Val	20	18.2	19.0
Met	1	0.9	1.0	Met	3	2.4	2.2
Ile	8	7.9	7.1	Ile	5	4.6	4.3
Leu	13	12.7	12.6	Leu	26	24.2	26.0
Phe	4	4.1	3.9	Phe	11	9.1	10.6
Lys	12	10.1	12.2	Lys	18	13.5	17.0

[a]All compositions normalized to Ala

No reconstitution or extraction is necessary for PVDF-bound samples when analyzed on the derivatizer / analyzer. After hydrolysis the PVDF pieces are placed directly on the glass frits for analysis *in situ* . The presence of EDTA, in this case in the electroblotting buffer, also improves the yields of certain amino acids as indicated in Table V. Note the upward trend of aspartic acid, glutamic acid and lysine values obtained from BSA as the concentration of EDTA present in the buffer increases from None to 0.1%. Hydrolysis conditions were those described for Table IV.

Table V. Effect of EDTA in Electroblot Buffer on BSA Composition

AA	Known	Amount of EDTA		
		None	0.05%	0.1%
Asp	***54***	***28.3***	***42.6***	***52.8***
Glu	***79***	***60.4***	***63.6***	***74.5***
Ser	28	18.7	19.3	23.0
Gly	16	17.8	19.3	19.8
His	17	13.5	10.6	11.6
Arg	23	27.6	25.9	23.1
Thr	34	29.1	26.1	26.7
Ala	46	46	46	46
Pro	28	25.5	23.9	25.8
Tyr	19	21.1	20.4	18.7
Val	36	37.1	35.6	35.8
Met	4	2.5	2.6	2.9
Ile	14	15.2	13.6	14.0
Leu	61	62.1	61.2	61.7
Phe	27	27.9	27.4	26.4
Lys	***59***	***30.0***	***47.4***	***51.8***

VI. CONCLUSIONS

The potential sensitivity which any pre-column chemistry offers can be realized only if the barriers of background contamination and sample loss are overcome. Greater care must be taken, not only in sample handling, but also in the choices of reagents, solutions, glassware and other laboratory paraphernalia which come into contact with the sample. The information presented here shows that results obtained from pre-column analysis methods are significantly improved when certain precautions are observed during sample isolation and hydrolysis. Choose buffers for sample purification which are compatible with the derivatization chemistry. In particular, avoid using metal phosphates with PITC. Avoid metal contamination as much as possible by using high quality water, reagents, glassware, etc. Use a vapor phase protocol with appropiate scavengers present for the hydrolysis of sub-microgram sample amounts. Always reconstitute hydrolyzates in a solution of 250 ppm EDTA.

VII. REFERENCES

LeGendre, N. and Matsudaira, P. (1088) Biotoohniquoc 6, 154
Yuen, S.W., Chui, A.H., Wilson, K.J. and Yuan, P.Y. (1988) Applied Biosystems Protein Sequencer User Bulletin No. 36, Foster City, CA
Dupont, D., Keim, P., Chui, A., Bozzini, M. and Wilson, K. (1988) Applied Biosystems Derivatizer-Analyzer Users Bulletin No. 2, Foster City, CA.
Swadesh, J.K., Thannhauser, T.W. and Scheraga, H.A. (1984) Anal. Biochem. 141, 397.
Matsubara, H. and Sasaki, R.M. (1969) Biochem. Biophys. Res. Comm. 35, 175.

29

AUTOMATIC HYDROLYSIS AND PTC AMINO ACID ANALYSIS A PROGRESS REPORT[1]

Karen A. West and John W. Crabb

W. Alton Jones Cell Science Center
10 Old Barn Road
Lake Placid, New York 12946

I. INTRODUCTION

Phenylthiocarbamyl (PTC) amino acid analysis has rapidly evolved into a practical method for picomole level quantitative amino acid analysis. First described in detail in 1984 (1), this phenylisothiocyanate (PITC) precolumn derivatization method utilizes reverse phase high performance liquid chromatography (RP-HPLC) to separate and quantify PTC-amino acid derivatives. Early efforts by Tarr (2,3) and by Waters Associates (4,5) resulted in commercialization of the method under the trade name PICO TAG amino acid analysis. Now most companies selling HPLC systems also describe how to perform PTC amino acid analysis using their respective instrumentation. Compared to other precolumn derivatization methods of amino acid analysis, the PTC method identifies all amino acids including proline and is generally less susceptible to incomplete derivatization and the generation of interfering compounds due to pH conditions and/or inappropriate molar ratios of reagent to sample. Although PTC derivatives are stable when dry and stored at -20°C, a slow, variable degradation of the derivatives occurs in solution at room temperature (3,5,6). As an approach to preventing time-dependent degradation and reducing variability, Applied Biosystems introduced instrumentation in October 1986 that automatically derivatizes each sample followed immediately by on line PTC

[1]Supported in part by USPHS grants EY06603, CA37589, DK38639 and NSF grant DMB-851 6111.

analysis. This article provides an overview of the performance of the Applied Biosystems automatic PTC amino acid analysis system including a progress report on the development of an automatic peptide/protein hydrolyzer.

II. MATERIALS AND METHODS

A. Chemicals and Glassware

Water was obtained from a Milli-Q apparatus (Millipore). Acetonitrile and methanol were from Burdick & Jackson or Applied Biosystems, concentrated HCl (Ultrex) was from J.T. Baker, N,N-diisopropylethylamine (DIEA), sodium acetate and phenylisothiocyanate were from Applied Biosystems and borane-DIEA complex was from Aldrich. An amino acid standard (Pierce H), mininert slide valves and 40 ml screw cap bottles were obtained from Pierce Chemical Co. A glassblower modified the 40 ml bottles for hydrolysis as described by Tarr (3). Pyrex hydrolysis tubes (6 x 50 mm) were obtained from Corning glass works (No. 9820-6) through VWR.

B. The Automatic Hydrolysis and PTC Amino Acid Analysis System

The Applied Biosystems automatic PTC amino acid analysis system is composed of the Model 420 Derivatizer, Model 130 Analyzer and Model 920 Computer. The hydrolyzer used for the present report consists of a station within the Model 420 which is composed of three temperature-controlled reaction chamber heads that clamp above and below glass sample frits. Our tests have been performed with a prototype that hydrolyzes one sample per hour; a more efficient prototype that simultaneously hydrolyzes three samples is under development. For automatic hydrolysis, 6N HCl containing 1% phenol is delivered to the hydrolyzer lower head (Fig. 1), the temperature is increased and a vapor phase hydrolysis carried out for 1 h at 155°C. Following hydrolysis the HCl vapor is flushed out with argon and the sample frit moved by turntable to the next station for derivatization. The Model 420 derivatizer sequentially performs the following tasks: a) an argon flush to remove oxygen; b) treatment with aqueous DIEA vapor to raise the pH for coupling; c) delivery of PITC followed by an argon flush to evaporate most of the heptane; d) a brief methanol delivery to facilitate coupling; e) another short DIEA vapor delivery then coupling proceeds for 20 minutes at room temperature; f) the derivatized sample is argon dried, extracted 2x with 200 µl of 30 mM sodium acetate buffer pH 5.1 and transferred to the Model 130 HPLC system where half of the sample (200 µl) is injected onto a 5µ narrow bore C18 column (2.1 x 250 mm); and g) following injection the entire derivatization system is flushed with methanol and argon dried.

All manual hydrolyses were by 6N HCl vapor containing 1% phenol at 150°C for 1 h (3). Autohydrolyses were according to the manufacturer's specifications; as an approach to preventing loss of Met by oxidation and loss of Asp and Glu from metal ions, a 30 µl aliquot of methanol containing 5% borane-DIEA, 0.5 mM EDTA was applied to frits with autohydrolysis samples.

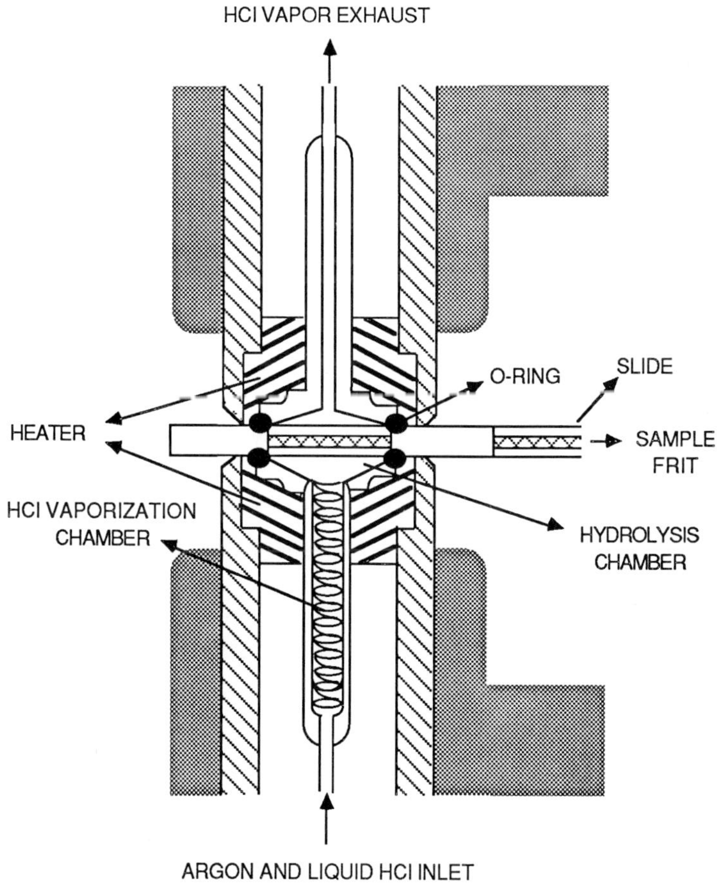

Fig. 1 Schematic cross-section of the hydrolyzer head.

III. RESULTS AND DISCUSSION

A. Resolution and Data Processing

A typical amino acid standard separation obtained with the Applied Biosystems instrumentation showing raw and enhanced data is presented in Fig. 2. The syringe drive Model 130 HPLC pumps provide a stable baseline with very little variation in retention times (Table I). In our work, the enhanced data has proven useful, for example, in quantifying carboxymethyl cysteine (CMC) in peptides containing one residue of CMC and 5 to 6 residues of either Asp or Glu, both of which elute adjacent to CMC (Fig. 2).

With or without automatic hydrolysis, on line statistical data analysis provides a valuable tool for monitoring and maintaining optimum instrument performance. We routinely assess instrument performance by measuring peak area reproducibility with the Model 920 computer. Without hydrolysis and without normalizing the data to an internal standard, the average peak area relative standard deviation for the 16 standard amino acids falls near 3% when our instrument is performing at its best (Table I).

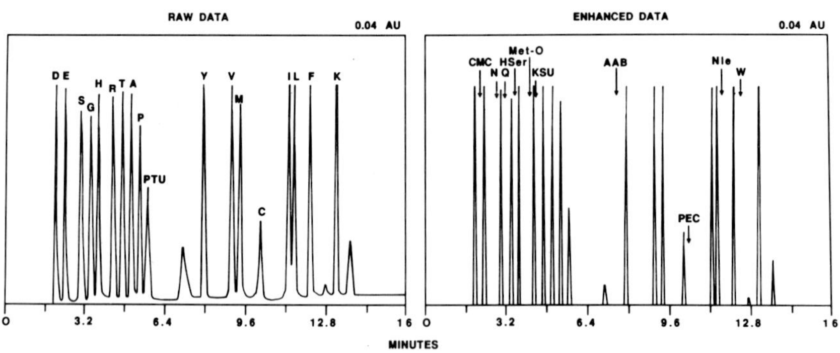

Fig. 2 Separation of PTC amino acid standards (150 pmol) using Applied Biosystems instrumentation at a flow rate of 300 µl/min and at 35°C. Solvent A was 50 mM sodium acetate pH 5.4 and solvent B was 70% acetonitrile. Chromatography conditions must be varied to optimize the resolution of some residues (e.g. CMC). Abbreviations: KSU, succinyl lysine; PEC, pyridylethylcysteine; CMC, carboxymethyl cysteine; MET-O, methionine sulfoxide; AAB, alpha aminobutyric acid; Nle, norleucine; HSer, homoserine.

B. Reproducibility

An automatic hydrolysis and PTC amino acid analysis system must provide reliable peak area reproducibility from protein hydrolysates. Table I shows that autohydrolysis of myoglobin and analysis at the 2 and 15 pmol protein level yielded average peak area relative standard deviations of about 10%, with glycine and proline exhibiting the most variation. The Pro and Gly variability appears to be associated in part with the sample frit pretreatment with the borane/DIEA/EDTA solution. A summary of the average reproducibility from manual and autohydrolysis of several small proteins relative to the amount analyzed is presented in Table II. These data indicate that for the proteins analyzed the present prototype hydrolyzer provides peak area reproducibility comparable to that obtained from manually hydrolyzed samples above the 10 pmol protein level.

Table I

REPRODUCIBILITY OF AUTOMATIC PTC AMINO ANALYSIS

Amino Acid	Amino Acid Standard Retention Time	Amino Acid Standard Peak Area	Autohydrolyzed Myoglobin Peak Area	
	Percent Relative Standard Deviation (%SD)			
ASP	1.36	2.9	10	9
GLX	0.58	3.1	9	11
SER	0.30	2.8	8	10
GLY	0.24	3.0	10	17
HIS	0.36	3.0	9	9
ARG	0.23	3.4	9	10
THR	0.21	2.8	7	8
ALA	0.13	2.5	8	9
PRO	0.16	2.7	32	14
TYR	0.07	2.8	7	5
VAL	0.07	2.7	9	9
MET	0.07	6.7	7	10
ILE	0.06	3.1	7	9
LEU	0.05	3.0	7	8
PHE	0.06	3.6	7	8
LYS	0.07	4.3	7	8
Average % SD	0.25	3.3	9.6	9.6
Amount Analyzed				
pmol Protein		-	15	2
pmol Amino Acid	150	150	30-285	4-38
Number of analyses (n)	9	9	6	7

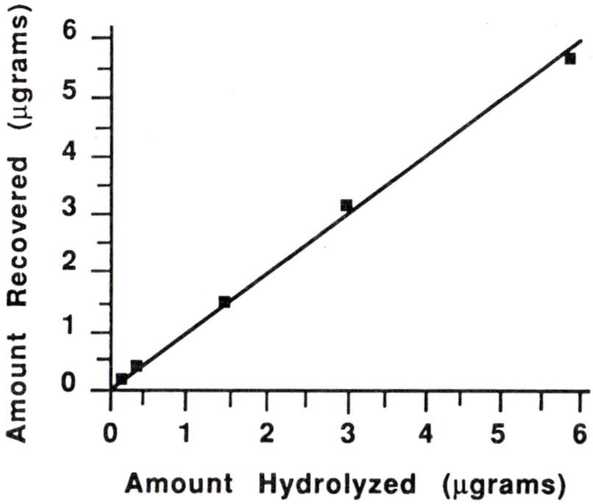

Fig. 3 Linearity of recovery of insulin following automatic hydrolysis. Each data point represents an average of three autohydrolyses. (These data were provided by D. Dupont, Applied Biosytems.)

D. Comparison of Automatic and Manual Hydrolysis

Average compositional data from several samples is presented in Table III, comparing results from automatic and manual hydrolysis. These preliminary data, generated from a broad size range of samples (12 to 1237 residues), indicate that overall average recovery (99%) and reproducibility (ca 16% standard deviation) obtained with the prototype automatic hydrolyzer were comparable to that obtained by manual hydrolysis. However, for this set of data, the compositional accuracy of the manually hydrolyzed samples was superior. The higher average errors associated with autohydrolyzed samples were due in large part to lower recoveries for Val, Ile and Met and inflated values for Gly, Ala and Pro. The incomplete hydrolysis of Val and Ile should be correctable by increased hydrolysis times. A remedy for the peaks generated as by-products of autohydrolysis is currently under investigation.

A promising application of autohydrolysis is in the quality control of solid phase peptide synthesis, particularly with synthesizers that automatically remove a peptide-resin sample from each coupling cycle. Table IV presents a comparison of automatic and manual hydrolysis and analysis of five synthetic resin-bound peptides generated from every other coupling cycle in the synthesis of an 11 residue peptide. The amount of the resin-bound peptide is not limiting and the results show that autohydrolysis yields satisfactory compositions.

TABLE III

COMPARISON OF AUTOMATIC AND MANUAL HYDROLYSIS IN AUTOMATED PTC AMINO ACID ANALYSIS[a]

Amino Acid	α-Phosphorylase[b] Kinase (M_r, 138,422)			β-Phosphorylase[c] Kinase (M_r, 125,177)			RNAase (M_r, 13,690)			Synthetic Peptide L-12-F (M_r, 1,395)			
	by sequence	Manual	Auto	by sequence	Manual	Auto	by sequence	Manual	Auto	by sequence	Manual	Auto	Auto
Asx	100	105	104	107	103	94	15	16	15	1	1	1	1
Glx	145	145	150	131	121	154	12	13	15	1	1	1	1
Ser	99	83	100	69	60	77	15	14	15	0	0	0	0
Gly	79	80	96	65	94	104	3	4	9	1	1	1	2
His	35	35	32	25	25	22	4	4	3	0	0	0	0
Arg	62	77	54	61	75	56	4	4	4	1	1	1	1
Thr	71	63	71	49	48	53	10	9	10	0	0	0	0
Ala	82	80	86	48	73	63	12	13	13	2	2	2	2
Pro	60	54	65	56	55	67	4	4	4	0	0	0	0
Tyr	45	44	46	49	48	49	6	5	6	1	1	1	1
Val	79	77	71	71	68	66	9	9	9	0	0	0	0
Met	36	36	25	21	20	13	4	2	3	1	1	1	1
Ile	69	65	59	70	60	64	3	2	2	0	0	0	0
Leu	142	147	145	127	119	146	2	2	4	2	2	2	2
Phe	38	41	42	42	40	51	3	3	4	1	1	1	1
Lys	63	67	67	68	65	66	10	10	9	1	1	1	0
Total Residues	1237	1199	1213	1092	1074	1145	116	114	125	12	12	12	12
pmol Analyzed		3	3		5	4		17	16		425	597	63
n		3	5		3	5		3	5		3	7	6
% SD Peak area		33	7		15	21		13	21		4	17	12
Average % ERROR		6	8		10	14		9	17		0	0	13
% Recovery auto/manual		100	110		100	80		100	93		100	94	117

a. Residues per mol by automatic phenylthiocarbamyl amino acid analysis using ABI Models 420, 130 and 920. All samples were RP-HPLC purified in aqueous TFA/acetonitrile solvents. Values are from several analyses (n), are averaged and rounded to the nearest integer and are uncorrected for partial destruction, incomplete hydrolysis or contaminations. Hydrolysis and peak area %SD were as described in Materials and Methods. Average % ERROR = 100| known value−measured value|÷ known value and averaged for all amino acids except Gly. % Recovery was evaluated by analyzing the same protein solution following both hydrolysis methods. b. Sequence data from reference 7. c. Sequence data from reference 8.

IV. CONCLUSIONS AND PROSPECTS

Independent of automatic hydrolysis, the Applied Biosystems automatic PTC amino acid analysis system provides an efficient approach to processing large numbers of samples and is more convenient and less labor intensive than any manual precolumn derivatization HPLC based amino acid analysis system that we have employed (e.g. PICO TAG/PTC, OPA). Retention time reproducibility is excellent and reasonable quantitation and compositional accuracy (~10% error) can be obtained around the 20 pmol amino acid level provided that instrument performance is monitored and maintained. When peak area average relative standard deviations creep into the 6-8% range for standard amino acids at the 100 pmol level, we start looking

TABLE IV

COMPARISON OF AUTOMATIC AND MANUAL HYDROLYSIS
QUALITY CONTROL AMINO ACID ANALYSIS OF SYNTHETIC RESIN-BOUND PEPTIDES

Amino Acid	3 residues Manual		Auto	5 residues Manual		Auto	7 residues Manual		Auto	9 residues Manual		Auto	11 residues Manual		Auto
Asx	1.3	(1)	0.9	1.4	(1)	1.5	1.6	(1)	1.4	2.7	(2)	1.7	2.6	(2)	2.7
Glx							1.1	(1)	0.9	1.0	(1)	1.0	1.0	(1)	1.1
Ser													0.7	(1)	0.8
Gly															
His															
Arg															
Thr										0.8	(1)	1.1	1.0	(1)	1.0
Ala															
Pro							1.1	(1)	1.1	1.0	(1)	1.2	1.2	(1)	1.1
Tyr													1.0	(1)	0.8
Val	0.9	(1)	1.1	1.8	(2)	1.6	2.1	(2)	1.4	2.0	(2)	1.4	1.6	(2)	2.0
Met															
Ile	0.8	(1)	1.0	0.9	(1)	0.7	1.0	(1)	0.9	1.0	(1)	1.0	0.8	(1)	0.9
Leu															
Phe				0.8	(1)	1.0	1.0	(1)	1.1	0.9	(1)	1.2	1.1	(1)	1.0
Lys															
pmol peptide analyzed	21		581	180		210	452		640	13		7222	170		439

Residues per resin-bound peptide by automatic phenylthiocarbamyl amino acid analysis using ABI Models 420, 130 and 920. Values are from a single analysis and are uncorrected. Numbers in parenthesis are the expected composition. The eleven residue peptide SYTNPEFVINV was synthesized with an ABI Model 430 Peptide Synthesizer and quality control amino acid analyses were performed on every other resin sample in order to monitor the efficency of the synthesis. Automatic and manual hydrolysis conditions were as described in Materials and Methods.

for potential problems such as leaks, inconsistent deliveries, dirty sample frits and old/contaminated chemicals (see Atherton, this volume).

The past months represent the earliest phase in the performance evaluation of the first Applied Biosystems automatic hydrolyzer. Our tests indicate that automatic hydrolysis can yield reasonable reproducibility and compositional accuracy with moderate amounts of protein. Improvements are still being implemented and we are optimistic that automatic hydrolysis will soon streamline the amino acid analysis procedure as well as enhance both sensitivity and accuracy.

V. REFERENCES

1. R.L. Heinrikson and S.C. Meredith. Anal. Biochem. 136, 65-74 (1984).
2. D.R. Koop, E.T. Morgan, G.E. Tarr and M.J. Coon. J. Biol. Chem. 257, 8472-8480 (1982).
3. G.E. Tarr in "Microcharacterization of Polypeptides, A Practical Manual", J.E. Shively, Ed., pp. 155-194, Humana Press, NJ, 1986.
4. B.A. Bidlingmeyer, S.A. Cohen and T.L. Tarvin. J. Chromatogr. 336, 93-104 (1984).
5. B.A. Bidlingmeyer, T.L. Tarvin and S.A. Cohen in "Methods in Protein Sequence Analysis 1986", K.A. Walsh, Ed., pp. 229-245, Humana Press, NJ, 1987.
6. T. Bergman, M. Carlquist and H. Jörnvall in "Advances Methods in Protein Microsequence Analysis" B. Wittmann-Liebold, Ed. pp 45-55, Springer-Verlag Berlin 1986.
7. N.F. Zander, H.E. Meyer, E. Hoffman-Posorske, J.W. Crabb, L.M.G. Heilmeyer Jr. and M.W. Kilimann. Proc. Natl. Acad. Sci. USA 85, 2929-2933 (1988).
8. M.W. Kilimann, N.F. Zander, C.C. Kuhn, J.W. Crabb, H.E. Meyer, and L.M.G. Heilmeyer, Jr. Proc. Natl. Acad. Sci. USA (1988) in press.

30

AMINO ACID ANALYSIS AT THE FEMTOMOLE LEVEL USING THE DIMETHYLAMINOAZOBENZENE SULFONYL CHLORIDE PRECOLUMN DERIVATIZATION METHOD : Potential and limitation

J.-Y. CHANG, R. KNECHT, P. JENOE
and S. VEKEMANS
Pharmaceuticals Research Laboratories, CIBA-GEIGY Ltd. Basel, CH-4002 Switzerland

I. INTRODUCTION

Amino acid analysis can be performed with two types of technologies. The conventional way is by first separating free amino acids by ion-exchange chromatography and then by detecting separated amino acids with post-column derivatization [1,2]. The contemporary way is by first derivatizing free amino acids and then by analysing derivatized amino acids by reverse phase HPLC. The advantage of the precolumn derivatization technique is its potential of high sensitivity and efficiency. Practically, any reagent which is capable of forming a covalent product with amino acids can be used for precolumn derivatization and many reagents have been proposed for this purpose [3-8]. However, in choosing a suitable reagent, many concerns need to be considered: (a) the ability to react with both primary and secondary amino acids; (b) the ability to yield a single derivative of each amino acid quantitatively and reproducibly; (c) the stability of derivatized amino acids; and (d) the sensitivity and wavelength of detecting amino acid derivatives.

For the past 7 years, our lab has been using the dimethylaminoazobenzene sulfonyl chloride (DABS-Cl, dabsyl chloride) precolumn derivatization technique for high sensitivity amino acid analysis [3,9]. The reproducibility and reliability of this technique have been thoroughly tested and procedures for routine analysis of nanogram amount of dabsylated hydrolysate are established. The DABS-Cl technique has the potential of analysing femtomole (picogram) amount of protein sample, but background contamination encountered during sample handling precludes routine analysis at this sensitivity level. These problems are addressed in this communication.

II. MATERIALS AND METHODS

The DABS-Cl technique consists of sample hydrolysis, dabsylation and HPLC quantitation. Ideally, these three operations should be fully automated. Automatic dabsylation is now being investigated in some labs (Dr. G. Begg, personal communication). In our lab, only HPLC quantitation of DABS-amino acids are currently performed automatically.

1. <u>Hydrolysis</u>:

Sample hydrolysis is the most crucial and difficult step in high sensitivity amino acid analysis because: (a) this is the step in which contaminants are most likely to be introduced and (b) this is also the step during which Met, Cys and Tyr will be partially destroyed if the vacuum system is not in optimal condition. Some measures are recommended to reduce the background and prevent the destruction of amino acids: (a) Vapor phase acid hydrolysis can reduce the background contamination by 75% [9]. A typical blank sample after vapor phase hydrolysis yields an average of only 0.5 to 3 pmoles of Asp, Glu, Ser and Gly. (b)

A good vacuum (lower than 0.1 mbar) in the hydrolysis vessel is absolutely essential for good yield of Met, Cys and Tyr. A good vacuum sealed within a liquid phase hydrolysis tube can usually be maintained throughout the hydrolysis. Unfortunately, none of the commercially available gas-phase hydrolysis vessels (including those from Pierce and Waters) have proven to be perfect after extended use. A slight leak of the vacuum usually occurs during hydrolysis, and this could lead to partial destruction of Met, Cys and Tyr. For this reason, it is important to hydrolyse both standard amino acid mixture and unknown samples simultaneously in the same vessel and quantitate the unknown samples based on the hydrolysed standard. So far, we have not found a substitute other than a good vacuum to prevent Met, Cys and Tyr destruction. Those who intend to automate hydrolysis process should bear this point in mind. The vapor phase hydrolysis can be performed with 6 N HCl at 110°C for 24 h or with TFA/6N HCl (1:2, v/v) at 150°C for 1 h. In addition, hydrolysis can be carried out with methane sulfonic acid for the purpose of Trp determination [10,11].

2. Derivatization:

Dabsylation is the most straight forward step in the DABS-Cl technique. The procedure requires minimum amount of labor and attention. Derivatized amino acid mixture, after dilution [9], are stable at room temperature up to one month. Nonetheless, automation of the dabsylation step could further increase the sensitivity and efficiency of the DABS-Cl method, and eliminate any possible manual errors.

3. Analysis of DABS-derivatives:

DABS-amino acids are completely separated on a reversed phase C-18 column from a variety of suppliers, including those from Merck, Beckman and Brownlee. Complete separation can be achieved within 14 min [12]. In our lab, we have been using

TABLE II

AUTOMATIC HYDROLYSIS AND PTC-AMINO ACID ANALYSIS
SUMMARY OF REPRODUCIBILITY RELATIVE TO AMOUNT ANALYZED

Sample	Amount Analyzed (pmol protein)	Number of Hydrolyses/Analyses	Overall Peak Area % SD
I. Automatic Hydrolysis			
Cytochrome *c*	109	8	17
RNase	45	9	7
CMC-Lysozyme	40	8	13
Cytochrome *c*	39	8	8
	$\bar{x}$ ~ 41	n = 25	$\bar{x}$ ~ 9
Myoglobin	19	8	11
Myoglobin	15	6	10
CMC-Lysozyme	10	10	24
RNase	7	9	15
Cytochrome *c*	6	7	12
Myoglobin	6	8	24
	$\bar{x}$ ~ 7	n = 34	$\bar{x}$ ~ 18
Myoglobin	2	7	10
II. Manual Hydrolysis			
RNase	6	10	12
Cytochrome *c*	6	10	10
RNase	2	10	12
	$\bar{x}$ ~ 5	n = 30	$\bar{x}$ ~ 11

Hydrolysis, PTC-amino acid analysis and calculations were as described in Material and Methods.

C. Linearity and Sensitivity

The response linearity of the system without automatic hydrolysis was investigated by injection of amino acid standards in amounts of 10-10,000 pmol. A linear response of peak area versus amount of amino acid analyzed was obtained between 10 and 5000 pmol (not shown). It is possible to detect PTC-amino acids at and below the 1 pmol level (1,3-6), however sample contamination and background level amino acids limit the useable analysis range in most laboratories. Presently in our laboratory, 20-30 pmol of amino acid analyzed is a practical lower operational limit for routine analyses.

Preliminary evaluation of the linearity of the system with automatic hydrolysis was performed by autohydrolysis of varying amounts of insulin. Figure 3 shows that a linear response of amount recovered versus amount hydrolyzed was obtained between 0.09-5.7 µg insulin (16-1000 pmol protein).

the Merck Lichrospher 100 CH-18/2 column (12.5 cm x 4 mm i.d.) for routine analysis. Separation is completed in 22 min and cycle time from injection to injection is about 30 min. Most columns have a life time of at least 1500 to 2000 injections. Complete separation of modified amino acid derivatives has also been reported [10].

III. RESULTS AND DISCUSSION

1. Sensitivity of the DABS-Cl method

The major attraction of the precolumn derivatization technique is its potential high sensitivity. DABS-amino acids (mono-derivative) have molar extinction coefficient of 34,000 at 420 nm. Because of the stable baseline obtained by monitoring at the visible wavelength, DABS-amino acids are readily detected at the femtomole level (Fig. 1) and has the potential to be detected at subfemtomole level [13]. Amino acid composition of picogram quantity of dabsylated protein hydrolysate can be accurately (± 10%) analysed (Figs. 2 & 3). However, this was achieved by hydrolysing and dabsylating a few hundred nanogram of protein and analysing only a portion of it. It is yet impossible to hydrolyse picogram amount of protein without risking contaminants (see Atherton, D, this volume). Despite strenuous purification and cleaning of the solvents and glasswares, there are 0.5 to 3 picomoles of Asp, Glu, Ser and Gly which remain difficult to remove. To offset this level of background contaminants down to less than 10% of the sample, one needs to hydrolyse at least 20 to 50 nanograms of protein. Thus, the major hurdle for the amino acid analysis at the femtomole level is not the detection limit, but the sample handling. To make femtomole amino acid analysis a routine technique, one need to (a) develope reliable methods for isolating picogram amount of proteins without introducing appreciable contaminants and (b) design a sample-handling and sample hydrolysis protocol which would reduce the background contaminant of the blank sample down to the low femtomole level.

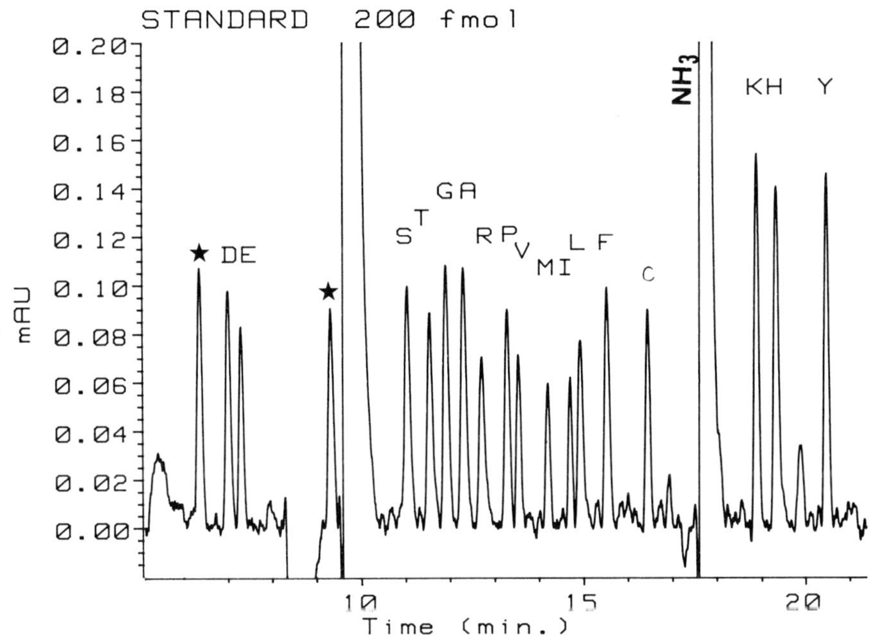

Figure 1 HPLC quantitation of DABS-amino acid standard (200 femtomoles each, except for Cys which is 100 femtomoles). The data was obtained from a Hewlett-Packard 1090 liquid chromatograph. The original chromatogram was automatically filtered and integrated to give this figure. Hydrolysed excess reagent was eluted at 10 min. Two major ghost peaks are marked by an asterisk. Solvent A was 25 mM sodium acetate, pH 6.45, containing 4% dimethylforamide. Solvent B was acetonitrile. The gradient was 18%B to 23%B in two min, 23%B to 30%B from 2 to 8 min, 30%B to 41%B from 8 to 13 min, 41%B to 51%B from 13 to 15 min, 51%B to 54%B from 15 to 17 min, 54%B to 64%B from 17 to 19 min, 64%B to 90%B from 19 to 20 min, and then 90%B to 18%B from 20 to 22 min. The column temperature was 40°C. The flow rate was 0.2 ml/min The detector wavelength was 456 nm. The column was Brownlee RP-18 Spheri-5 (10 cm x 2.1 mm).

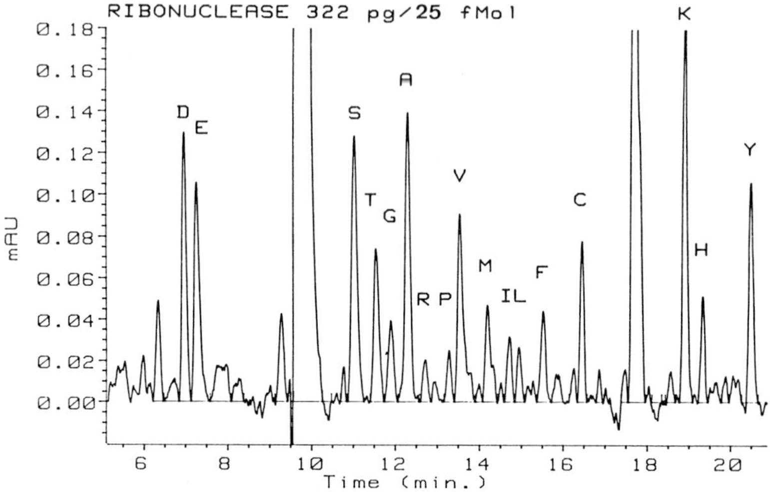

Figure 2 Chromatogram from the amino acid analysis of ribonuclease at the femtomole level. Five hundred nanograms of ribonuclease were hydrolysed, dabsylated and diluted with the dilution buffer, and 25 femtomoles (322 picograms) of the sample were injected for analysis. Chromatographic conditions are described in Fig. 1

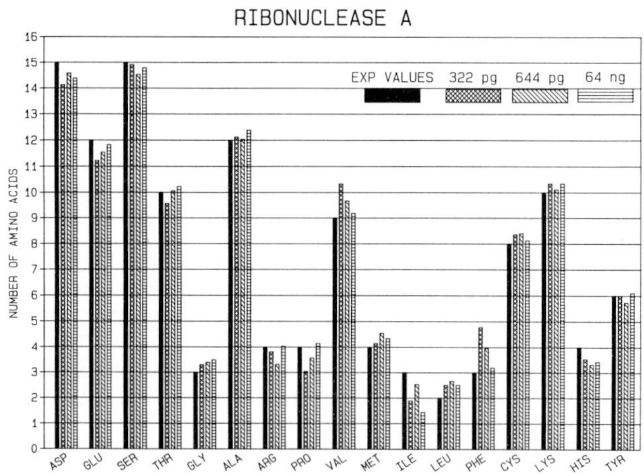

Figure 3 Reproducibility and accuracy of amino acid quantitation at the femtomole level. Five hundred nanograms of ribonuclease were hydrolysed, dabsylated and 322 pg, 644 pg and 64 ng were analysed. Conditions see Fig.1 and Fig. 2.

2. Reliability and accuracy of the DABS-Cl method

The major concern of the precolumn derivatization approach is its reliability and accuracy as compared to the conventional amino acid analyser. The reliability of a precolumn derivatization technique depends critically upon the reproducibility of derivatization and the stability of derivatized amino acids (see A. Smith et al., this volume). Though all reported reagents have been shown to react with amino acid in a reproducible manner [3-9], the stability of their amino acid derivatives varies significantly (20% destruction of certain amino acid derivatives can vary from seconds to years). The sulfonamide bond of DABS and DNS derivatives are comparably the most stable ones. It has been demonstrated that all DABS-amino acids can be left at room temperature in the dilution buffer (ethanol/phosphate buffer, pH 7.0, 1/1, v/v) over a period of one month without any detectable degradation (< 2%). The azo-dye chromophore is also extremely stable to day light and devoid of the fluorescence quenching problem encountered by the DNS-Cl method [6]. These characteristics ensure the reliability and accuracy of the DABS-Cl method which have been demonstrated by several independent labs [9,10,12]. Fig. 4 shows the reproducibility and accuracy of 4 interferon samples which were hydrolysed, dabsylated and analysed independently.

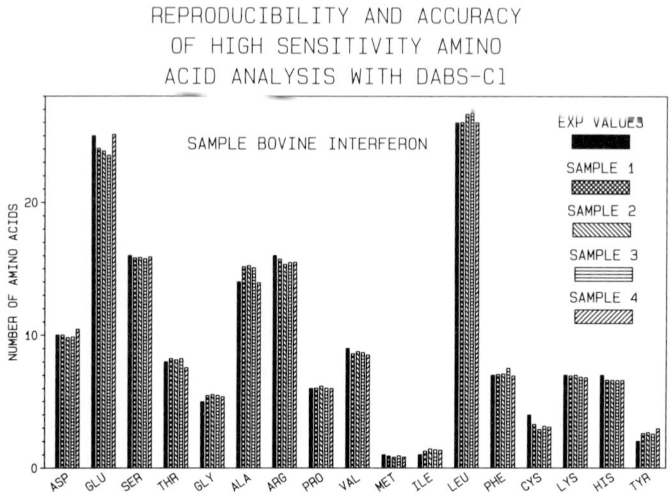

Fig. 4 Reproducibility of the DABS-Cl method. Four bovine interferons (1 ug each) were hydrolysed and dabsylated independently. Five percent of each derivatized sample were analysed. The expected values are shown at the left column of each amino acid.

3. Limitations of the DABS-Cl method

Despite its advantages, the DABS-Cl method possesses limitations which are not shared by the conventional analyser. The DABS-Cl technique has restricted tolerance of salt contaminant in the sample. Large amount of salts present in the sample will possibly alter the pH of the dabsylation buffer and obstruct the reproducibility of the derivatization. Our preliminary studies have shown that when 1 ug of protein was hydrolysed and dabsylated, the approximate limits of salt tolerance were: urea, 10 ug; sucrose, 50 ug; sodium dodecyl sulfate, 100 ug; mannitol, 100 ug; benzethonium chloride, 100 ug; sodium chloride, 1000 ug; sodium acetate, 2000 ug. Interferences of these salts were investigated independently. When concentration of salts are above these limits, deviations of amino acid recoveries could exceed 15%. In applying the DABS-Cl method, one must also know the approximate amount of sample in hand. The reagent must be at least four fold over the total amino acids, otherwise incomplete derivatization will occur. For instance, when dabsylation was carried out with 20 ul of sodium bicarbonate buffer (50 mM, pH 8.3) and 40 ul of DABS-Cl solution (4 mM in acetonitrile), the sample should be kept within 0.05 to 4 ug. If the integrated excess reagent peak is lower than 75% of the total reagent added, experiment must be repeated. Fortunately, too much of the sample is never a serious problem in the high sensitivity amino acid analysis.

IV. REFERENCES

1. Spackmann, D. H., Stein, W. H. and Moore, S. (1958) Anal. Chem. 30, 1190-1196.

2. Bensen, J. R. and Hare, P. E. (1975) Proc. Natl. Acad. Sci. USA 72, 619-622.

3. Chang, J.-Y., Knecht, R. and Braun, D. G. (1983) Methods Enzymol. 91, 41-48.

4. Lindroth, P. and Mopper, K. (1979) Anal. Chem. 51, 1667-1674.

5. Heinrikson, R. L. and Meredith, S. C. (1984) Anal. Biochem. 136, 65-74.

6. Wilkinson J. M. (1978) J. Chromatogr. Sci. 16, 547-552.

7. Einarsson, S., Josefsson, B. and Lagerkvist, S. (1983) J. Chromatogr. 282, 609-618.

8. Watanabe, Y. and Imai, K. (1981) Anal. Biochem. 116, 471-472.

9. Knecht, R. and Chang, J.-Y. (1986) Anal. Chem. 58, 2375-2379.

10. Vendrell, J. and Aviles, F. X. (1986) J. Chromatogr. 358, 401-413.

11. Simpson, R. J., Neuberger, M. R. and Liu, T. Y. (1976) J. Biol. Chem. 251, 1936-1940.

12. Hughes, G. J., Frutiger, S. and Fonck, C. (1987) J. Chromatogr. 389, 327-330.

13. Nolan, T. G. and Dovichi, N. J. (1987) Anal. Chem. 59, 2803-2805.

31

DETECTION OF NORLEUCINE INCORPORATION INTO RECOMBINANT PROTEINS SYNTHESIZED IN E. COLI

Bernard N. Violand, Ned R. Siegel, Gregg Bogosian, Wesley E. Workman and James F. Kane

Monsanto Agricultural Co., Animal Sciences Division
St. Louis, MO 63198

I. INTRODUCTION.

It has been well-documented that the methionine analog norleucine can substitute for methionine in bacterial proteins (1,2,3). This has been demonstrated by growing cells in the presence of limiting methionine and adding norleucine. Norleucine can substitute for methionine in proteins both at the amino-terminus and at internal residues (4,5). Norleucine containing proteins resemble their wild-type counterparts in terms of structure and activity as demonstrated for Staphylococcal aureus staphylococcal nuclease (6), E. coli beta-galactosidase (7) and E. coli adenylate kinase (8).

This report describes the isolation and characterization of norleucine containing species of recombinant methionyl bovine somatotropin, Sometribove. The norleucine discovered in Sometribove was endogenously synthesized by E. coli strain W3110G containing the recombinant plasmid pBGH1. The biosynthetic route by which norleucine is produced in this strain of E. coli is discussed in detail in another report (9). Sometribove containing only one norleucine residue was separated from normal Sometribove by isoelectric focusing or RP-HPLC. The sites of norleucine incorporation were determined by tryptic peptide mapping of the altered bovine somatotropin followed by amino acid analysis, amino acid sequencing and FAB-MS of the modified peptides. Similar norleucine incorporation has also been shown to occur in recombinant porcine somatotropin.

II. RESULTS AND DISCUSSION.

A. Isolation and Characterization of Norleucine Containing Species of Sometribove.

1. Sometribove containing norleucine at the amino-terminus.

Figure 1 shows an isoelectric focusing (IEF) gel of

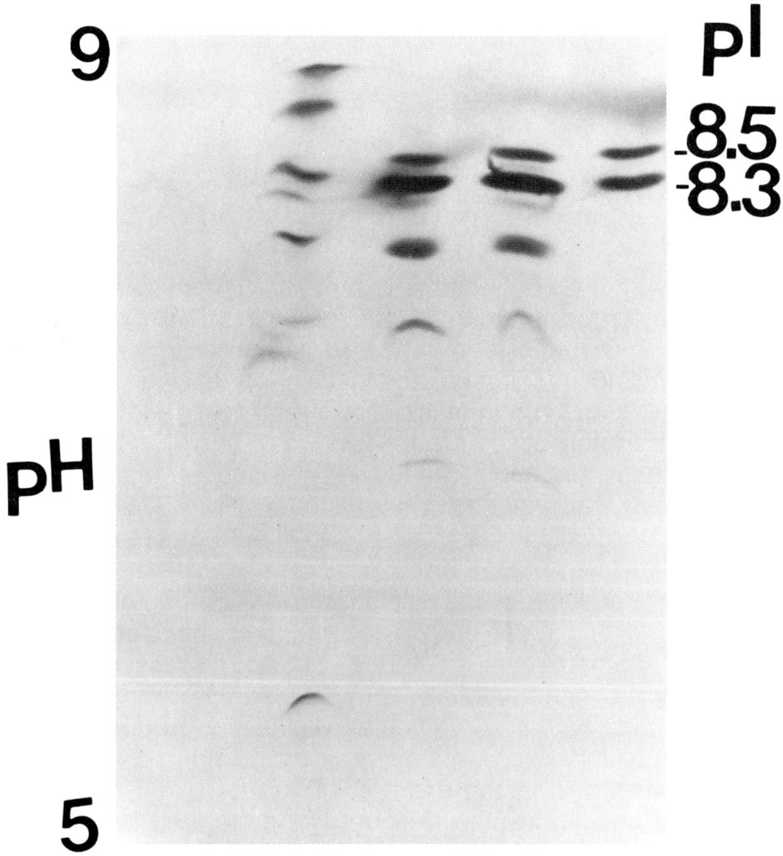

Figure 1. Isoelectric focusing gel (pH 5 to 9) of Sometribove (lanes 2,3) and partially purified 8.5 pI form of Sometribove (lane 4). Lane 1 is Pharmacia IEF standards.

Sometribove. The normal pI of this protein under the conditions used for IEF is 8.3. Preparative IEF using an LKB ampholine electrofocusing column was used to prepare Sometribove which contained approximately 50% normal pI and 50% of an 8.5 pI species (lane 4).

Peptide mapping using trypsin digestion followed by RP-HPLC separation of the generated peptides was used to analyze the partially purified 8.5 pI species of Sometribove (Figure 2). All the tryptic peptides appeared normal except for the amino-terminal peptide (peak 1) which was reduced in intensity and a new later-eluting peptide (peak 2) of almost equal intensity was observed. The normal amino-terminal tryptic peptide consists of 17 amino acids, residues -1 to 16. The amino-terminal methionine of Sometribove is numbered -1 followed by residues 1,2 etc. as shown in Figure 3. The normal and modified peptides were isolated by RP-HPLC and subjected to amino acid analysis. Identical results were obtained for the two peptides except that in

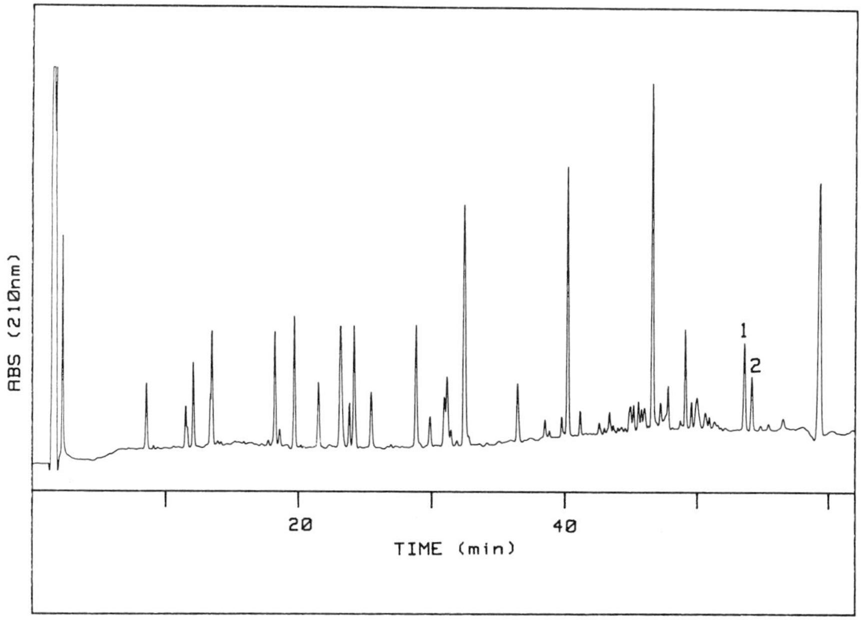

Figure 2. RP-HPLC tryptic peptide map of partially purified 8.5 pI form of Sometribove shown in lane 4 of Figure 1. Peptide 1-Normal amino-terminal peptide, residues -1 to 16. Peptide 2-Amino-terminal peptide containing norleucine at residue -1.

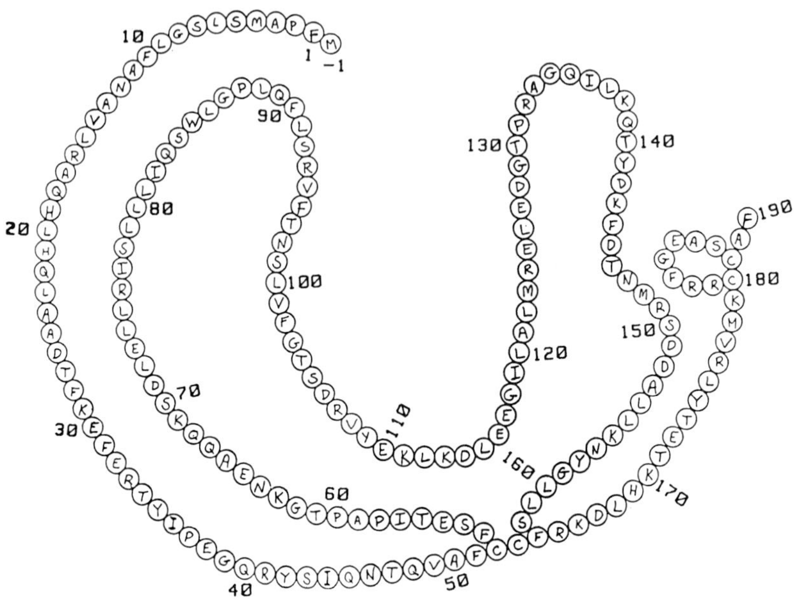

Figure 3. Amino acid sequence of recombinant methionyl bovine somatotropin (Sometribove). The methionines are at residues -1, 4, 123, 148 and 178.

the modified peptide the methionine was reduced in intensity and a new amino acid was found which eluted on the descending limb of leucine (Figure 4). The retention time of this new amino acid corresponded exactly to that of norleucine. Table I shows that the modified peptide contained 1 methionine and 1 norleucine.

Since the amino-terminal tryptic peptide of Sometribove normally contains methionines at residues -1 and 4, amino acid sequencing was used to determine which methionine was replaced by norleucine. This result showed that the norleucine was incorporated at the amino-terminal residue and methionine was at residue 4. The sequence obtained for this peptide was:

-1 4 10
nLeu-F-P-A-M-S-L-S-G-L-F-A-N-A-V-L-R

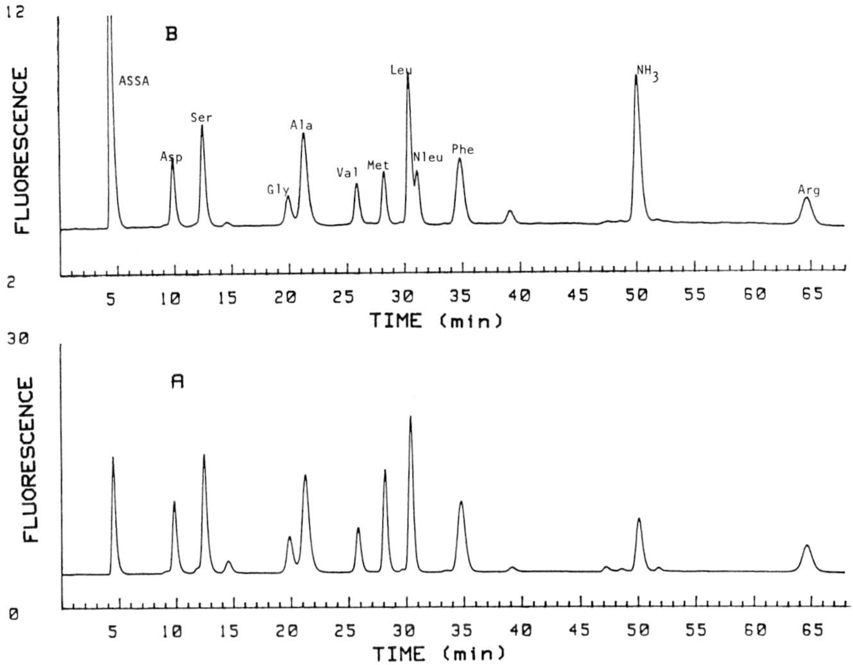

Figure 4. Amino acid analyses of normal (A) and modified (B) amino-terminal peptides.

TABLE I. Amino Acid Compostion of Normal and Modified Amino-Terminal Peptides of Sometribove.

AMINO ACID	THEORETICAL	NORMAL PEPTIDE	MODIFIED PEPTIDE
Asx	1	1	1
Ser	2	2	2
Gly	1	1	1
Ala	3	3	3
Val	1	1	1
Leu	3	3	3
Phe	2	2	2
Arg	1	1	1
Pro	1	n.d.	n.d.
Met	2	2	1
nLeu	0	0	1

n.d.-not determined

Masses of the normal and altered amino-terminal peptides were determined by FAB-MS to be 1825 and 1807 daltons, respectively. This is the exact difference expected between these two peptides since a methylene group (14 daltons) replaces a sulfur (32 daltons) during this amino acid substitution. This data confirmed the amino acid analyses and sequencing data obtained for these two peptides.

The only chemical modification present in the 8.5 pI form of Sometribove was the substitution of norleucine for methionine at the amino-terminus. Both of these amino acids contain uncharged side chains but since this substitution occurs at the amino-terminus the pK_b of the alpha-amino group is altered which could affect the overall pI of this protein. It has been previously reported that the amino-terminal amino acid does affect the overall pI of natural bovine somatotropin (10).

2. RP-HPLC separation of Sometribove containing norleucine at residue 123.

The above analyses demonstrated that norleucine could substitute for methionine during biosynthesis of Sometribove. However, IEF has the disadvantage that it is not quantitative and therefore an HPLC method was developed which was used to quantitate the % of norleucine substitution.

RP-HPLC using a Vydac C-18 column with TFA and acetonitrile as the mobile phase resolved Sometribove into three distinct peaks (Figure 5). These proteins were isolated and characterized by tryptic peptide mapping. The first component was comprised of normal Sometribove. The second-eluting component was determined to be N-formyl-methionyl-Sometribove which presumably was present because of incomplete removal of the N-formyl group during synthesis of this protein. The third component, which consisted of 5% of the total Sometribove for this preparation, was subjected to tryptic peptide mapping as shown in Figure 6. This altered molecule differed from normal Sometribove because two peptides were absent in the map: one of amino acid residues 112-124 and the other of residues 114-124. Two new later-eluting peptides were observed in place of the above two peptides and they were isolated for characterization. Amino acid analysis of the two new peptides revealed that no methionine, the normal amino acid at residue 123, was present but one mole of norleucine per mole of peptide was present. Amino acid sequencing results demonstrated that

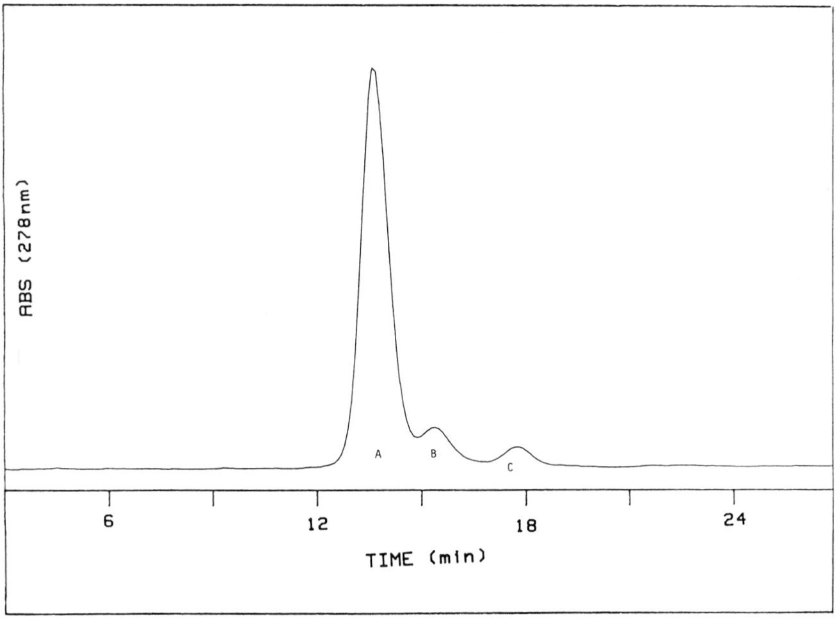

Figure 5. RP-HPLC analysis of intact Sometribove. Peak A-Normal Sometribove. Peak B-N-formylated Sometribove. Peak C-Sometribove with norleucine at residue 123.

both new peptides contained norleucine in place of the normal methionine at residue 123. The sequences for these two peptides were:

112 124	Peptide
L-K-D-L-E-E-G-I-L-A-L-nLeu-R	3
D-L-E-E-G-I-L-A-L-nLeu-R	4

Two peptides were obtained for this portion of Sometribove because 100% cleavage of lysine-113 by trypsin does not occur under the conditions used for this digestion. The only chemical difference between norleucine and methionine is the replacement of a sulfur by a methylene group, yet this substitution at residue 123 results in a dramatic increase in the retention time of this modified Sometribove. Since there are five methionine residues in Sometribove there are four other potential mono-norleucine species which could be synthesized, but none of these species were resolvable from normal Sometribove by RP-HPLC.

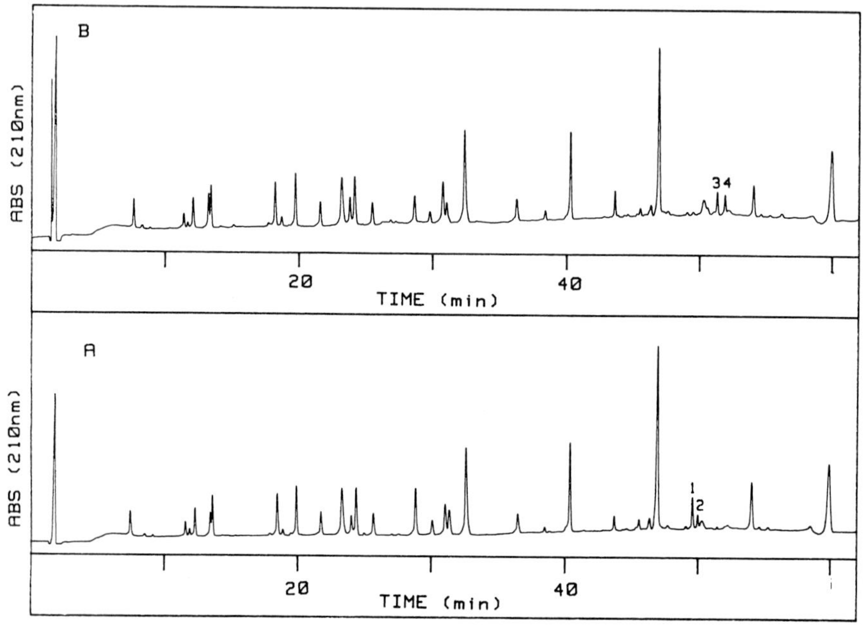

Figure 6. RP-HPLC tryptic peptide maps of normal Sometribove (A) and peak C (B) from Figure 5. Peak 1-Normal residues 112-124. Peak 2-Normal residues 114-124. Peak 3-Residues 112-124 with norleucine 123. Peak 4-Residues 114-124 with norleucine 123.

This RP-HPLC method can be used for analyzing Sometribove both in the absence and presence of a disulfide reducing agent, so that it is not necessary to have the disulfide bonds intact to obtain this high resolution. This allows the analysis of Sometribove directly from refractile bodies without requiring an oxidation/folding step.

Masses of the normal and norleucine containing peptides of residues 112-124 were determined by FAB-MS to be 1500 daltons and 1482 daltons, respectively. This 18 dalton difference confirmed the sequencing data obtained for these peptides.

B. Amino Acid Sequencing of Intact Sometribove.

The above data showed that two of the five methionine residues in Sometribove could be replaced by norleucine, and these two mono-norleucine species could be resolved from the remainder of the protein either by IEF or RP-HPLC. The substitution of norleucine for methionine should occur randomly in Sometribove, therefore all methionine sites in Sometribove should be replaced by norleucine to the same extent. Since there is a methionine at residue 4 in Sometribove the % of norleucine substitution at this residue, along with residue -1, could be determined by amino acid sequencing of the intact protein. The % of norleucine substitution at residues -1 and 4 determined by sequencing was compared to the % obtained by RP-HPLC analysis (detects norleucine at residue 123) as shown in Table II. Within experimental error, this data shows that norleucine substitution was equivalent at these three residues. No method is available to easily quantitate the amount of norleucine at residues 148 and 178, but amino acid analysis of the entire protein was used to determine the overall % norleucine. This analysis gave a similar result (5.4%) to that obtained in Table II demonstrating that norleucine substitution occurs randomly in Sometribove.

C. Analysis of Sometribove Synthesized in the Presence of Exogenously Added Methionine.

The effect of adding excess methionine to the fermentor during the synthesis of Sometribove was investigated to determine if norleucine-free Sometribove could be produced. Figure 7 shows a RP-HPLC chromatogram of Sometribove synthesized in the presence of added methionine. The norleucine-123 species of Sometribove was absent in this chromatogram (Compare to Figure 5). This shows that Sometribove containing norleucine at residue 123 was not synthesized under these fermentation conditions. To confirm the absence of norleucine the Sometribove was analyzed by amino acid sequencing. No norleucine was detected at

TABLE II. Per Cent of Norleucine in Sometribove Determined by RP-HPLC vs Sequencing.

Method	Molecule Assayed	% nLeu/Met+nLeu
RP-HPLC	nLeu(123)	5.0
Sequencing	nLeu(-1)	5.0
Sequencing	nLeu(4)	5.9

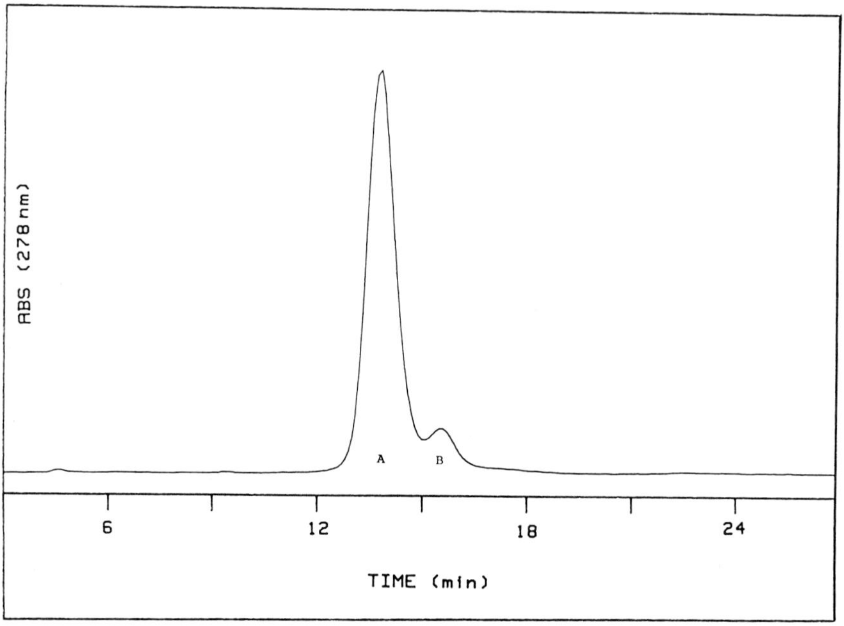

Figure 7. RP-HPLC analysis of Sometribove synthesized in the presence of added methionine or 2-hydroxy-4-thiobutanoic acid. Notice the absence of Sometribove containing norleucine-123. (Compare to Figure 5).

residues -1 or 4 confirming the absence of norleucine when methionine was added to the fermentor. The 8.5 pI species of Sometribove, which contains norleucine at residue -1, was not present when this sample was analyzed by IEF (not shown). Amino acid analysis of the entire protein also showed the absence of norleucine. Norleucine-free Sometribove was also produced when the methionine precursor, 2-hydroxy-4-methylthiobutanoic acid, was added to the fermentor instead of methionine. This data demonstrated that Sometribove devoid of norleucine could be produced by adding methionine or 2-hydroxy-4-methylthiobutanoic acid during the synthesis of this protein.

III. SUMMARY.

This report demonstrates that norleucine was incorporated into recombinant methionyl bovine somatotropin during its synthesis in E. coli. This incorporation significantly altered either the pI or hydrophobicity of this protein depending on which methionine was replaced. The amount of norleucine incorporation was easily quantified by either RP-HPLC or amino acid sequencing of the intact protein. Incorporation of norleucine into Sometribove could be prevented by the addition of methionine to the fermentor during synthesis of this protein.

ACKNOWLEDGEMENTS.

We gratefully acknowledge the FAB-MS data which was generated by E. Kolodjiez and P. Toren of the Monsanto Physical Sciences center. We would also like to acknowledge the excellent technical assistance of M.R. Schlittler and A. Johnson.

REFERENCES.

1. Munier, R., and Cohen, G.N. (1956) Biochim. Biophys. Acta 21, 592-593.
2. Cohen, G.N., and Munier, R. (1959) Biochim. Biophys. Acta 31, 347-356.
3. Munier, R., and Cohen, G.N. (1959) Biochim. Biophys. Acta 31, 378-391.
4. Trupin, J., Dickerman, H., Nirenberg, M., and Weissback, H. (1966) Biochem. Biophys. Res. Comm. 24, 50-55.
5. Kerwar, S.S., and Weissback, H. (1970) Arch. Biochem. Biophys. 141, 525-532.
6. Anfinsen, C.B., and Corley, L.G. (1969) J. Biol. Chem. 244, 5149-5152.
7. Naider, F., Bohak, A., and Yariv, J. (1972) Biochem. 11, 3202-3208.
8. Gilles, A., Marlieres, P., Rose, T., Sarfati, R., Longin, R., Meier, A., Fermandjian, S., Monnot, M., Cohen, G.N., and Barzu, O. (1988) J. Biol. Chem. 263, 8204-8209.

9. Bogosian, G., Violand, B.N., Dorward-King, E.J., Workman, W.E., Hartley, D.L., and Kane, J.F., J. Biol. Chem., in press.

10. Ellis, S., Lorenson, M., Grinderland, R.E., and Callahan, P.X., (1972) "Growth and Growth Hormone" (eds. Pecile, A. and Mueller, E. E.) p.55, Exerpta Medica, Amsterdam.

32

OCCURRENCE OF METHIONINE SULFOXIDE DURING PRODUCTION OF RECOMBINANT HUMAN INSULIN-LIKE GROWTH FACTOR I (IGF-I)

Maris G.N. Hartmanis

KabiGen AB
S-112 87 Stockholm, Sweden

Åke Engström[1]

Department of Immunology
Uppsala University
S-751 23 Uppsala, Sweden

I. INTRODUCTION

During the production of recombinant proteins and peptides in heterologous systems unwanted modifications can occur. Some of the modifications are not always easy to detect and characterize. Here we present the use of mass spectrometry for characterization of an amino acid modification that occurred during production of recombinant human IGF-I. An essential component in this analysis is the possibility to achieve accurate molecular weight determinations of intact peptides.

Purified human recombinant IGF-I produced in Escherichia coli was shown to be heterogeneous by a high resolution HPLC method. This method revealed two forms of IGF-I that were indistinguishable by

---

[1]Supported by a grant from the Swedish Board for Technical Development.

biological activity, amino terminal sequence analysis, isoelectric point, SDS gel electrophoresis and amino acid analysis. By the combined use of plasma desorption mass spectrometry (PDMS), enzymatic fragmentation and HPLC the modified IGF-I was shown to contain a methionine sulfoxide at the position of the methionine.

II. MATERIALS AND METHODS

KabiGen's EcoSec expression system was used for production of recombinant human IGF-I (1). IGF-I is produced in E. coli as a secreted fusion protein, where the fusion partner is a synthetic IgG-binding peptide derived from Staphylococcus aureus protein A.

After immunoaffinity chromatography on IgG-Sepharose, the fusion protein is cleaved at a hydroxylamine sensitive asparagine-glycine bond. After cleavage with 2M hydroxylamine, pH 9.2, for 5h at 45°C, IGF-I is purified using standard chromatography techniques.

Samples for PDMS were prepared by drying 1-5 μg of peptide dissolved in 0.05% trifluoroacetic acid containing 50% ethanol onto aluminum foils covered with nitrocellulose. In order to remove possible interfering cations, the samples were washed with 0.1% trifluoroacetic acid and reanalyzed.

PDMS (2) was carried out using a BIO ION 20 instrument (BioIon Nordic AB, Uppsala, Sweden). Spectra of the intact IGF-I molecules were measured at BioIon Nordic AB using an instrument equipped with a long flight tube (3).

Prior to enzymatic digestion, IGF-I was treated with dithioerythritol and reacted with 4-vinylpyridine.

III. RESULTS

After a final purification step using reverse phase HPLC, human recombinant IGF-I, Fig. 1, was found to be homogeneous as determined by a variety of analytical techniques. However, use of a high resolution reverse phase HPLC method on the purified material revealed two forms of IGF-I, Fig. 2.

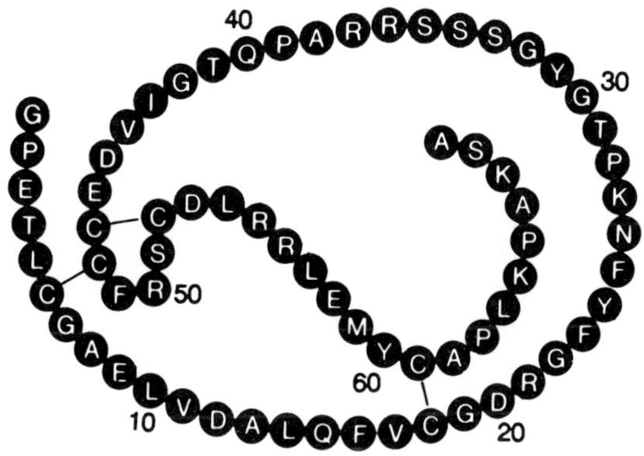

Fig. 1. Amino acid sequence of human insulin-like growth factor I (IGF-I).

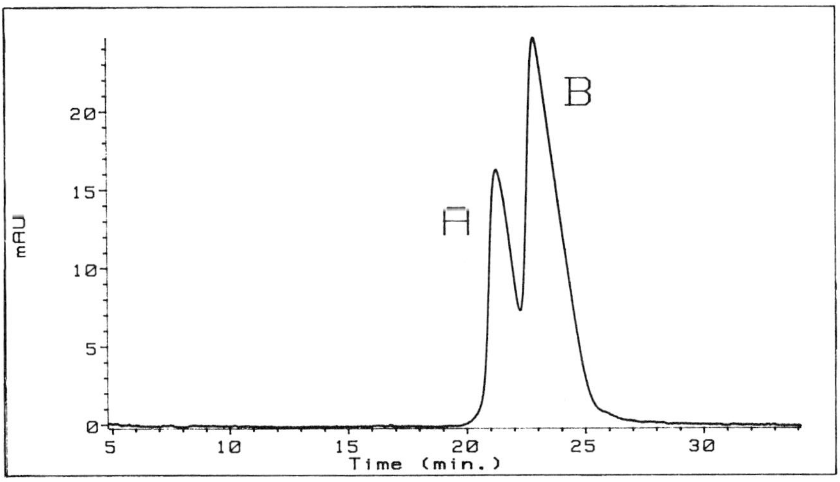

Fig. 2. Reverse phase HPLC separation of purified IGF-I into two forms, A and B. The separation was carried out using a 5 μm Vydac C_{18} (4.6 x 250 mm) column. A linear gradient was run from 32 to 35% acetonitrile in 0.25% pentafluoropropionic acid in water for 30 min at a flow rate of 1 ml/min. The absorbance was monitored at 280 nm.

Efforts to distinguish between the two components by different analytical techniques were unsuccessful. There were no detectable differences between the two forms in specific activity, amino terminal sequence, electrophoretic mobility, isoelectric point or amino acid composition after acid hydrolysis. In order to detect possible minor chemical modifications of the recombinant IGF-I, the purified fractions A and B, Fig. 2, were subjected to PDMS analysis. The results showed a mass difference of approximately 15 mass units between fractions A and B, Fig. 3. Fraction B exhibits a mass corresponding to the calculated molecular mass of native IGF-I, and fraction A has a mass which is higher than expected. We concluded that fraction B was the correct and unmodified IGF-I and that fraction A had undergone some sort of modification that made it slightly less hydrophobic in the HPLC system used to separate the two forms, Fig. 2.

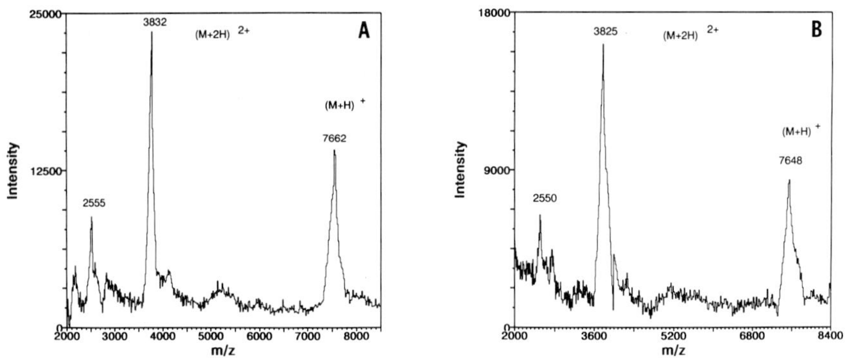

Fig. 3. PDMS analysis of the purified fractions A (A) and B (B) from Fig. 2. The calculated molecular weight of native IGF-I is 7648 Daltons.

In order to determine the position and nature of the modification, fraction A was reduced with dithioerythritol, treated with 4-vinylpyridine and digested with Staphylococcus aureus V8 protease, which cleaves at glutamic acid residues. The digest was then analyzed with PDMS, Fig. 4.

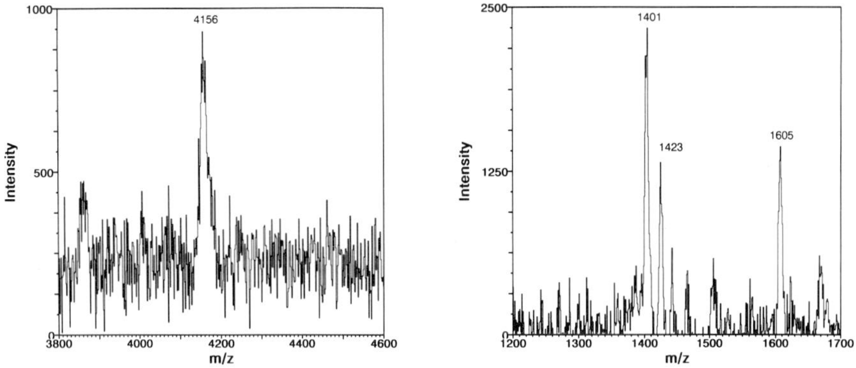

Fig. 4. Two regions of the PDMS spectrum of V8 digested fraction A of IGF-I after reduction with dithioerythritol and treatment with 4-vinylpyridine. A mass of 4156 corresponds to IGF-I fragment 10-46. A mass of 1605 corresponds to IGF-I fragment 47-58, with only one of the three cysteines modified with 4-vinylpyridine. The masses of 1401 and 1423 are 16 mass units higher than the expected masses of the hydrogen and sodium forms of IGF-I fragment 59-70.

In the spectra we detected the masses corresponding to fragment 10-46 (4156 mass units) and fragment 47-58 with only one of the three cysteines reacted with 4-vinylpyridine (1605 mass units). The other two masses in Fig. 4, at 1401 and 1423 mass units, are 16 mass units higher than expected for the hydrogen and sodium forms of fragment 59-70. The IGF-I fragment 59-70 contains the only methionine residue in the molecule, Fig. 1, and an increase in mass by 16 mass units indicates a formation of methionine sulfoxide. To confirm this oxidation, we separated the V8 fragments by reverse phase HPLC and identified the fraction containing fragment 59-70 (1400 mass units) by mass spectrometry, Fig. 5A. After treatment of this fragment with 2-mercaptoethanol for 48 h, molecules with the expected mass of the methionine form of fragment 59-70 (1384 mass units) were detected, Fig. 5B.

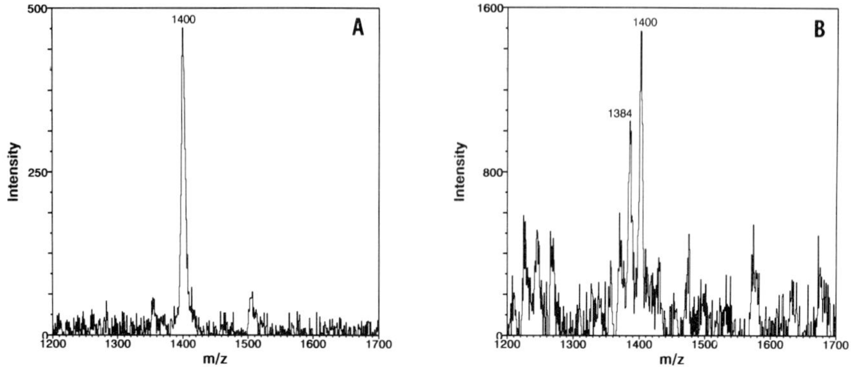

Fig. 5. PDMS analysis of IGF-I fragment 59-70 from fraction A after purification on reverse phase HPLC before (A) and after (B) reduction. The peptide was reduced with 5% 2-mercaptoethanol in 0.2 M ammonium bicarbonate buffer, pH 8.4, for 48 h. A mass of 1384 indicates a reduction of the methionine sulfoxide at position 59 to methionine.

The existence of a methionine sulfoxide in fraction A of IGF-I, Fig. 2, was further shown by Edman degradation of cyanogen bromide treated peptides. The sequence analysis of the chemical digest showed, in the case of fraction B, both the amino terminal sequence and the sequence of fragment 60-70. Analysis of the cyanogen bromide treated fraction A showed only the native amino terminal sequence.

Additional support for the presence of methionine sulfoxide in fraction A was obtained by mild oxidation of fraction B with N-chlorosuccinimide. The oxidation produced a component which coeluted with fraction A in the HPLC system described in Fig. 2.

The oxidation of methionine to methionine sulfoxide in IGF-I could largely be eliminated by carrying out the chemical cleavage procedure with hydroxylamine under strictly anaerobic conditions, using argon or nitrogen. However, approximately 5% of the total IGF-I produced is oxidized during the E. coli fermentation.

IV. DISCUSSION

The existence of a single methionine sulfoxide in human recombinant IGF-I produced in E. coli could be shown by a combination of analytical techniques. Mass spectrometry of the purified intact peptide variants together with the separation capability of a high resolution chromatography system were the crucial steps in this procedure. The ^{252}Cf PDMS time of flight instrument used can neither resolve the oxidized peptide from the native peptide in a mixture nor produce sequence ions. The combined use of techniques, however, circumvented this problem.

For a long time mass spectrometry has been the "specialists" tool in biochemistry and protein chemistry, but the introduction of modern, less expensive and easy to use instruments may give the PDMS technique a place in many biochemistry laboratories.

V. ACKNOWLEDGMENTS

We thank Dr. Anthony Craig for carrying out the analysis of the intact IGF-I molecules and Mr. Göran Forsberg for the N-chlorosuccinimide experiments.

VI. REFERENCES

1. Moks, T., Abrahmsén, L., Österlöf, B., Josephson, S., Östling, M., Enfors, S-O., Persson, I., Nilsson, B., and Uhlén, M. (1987) Bio/Technology 5, 379.
2. Sundqvist, B., and Macfarlane, R.D. (1985) Mass Spectrom. Rev., 4, 421.
3. Kamensky, I., and Craig, A.G. (1987) Anal. Instr., 16, 71.

Section IV

Introduction

Kenneth J. Wilson

SECTION IV. HPLC TECHNIQUES APPLIED TO PROTEINS

The articles presented in this section cover a wide selection of the chromatographic techniques used in protein isolation/characterization. They are arranged in the order that one would use them during the analysis of a given sample . . . starting with isolation procedures and finishing with those employed for characterization.

Preparative isolations using a 'specialized' support are covered in Nau for immunoglobins. Similarly, the new hydrophilic, polymeric material discussed by Kitagawa appears to provide desired prerequisites for protein isolation: high recovery, moderate elution conditions and selection of surface chemistry.

Separations of very hydrophilic peptides have been a problem for years and is addressed in the article by Nagarajan et al. The choice of ion pairing agent and pH, as well as the chromatographic support (not covered in this article), were varied to determine effects ultimately to achieve the desired separation.

Analysis of a posttranslational modification, specifically glycosylation, is the topic of the Olechno, et al. article. The need to carryout such characterizarions on native and recombinant products has been, and will continue to be, a major undertaking in biotechology related areas.

The remaining articles deal with high-sensitivity peptide mapping, the problems associated with current instrumentation, sample handling and choice of column sizes and support material. Stone et al. have developed 'routine' methods for preparing samples for enzymatic digestion and mapping. Cheap and easily packed columns for the chromatographic separations were covered by Southan, and use of narrow bore columns and diode array detection was presented by Burgoyne et al.

Although many of the topics in this section have been presented in the past, these reviews serve as up-to-date information on the 'how to do' when the need arises.

33

TECHNIQUES IN PROTEIN CHEMISTRY OPTIMIZATION OF MONOCLONAL ANTIBODY PURIFICATION

David R. Nau, Ph.D.

J.T.BAKER INC.
222 Red School Lane
Phillipsburg, NJ 08865

I. INTRODUCTION

Antibody purification is one of the oldest challenges in protein biochemistry. Even today, the development of rapid and economical purification schemes for antibodies remains a major problem. Indeed, it has been suggested that protein purification is the Achilles heel of modern biotechnology.

For protein chemists or process engineers faced with the task of purifying numerous antibodies from a number of sources, the most practical approach to purification would involve the use of one or two highly selective chromatographic matrices which have universal applicability, with minor adjustments in purification protocols, to attain any level of purity from laboratory to process-scale.

II. BAKERBOND ABx*

The requirement for an economical chromatographic matrix able to rapidly purify large quantities of antibodies to homogeneity compelled us to investigate synthetic approaches in an attempt to construct a chromatographic surface that would bind antibodies more selectively than conventional ion exchange matrices. Using mixed mode interactions as the basis, silica gel as the most advantageous support, and a hydrophilic polymeric coverage to increase stability, eliminate non-specific binding, and maximize recovery and capacity, we developed BAKERBOND ABx*, antibody exchanger (1,2).

The major advantage of ABx is that it binds all immunoglobulins rather selectively, while exhibiting little or no affinity for albumins, transferrins, proteases, insulin, lipopolysaccharides or pH indicator dyes, which elute in the void volume, leaving the antibody(s) and a relatively small number of contaminants bound to the matrix (1,2). Therefore, ABx may be used to separate immunoglobulins from crude antibody preparations in a single step, at purities greater than 75%. Because most proteins are not bound, the majority of the ABx ligands are available to bind immunoglobulins; this substantially increases the capacity of ABx to bind antibodies present in crude samples. To date, literally hundreds of antibodies have been purified on ABx in this and other laboratories with near quantitative recovery (1,2).

III. OPTIMIZATION OF ABx SEPARATIONS VIA MOBILE PHASE MANIPULATION AND GRADIENT PROFILE

Due to their complex structure, individual proteins have numerous sites at which physical interactions may occur. Therefore, bonded phases such as ABx which operate via multiple chromatographic adsorption mechanisms permit changes in selectivity in the presence of various mobile phases which are not encountered with more conventional supports.

Early experiments with ABx indicated that elution with various salts produced distinct elution profiles, and often, the monoclonal and other bound proteins actually reversed their relative order of elution (figure 1). In general, the purity of the immunoglobulin fractions and the resolution between the antibody(s) and the major bound, non-immunoglobulin peak progressively increase as the anionic species of the elution buffer is changed from phosphate, to acetate, to chloride (halide), and then, to sulfate (figure 1). That is, elution with phosphate often produces antibodies of the lowest purity, while elution with sulfate typically produces higher purities (1,2).

As indicated in the previous example (figure 1), gradient elution of antibodies from ABx often provides adequate resolution and purity. However, in most cases, a more effective elution method is the use of step gradients in conjunction with optimized mobile phase conditions. In many production environments step elution is used extensively, not only due to economic reasons (simpler equipment and reduced buffer costs), but also because resolution can be increased dramatically (figure 2). However, most chromatographers make

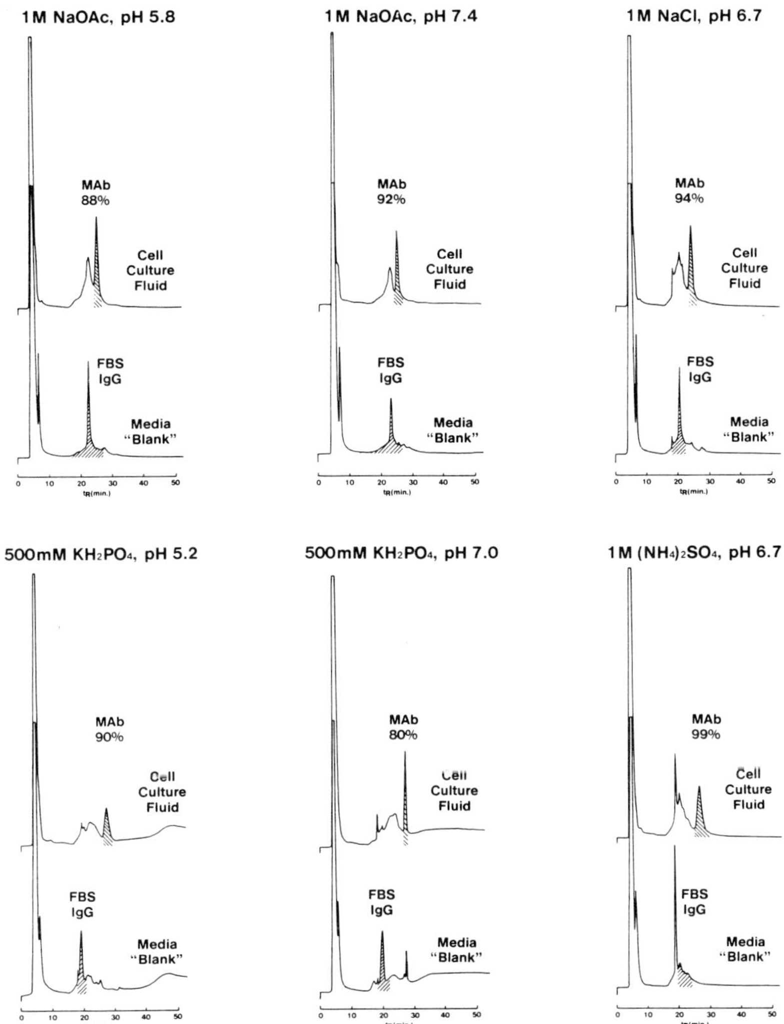

Figure 1. Effect of elution buffer on the purification of a mouse monoclonal antibody from a cell culture ultrafiltrate with 5 micron BAKERBOND ABx.

Chromatography was conducted on a column (7.75 x 100 mm) containing 5 micron ABx with an initial (A) buffer of 10 mM MOPSO plus 15 mM MES, pH 5.6, and a final elution (B) buffer of either, 1 M NaOAc, pH 5.8, or, 1 M NaOAc, pH 7.0, or, 1 M NaCl plus 20 mM NaOAc, pH 6.7, or, 500 mM KH_2PO_4, pH 5.2, or, 500 mM KH_2PO_4, pH 7.4, or, 500 mM $(NH_4)_2SO_4$ plus 20 mM NaOAc, pH 6.7. The gradient consisted of 100% A buffer for 4 min, followed by a linear gradient from 100% A to 100% B buffer over 26 min, and the flow rate was 0.7 mL/min for 4 min, then changed to 1.0 mL/min; the back pressure was 200 psi. Proteins were detected at 280 mn and 2.0 AUFS. The sample consisted of either, 0.4 mL of fetal bovine serum containing a hormone/peptide supplement ("media blank"), or 0.3 mL of cell culture ultrafiltrate ("culture fluid") diluted to 1.0 mL with buffer A. The monoclonal antibody (cross-hatched peaks labeled "MAb") eluted as the last major peak and was resolved from most protein contaminants, including the host (fetal bovine serum polyclonal) IgG (cross-hatched peaks labeled "FBS IgG"). The purity of the monoclonal (by SDS PAGE and size exclusion chromatography) is given in each chromatogram.

---

little use of this more traditional method, perhaps because gradient elution is a simpler technique. Indeed, one of the disadvantages of step elution is that more extensive method development is typically required to optimize the separation (figure 2). Another problem with step gradient elution is the possibility of creating false peaks or split peaks; these are the result of multiple or drastic changes in ionic strength and/or pH, or, the use of excessive column overload without proper loading studies during methods development. In order to determine the proper step gradient to employ, a first approximation is to run a long gradient, determine the ionic strength at which the monoclonal elutes, and in a subsequent chromatographic run, and use a step which is half of this concentration. This step can be held constant for a short period in order to elute the contaminating proteins, and then, followed by a linear gradient or a second step gradient to elute the monoclonal (figure 2).

Although step gradients of a wide range of ionic strengths are often capable of providing adequate resolution, the most efficient step gradient lies between two extremes (figure 2). At one extreme, the ionic strength is too low to enable the full elution of the contaminating proteins in a timely manner (figure 2). (Although one of the advantages of step gradients is to enhance resolution, another is to reduce run time, since recovery as well as process economics are often proportional to residence time on the column). On the other hand, if the ionic strength of the step is too high, the separation is rapid, but the monoclonal partially co-elutes with the contaminating proteins (figure 2). Obviously then, the ionic strength of the step should be between these two extremes; high enough to elute contaminants quickly, but neither at the expense of creating false peaks or split peaks, nor by eluting the monoclonal so quickly that

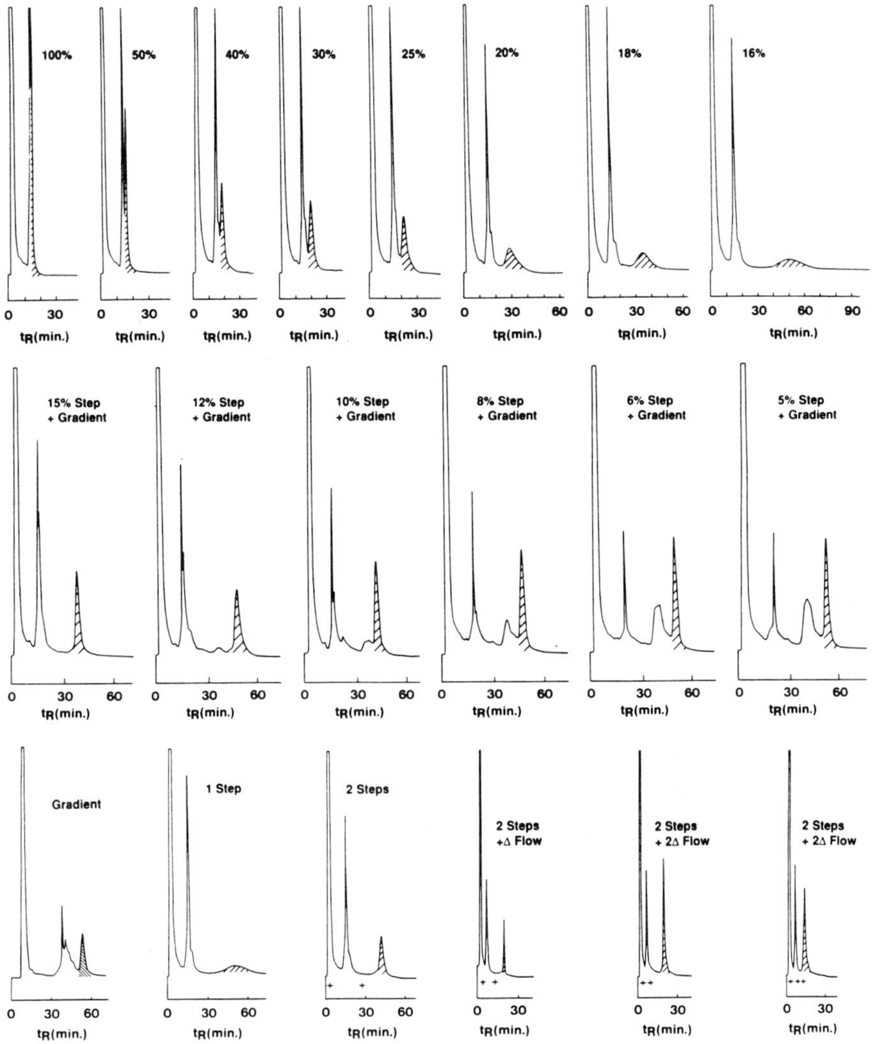

Figure 2. Effects of step gradients on the resolution of a monoclonal antibody from a fetal bovine serum-supplemented cell culture fluid on 15 micron BAKERBOND ABx.

Chromatography was conducted on a column (7.75 x 100 mm) packed with 15 micron ABx; the equilibration (A) buffer was 10 mM MOPSO plus 15 mM MES, pH 5.6, and the elution buffer was 500 mM $(NH_4)_2SO_4$ plus 20 mM NaOAc, pH 6.7. The flow rate was 1.0 mL/min (except as noted below) at a back pressure of 50 psi, and proteins were detected at 280 nm and 1.0 AUFS. The sample consisted of 0.4 mL of a serum-supplemented cell culture ultrafiltrate diluted to 1.0 mL with buffer A. An initial isocratic gradient at 100% A buffer was held for 4 min, followed directly by either a single step gradient to various percentages of the B buffer (top row; 100% B through 16% B), or (bottom row) lower ionic strength step gradients with a concomitant linear gradient to 100% B buffer over 30 min (i.e., 100% A, step to 15% B, then 15% B to 100% B over 30 min, etc.), or, (in the bottom row) either a linear gradient (see Figure 1), or, an initial step gradient to 16% B buffer, with or without a second step gradient to 50% B buffer and changes in flow rate. In these lower chromatograms, the first + denotes a step gradient to 16% B buffer, with or without an increase in flow rate to 2.5 mL/min; the second + denotes a second step gradient to 50% B buffer; and the third + sign denotes a decrease in flow rate back to 1.0 mL/min.

---

it co-elutes, and reduces product purity. Ideally, there is an ionic strength at which all the contaminating proteins elute rather quickly, yet the monoclonal is still bound tightly (figure 2).

Typically, it is advantageous to employ a second step gradient or rapid linear gradient following the elution of the contaminating proteins to quickly elute the monoclonal (figure 2). This not only reduces run time, but also, elutes the product into a smaller volume of a more appropriate buffer (e.g., ammonium acetate for lyophilization, concentrated ammonium sulfate for binding to a HIC matrix, PBS, an assay buffer, a formulation buffer, or any buffer which enhances the stability of the monoclonal). Finally, it should be mentioned that the manipulation of flow rate can be used to further reduce run time, and to increase the specific activity of the final product (figure 2). As a result of the high degree of resolution which can be obtained with step gradients, 40 micron ABx may be used to achieve to large scale , high performance separations in an economical manner (1,2).

IV. HYDROPHOBIC INTERACTION CHROMATOGRAPHY (HIC)

Previously, we had reported that HIC may be useful in the purification of some monoclonal antibodies from mouse ascites fluids (2). However, subsequent research in a number of laboratories indicated that, in general, HIC is not a sufficiently specific method for the one-step purification of monoclonal antibodies (2). These studies suggested that, like anion exchange chromatography, HIC is not particularly desirable as an initial purification step; virtually all the proteins in antibody preparations are bound, and as such, the number of ligands which are available to bind antibodies are dramatically reduced (1,2). Furthermore, individual antibodies elute as broad peaks and at a wide range of retention times within the gradient, hindering method development and identification of the antibody peak, and decreasing the possibility of obtaining homogeneous antibody (1,2). These considerations, as well as the apparent need to use large amounts of ammonium sulfate to bind immunoglobulins, were believed to limit the effectiveness of HIC as an economical initial step in preparative antibody purification schemes.

However, recent work in this laboratory suggests that the lack of specificity for antibody binding on HIC supports which was observed in these preliminary investigations may have simply resulted from the use of sub-optimal binding buffer conditions (1,2). With BAKERBOND HI-Propyl*, it is possible to use mild binding conditions (700 mM ammonium sulfate) to purify human or mouse IgG monoclonal antibodies from fetal bovine serum-supplemented cell culture fluids to near homogeneity (and 95% free of bovine IgG) with near quantitative recovery of immunoglobulin mass (figure 3).

HI-Propyl is also a simple, efficient, economical, and direct second purification step for IgG following an initial step with ABx (figure 3). One advantage of this approach is that antibodies eluted from ABx may be loaded directly onto HI-Propyl (1). Furthermore, with these optimized buffer conditions, virtually all of the contaminating fetal bovine serum proteins which do bind to ABx do not bind to HI-Propyl, and flow through in the void volume (figure 3). Therefore, antibodies eluted from ABx, either in the column or batch mode, may be selectively bound to HI-Propyl and eluted in a substantially purified form, free of bovine immunoglobulins and other fetal bovine serum proteins (figure 1,2). Another advantage of downstream purification steps with HI-Propyl is that a significant amount of nucleic acid and pyrogen can also be stripped from monoclonals, because under these optimized buffer conditions nucleic acids and lipopolysaccharides exhibit little or no binding (figure 3). Although bacterial endotoxins do not bind to ABx, some DNA and RNA co-elute with immunoglobulins; furthermore, nucleic acids and pyrogens do exhibit a significant amount of non-specific binding to antibodies. However, like hydrophobic interaction chromatography with HI-Propyl, anion exchange chromatography with BAKERBOND MAb* can be used to depyrogenate, remove nucleic acid and further purify monoclonal antibodies eluted from ABx (1,2).

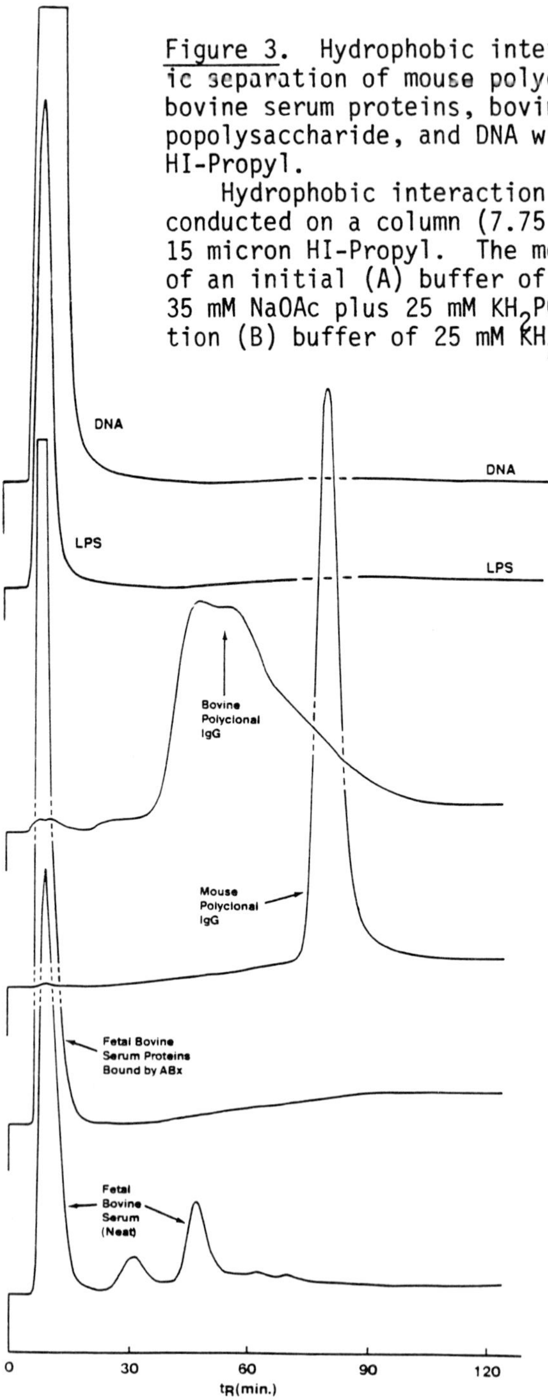

Figure 3. Hydrophobic interaction chromatographic separation of mouse polyclonal IgG from fetal bovine serum proteins, bovine polyclonal IgG, lipopolysaccharide, and DNA with BAKERBOND HI-Propyl.

Hydrophobic interaction chromatography was conducted on a column (7.75 x 100 mm) packed with 15 micron HI-Propyl. The mobile phase consisted of an initial (A) buffer of 700 mM $(NH_4)_2SO_4$ plus 35 mM NaOAc plus 25 mM KH_2PO_4, pH 7.2 and an elution (B) buffer of 25 mM KH_2PO_4, pH 7.2, with a linear gradient from 100% A to 100% B buffer over 1 hr, at a flow rate of 0.5 mL/mn, with a back pressure of 25 psi. The sample consisted of either 2 mg of commercially-purified bovine or mouse immunoglobulin G, human sperm DNA, E. coli lipopolysaccharide, 0.5 mL of neat fetal bovine serum, or 0.5 mL of fetal bovine serum proteins which bind to ABx, each diluted to 2 ml with buffer A and 2.1 M $(NH_4)_2SO_4$ buffer to give a salt concentration equal to buffer A. Proteins were detected by UV absorbance at 280 nm; DNA was detected at 260 nm.

These techniques are able to produce electrophoretically pure monoclonal antibody and bring the levels of pyrogen and nucleic acid contamination well below those required for *in vivo* use (i.e., 2 pg DNA per dose and 0.05 ng lipoplysaccharide per mL). Since the mechanisms of separation on these complementary bonded phases are each entirely different, ABx plus MAb or HI-Propyl offer selectivity and resolution which is greatly enhanced relative to any single chromatographic matrix alone (1,2).

V. SUMMARY

Several factors are of utmost importance in the preparative-scale purification of antibodies. These include longevity, capacity, recovery, hygiene and economy. BAKERBOND ABx, BAKERBOND MAb and BAKERBOND HI-Propyl have been designed with these factors in mind.

As new types of monoclonal antibodies are purified and as purity requirements increase, it becomes increasingly apparent that the molecular diversity inherent within the immunoglobulin family makes a universal single step purification method less likely. However, the development of dedicated chromatographic matrices such as ABx, which utilize composition surface interactions to maximize selective antibody binding and resolution, represents a major step in the right direction.

VI. REFERENCES

1. Nau, D. R. (1988) in *The Role of HPLC in Biotechnology* (W. Hancock, Ed.) Wiley, N.Y. (in press).

2. Nau, D. R. (1988) in *Monoclonal Antibodies and Nucleic Acid Probes* (G. Prakash and B. Swaminathan, Eds.), Marcel Dekker, N.Y.

*BAKERBOND, BAKERBOND ABx, BAKERBOND MAb, and BAKERBOND HI-Propyl are all trademarks of J.T.Baker Inc., Phillipsburg, NJ. BAKERBOND ABx is registered under U. S. Patent No. 4,606,825.

34

NEW HYDROPHILIC POLYMER FOR PROTEIN SEPARATIONS BY HPLC

Naotaka Kitagawa

Interaction Chemicals Inc.
Mountain View, California 94043

I. INTRODUCTION

Non-porous spherical ion-exchange materials for fast high performance liquid chromatography of proteins have recently become the object of public attention.[1-4] It is generally recognized that use of non-porous materials eliminates the intraparticulate diffusion that is responsible for band-broadening, losses in efficiency and resolution.[4] Another distinctive characteristic of non-porous materials is that they can easily be made rigid and withstand high flow rates and high pressures. Thus, high speed analyses are possible. However, non-porous materials have limited loading capacities because of their small surface area. Consequently, smaller particle sizes must be utilized to compensate for these reduced capacities and this results in increased column pressures, an unwanted property. Because of the possibility of larger sample loadings offered by porous materials, these still dominate the field of HPLC analysis.

If one could design a macroporous polymeric material that withstands high flow rates and high pressures, the material would permit high loading even during high speed separations. This is particularly true for surface-mediated separations such as ion-exchange chromatography and hydrophobic interaction chromatography, because proper pore distributions and particle sizes will eliminate or minimize the stagnant mobile phase mass transfer effects.[5]

In this paper, two new HPLC columns, packed with Hydrophase HP-CM2 and HP-Butyl polymers, are reported to be suitable for high loadings and high speed analyses in protein separations.

A hydrophilic, macroporous, synthetic polymer, Hydrophase HP-100, whose characteristics have been reported previously[6], was modified to contain carboxy methyl groups for the HP-CM2 polymer (a weak cation separation medium) and to contain n-Butyl groups for the HP-Butyl polymer (a hydrophobic interaction separation medium). Both columns exhibited not only high loading capacities for protein molecules but also durability against high flow rates through various buffers and salt gradients. The mass transfer characteristics of the packing materials at high flow rates seem to be exceptionally good in separating protein mixtures.

II. MATERIALS AND METHODS

Columns

Hydrophase HP-CM2 and HP-Butyl polymers sized in 10 um were obtained from Interaction Chemicals, Inc. (Mountain View, CA, U.S.A.). The polymers were packed in 7.8 x 100 mm stainless steel columns equipped with titanium fruits by means of a Shandon packer using an aqueous packing medium.

HPLC

A high pressure eluent-delivery system (EM Science, Cherry Hill, NJ, U.S.A.) was equipped with an L-6000 Pump, an L-6200 Intelligent Pump, an L-4000 UV Detector, a D-2000 Chromato-Integrator (Hitachi, San Jose, CA, U.S.A.) and a 7161 sample injector (Rheodyne, Cotati, CA, U.S.A.).

CHEMICALS

Deionized glass distilled water was used for preparation of buffers and sample protein solutions. All reagents were of analytical grade. Sodium chloride (Curtin Matheson, Florence, KY, U.S.A.), hydrochloric acid and sodium hydroxide (Fisher Scientific, Santa Clara, CA, U.S.A.) were used. Bis(2-hydroxymethyl)imino-tris-(hydroxymethyl)-methane (Bis-Tris) and 2-(N-morpholino)ethane sulfonic acid (MES) were obtained from Research Organics Inc. (Cleveland, OH, U.S.A.). Sodium phosphate monohydrate and ammonium sulfate were obtained from EM Science (Cherry Hill, NJ, U.S.A.). Cytochrome c (horse heart), hemoglobin (Hb) (bovine blood), ovalbumin (chicken egg), lysozyme (chicken egg white), alpha-chymotrypsinogen (bovine pancreas), ribonuclease A (bovine pancreas), insulin chain A oxidized (bovine insulin), carbonic anhydrase (bovine erythrocytes), and transferrin (bovine) were obtained from Sigma (St. Louis, MO, U.S.A.). HbA/F, HbF, HbA/C, HbA/F and HbAFSC were obtained from Isolab (Akron, OH, U.S.A.). Sodium azide from Aldrich (Milwaukee, WI, U.S.A.), 0.25 g/L, was used to prevent bacterial growth.

All mobile phases were prepared at room temperature, degassed under vacuum and filtered through 0.45 um membranes prior to use. Protein solutions were freshly prepared and stored at 4°C when not in use. All experiments were run at room temperature.

III. RESULTS AND DISCUSSION

The present work was performed to examine a macroporous weak cation-exchange polymer and a hydrophobic interaction chromatographic polymer for high sample loadability and also for high speed separations.

pK of Hydrophase HP-CM2 Polymer

pK of HP-CM2 was obtained from a titration profile with 0.1 N NaOH solution. It was determined to be approximately 7.00.

Ion-Exchange Capacity of Hydrophase HP-CM2 Polymer

Wet H^+ capacity of HP-CM2 was determined by titrating the polymer with 0.1 N NaOH solution to the phenolphthalein end point for the cation-exchange capacity. The capacity of the polymer was approximately 136 meq/wet-gram.

Column Capacity of Hydrophase HP-Butyl

The protein-binding capacity of 7.8 x 100 mm columns was determined by sequential injections of ovalbumin solution until the protein was detected in the break-through volume. The column capacity was thereby determined to be approximately 190 mg/column.

Observed Plate Number of Hydrophase HP-CM2 Column

Using 3-aminophenol, an efficiency of 14,000 plates per meter with an asymmetry factor of 1.09 was recorded using 15 mM Bis-Tris buffer isocratically at pH 6.5 and a flow rate of 0.5 mL/min.

Hydrophase HP-CM2 in Cationic Ion-exchange Modes

Protein separations of ovalbumin, myoglobin, hemoglobin and cytochrome c were achieved using various gradients and flow rates and the results were shown in Figs. 1, 2, and 3. It should be noted that the higher flow rates did not alter selectivity or resolution. It seems that separation patterns at high flow rates resemble that of non-porous packing materials.

Figure 1

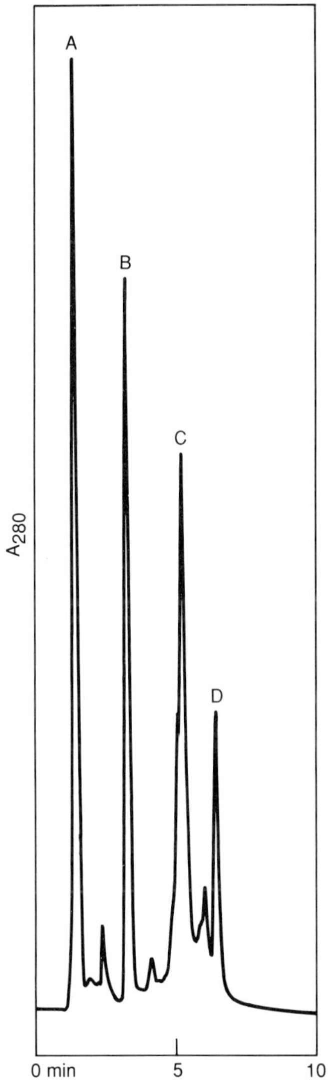

Figure 2

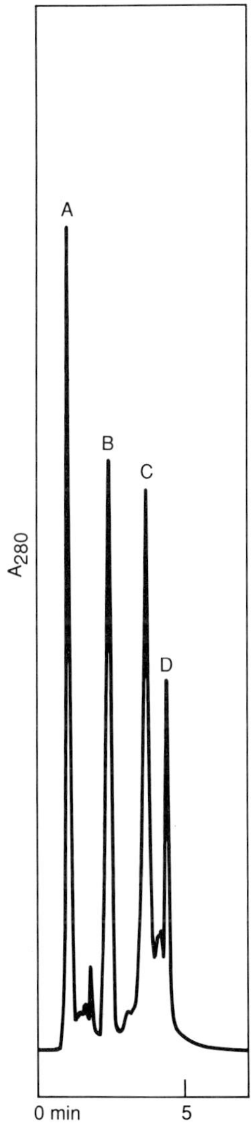

Figure 3

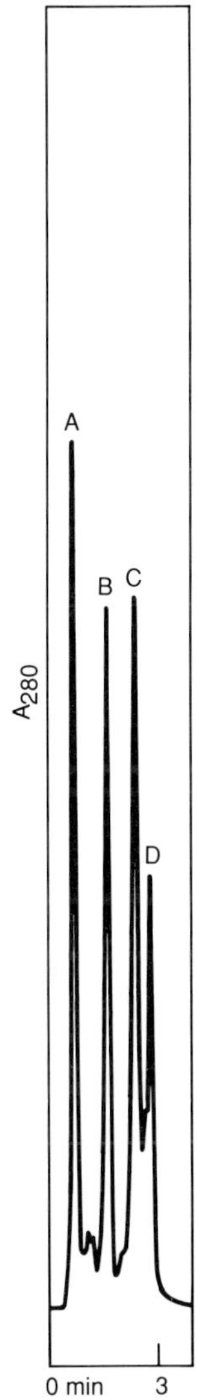

Fig. 1, 2 & 3. High speed protein separations on Hydrophase HP-CM2. Elution order is: A)ovalbumin B)myoglobin C)hemoglobin D)cytochrome c. Elution conditions were: Fig. 1) a 5-min linear gradient from 10 mM Bis-Tris (pH 5.9) to 0.2 M sodium chloride in 10 mM Bis-Tris (pH 5.9) at a flow rate of 1.5 mL/min; Fig. 2) a 3-min linear gradient from 10 mM Bis-Tris (pH 5.9) at a flow rate of 2.0 mL/min; Fig. 3) a 1-min linear gradient from 10 mM Bis-Tris (pH 5.9) at a flow rate of 3mL/ min. Detected at UV 280 nm.

Hemoglobin variants separation involving HbF HbA, HbS, and HbC were achieved using a 60-min linear gradient (Fig. 4) and a 5-min linear gradient (Fig. 5). Comparing these two figures, it is interesting to note that resolution in the two chromatograms was not significantly changed despite the 75% reduction in analysis time (Fig. 4).

Separation of ovalbumin, alpha-chymotrypsinogen A, and lysozyme was achieved using a 30 minute linear gradient (Fig. 6).

Separation of insulin chain A, carbonic anhydrase, transferrin, and cytochrome c was achieved in 11 minutes using a 15 minute linear gradient (Fig. 7).

<u>Hydrophase HP-Butyl in Hydrophobic Interaction Modes</u>

Separation of cytochrome c, myoglobin, ribonuclease, ovalbumin, and alpha-chymotrypsinogen was achieved in 9 minutes (Fig. 8).

Separation of cytochrome c, myoglobin, and lysozyme was achieved in 7 minutes (Fig. 9) and in less than 4 minutes (Fig. 10). It is interesting to note that results similar to that obtained in ion-exchange (IEX) modes at a high flow rate were also obtained using hydrophobic interaction chromatography (HIC) modes.

IV. CONCLUSIONS

A weak cation-exchange stationary phase (carboxymethyl groups), Hydrophase HP-CM2, and a hydrophobic interaction stationary phase (n-butyl groups), Hydrophase HP-Butyl, were examined in 7.8 x 100 mm steel columns for protein separations. Because of the macroporous nature of the packing materials, the columns exhibited high capacities and high resolution.

In order to take full advantage of column durability when used at high flow rates, short gradient profiles were employed to reduce analysis times for protein separations. The resulting chromatograms exhibited both high selectivity and high resolution. The columns can be used for both semi-preparative work and analytical on-line monitoring of regular preparative work.

It is significant that 7.8 x 100 mm columns can separate complex mixtures of proteins in only a few minutes. It is also significant that these sample mixtures of proteins contained not micro-gram but miligram quantities of proteins. We expect that if a proper column configuration is chosen and a proper reduction of the dead volume is achieved one minute separations of many protein mixtures may be possible with

Figure 4

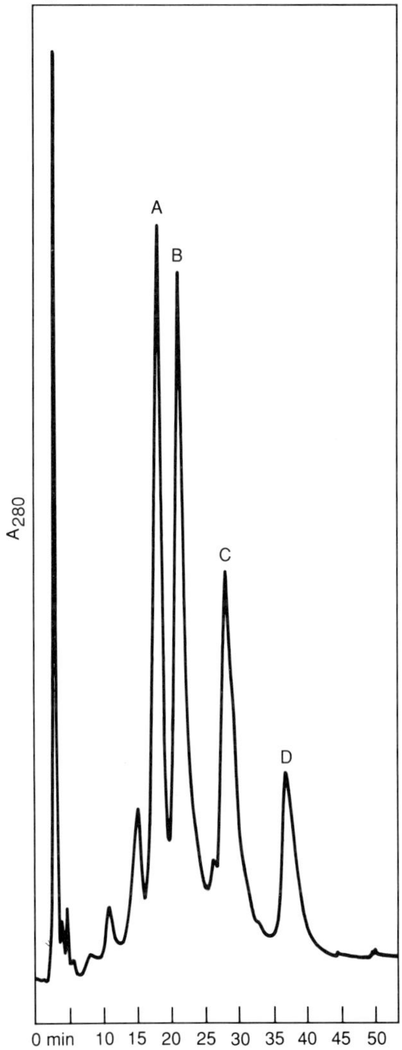

Figure 5

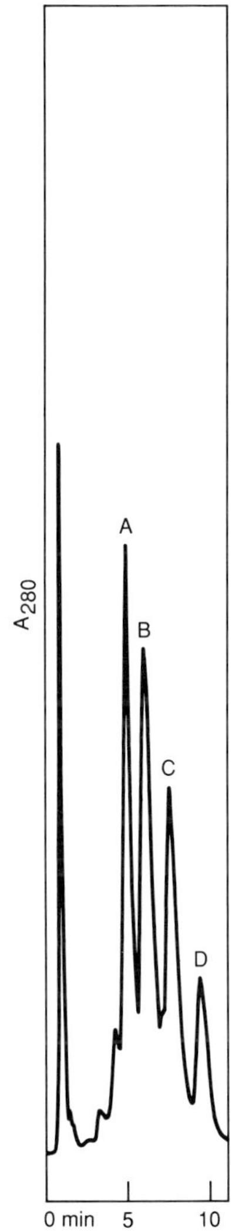

Fig. 4 & 5. Hb variants separations on Hydrophase HP-CM2. Elution order is: A)HbF B)HbA C)HbS D)HbC. Elution conditions were Fig. 4) a 60-min linear gradient from 20 mM Bis-Tris (pH 6.3) to 0.13 M sodium chloride in 20 mM Bis-Tris (pH 6.3) at a flow rate of 1.0 mL/min; Fig.5) a 5-min linear gradient from 20 mM Bis-Tris (pH 6.0) to 0.1 M sodium chloride in 20 mM Bis-Tris (pH 6.1) at a flow rate of 3 mL/min. Detected at UV 280 nm.

Figure 6

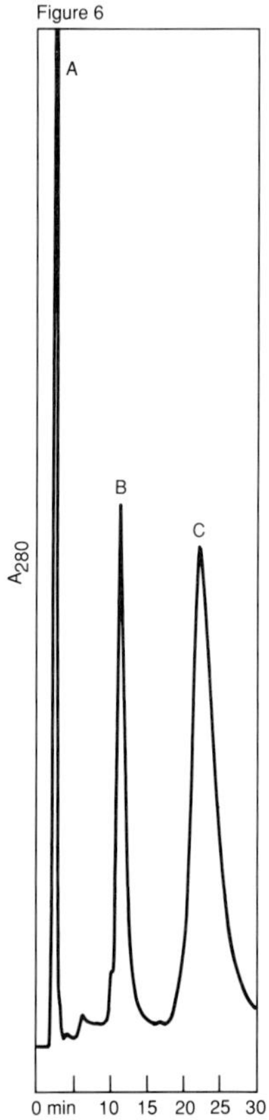

Figure 7

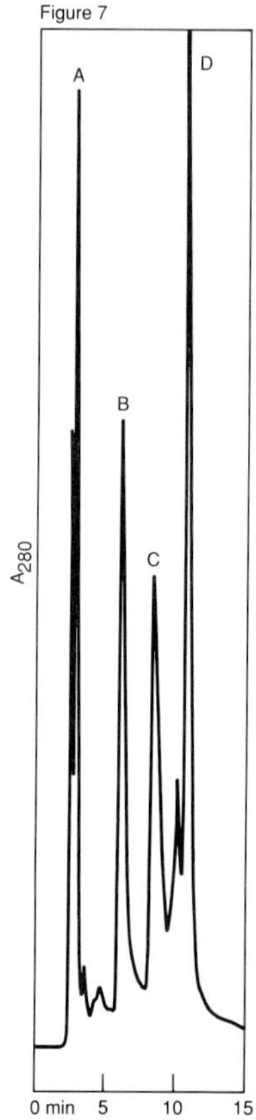

Figure 8

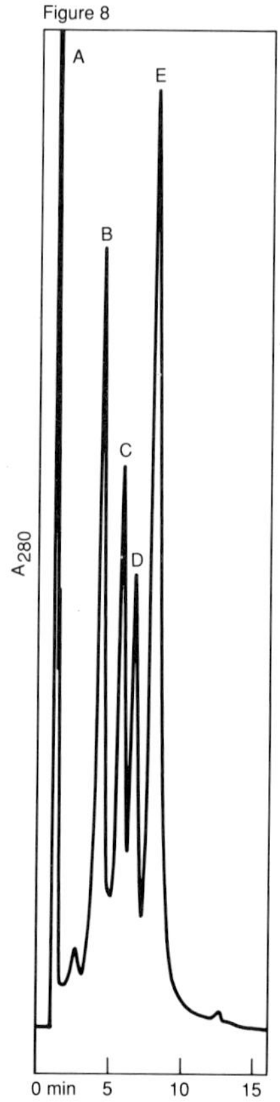

Fig. 6. Separation of proteins on Hydrophase HP-CM2. Elution order is: A)ovalbumin B)alpha-chymotripsinogen-A C)lysozyme. Elution condition was a 30-min linear gradient from 15 mM Bis-Tris (pH 6.7) to 0.5 M sodium chloride in 15 mM Bis-Tris (pH 6.7) at a flow rate of 1.0 mL/min. Detected at UV 280 nm.

Fig. 7. Separation of proteins on Hydrophase HP-CM2. Elution order is: A)insulin B)carbonic anhydrase C)transferrin D)cytochrome c. Elution condition was a 15-min linear gradient from 20 mM Bis-Tris (pH 6.0) to 0.3 M sodium chloride in 20 mM Bis-Tris (pH 6.0) at a flow rate of 1.0 mL/min. Detected at UV 280 nm.

Fig. 8 High speed separation of proteins on Hydrophase HP-Butyl. Elution order is: A)cytochrome c B)myoglobin C)ribonuclease D)ovalbumin E)alpha-chymotripsinogen. Elution condition was a 5-min linear gradient from 1.7 M ammonium sulfate in 0.1 M sodium phosphate (pH 7.0) to 1.7 M ammonium phosphate (pH 7.0) at a flow rate of 2.0 mL/min. Detected at UV 280 nm.

Figure 9

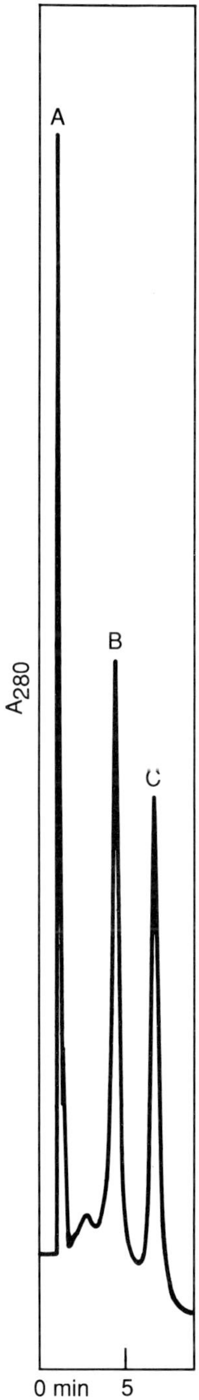

Figure 10

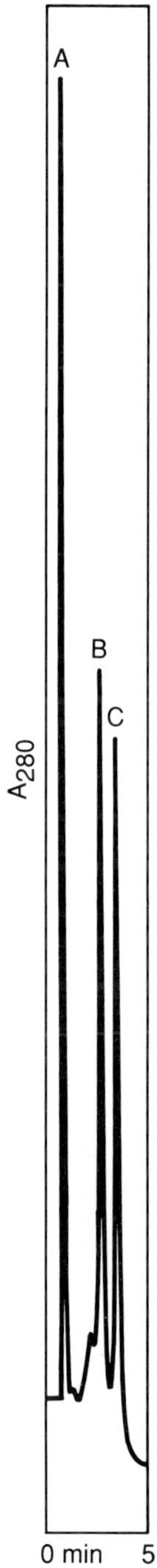

macroporous packing materials. We are continuing investigations in these areas.

ACKNOWLEDGEMENTS

The author expresses his thanks to his wife, Yoko, for her encouragement and Dr. James R. Benson of Interaction Chemicals for his support.

V. REFERENCES

1. D. J. Burke, et al., J. Chromatogr. 353 (1986) 425-437

2. D. J. Burke, et al., J. Chromatogr., 359 (1986) 533-540

3. Y. Kato, et al., J.Chromatogr., 398 (1987) 327-339

4. M. A. Rounds & F. E. Regnier, J. Chromatogr., 443 (1988) 73-83

5. L. R. Snyder & J. J. Kirkland, Introduction to Modern Liquid Chromatography, Wiley-Interscience, New York, 1979

6. N. Kitagawa, J. Chromatogr., 443 (1988) 133-141

35

EFFECTS OF PH AND ION PAIRING AGENTS ON THE REVERSE PHASE HPLC OF HYDROPHILIC PEPTIDES: A CASE STUDY OF PENTIGETIDE AND ITS FRAGMENTS

Gobi R. Nagarajan
Jo-Lynne S. Boone[1]
Thomas J. Stolzer
Steven P. Richieri

Immunetech Pharmaceuticals, San Diego, California 92121, USA

I. ABSTRACT

The reverse phase HPLC analysis of a mixture of similar hydrophilic peptides was examined with respect to the effect of pH and ion pairing agents on retention and resolution. The model series of peptides chosen for this study were pentigetide, β-Asp[1]-pentigetide, D-Ser[2]-pentigetide, SDPR, DPR, PR and DSD. Net acidic fragments were retarded by low pH. Basic and neutral peptides had increased retention in the presence of hexanesulfonic acid. Trifluoroacetic acid was found to retard all of the peptides. From these studies it is clear that the effects of both pH and ion pairing agents are significant factors to be considered when optimizing the separation of a homologous peptide mixture.

II. INTRODUCTION

Over the past two decades, development and application of high performance liquid chromatography (HPLC) have revolutionized the analysis, characterization and isolation of peptides and proteins.

---

[1]Author to whom correspondence may be addressed.

The reverse phase HPLC of polar peptides represents a difficult problem in terms of both retention and resolution. Pentigetide, a clinically important compound (1-3), is an especially good example.

Pentigetide is a polar peptide (Asp-Ser-Asp-Pro-Arg; DSDPR) derived from Immunoglobulin E (IgE) Fc region, with two acidic side chains, a basic side chain and free termini. As can be seen from the structure all of the likely metabolic fragments and synthetic impurity peptides would be expected to be hydrophilic with similar chromatographic properties. Although very similar hydrophilically, they differ by the ratio of acidic and basic residues.

In the following paper, we have shown that the hydrophilic fragments and impurities can be separated and identified by reverse phase HPLC through the judicious use of pH and ion pairing agents.

III. MATERIALS AND METHODS

Reagents (HPLC or ACS grade): water; freshly distilled trifluoroacetic acid (TFA) (Pierce); 1-hexanesulfonic acid, sodium salt (HSA) (Eastman Kodak Co.); sodium phosphate monobasic, sodium phosphate dibasic, o-phosphoric acid and acetonitrile (Fisher Scientific).

The analytical HPLC system set at 214 nm consists of a 421A system controller, two 114M pumps (1 mL/minute), a 340 organizer with 1.2 mL mixer, a 160 uv detector (all Beckman), a SIL-6A autoinjector with a SCL-6A controller, a CTO-6A column oven (35° C), a C-R3A integrator (all Shimadzu) and an Adsorbosphere C_{18} 5μm, 25 cm x $^{1}/4$ " column (Alltech).

Peptides: Pentigetide (DSDPR), β-Asp^{1}-pentigetide (β-DSDPR), D-Ser^{2}-pentigetide [D(D-Ser)DPR], SDPR, DPR and PR were synthesized in solution via mixed anhydride method (isobutyl chloroformate) by stepwise condensation starting from N$^{\omega}$-nitro-L-arginine benzyl ester. The β-carboxyl group of aspartic acid was protected as benzyl ester, and serine hydroxyl as benzyl ether. After each coupling, the Boc protecting group was removed by TFA. The nitro and benzyl groups were removed in the final step by catalytic hydrogenation (Pd black, 1 atm.). DSD was synthesized by the solid phase method of Merrifield (4). The crude peptides were purified on a DEAE-Sephadex A-25 column (10mM-300mM, NH_4HCO_3 gradient) and/or a Vydac C_{18} column (H_2O/60% aq. CH_3CN gradient). The peptides were pure by analytical HPLC, amino acid analysis and TLC.

A. Column Acceptability Requirements

All columns used for this study had a resolution factor (R) of $\geq$ 1.2 between pentigetide and β-Asp^1-pentigetide in solvent system 4 (described below) as defined by $R = 2\,(t_{R\beta} - t_R)/(W_\beta + W)$, where $t_{R\beta}$ and t_R are the retention times and W_β and W are the basewidths at baseline of β-Asp^1-pentigetide and pentigetide respectively (5).

B. Experimental Procedure

In order to study retention effects, duplicate 20 µl injections of each peptide (1 mg/mL in 0.1 M sodium phosphate, pH 7.4) were made and followed with a blank run (0.1 M sodium phosphate, pH 7.4). Resolution was studied using peptide mixtures with varying relative amounts of the component peptides.

Five binary gradient solvent systems were evaluated in this study, with each pair of mobile phase buffers consisting of an aqueous Buffer A (described below) and a Buffer B containing CH_3CN in Buffer A. The following compositions of Buffer A were tested:

1. 0.1% TFA, pH 2.2
2. 0.1 M sodium phosphate, 5 mM HSA, pH 4.5
3. 0.1 M sodium phosphate, pH 3.3
4. 0.1 M sodium phosphate, pH 4.5
5. 0.1 M sodium phosphate, pH 7.4

For solvent systems 1-4, Buffer B was 60% CH_3CN in Buffer A; for solvent system 5, Buffer B was 30% CH_3CN in Buffer A. The gradient used for solvent systems 1-4 was 7 minutes isocratic 100/0 A/B, followed by a linear gradient over 15 minutes from 100/0 A/B to 95/5 A/B, then isocratic 95/5 A/B for 8 minutes. The gradient for solvent system 5 was identical except the final concentration was 9/1 A/B.

IV. RESULTS AND DISCUSSION

Pentigetide has been shown to inhibit hypersensitivity in the Prausnitz-Kunstner reaction, indicating its possible therapeutic value for allergic diseases (1). Subsequent clinical studies of pentigetide have demonstrated that it has therapeutic activity in humans for treatment of allergic rhinitis (2, 3). Metabolic (6) and clinical studies of pentigetide required development of a method to measure shortened metabolic fragments and likely synthetic impurities. This need stimulated our interest in studying the effects of pH and ion pairing on resolution and retention. In order to develop the proper analytical system the desired standards β-DSDPR, D(D-Ser)DPR, SDPR, DPR, PR and DSD, of pentigetide were synthesized and purified by standard chemical techniques.

Preliminary studies with various gradient and solvent systems indicated that mixtures of pentigetide with likely fragments and synthetic impurities did not resolve. Our expectation that the extent of peptide protonation would change chromatographic properties on reverse phase HPLC prompted an evaluation of the effects of mobile phase pH on resolution. In the current case because the peptides in the mixture had different numbers of acidic and basic residues, it seemed likely that the resolution and retention would be sensitive to pH. The HPLC chromatograms of pentigetide and a mixture of fragments and synthetic impurities are shown in Figures 1 & 2. At pH 3.3 (solvent system 3) retention was satisfactory but the resolution was unacceptable (Fig. 1a).

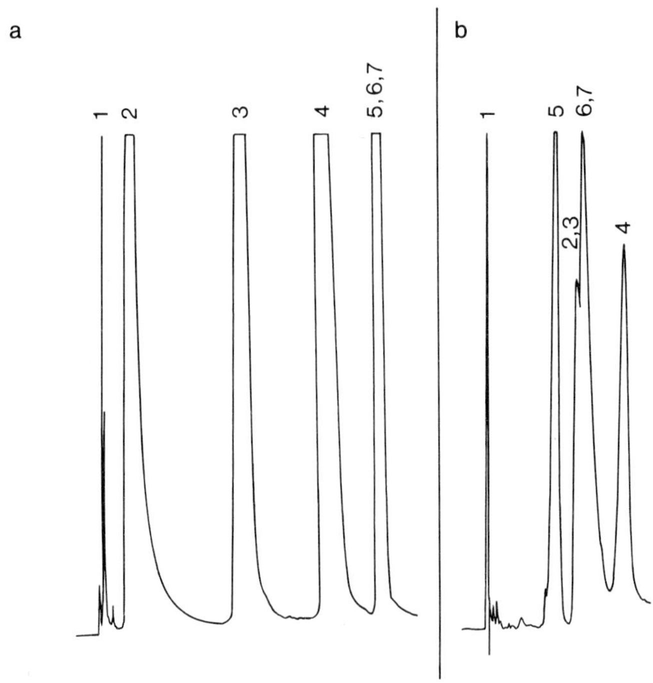

Figure 1. Chromatogram of pentigetide and analogs in (a) solvent system 3, (b) solvent system 5. (1=DSD, 2=PR, 3=DPR, 4=SDPR, 5=DSDPR, 6=β-DSDPR, 7=D(D-Ser)DPR).

When the pH was increased to 7.4 (solvent system 5), the retention times and elution order changed, and the resolution was generally poorer (Fig. 1b). An evaluation of these results suggested that an intermediate pH would be necessary to give the best combination of resolution and retention. Figure 2 is a chromatogram of the same peptide mixture illustrated in Figure 1, at the intermediate pH of 4.5 (solvent system 4). This pH resolves all of the pentigetide components. The difference in retention and resolution shown in Figures 1 and 2 can be explained by pH differences since the effective mobile phases and gradients are identical except for their pH.

An analysis of the chromatograms shows that the capacity factor, k', is increased as pH decreases for acidic peptides. The trend for neutral peptides is not clear from this study. The k' of the one basic peptide in the mixture increases with increasing pH. Our data for acidic peptides can be explained by the notion that at low pH, acidic groups are protonated and the peptide is rendered more hydrophobic, causing the observed increase in k'. Likewise it is reasonable that basic peptides have an increased k' at higher pH.

In an attempt to further enhance resolution, three ion pairing agents were compared. Table I compares the effects of phosphate, HSA and TFA on k'. A comparison of HSA and phosphate (solvent systems 2 and 4 respectively; Table I) shows that acidic peptides are retained less

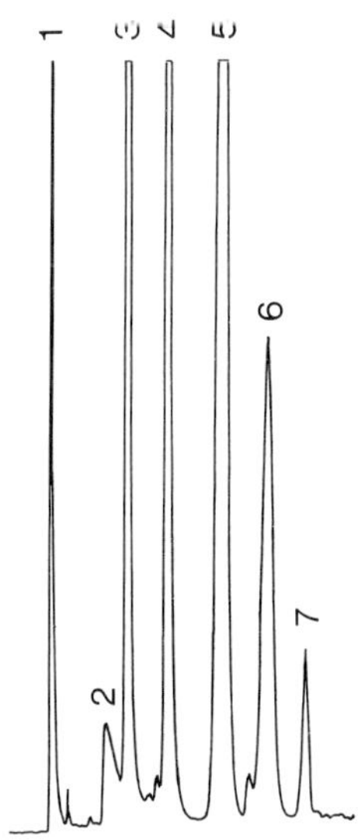

Figure 2. Chromatogram of pentigetide and analogs in solvent system 4. Same peak numbering applies as in Figure 1.

in HSA than in phosphate buffer, while basic and neutral peptides are retained longer. The pH in both systems is the same. The effects of HSA on retention agree with a previously described mechanism in which HSA intercalates into the hydrophobic stationary phase and ion pairs with positively charged peptide functionalities (7-9).

TABLE I. Effects of ion pairing agents (PO_4, HSA, TFA) on k'

Compound	AA Ratios Acidic/Basic	k' HSA[a]	PO_4[b]	TFA[c]	PO_4[d]
DSD	3 / 1	0.11	---	1.08	0.11
DSDPR	3 / 2	2.90	4.54	13.20	7.40
β-DSDPR		3.46	6.80	10.26	7.35
D(D-Ser)DPR		3.01	4.69	12.56	8.02
SDPR	2 / 2	5.47	2.89	12.98	5.50
DPR		3.89	1.95	9.03	3.44
PR	1 / 2	7.4	1.22	5.94	1.05

[a]Solvent system 2
[b]Solvent system 4
[c]Solvent system 1
[d]Solvent system 3

The most commonly used ion pairing agent TFA was also examined. Here, however, there is a difference in both pH and ion pairing agent. The results from a comparison of TFA and phosphate (solvent systems 1 and 3 respectively) are shown in Table I. Longer retention was observed for all peptides in TFA than in phosphate buffer,with the most pronounced effect on the basic peptide PR. Based on the results above, the longer retention for acidic and neutral peptides may be attributed simply to the lower pH in the TFA system (pH 2.2 *vs.* pH 3.3 in the phosphate system). Recalling the decreased retention with decreasing pH shown above for the basic fragment PR, the result in TFA cannot be due to pH alone and presumably is a combination of pH and classical ion pairing.

In summary, when designing a separation system for a mixture of similar hydrophilic peptides, the effects of pH and solvent system should be considered. The peptidic resolution and retention have been shown to be sensitive to mobile phase pH, ion pairing agents and to the number of acidic and basic groups present in the peptide. The judicious choice of pH and ion pairing agents will allow the separation of a complex mixture of similar peptides.

V. ACKNOWLEDGMENTS

The authors wish to thank Ms. D. Pearson for artwork and typing.

VI. REFERENCES

1. R.N. Hamburger (1975) Science 189:389.

2. G.A. Cohen, R.D. O'Connor, and R.N. Hamburger (1984) Ann. Allergy 52:83.

3. G.S. Hahn (1986) Nature 324:283.

4. R.B. Merrifield (1963) J. Am. Chem. Soc. 85: 2149.

5. E.L. Johnson and R. Stevenson (1978) "Basic Liquid Chromatography," p. 16. Varian, Palo Alto.

6. P.P. Benham, G. Nagarajan, S.L. Chiang, *et al.* (1985) Fed. Proc., USA 44, No. 3, Abstract No. 1014.

7. W.S. Hancock, C.A. Bishop, L.J. Meyer, *et al.* (1978) J. Chromatogr. 161:291.

8. M. T. W. Hearn *In* Chromatographic Series "Ion Pair Chromatography" (M.T.W. Hearn, ed.), Vol. 31, p. 207. Marcel Dekker, New York.

9. S.S. Yang and R.K. Gilpin (1987) Anal. Chem. 59:2750.

36

CHROMATOGRAPHIC ANALYSIS OF GLYCOPROTEIN DERIVED CARBOHYDRATES: PELLICULAR ION EXCHANGE RESINS AND PULSED ELECTROCHEMICAL DETECTION.[1]

Joseph D. Olechno, Steven R. Carter, William T. Edwards, Dennis G. Gillen, Dionex Corporation, Sunnyvale, CA 94086
R. Reid Townsend, Y. C. Lee, Mark R. Hardy,
Department of Biology,
The Johns Hopkins University, Baltimore, MD, 21218

I. INTRODUCTION

It is now well recognized that the fate and function of glycoproteins is intimately associated with their covalently attached oligosaccharide chains (for review see reference 1). Co-translational transfer of core oligosaccharide chains to the nascent polypeptide is required for proper folding and subunit assembly in the endoplasmic reticulum (2). After processing of the core oligosaccharides and specific addition of sugars in the Golgi apparatus, certain completed carbohydrate structures (e.g., mannose-6-phosphate containing oligosaccharide chains) direct the glycoproteins to intracellular destinations such as the lysosomes (3). The activity of integral membrane receptors, such as the insulin receptor, can be modulated according to the complement of carbohydrate structures (4). The binding affinity of

---

[1] RRT was supported by National Institutes of Health Research Grant DK31376.
MRH is the recipient of a Postdoctoral Carbohydrate Research Fellowship from Dionex Corporation.
YCL is supported by National Institutes of Health Research Grant DK9970.

interaction of tissue plasminogen activator, tPA, with fibrin) has been related to the extent of protein glycosylation (5).

The carbohydrate portion of glycoproteins can affect their antigenicity by either "masking" peptide epitopes (6) or by participating as epitopes (7). In addition, oligosaccharide chains serve as signals for lectin-mediated cell-cell adhesion (8) and for recognition of glycoproteins by specific cell types (9). Complex carbohydrates act as blood group determinants in mucus glycoproteins (10). The carbohydrate chains of mucus glycoproteins act as barriers against proteolytic digestion of the gastrointerogical tract (11). Glycoproteins have even been shown to act as antifreeze in fish that live under the antarctic ice (12).

The nature of carbohydrate groups in genetically engineered products are drawing an enormous amount of attention. For example, erythropoietin produced in different cells possess different carbohydrate structures (13,14).

All of the above biological interactions are dependent upon the interaction of protein moieties with the three-dimensional structures of oligosaccharide chains on glycoproteins. Tertiary structure and activity of proteins may be dramatically altered by changing a single amino acid residue as is seen in normal and sickle cell hemoglobin. The change of a single sugar residue in a glycoprotein derived oligosaccharide can also change structure or biological activity drastically as seen in blood group determinants (10). Carbohydrate structures can also be drastically altered even without changing a single sugar residue, by changing the position of inter-sugar linkage, anomeric configuration and branching position. Most glycoproteins have oligosaccharide chains at different peptide loci and frequently an array of carbohydrate structures is present at each of these glycosylation sites (15) (microheterogeneity).

Structural elucidation of carbohydrates on glycoproteins generally involves i.) release of the oligosaccharides from the protein, ii.) purification to produce homogenous fractions, and iii.) structural studies, using primarily mass spectrometry and high resolution ^{13}C(16) and ^{1}H NMR (17). The release of oligosaccharides can be accomplished through either chemical techniques (e.g., hydrazinolysis [18,19], reductive alkaline hydrolysis [20], etc.) or, alternatively, by using endoglycosidases (and other enzymes) which cleave a wide spectrum of oligosaccharide structures from glycoproteins (21). High performance liquid chromatographic (HPLC) methods using amine- (22) and alkyl (21) bonded stationary phases have been used extensively to separate oligosaccharides according to size, monosaccharide composition and, to some extent, linkage position of individual sugars and branching patterns (23-26). Recently, a novel approach has been introduced which can effectively separate oligosaccharide positional isomers (27-29, 35). For the first time large oligosaccharides, which differ only in a single linkage position can be resolved.

A key feature of this method is the separation of oligosaccharides, either neutral or acidic, as their oxyanions using a Dionex pellicular anion exchange column under highly alkaline conditions (pH=12-14) (Table 1). The enhanced chromatographic selectivity for positional isomers can be attributed to both the differential pKa values of the sugar hydroxyl groups and the accessibility of the oxyanions (which is related to the three dimensional structure of the oligosaccharides) to the quaternary amine stationary phase.

Critical to the analysis of carbohydrates using liquid chromatography is the availability of a sensitive technique for detection. While refractive index detectors have been used for the detection of non-derivatized carbohydrates (23), this technique is relatively insensitive (μmoles per injection) and can be plagued by difficulties arising from temperature instability, gradient incompatibility and non-carbohydrate sample contamination. UV

TABLE 1. Neutral Sugar pKa's.

Sugar	pKa
Galactose	12.39
Glucose	12.28
Xylose	12.15
Mannose	12.08
Fructose	12.03
Sorbitol	13.60

Common neutral sugars such as hexoses are ionized only at high pH. Even epimers have sufficiently different pKa's for resolution. Data taken from Lange's Handbook of Chemistry(13th Edition).

absorbance (190-220 nm) can be used for the detection of some carbohydrates (especially those containing amide bonds) in the 5-50 nmole per injection range. However, high sensitivity was attained only after derivatizing the carbohydrate with radiolabels (30), chromophores (31,32) or fluorophores (21). Pulsed electrochemical detection (33-36) can now be used to monitor underivatized oligosaccharides in the picomole range (10-100 pmol). Coupling the unprecedented resolution of pellicular anion exchange resins with pulsed electrochemical detection provides the ability to produce high resolution oligosaccharide profiles of either glycoproteins (FIGURE 1) or individual glycosylation sites (FIGURE 2) (37). These maps are analogous to the familiar maps obtained by reverse phase HPLC after tryptic digestion of proteins.

FIGURE 1. Oligosaccharide Profile from a Glycoprotein

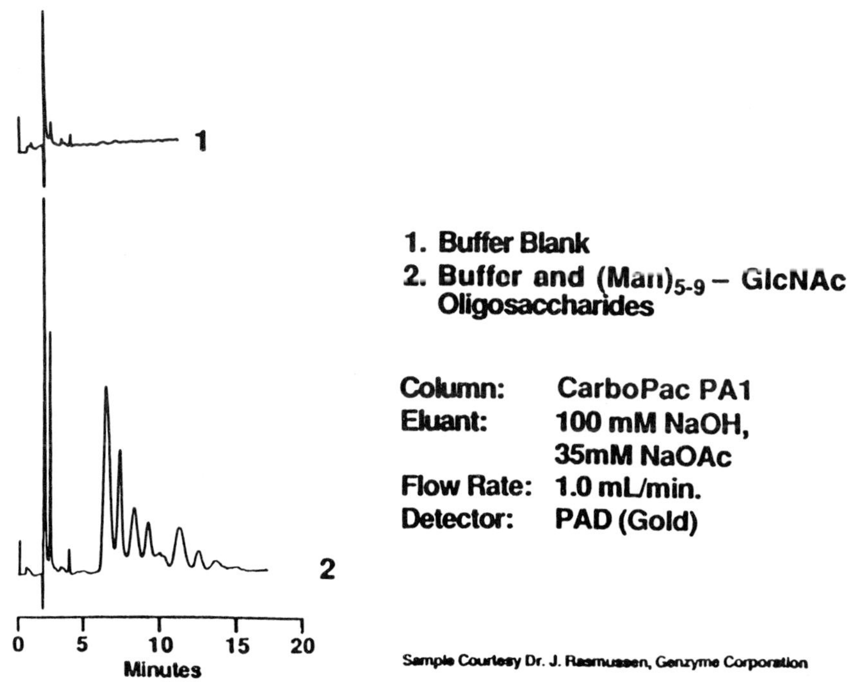

Glycoproteins such as Ribonuclease B, shown here, can be treated with enzymes to release their oligosaccharide chains. After clean-up with a reverse phase cartridge, the carbohydrates can be chromatographed with an ionic strength gradient. Changes in microheterogeneity between batches of glycoprotein may cause a change in the number and ratio of peaks.

FIGURE 2. Separation of Glycopeptide Structural Isomers

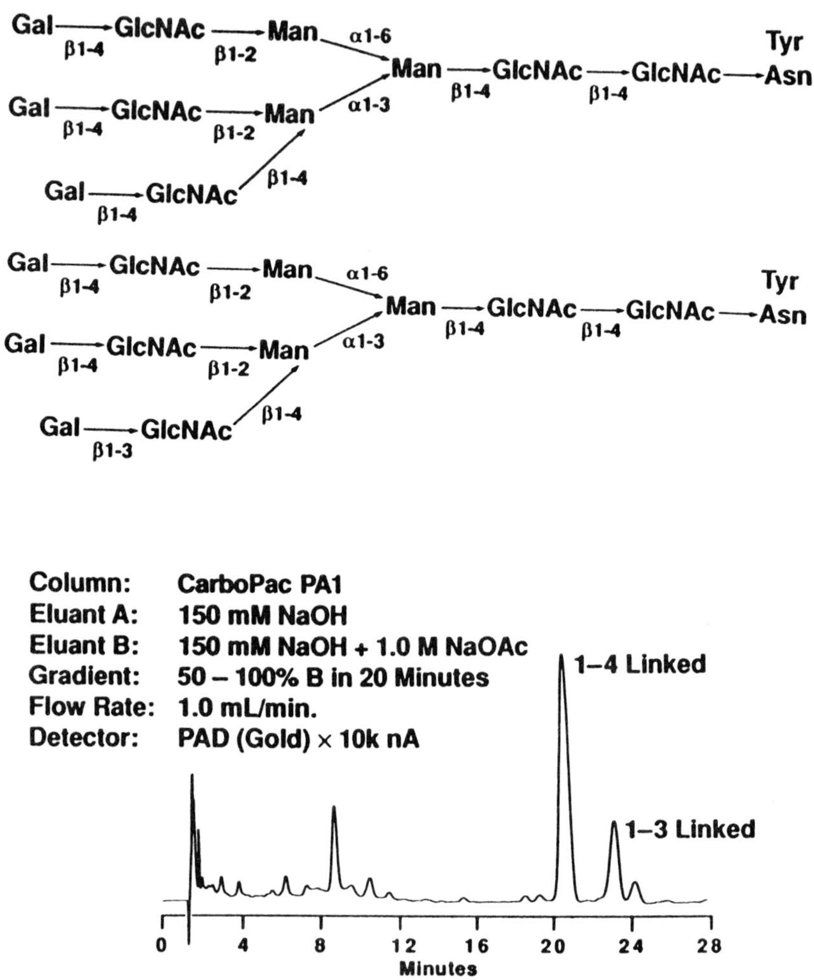

Above. The structures of the two glycopeptides are identical except for a single non-reducing terminus, (β1-3 vs β1-4).
Below. The two peptides are easily resolved. Unpublished data show resolution is retained even with changes to the peptide fragment.

The technique of pellicular anion exchange chromatography and pulsed electrochemical detection can also be used for the carbohydrate compositional analysis (38) of both individual oligosaccharides or of entire glycoproteins (see FIGURE 3 for carbohydrate compositional analysis).

II.PELLICULAR ANION EXCHANGE CHROMATOGRAPHY

While anion exchange chromatography has been used for the separation of sialylated, phosphorylated or sulfated carbohydrates (39), the technique was not commonly employed for the analysis of neutral oligosaccharides. This was due, in part, to the fact that column packings were not compatible with mobile phases sufficiently alkaline to ionize neutral carbohydrates. Instead, neutral mono- and oligosaccharides have been resolved by anion exchange chromatography in borate buffers (38). Borate complexes with the cis-diols of the carbohydrate to form a charged compound. These borate complexes tend to dissociate and reassociate continuously during chromatography to produce broad peaks and poor resolution.

Table 1 shows the pKa's of a number of neutral monosaccharides. At pH 13 all of the listed carbohydrates are ionized and can be separated in a way analogous to the separation of chloride and nitrate under neutral conditions. These values would be of no practical importance if one were restricted to alkali-labile silica based columns. Polystyrene-divinylbenzene anion exchange columns now allow the chromatographic separations of these analytes as oxyanions under strongly alkaline conditions.

Alkali compatible columns do not alone solve the problem of carbohydrate separations. Traditional porous polystyrene-divinylbenzene anion exchange columns can separate carbohydrates but tend to produce broad peaks. Peak broadening is due primarily to the time involved in the diffusion of the analyte into and out of the porous packing material. Smaller porous beads increase the efficiency of the column but at the cost of increased operating pressure.

Dionex MicroBead pellicular anion exchange column packing material uses a 10 micron, surface sulfonated, polystyrene-divinylbenzene bead as a support particle. Small (0.1 micron) particles of anion exchange material are bound in a mono-layer to the surface of the 10 micron bead. The small beads are held in place by strong electrostatic forces that are not neutralized even at the extremes of pH. The MicroBeads have

FIGURE 3. Carbohydrate Compositional Analysis

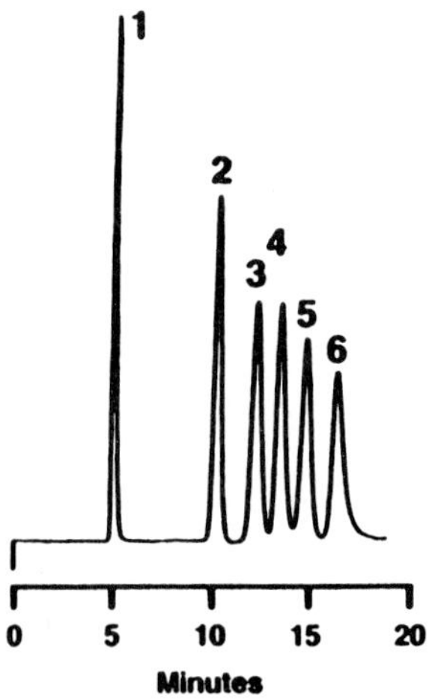

Column: CarboPac PA1
Eluant: 15 mM NaOH
Flow Rate: 1 mL/min.
Detector: PAD (Gold)
Postcolumn addition of 0.3 M NaOH

	nmoles
1. Fucose	4
2. Galactosamine	5
3. Glucosamine	5
4. Galactose	5
5. Glucose	5
6. Mannose	5

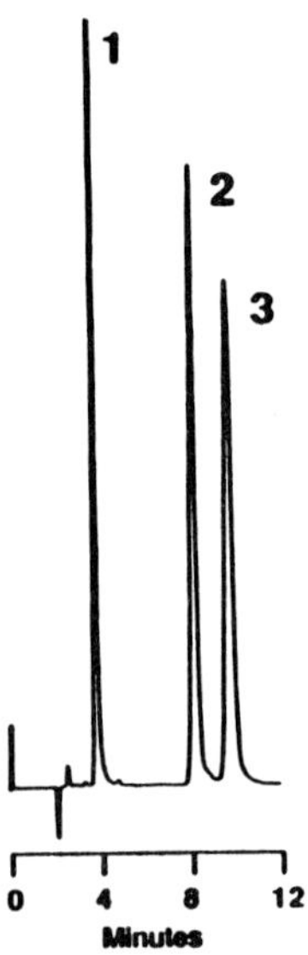

Peaks	µg
1. N-Acetylneuraminic Acid	2.5
2. Glucuronic Acid	2.0*
3. N-Glycolylneuraminic Acid	2.5

* Internal standard

Column: CarboPac PA1
Eluant: 100 mM NaOH, 150 mM NaOAc
Flow Rate: 1.0 mL/min.
Detector: PAD (Gold)

Above. Theses are the sugars released from an acid hydrolysis of a mammalian glycoprotein. The acetyl groups are hydrolyzed to yield the free aminosugars. Sialic acids are destroyed under these conditions.
Below. Milder hydrolytic conditions can release the sialic acid residues from glycoproteins.

10^{-6} times the internal volume of a typical 10 micron bead. This improves the surface to volume ratio providing mass transfer that is on the order of one hundred times faster than for traditional porous 10 micron particles. The support beads are stable to changes in ionic strength and pH and do not swell or shrink under chromatographic conditions which can cause chromatographic artifacts commonly associated with bead size change.

For monosaccharides of comparable structures, the carbohydrates elute in the expected order, i.e., highest pKa (weakest acid) first. The elution of complex oligosaccharides is less predictable but can be approximated with a number of general rules (27-29).

1. Larger oligosaccharides tend to elute after smaller oligosaccharides. Oligomeric series of polysaccharides can easily be generated with each peak corresponding to a component one sugar unit longer than the previous peak.

2. Adding a fucose α1-2, α1-3 or α1-4 to an oligosaccharide markedly reduces (10-30 minutes) its retention time.

3. Adding a sialic acid (or phosphate or sulfate ester group) significantly increases the retention time of the oligosaccharide. In this case, the carbohydrate is a much stronger anion and is retained correspondingly longer.

4. Oligosaccharide alditols elute significantly earlier than their non-reduced parent compounds.

5. Certain sequences in oligosaccharides (e.g., Gal β1-3 GlcNAc) assume conformations which optimally position the oxyanions resulting in stronger interaction with the column and increased retention.

Carbohydrates can undergo a series of well characterized base catalyzed reactions at elevated pH (40). To determine whether the high alkalinity required to produce carbohydrate oxyanions catalyzes analyte decomposition, a number of different groups have monitored the effects of high pH chromatography on simple monosaccharides and complex oligosaccharides.

The best known degradation of carbohydrates is the Lobry de Bruyn-van Ekenstein transformation (41-44) which includes the keto-enol tautomerization and epimerization adjacent to the carbonyl of reducing sugars. For example, glucose is transformed first to fructose and mannose and eventually to many different hexoses. Occasionally, the description of this transformation loosely includes the oxidative decomposition of the starting material to carbon dioxide and smaller sugars. However, at room temperature, studies show that with 30 mM NaOH (an eluant commonly used for anion exchange chromatography of carbohydrates) (45) this reaction is slow with less than five percent of a glucose sample transformed to mannose in a period of six hours. With most chromatography taking less than 30 minutes, this reaction, in general, is not a problem.

There have been no reports of deacylation of N-acylated amino sugars. This reaction is, if anything, even slower than the Lobry de Bruyn-van Ekenstein transformation. On the other hand, O-acylations tend to degrade under the alkaline conditions of chromatography.

Reducing carbohydrates which are substituted in the 3 position can undergo a beta elimination (called "peeling" or "stripping") under alkaline conditions (46). On the other hand, oligosaccharides ending in alditols or glycosides do not normally undergo beta elimination. When the group at position 3 is a good leaving group such as sulfate or phosphate, this reaction is very facile but it occurs only slowly, if at all, when the group at position 3 is another carbohydrate. Broad, asymmetrical peaks are a ready indicator of this reaction during chromatography. Sulfated and phosphorylated carbohydrates can be resolved under non-degrading, neutral pH conditions due to the strong ionic character of the analyte. Of those tested so far, most oligosaccharides 3-O linked at the reducing terminus appear to undergo negligible decomposition under standard analytical conditions.

III. PULSED AMPEROMETRIC DETECTION

Single potential electrochemical detection of carbohydrates has not proved generally applicable because of rapid working electrode surface fouling. Carbohydrates passing through the electrochemical cell are adsorbed onto the surface of the electrode and partially oxidized. While this oxidation produces a current that allows detection of the carbohydrate, the absorbed partially oxidized materials form a layer preventing new analyte from contact with the electrode surface and subsequent oxidation. Techniques to overcome this problem have been explored including various automatic mechanical cleaners. With mechanical cleaners, the dead volume is increased significantly producing broad peaks. Alternative techniques may loose both sensitivity and linearity while still requiring manual electrode polishing and cell reequilibration. All methods of mechanical electrode cleansing have proved to be less than optimal.

In pulsed amperometry the voltage potential of the working electrode is automatically cycled to three different values (See FIGURE 4). During the first stage, the analytical potential (E_1, t_1), carbohydrates in the eluant stream are oxidized. At this voltage, carbohydrates are adsorbed to the partially electron deficient gold surface. It is during this time that signal from oxidation of the analytes is collected. During the second pulse, (E_2, t_2), signal is no longer collected. Any carbohydrates which were adsorbed onto the surface of the electrode during the first stage are oxidized to water soluble organic acids. The gold oxide which forms on the surface of the electrode is no longer electron deficient and is no longer adsorptive. In the third pulse, (E_3, t_3), the gold oxide is reduced to provide a fresh clean gold surface. Carbohydrates do not adsorb on to the surface during this step, probably because of the charge on the working electrode. This three pulse cycle continuously repeats, up to five times per second, depending upon t_n. The voltages and pulse durations can both be adjusted to detect other analytes or to reduce the background due to sample matrix interferences and improve the analyte signal-to-noise ratio.

The linearity of pulsed amperometric detection has been shown to be greater than three orders of magnitude for numerous monosaccharides (47). Detection limits for both mono- and oligosaccharides have been reported at the low picomole levels (27, 37). Pulsing reduces the time required for baseline stabilization compared to non-pulsed detectors. While different carbohydrates give different molar electrochemical responses, the variations are generally small within a related group.

FIGURE 4. Schematic of Pulsed Amperometric Detector Potentials

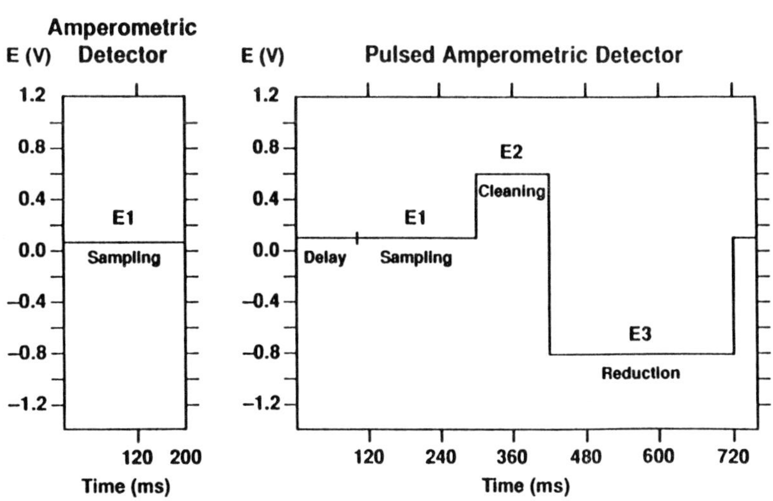

By pulsing the potential, the surface is continuously cleaned, yielding good linearity and high sensitivity.

IV. CONCLUSIONS

High performance, MicroBead, pellicular anion exchange resins can deliver superior chromatographic resolution of both mono- and oligosaccharides with little, if any, decomposition due to the alkalinity of the eluants. This technique can be used for acidic, neutral and amino sugars. Complicated glycoprotein derived oligosaccharides, both neutral and anionic, can be separated with unprecedented resolution.

Pulsed amperometry is a sensitive, gradient compatible alternative to carbohydrate detection by both refractive index and UV absorbance. The sensitivity achieved by this technique matches or exceeds that obtained by producing UV absorbing or fluorescent compounds and approaches that obtained by radiolabelling. This technique is well suited for the analysis of picomole amounts of either monosaccharides derived from glycoprotein hydrolysates or for oligosaccharides and glycopeptides derived from glycoproteins.

Together, these techniques provide the analytical biochemist with the tools necessary for the analysis of glycoprotein derived carbohydrates.

V. REFERENCES

1. Rademacher, T.W., Parekh, and Dwek, R.A. (1988) Ann. Rev. Biochem. 57, 785-838.
2. Matzuk, M.M. and Boime, I. (1988) J. Cell Biol. 106, 1049-59.
3. Sahagian, G.G. (1987) Vertebrate Lectins. (K. Olden and J. B. Parent, eds.) Van Nostrand Reinhold pp. 46-64.
4. Podskalny, J.M., Rouiller, D.G., Grunberger, G., Baxter, R.C. McElduff, A., and Gorden, P. (1986) J. Biol. Chem. 261, 14076-81.
5. Wilheim, J., Kalyan, N., Lee, S.G., Hum, W.-T., and Hung, P.P. (1988) FASEB J. 2, A344.
6. Bruck, C., Protetelle, D., Burny, A., and Zavada, J. (1982) Virology, 122, 353-62.
7. Galili, U., Clark, M.R., Shohet, S.B., Buehler, J., and Macher, B.A. (1987) Proc. Natl. Acad. Sci. 84, 1369-73.
8. White, S.L. Vertebrate Lectins. (K. Olden and J.B. Parent, eds.) Van Nostrand Reinhold. pp. 46-64.
9. Ashwell, G. and Harford, J. (1982) Ann. Rev. Biochem. 51,

531-54.
10. Hounsell, E. and Feizi, T. (1982) Medical Biology, 60, 227-36.
11. Wold, F. (1981) A. Rev. Biochem., 50, 783-814.
12. DeVries, A. , Komatsu, S., Feeney, R. (1970) J. Biol. Chem., 245, 2901-08.
13. Tsuda, E., Goto, M., Murakami, A., Akai, K., Ueda, M., Kawanishi, G., Takahashi, N., Sasaki, R., Chiba, H., Ishihara, H., Mori, M., Tejima, S., Endo, S. and Arata, Y. (1987) Biochemistry, 27, 5446-5654.
14. Sasaki, H., Bothner, B., Dell, A. and Fukuda, M. (1987) J. Biol. Chem., 262, 12059-76.
15. Slomiany, B. L., Zdebska, E. and Slomiany, A. (1984) J. Biol. Chem., 259, 2863-69.
16. Dill, K. and Allerhand, A. (1979) J. Biol. Chem., 254, 4224-31.
17. Van Halbeek, H., Dorland, L., Vliegenthart, J. F. G., Kochetkov, N. K., Arbatsky, N. P. and Derevitskaya, V. A. (1982) Eur. J. Biochem., 127, 21-29.
18. Bayard, B. and Montreuill, J. (1974) Colloq. Int. CNRS., 221, 209-218.
19. Dimitriev, B., A. Knirel, Y. A. and Kochetkov, N. K., (1973) Carbohydr. Res., 29, 451-7.
20. Iyer, R. N. and Carlson, D. M. (1971) Arch. Biochem. Biophys., 142, 101.
21. Tomiya, N., Kurono, M., Ishiihara, H., Tejima, S., Endo, S., Arata, Y., and Takahashi, N. (1987) Anal. Biochem. 163, 89-99.
22. Blanken, W.M., Bergh, M.L.E., Koppen, P.L., and Van den Eijnden, D.H. (1985) Anal. Biochem. 145, 322-30.
23. Clarke, P. I., Narasimhan, S., Williams, J. M. and Clamp, J. R. (1983) Carbohydr. Res., 118, 147-55.
24. Parente, J., Strecker, G., Leroy, Y., Montreuill, J. and Fournet, B. (1982) J. Chromatogr., 249, 199-204.
25. Shaw, P. and Wilson, C. (1982) J. Chrom. Sci., 20, 209.
26. Dube, V. A. and Bush, C. A. (1984) Biochim. Biophys. Acta, 802, 29.
27. Hardy, M.R. and Townsend, R.R. (1988) Proc. Natl. Acad. Sci. 85, 3289-93.
28. Townsend, R.R., Hardy, M.R., Hindsgaul, O., and Lee, Y.C. (1988) Anal. Biochem., in press.
29. Townsend, R.R. and Hardy, M.R. manuscript submitted to Carbohydr. Res.
30. Takasaki, S., Mizuochi, T., and Kobata, A. (1982) Methods in Enzymology (Ginsberg, V.,ed.) Vol. 83, pp. 263-68 Academic Press, New York.

31. Wang, W.T., LeDonne, N.C., Jr., Ackerman, B., and Sweeley, C.C. (1984) Anal. Biochem. 141, 366-81.
32. Daniel, P.F., DeFeudis, D.F., Lott, I.T., and McCluer, R.H. (1981) Carbohydr. Res. 97, 161-80.
33. Rocklin, R. D. and Pohl, C. A., (1983) J. Liq. Chromatogr., 6, 1577-90.
34. Hughes, S. and Johnson, D. C., (1981) Anal. Chim. Acta, 132, 11.
35. Olechno, J. D., Carter, S. R., Edwards, W. T., and Gillen, D. G. (1987) American Biotechnology Laboratory 5, 38-50.
36. Johnson, D. C. (1986) Nature, 321, 451-456.
37. Chen, L-.M., Yet, M-.G. and Shao, M.-C. (1988) FASEB J., 2, 2819-24.
38. Hardy, M. R., Townsend, R. R. and Lee, Y. C. (1988) Anal. Biochem., 170, 54-62.
39. Mellis, S. J. and Baenziger, J. U. (1983) Anal. Biochem., 134, 442-49.
40. Ferrier, R. and Collins, P.M. (1972) Monosaccharide Chemistry, Penguin Books, London.
41. Lobry de Bruyn, C. A., (1895) Rec. Trav. Chim., 14, 150.
42. Lobry de Bruyn, C. A. and van Ekenstein, W. A., (1895) Rec. Trav. Chim., 14, 195.
43. Lobry de Bruyn, C. A. and van Ekenstein, W. A., (1897) Rec. Trav. Chim., 16, 262.
44. Johansson, M.H. and Samuelson, O., (1976) Chem. Scr., 9, 151.
45. Wang, W.-T. and Zopf, D., Carbohydr. Res., in press.
46. Hough, L., Jones, J. V. S. and Ko, A., (1974) Colloq. Inst. CNRS., 221, 255-61.
47. Olechno, J. D. and Carter, S. R., unpublished data.

37

ENZYMATIC DIGESTION OF PROTEINS AND HPLC PEPTIDE ISOLATION IN THE SUB-NANOMOLE RANGE

Kathryn L. Stone
Mary B. LoPresti
Nancy D. Williams
J. Myron Crawford
Raymond DeAngelis
Kenneth R. Williams

Yale University School of Medicine Protein and Nucleic Acid Chemistry Facility
Yale University
New Haven, CT

I. INTRODUCTION

Although it is now possible to routinely obtain useful amino acid sequence data from 100-200 picomole amounts of protein (1), this direct approach often cannot be used because approximately 80-90% of all eukaryotic proteins have blocked NH_2-termini (2,3). An alternative approach that is frequently taken is to enzymatically digest the blocked protein and then isolate and sequence one or more of the resulting peptides. While in theory this latter approach should not require substantially larger amounts of protein than are needed for direct sequencing of the intact protein, in practice, almost an order of magnitude more protein is usually taken if the goal is to isolate and sequence a peptide (1). Based on our research, the major reasons for needing more than a nanomole of protein for an enzymatic digestion are incomplete denaturation of the protein, loss of the denatured protein prior to digestion and the presence of detergents such as SDS that interfere with both the enzymatic digestion and the subsequent HPLC. In this chapter a simple carboxamidomethylation/trypsin digestion procedure will be given that avoids loss of the denatured protein and that has now been successfully used on over 50 different proteins that have been submitted to the Yale University School of Medicine Protein and Nucleic Acid Chemistry Facility. Only minor modifications in this procedure are required in the case of proteins that have been electroeluted from SDS polyacrylamide gels or that have been electroblotted onto polyvinylidene difluoride (PVDF) membranes (4). Practical suggestions will also be given on isolating the resulting peptides on conventional and narrowbore HPLC columns. Finally, several comparative HPLC profiles will be shown that were obtained on three different HPLC systems.

II. EXPERIMENTAL PROCEDURES

A. Sample Preparation

If the protein to be digested contains less than 0.2 millimoles of monovalent salt and is free of SDS then, in many cases it can simply be transferred to a 1.5 ml eppendorf tube and taken to dryness in a Speedvac. If the amount of non-volatile salt in the sample is excessive then the protein can usually be precipitated in good yield by adding a final concentration of 10% trichloroacetic acid (TCA) followed by incubating on ice for 60 min. Residual TCA is removed by washing with acetone. In general, we usually concentrate the protein in a Speedvac until the protein concentration is above 0.1 mg/ml and, if necessary, use dialysis to decrease the glycerol concentration to 10% (w/v) or less prior to carrying out the TCA precipitation. Samples that have been electroeluted from SDS polyacrylamide gels are usually dialyzed versus 0.05% SDS, 5 mM NH_4HCO_3 and then up to 1 ml of this solution is concentrated to a final volume of 50 μl in a Speedvac. The SDS is then extracted by adding 450 μl cold acetone and incubating for 3 hrs at-20°C. If the protein has been electroblotted onto a PVDF membrane then, based on the proteins that we have so far tested, it can be eluted in high yield (typically 75%) by following a procedure similar to that suggested by Montelaro (5). In this instance the Coomassie-blue stained protein is eluted from the PVDF membrane by cutting out the appropriate band and submerging it in 200 μl 40% acetonitrile in water. After incubating at 37°C for 3 hrs the supernatant is removed and the PVDF membrane is re-extracted with 200 μl 40% acetonitrile in 0.05% trifluoroacetic acid. After incubating at 50°C for 20 min, the two supernatants are combined, and then evaporated to dryness in a Speedvac. In all cases the final dried protein sample is redissolved in 50 μl 8 M urea, 0.4 M NH_4HCO_3 and then carboxamidomethylated and digested with trypsin as described below.

B. Carboxamidomethylation and Trypsin Digestion

The experiments that will be described, which examined the effect of various parameters on the extent of trypsin digestion, were carried out on 50 pmol aliquots of transferrin or bovine serum albumin that had been dried _in vacuo_ in 1.5 ml eppendorf tubes. After redissolving in 50 μl 8 M urea, 0.4 M NH_4HCO_3, pH 8.0, the proteins were reduced by adding 5 μl 45 mM dithiothreitol and then incubating at 50°C for 15 min. After cooling to room temperature, 5 μl 100 mM iodoacetamide was added and the reaction mixture was incubated for 15 minutes prior to adding 140 μl water. Trypsin (TPCK-treated from Cooper Biomedical) was then added in a volume of 5 μl at an enzyme/substrate (weight/weight) ratio of 1/25. After 24 hrs at 37°C the digest was stopped by freezing or by injecting onto an HPLC. If an unknown protein is being digested then 10% is usually removed after solubilizing it in 50 μl 8 M urea, 0.4 M NH_4HCO_3 so that the protein concentration can be verified by ion exchange amino acid analysis prior to carrying out the digest. Our previous studies demonstrate that a final protein concentration of about 25 μg/ml is needed to ensure reasonably complete digestion (6). If the final protein concentration is anticipated to fall below this figure, as in the case of 100 picomole amounts of low molecular weight proteins, then the volume of 8 M urea, 0.4 M NH_4HCO_3 (and all other volumes) should be reduced accordingly. Since we have previously demonstrated that there can be significant variation in the purity of trypsin preparations obtained from different vendors, it is recommended that any new preparation be pre-tested before use by carrying out a digest on insulin B-chain as previously described (7). Most of the comparative HPLC experiments that follow utilized 50 picomole aliquots derived from a large scale trypsin digest of 3.0 mg of carboxamidomethylated transferrin. After extensive dialysis, 8.7 mg of carboxamidomethylated transferrin was dried _in vacuo_ and then re-dissolved in 1.0 ml 8 M urea. Approximately 0.35 ml of this transferrin solution was heated at 65°C for 10 min prior to adding 0.92 ml 0.1 M NH_4HCO_3 and 0.12 ml 1.0 mg/ml trypsin. After incubating at

37°C for 24 hrs, suitable dilutions of this digest were made in a total volume of 20 μl 0.05% trifluoroacetic acid. Digests were run on either a 2.1 mm x 25 cm Vydac C-18 column or a 2.0 mm x 15 cm Delta Pak C-18 column, as described in the figure legends. These two columns provide similar peptide resolution (unpublished results), and were used interchangeably.

C. Analytical and Narrow-Bore HPLC

With the exception of three chromatograms (see Figures 6 and 7) that were obtained on either a Waters Associates HPLC System that has been previously described (7), or on an Applied Biosystems Model 130 HPLC System, all of the HPLC experiments were carried out on a Hewlett Packard Model 1090 HPLC that was equipped with a diode array detector and a 200 μl injection loop. Data was in all cases acquired on a Nelson Analytical Model 4416X Multi-Instrument Data System. In all instances the following gradient was used at the flow rates indicated in each figure legend.

0-63 min	2-37%B
63-95 min	37-75%B
95-105 min	75-98%B

The gradient was followed by a 2.5 ml isocratic wash at 98% B prior to re-equilibrating with approximately 15 column volumes buffer A which is 0.060% trifluoroacetic acid. Buffer B is 0.056% trifluoroacetic acid in 80% acetonitrile. Small adjustments, on the order of ±0.002%, were made in the concentration of trifluoroacetic acid in either buffer as appropriate to balance the apparent absorbances at 210 nm.

III. RESULTS

A. Important Variables Affecting the Extent of Trypsin Digestion

Based on our experience, one of the major reasons for enzymatic digestions not working is simply over-estimation of the actual protein concentration due to losses incurred in the final sample preparation steps or the inaccuracy of colorimetric and dye-binding protein assays. For this reason it is essential to obtain an accurate amino acid analysis either immediately before drying the sample in the eppendorf tube in which it will be digested or preferably, after redissolving the sample in 8 M urea, 0.4 M NH_4HCO_3. A second major problem is the presence of detergents. Fig. 1a demonstrates that 0.05% SDS is sufficient to completely prevent trypsin digestion, possibly by denaturing the trypsin. All absorbance peaks present in the 0.05% SDS panel in Fig. 1a are due to contaminants in the SDS and not to a limited extent of cleavage. As shown in Fig. 1b, SDS also interferes with reversed phase HPLC. SDS concentrations above 0.005%, that is 10 μg SDS in our usual 0.2 ml injection volume, result in a general loss in resolution and later elution of peptides. Acetone precipitation provides a relatively simple means of reducing the SDS concentration to a level that no longer interferes with either the enzymatic digestion or the subsequent HPLC. In a previous study it was demonstrated that this approach was quite effective even in the rather extreme instance where a mixture containing 50 picomoles of protein and 500 μg SDS was extracted (6).

Fig. 2 demonstrates that heating in 8 M urea is not sufficient to adequately denature bovine serum albumin. As shown in this figure, prior carboxamidomethylation markedly improves the extent of digestion, presumably by more fully and irreversibly denaturing the protein. Since under the conditions given in Experimental Procedures the excess reagents left over from the carboxamidomethylation reaction do not interfere with the subsequent trypsin digestion, there is no need for an intermediate gel filtration or dialysis step. Following carboxamidomethylation, the urea concentration must be reduced to 2 M prior to carrying out the trypsin digestion by diluting with water as described in Experimental Procedures. The maximum urea concentration that can be tolerated is about 2 M. At higher concentrations, the extent of trypsin digestion is poor.

If the final purification step depends upon electroeluting the protein from an SDS polyacrylamide gel, then an alternative approach that can be used is to blot the sample onto polyvinylidene difluoride (PVDF) membranes, stain with Coomassie Blue and then elute the protein with 40% acetonitrile as described in Experimental Procedures. Fig. 3 shows the results of an experiment where 250 picomoles of myoglobin that had been electroblotted onto PVDF was subjected to trypsin digestion and the resulting HPLC chromatogram compared with a control digest carried out on the same amount of protein that had not been eluted from PVDF. Amino acid analyses, carried out on duplicate strips

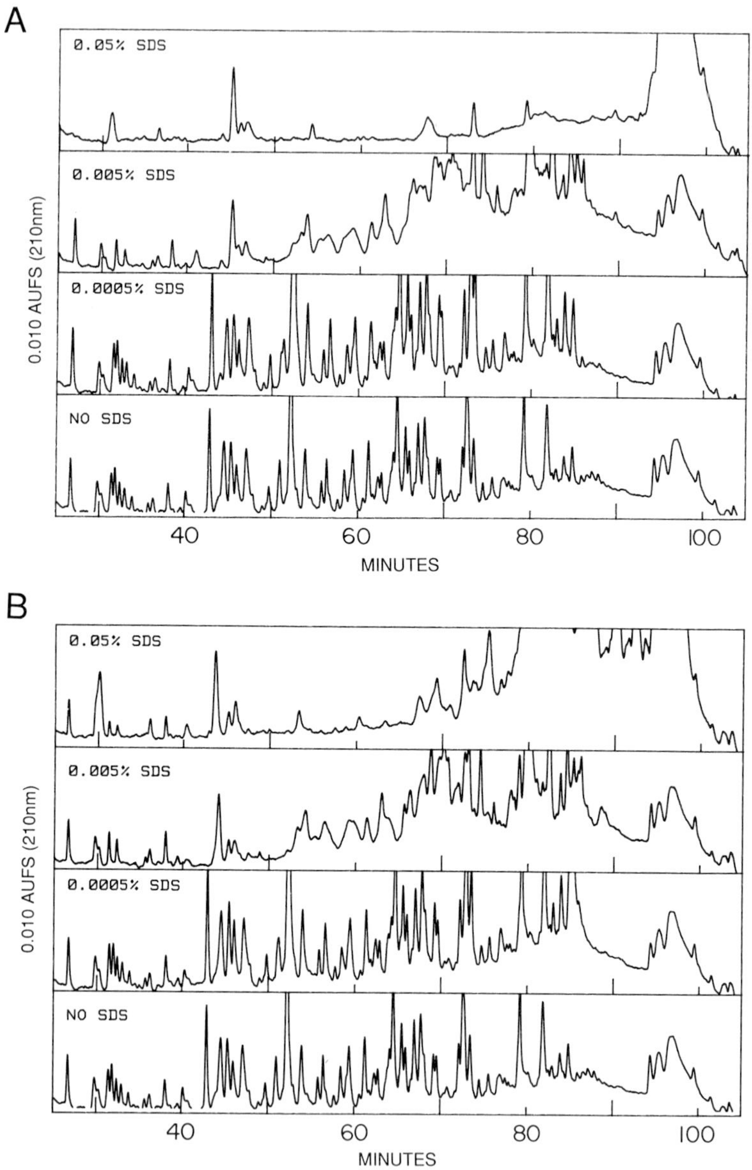

Fig. 1. Effect of SDS on trypsin digestion and reversed-phase HPLC. SDS was added at the indicated final concentrations either before (A) or after (B) trypsin digestion of several 50 picomole aliquots of transferrin. Each digest was then subjected to reversed-phase HPLC at a flow rate of 0.15 ml/min on a 2.1 mm x 25 cm Vydac C-18 column as described in Experimental Procedures.

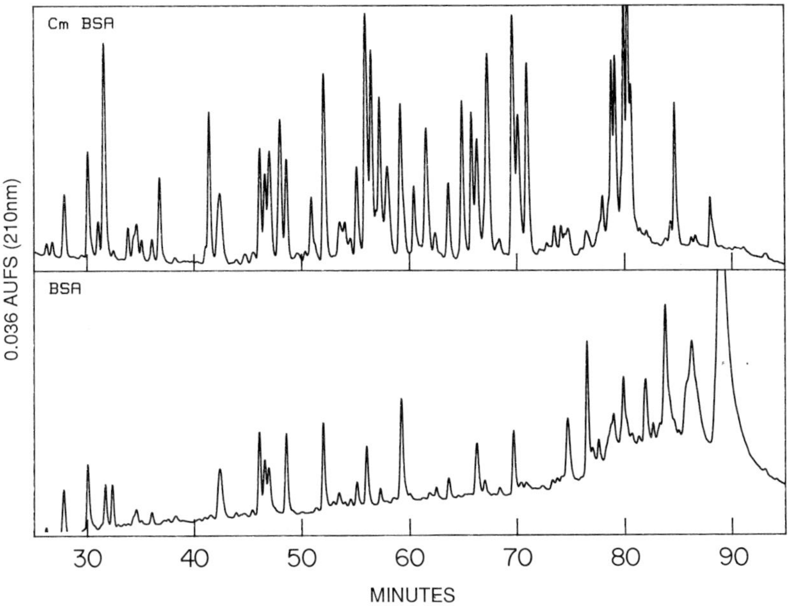

Fig. 2. Effect of carboxamidomethylation on trypsin digestion. Two 50 picomole amounts of bovine serum albumin were digested with trypsin either without (lower panel) or with (upper panel) prior carboxamidomethylation. Each aliquot was then injected onto a 2.1 mm x 25 cm Vydac C-18 column that was eluted at a flow rate of 0.15 ml/min as described in Experimental Procedures. The large peak eluting at about 89 min in the lower panel corresponds with the elution position of intact serum albumin.

of PVDF, indicated that only about 28% of the 1.2 nanomole myoglobin originally applied to the SDS polyacrylamide gel had been successfully transferred to the PVDF membrane. Again based on amino acid analysis of duplicate PVDF strips, approximately 80% of the 340 picomoles of myoglobin that had been successfully transferred to the PVDF membrane was eluted using the method described in Experimental Procedures. As shown in Fig. 3 the electroblotted/eluted myoglobin is amenable to trypsin digestion. The differences in the chromatograms for the control versus the electroblotted/eluted myoglobin probably result from either different final extents of cleavage or covalent modification of the myoglobin during SDS polyacrylamide gel electrophoresis.

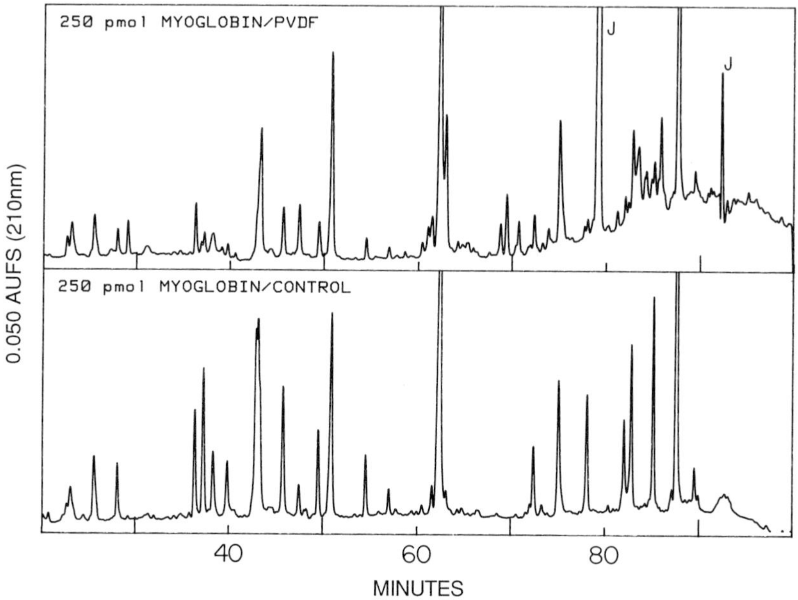

Fig. 3. HPLC separation of tryptic peptides from 250 picomoles myoglobin either before (lower panel) or after (upper panel) SDS polyacrylamide gel electrophoresis, electroblotting and elution from a PVDF membrane. The reversed-phase HPLC was carried out on 2.1 mm x 25 cm Vydac C-18 column that was eluted at a flow rate of 0.15 ml/min as described in Experimental Procedures. The two peaks in the upper chromatogram marked "J" are artifact peaks arising from the PVDF membrane and/or the SDS polyacrylamide gel.

B. HPLC Isolation of Sub-nanomole Amounts of Peptides

In devising appropriate strategies for the HPLC fractionation of enzymatic digests of sub-nanomole amounts of proteins there are at least three parameters that must be optimized. These include the analysis time, peak resolution, and sensitivity of detection. In terms of run time we have arbitrarily chosen a turn-around time of about 3 hours which is short enough to permit 2 collection runs to be easily completed per day and, taking column re-equilibration into account, allows for at least a 105 min gradient. Although peptide resolution is indirectly proportional to gradient steepness, near maximum resolution (with a constant run time of about 105 min on a narrowbore column) appears to be attained with the use of gradient volumes that correspond to approximately 50 column volumes (7). In the case of the 15 cm columns used in this study this would correspond to gradient volumes of about 24 and 90 ml respectively for the 2.0 mm and 3.9 mm inner diameter columns. These gradient volumes can be approximately reached within 105 min if flow rates of 0.2 and 1.0 ml/min respectively are used. An opposing consideration is peak sensitivity. Since peak volume is directly related to the flow rate (6-8), maximum sensitivity requires minimal flow rates. This phenomenon is demonstrated in Fig. 4 where the sensitivity of detection on a conventional 3.9 mm ID column has been increased nearly four-fold by decreasing the flow rate (and hence the total gradient volume since the gradient time was kept constant) from 1.0 to 0.2 ml/min. As shown by a comparison of the top two panels in Fig. 4 this substantial reduction in gradient volume is accompanied by a

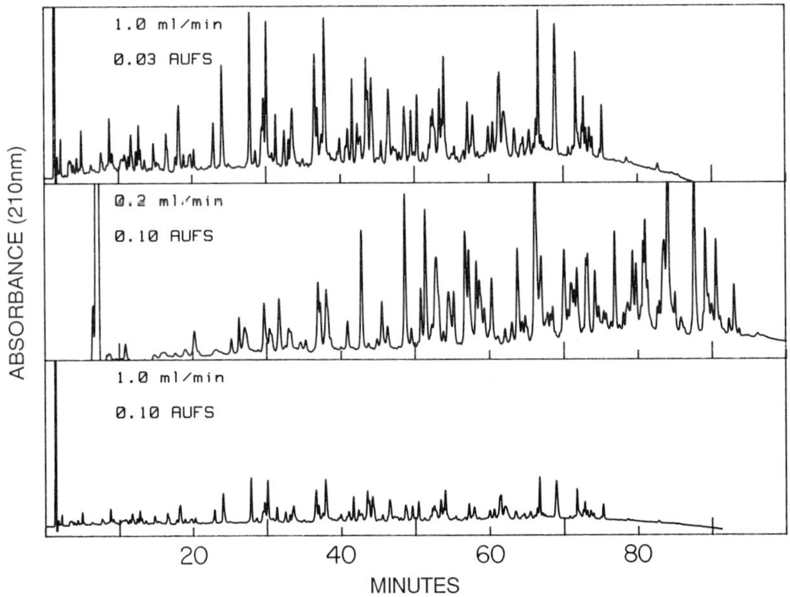

Fig. 4. Effect of decreased flow rate on the reversed phase HPLC separation of tryptic peptides from 250 picomoles transferrin on a 3.9 mm x 15 cm Delta Pak C-18 column that was eluted at the indicated flow rates as described in Experimental Procedures. The top and bottom chromatograms are identical except for the use of different full scale absorbance settings as shown.

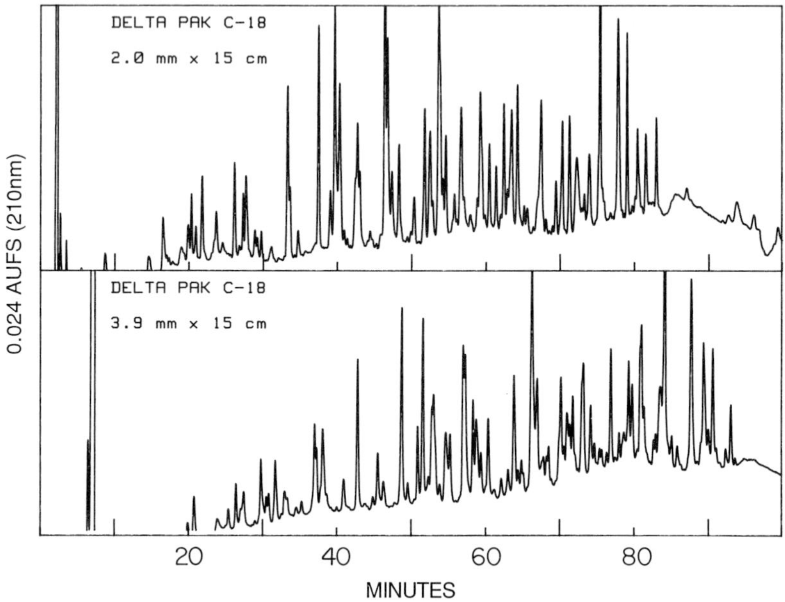

Fig. 5. Reversed-phase HPLC of tryptic peptides from 50 picomoles transferrin on either a 2.0 mm x 15 cm (top chromatogram) or a 3.9 mm x 15 cm (bottom chromatogram) Delta Pak C-18 column that was eluted at 0.2 ml/min as described in Experimental Procedures.

relatively minor decrease in resolution that in many cases may be tolerable. That is, 116 peaks were detected by the Nelson Analytical Data System at 1.0 ml/min compared to the 100 peaks that were detected at 0.2 ml/min.

Fig. 5 demonstrates that even as little as 50 picomoles of an enzymatic digest can easily be detected at 0.2 ml/min on either a conventional (3.9 mm ID) or narrow-bore (2.0 mm ID) column. This figure clearly demonstrates that under the conditions used, column diameter has very little influence on sensitivity. In terms of resolution, more peaks were actually detected on the 2.0 mm ID narrow-bore (106 peaks) rather than the conventional 3.9 mm ID column (93 peaks). If the two columns are run at more similar linear flow velocities and ratios of sample load to column volume (i.e./ 50 picomoles and 0.2 ml/min for the 2.0 mm ID column compared to 250 picomoles and 1.0 ml/min for the 3.9 mm ID column) then slightly better resolution is actually obtained on the 3.9 mm ID column. That is, under these conditions 116 peaks were detected on the 3.9 mm ID column compared to the 106 peaks detected on the 2.0 mm ID column.

Several comparisons were carried out on both columns at two different sample loads to serve as a guide in selecting optimum flow rates. In general, it would appear that 250 picomoles might serve as an approximate estimate of the minimum amount of a digest that can be optimally fractionated on a 3.9 mm ID column. Little loss in resolution occurs on decreasing the flow rate from 1.0 to 0.4 ml/min which substantially reduces the resulting peak volume, thus increasing sensitivity of detection and the ease with which a reasonably large percentage of the resulting peaks can be directly loaded onto a commercial amino acid sequencer. Below 250 picomoles it would appear that a narrow-bore column provides the best approach. Using our standard 105 min program nearly 50 column volumes can be pumped through the column at a rate of 0.2 ml/min thus providing excellent resolution and sensitivity. Although little loss in resolution occurs as the flow rate is decreased to 0.1 ml/min, at this flow rate the "delay time" between when a peak is detected and when it actually appears at the fraction collector becomes too large for convenient use of automatic peak detection. Hence the delay in the Waters Associates System described in reference 7 is about 70 μl which at a flow rate of 0.1 ml/min corresponds to 42 seconds. Under these conditions, if a second peak begins to elute during the 42 seconds that the fraction collector is waiting for the first peak to appear, then in the case of the ISCO Model 2150 Peak Detector the two peaks will be pooled together. This phenomenon seldom occurs at 0.2 ml/min where the delay is about 21 sec.

Finally, we have carried out some comparative studies using different HPLC systems. Fig. 6 shows that there is relatively little difference in resolution at 0.15 ml/min among a Waters Associates, an Applied Biosystems 130 and the HP 1090 HPLC System that we tested. The exceptionally long gradient delay on the Waters HPLC System has minimal effect on resolution and can be alleviated by inserting the WISP sample injector in the "A" buffer line just prior to the mixing tee (7). Of the three systems tested, the Applied Biosystems was the only HPLC equipped with a manual sample injector. At least in the case of the three particular units that were tested, the HP 1090 appears to provide optimum resolution with minimal gradient delay. Decreasing the flow rate further to 50 μl/min, which only corresponds to a total gradient volume of about 6 column volumes using the conditions given in Fig. 7, resulted in a substantial loss in resolution on both the HP 1090 and Applied Biosystems HPLC Systems due largely to the steepness of the resulting gradient. Under these conditions it would appear that the Applied Biosystems Model 130 provides a somewhat shorter gradient delay but slightly worse resolution than the HP 1090 System. At such low flow rates better resolution might actually be obtained on a 1 mm ID column.

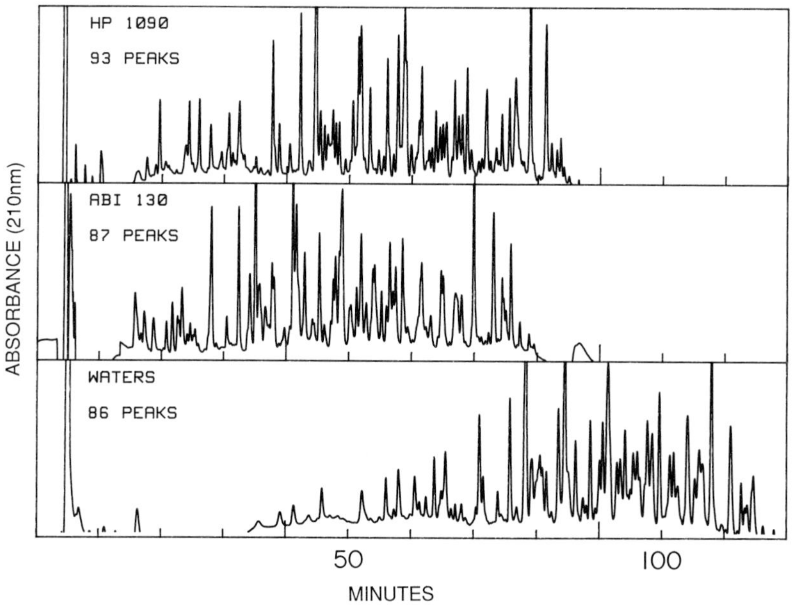

Fig. 6. HPLC separation of tryptic peptides from 50 picomole aliquots of transferrin that were chromatographed on the three different HPLC systems indicated above. In all cases a 2.1 mm x 25 cm Vydac C-18 column was used that was eluted at a flow rate of 0.15 ml/min as described in Experimental Procedures. The indicated number of absorbance peaks were detected by a Nelson Analytical Model 4416X Multi-Instrument Data System and the 210 nm full scale absorbance settings used were 0.036 in the case of the Applied Biosystems and Waters Associates HPLC Systems and 0.022 in the case of the Hewlett Packard System.

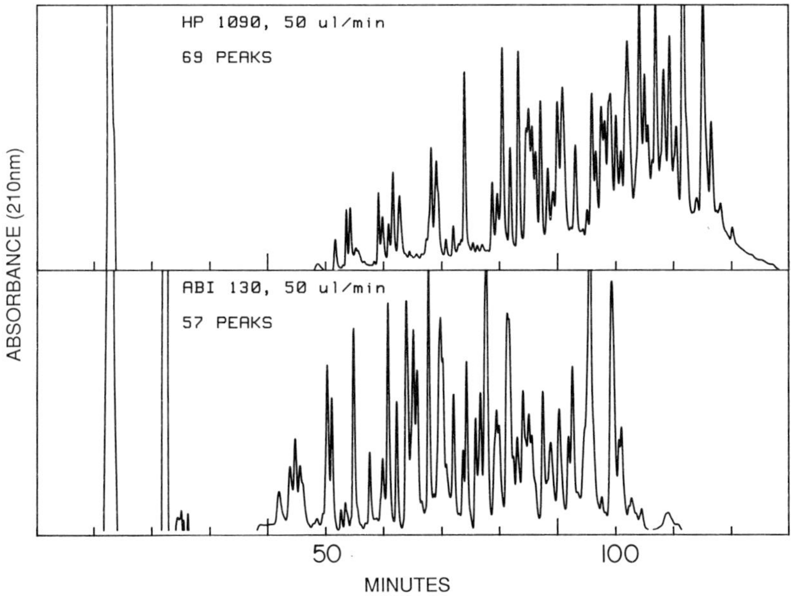

Fig. 7. HPLC separation of tryptic peptides from 50 picomole aliquots of transferrin that were chromatographed on the indicated HPLC systems at a flow rate of 50 μl/min. In both cases a 2.1 mm x 25 cm Vydac C 18 column was used and the indicated number of absorbance peaks were detected by a Nelson Analytical Model 4416X Multi-Instrument Data System. The 210 nm full scale absorbance settings used were 0.070 and 0.066 for the Applied Biosystems and Hewlett Packard HPLC Systems respectively.

IV. CONCLUSIONS

Based on our studies, the relatively straightforward procedures that have been presented can be used to routinely generate and isolate tryptic peptides from as little as 50 picomoles of protein in amounts sufficient to permit complete amino acid sequencing. If this amount is increased to above about 100 picomoles, then the resulting peptides can be repurified prior to sequence analysis using a different chromatographic support (7). Experiments with synthetic peptides confirm that no detectable NH_2-terminal blocking occurs (as a result of cyanate formation from the urea) during the course of the carboxamidomethylation or subsequent trypsin digestion (6). Although decreasing the flow rate from a typical value of 1.0 ml/min to 0.4 ml/min permits a substantial increase in sensitivity on a 3.9-4.6 mm ID column, maximum resolution and sensitivity in the 50-250 picomole range requires the use of narrow-bore columns. Providing that the flow rates are kept at or above 0.15 ml/min then the resulting peptides can be automatically and individually collected in 1.5 ml eppendorf tubes (which can be conveniently positioned on the top of 13 x 100 mm test tubes) using commercially available peak detectors and fraction collectors (7). When we have stored the resulting peptides at 5°C in their original, tightly capped eppendorf tubes, we have not, in general, detected any significant losses of peptides due to adsorption and have on numerous occasions successfully sequenced 50-100 picomole amounts of peptides that have been stored in this manner for over a year. Overall, successful isolation and sequencing of tryptic peptides from sub-nanomole amounts of protein appears to be most dependent upon such basic parameters as accurate estimation of the protein concentration, complete denaturation of the substrate, removal of excess detergents such as SDS, and a reliable measure of the "delay" in the HPLC system.

V. ACKNOWLEDGEMENTS

The authors wish to thank the Sep/a/ra/tions Group as well as the Waters Chromatography Division, Millipore Corporation for providing some of the HPLC columns used in this study and Rosella Gianella of Applied Biosystems Inc. for arranging for the temporary loan of a Model 130 HPLC System.

VI. REFERENCES

1. K.R. Williams, D. Atherton, A.V. Fowler, R. Kutny, A.J. Smith, R.L. Niece (1988). in FASEB J. in press.

2. J.L. Brown and W.K. Roberts (1976), J. Biol. Chem. 251:1009-1014.

3. J.L. Brown (1979). J. Biol. Chem. 254: 1447-1449.

4. P. Matsudaira (1987). J. Biol. Chem. 262: 10035-10038.

5. R.C. Montelaro (1987). Electrophoresis 8: 432-438.

6. K.L. Stone, M.B. LoPresti, K.R. Williams (1988) in "Focus on Laboratory Methodology in Biochemistry" (C. Fini, ed) CRC Press, in press.

7. K.L. Stone and K.R. Williams (1988) in "Macromolecular Sequencing and Synthesis (D.L. Schlesinger, ed) p. 7-24 Alan R. Liss Inc., New York.

8. T.D. Schlabach and K.J. Wilson (1987). J. Chromtography, 385, 65-74.

38

THE USE OF GLASS CAPILLARY TUBES AS DISPOSABLE MICROBORE COLUMNS FOR RP-HPLC OF PROTEINS AND PEPTIDES

Christopher Southan

Department of Medicinal Chemistry, Smith Kline & French Research Ltd Welwyn, Hertfordshire, AL6 9AR, UK

1. INTRODUCTION

For both analytical and micro-preparative HPLC in protein chemistry microbore columns of 1 mm I D have several advantages over 4.6 mm or 2 mm (1). With appropriate instrumentation the reduced solvent consumption and increased detection sensitivity are both proportional to the column cross-sectional area. This allows numerous analytical separation optimising runs to be performed using minimum amounts of sample. If RP-HPLC is used as the second of a 2-dimensional technique small aliquots of fractions from the first separation method may be sufficient for analysis and if an organic modifier is present in the first dimension fractions this can be diluted out for re-chromatography under the same or different reverse-phase conditions. The main advantage for micro-preparative runs and de-salting steps is that components are eluted in small volumes (50-200 μl) at high concentration factors (2). The collected protein or peptide can then be used directly, or even diluted out, for procedures such as sequencing, amino acid analysis, mass-spectrometry, derivatisation, and cleavage experiments. Losses are reduced by avoiding a post-collection concentration step.

Although an increasing number of HPLC instruments now offer adaptions for flow rates down to 50 - 100 μl/min the available range of commercial 1 mm columns remains limited both in lengths and packing types. This is partly due to the difficulties of slurry-packing such small columns (3). Consequently, their purchasing costs can approach that of 4.6 mm columns containing 20 times the amount of packing material. This expense factor is compounded by the shorter lifespan of 1 mm columns, the lack of any cartridge-based system for this diameter, and the tendency of guard columns to impair resolution. In an attempt to overcome some of these limitations this work describes the construction and evaluation of disposable dry-packed, microbore columns made from standard glass capillary tubes.

II. EXPERIMENTAL

A. Column construction

The 50 µl borosilicate-glass capillary tubes with green color-code rings, (127 mm, 1.52 mm OD, 0.92 mm ID) are available from many laboratory suppliers (Camlab, Cambridge, UK). The columns are constructed simply by securing the glass tubes into 1/16" plastic unions (Upchurch, CTFE part No. U243K, or Alltech, Kel-F part No. 32140) using standard fingertight fittings (Upchurch Kel-F F200 male nuts with F142 CTFE ferrules). To support the column bed the orifice in the lower union was plugged with a small piece of plastic frit material cut from inside a Sep-Pak cartridge (Millipore/Waters, Harrow, UK). A diagram of the column construction is shown below (Figure 1).

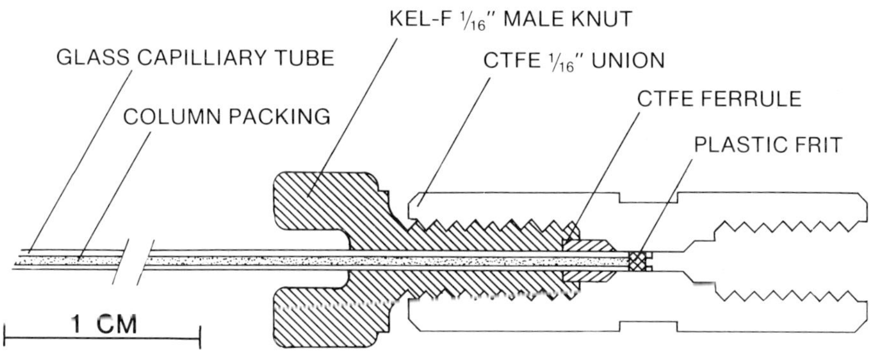

Fig. 1. A diagram showing the lower end assembly.
The upper one is identical except that the plastic frit is omitted.

B. Packing

The column material needed to be dry and finely dispersed. A filling funnel was made by cutting back a 1 ml disposable plastic pippete tip to give a push-fit on to the top of the glass capillary. With the bottom of the tube sealed with parafilm, rapid packing was induced by careful vibration on a vortex mixer. For the comparative evaluations described in this work columns were filled with Aquapore RP 300 7-µm C_8 or Speri-5 5-µm C_{18}, emptied from a 30 mm x 4.6 mm cartridge (Applied Biosystems, Warrington, UK). After packing, the column was assembled as described above. This was fitted into the HPLC instrument in the usual way, although the use of fingertight fittings with plastic HPLC tubing (Tefzel, Alltech, Lancashire, UK) was preferable to steel in being less likely to exert twisting forces on the glass tube. Using flow rates of between 100 and 200 μl/min the column was washed with 50% acetonitrile to wet, bed down, and wash the packing material. Any leaks could be eliminated by further gradual tightening of the male nuts, allowing operating pressures of up to 1500 p.s.i. After the column material had bedded down to a constant volume, i.e. compressed by approximately 20%, the top fitting was removed and the tube cut cleanly at, or below, the packing surface with a glass tube-cutter. The column was then re-assembled and connected to the UV detector. After prolonged use the top fitting could be dismantled to check for void formation and the tubing re-cut if necessary.

C. HPLC Equipment and Conditions

Two HPLC instruments were used for column evaluation. The first was from Gilson (Wisconsin, USA) and consisted of the 802 controller, two 302 pumps with 5 ml heads, and a 116 dual wavelength UV detector, all controlled by an IBM PC with Gilson 712 software. Microbore adaptations consisted of a 65 μl mixer and 6 μl flow-cell in the detector. The second instrument was a Beckman System Gold with IBM-controlled software, a 167 dual-wavelength detector, and a microbore mixer connected directly to the injector.

Packing comparisons were performed with both C_8 and C_{18} material. A 50 mm x 1 mm Aquapore RP-300 7-μm C_8 steel column, (Applied Biosystems, Warrington, UK) was compared with a 50 mm glass capillary containing the same packing from a sacrificed 4.6 x 30 mm cartridge. Solvent A was 0.06% trifluoroacetic acid (TFA) in water, and solvent B, 0.04% TFA in acetonitrile. Using the Gilson instrument with a 200 μl loop, chromatography was carried out at room temperature, 100 μl/min flow-rate, a gradient of 0% to 60% B in 30 min, and absorption at 215 nm.

The same conditions were used for a rapid de-salting application using a 30 mm glass capillary. Using the manual flow-programming mode the following sequence was employed: 1) inject the first sample and wait for the absorbence from the break-through peak to return to baseline; 2) switch the loop out and ramp 0 to 80% B in 2 min; 3) when the absorbance returns to baseline ramp 80 to 0% B in 2 min; 4) after re-equilibrium at 0% B the loop is switched in and the next sample can be injected.

The C_{18} columns were used with a different solvent system suitable for small peptides. A 50 mm x 1 mm Applied Biosystems column was compared with a 50 mm glass capillary, both containing Spheri-5 RP 18, 5 μm packing. From a 0.05 M ammonium acetate stock solution brought to pH 6.0 with dilute phosphoric acid, solvent A was prepared by 1:1 dilution with water and B with 1:1 acetonitrile. Using the Beckman instrument with a 50 μl loop chromatography was carried out at room temperature, 150 μl/min flow-rate, a gradient of 0% to 40% in 20 min, and absorption at 215 nm.

D. Standard samples

The four peptides released by thrombin from fibrinogen were used as an easily-prepared standard peptide solution and for the purification of fibrinopeptide A, (FPA) as described (4). For a standard protein the enzyme, Glutathione transferase (GSHT), type 1:1, (EC 2,5,1,18) Mr 25k, was also purified by HPLC as described (5). A peptide-plus-protein standard was prepared by mixing FPA and GSHT to give a full-scale deflection for each peak from a 10 µl injection under the conditions described.

III. RESULTS

The chromatograms in Figure 2 show the performance of the dry-packed glass capillary compared to a commercial slurry-packed steel column of similar dimensions and containing the same C8 packing material.

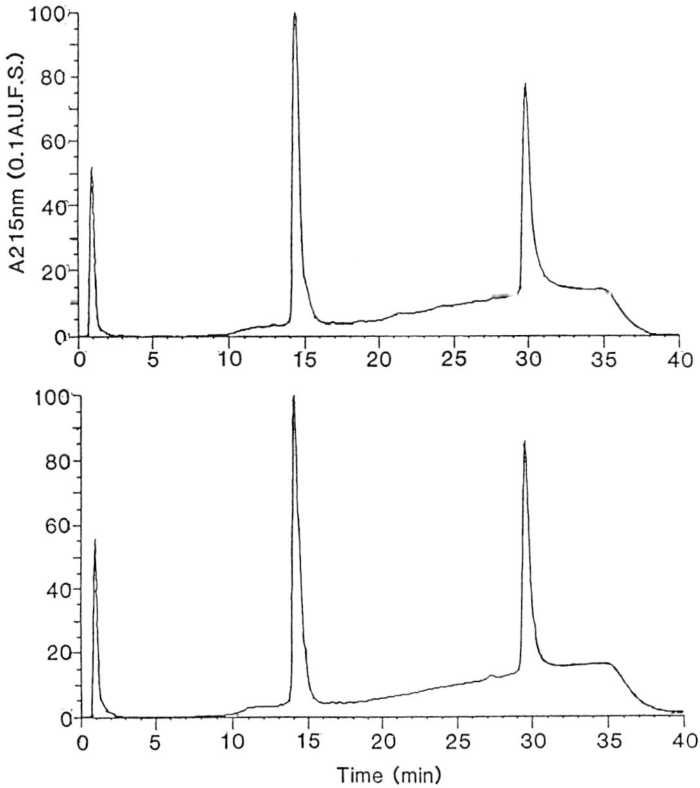

Figure 2. Comparison of the glass capillary (top) with an Applied Biosystems 50 mm x 1 mm column (bottom).
Both were packed with 7-µm Aquapore RP 300. The peak eluting at 14.5 min is the peptide, FPA, and the one at 30 min the protein, GST 1:

The separation pattern is similar although some peak tailing has occurred with the glass capillary. Table 1 shows some quantitative parameters compared for the two columns, retention times, peak heights, and peak areas. Within the determined limits of error peak areas were the same for both columns, implying equivalent recoveries for the eluted compounds. The performance comparison is extended to a different packing type and particle size in figure 3 (overleaf). The separation of a group of 5 fibrinopeptides is shown. The capillary column (b) again shows some peak tailing but the resolution compares well with the commercial column. In figure 3(c) a 3 cm version of the capillary column has been used as a guard column with only a slight deterioration in resolution. The use of a 3 cm packed capillary is also shown in figure 4 (overleaf) as applied to micro-scale de-salting. By using very short gradients for elution a total cycle time, including re-equilibration, of less than 10 min was achieved as well as very high concentration factors for the eluted compounds.

TABLE 1. Comparative separation parameters.
The errors given correspond to ± the standard deviation for five runs.

Column	Peak	Retention Time	Peak Height	Peak Area
metal	peptide	14.21 ± 0.18	1002 ± 19	517 ; 9.5
glass	peptide	14.41 ± 0.05	970 ± 21	517 ± 6.9
metal	protein	29.57 ± 0.29	748 ± 64	369 ; 13
glass	protein	29.83 ± 0.04	634 ± 22	362 ± 10

IV DISCUSSION

The separations shown in this work illustrate that effective microbore columns can be constructed out of disposable glass tubes. Although inferior to a commercial slurry-packed steel column the separation performance of the dry-packed capillaries was remarkable considering their low cost, simple construction and rapid assembly. With large particle packings (≤10μm) short columns can have low enough back-pressure for syringe-loading to achieve very large concentration factors. Although testing has so far been limited to reverse-phase material, in principal any type of chromatographic column packing could be used including those salvaged from discarded HPLC or FPLC columns. The absence of metal surfaces also implies increased "biocompatibility" over conventional steel columns and frits.

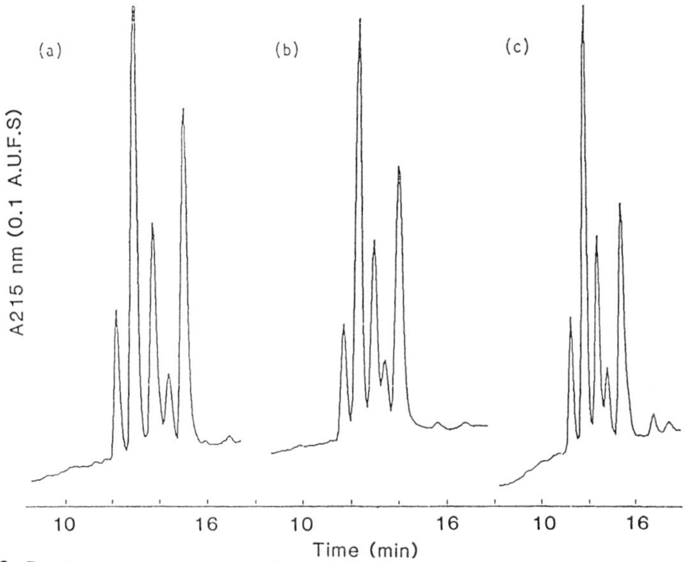

Figure 3. Performance comparison with C 18 5μm packing and separation of a peptide mixture.
Chromatogram (a) is the commercial column, (b) the 50 mm capillary and (c) a 30 mm capillary coupled as a guard column to the commercial 50 mm column.

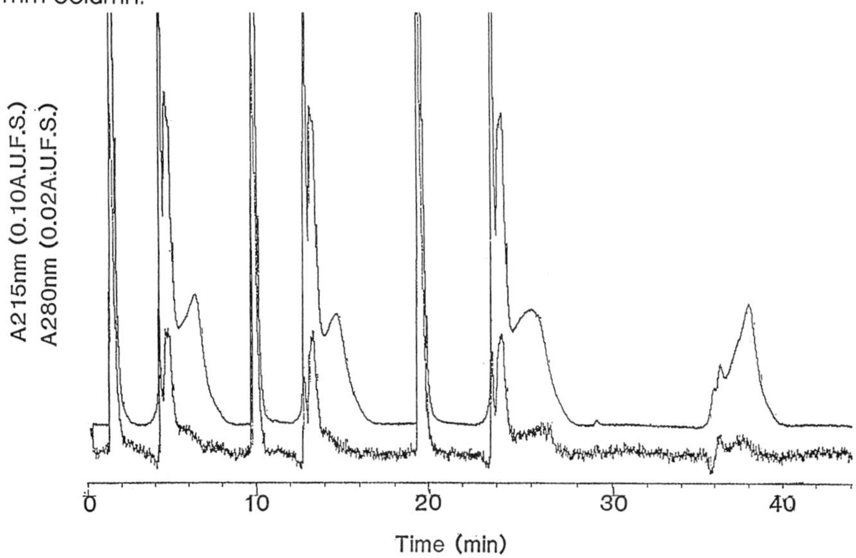

Figure 4. Rapid de-salting of a peptide-protein mixture using a 30 C8 mm glass capillary. The HPLC conditions were the same as Figure 2.
The upper trace was 215m, the lower one 280 mm. Three cycles of de-salting are shown with injections at 1, 10 and 20 min. At 35 min a blank gradient was run to check for complete elution from the previous runs.

The combined advantages of flexibility of design, ease of construction, and disposability suggest a wide range of possible applications. An obvious one is for use under conditions known to be deleterious to column packings. This would include: unfiltered solutions, samples with strong "memory" effects or non-elutable components, extremes of pH, radioactive material, polybuffers, and detergents. The columns can be used to test and compare untried packing materials for their suitability before purchasing commercial versions, or may even be adequate to substitute for them.

V REFERENCES

1. K.L. Stone and K.R. Williams (1988) In: "Macromolecular Sequencing and Synthesis" (D.H. Schlesinger, Ed.) pp 7-24, Alan R. Liss, Inc., New York.

2. E.C. Niece, C.J. Lloyd and A.W. Burgess, (1984) J. Chromatogr., 296, 153-170.

3. D.C. Shelly and T.J. Edkins (1987) J. Chromatogr. 411:185-199.

4. C. Southan, (1987) J. Chromatogr., 397: 399-404.

5. A. Ostlund-Farrants, D.J. Meyer, B. Coles, C. Southan, A. Aitken, P.J. Johnson, and B. Ketterer, (1987) Biochem. J., 245: 423-428.

39

ANALYSIS OF PEPTIDES AT LOW LEVELS USING A CHROMATOGRAPHY SYSTEM OPTIMIZED FOR NARROW-BORE COLUMNS

Robert Burgoyne
Catherine Stacey
Patricia Young
Nancy Astephen
Michael Merion

Waters Chromatography Division
Millipore Corporation
Milford, Massachusetts 01757
U.S.A.

I. INTRODUCTION

The elucidation of protein structure is a key step in the understanding of biochemistry and molecular biology. A limiting factor in protein primary structure determination is the availability of sufficient quantities of protein (1). Isolation to homogeneity using a variety of techniques including chromatography and electrophoresis often provide as little as 50 pmoles protein for structure analysis. The purified protein is then digested according to specific chemical and enzymatic cleavage techniques to produce numerous peptide fragments (2). Recent efforts have addressed the development of techniques to permit routine digestions of 50 pmoles protein (3). These efforts have also stimulated work in the separation of the complex peptide mixtures, a process called peptide mapping. The separation technique of choice is high performance liquid chromatography (HPLC). Mapping may be used for two purposes: 1) to purify individual peptides for subsequent amino acid composition or sequence analysis or 2) to identify key marker peaks in a complex peptide map as indicators of protein structural integrity.

Advances in amino acid compositional determination and sequence analysis have dramatically reduced the amount of peptide required to generate structural information. The utilization of pre-column derivatization techniques such as the Waters Pico-Tag[R] System enables accurate compositional analysis using only tens of picomoles of a starting sample (4). Similar quantities of peptide can also yield amino acid sequence information using either gas phase or covalent sequencing techniques (5,6). However, in order for these techniques to be useful, high sensitivity, high resolution separations that yield a purified peptide in fraction volumes less than 100ul must first be available. Peptide mapping has also become an important tool in the quality control of protein production that utilizes recombinant DNA techniques (7). Minor changes in amino acid composition, side chain modifications, and oxidation state of side chain groups can often be identified by shifts in retention time of specific peptides. Additional information can also be obtained through on-line spectral analysis of individual peptides using a photodiode array HPLC detector. For example, manipulation of spectral data can help to identify the presence of specific aromatic amino acid residues, such as tryptophan (8). Since the tryptophan codon is nondegenerate, its position in a peptide sequence provides useful information about the DNA code. A partial sequence of a protein covering a tryptophan region can be keyed to the DNA sequence used to encode that portion of the protein. Oligonucleotide probes can be synthesized which enable cloning of the gene which codes for the protein.

In this paper we describe an HPLC system called the Peptide Analyzer[TM] that has been optimized for the high resolution analysis and microisolation of peptides obtained from complex peptide mixtures. The system features two user-selectable flow paths for microbore and analytical separations. A unique solvent delivery system provides accurate and reproducible gradient delivery for separating complex peptide mixtures. Photodiode array detection affords high sensitivity detection while maximizing the amount of spectral information obtained during a single chromatographic run. This system has been evaluated using 1,2, and 3.9 mm internal diameter reverse phase columns. High quality peptide maps can be obtained from as low as a few picomoles, and up to nanomoles of original protein digest. As low as 1 picomole sensitivity can be obtained at 214 nm, with post-run spectral analysis that can positively identify peptides containing tryptophan using as little as 25 pmoles of sample. Microisolation of tens of picomoles purified peptides recovered in peak volumes of 50-75 ul is demonstrated. This capability is key for facilitating high sensitivity amino acid compositional determinations and sequence analysis.

II. MATERIALS AND METHODS

A. Chromatographic Instrumentation

All mapping was accomplished using a Peptide Analyzer (Waters) consisting of a modified 600E high performance gradient pump, 600E PowerLine™ Controller, Model 7000 Switching Valve (Rheodyne), U6K Variable Volume Injector and 990+ Photodiode Array Detector. System reproducibility studies were conducted using a 712 WISP Sample Processor (Waters) in place of the manual injector. Analytical and narrow-bore separations were done on Delta-Pak™ C18 (300A, 5um) columns having 3.9mm x 15cm and 2.0mm x 15cm dimensions, respectively. Microbore chromatography utilized a 1.0mm x 25cm Aquapore RP-300 C8 (300A, 7um) column (Brownlee).

Amino acid compositional analyses was conducted using a Waters Pico-Tag System consisting of two Model 510 pumps, Model 440 Absorbance Detector, Temperature Control Module, Refrigerated 712 WISP Sample Processor and Model 840 Data and Chromatography Control Station equipped with Version 6.2 software (Expert ™).

The system was equipped with a Pico-Tag Column, 3.9mm x 30cm, operated at 46°C with Pico-Tag Eluents A and B (Waters). The separation was achieved using a 21-min non-linear gradient (curve 5) to 44% eluent B at 1.0 ml/min followed by a 0.5 min linear gradient to 100%B with concurrent increase in flow to 1.5 ml/min. PTC-derivatized amino acids were detected at 254nm. A Pico-Tag Workstation (Waters) was used for sample hydrolysis and derivatization.

B. Samples and Reagents

Trifluoroacetic acid and 6N hydrochloric acid were Pierce Sequanal Grade. Acetonitrile, sodium acetate and ammonium acetate were Baker HPLC Grade. Water was 18 megohm produced by a Milli-Q™ System (Millipore). Adrenocorticotropic hormone fragments, ACTH 1-4, ACTH 1-10, ACTH 4-10, ACTH 11-24, neurotensin acetate salt, oxytocin acetate salt, transferrin acetate salt, and trypsin bovine pancreas were all purchased from Sigma Chemical Co.

C. Protocols

All peptide maps were generated using mobile phases containing 0.1% (w/v) TFA in neat or premixed aqueous and acetonitrile solutions. Liter quantities of mobile phases were prepared on a daily basis to minimize UV absorbing TFA breakdown products (9). Milli-Q water and acetonitrile, premixed if required, were degassed with helium for ten minutes prior to introduction of the total contents of a 1.0 gram ampoule of TFA. After thorough stirring, the Peptide Analyzer was primed with the eluents and helium sparging was reduced to 20% for continuous operation.

A tryptic digest of transferrin was produced by the method of Gooderham (11). The sample was centrifuged at 5000 rpm for five minutes in a Sorvall RC-5B Refrigerated Superspeed Centrifuge (DuPont). The supernatant fluid was filtered through a 0.45um Millex-HV filter (Millipore) and stored at 4°C.

Retention time reproducibility was determined by nine repetitive injections of 4 ACTH peptides made using a variety of flow and column combinations. In all cases Buffer A was 0.1% TFA/H_2O and Buffer B was 0.1% TFA/Acetonitrile. The following elution conditions were used: 1 ml/min.; 250 pmoles of each of the peptides were separated on a 3.9mm x 15cm Delta-Pak C18 column. Gradient: 0%B to 100%B over 30 minutes, Curve 6. 250ul/min.; 50 pmoles of each of the peptides were separated on a 2mm x 15cm Delta-Pak C18 column. Gradient 0%B to 100%B over 45 minutes, Curve 6. 100 ul/min.; 50 pmoles of each of the peptides were separated on a 2mm x 15cm Delta-Pak C18 column. Gradient: 0%B to 100%B over 60 minutes, Curve 6. 50 ul/min.; 25 pmoles of each of the peptides were separated on a 1mm x 25cm Aquapore RP-300 column. Gradient: 0%B to 100%B over 120 minutes, Curve 6.

Second derivative absorbance spectra were obtained using standard post-run analysis software contained in the 990+ Photodiode Array Detector. Focus was placed on the identification of tryptophan-containing peptides. This capability was demonstrated using ACTH 4-10 which has the sequence: Met-Glu-His-Phe-Arg-Trp-Gly. The peptide was suspended in 0.1% (w/v) aqueous TFA and chromatographed on a 2mm I.D. x 15cm Delta-Pak C18 column. Data collection across the peak of interest is maximized by monitoring spectra across a 200-300nm range with 1.4nm spectral resolution. Sensitivity is enhanced by applying a baseline subtraction routine prior to generating first and second derivative spectra. Oxytocin and neurotensin purified for amino acid analysis were suspended in 0.1% aqueous TFA prior to injection into the Peptide Analyzer. Two loading levels of each peptide, 25 and 100 picomoles, were chromatographed and were collected in 50-75ul fraction volumes into pyrolyzed 6 x 50mm sample tubes (Waters). Fractions were dried under vacuum and subsequently hydrolyzed using 6N hydrochloric acid containing 0.5% phenol for 22 hours at 110°C. The hydrolysis products were dried under vacuum. Following PITC derivatization using standard Waters Pico-Tag protocol, amino acid analyses were conducted using 80% and 40% of the 25 pmole and 100 pmole samples, respectively.

III. RESULTS

A. System Fluidics

A schematic diagram of the Peptide Analyzer fluid path is seen in Figure 1. Four solvent inlets are initially fed into a gradient proportioning valve. The outlet of the valve goes to the solvent inlet manifold, and is divided into two paths, each feeding a 100ul pump head. The flow from the dual reciprocating pump heads is then combined within a reference valve, and the outlet goes to the pressure transducer. From the transducer, the solvent passes through a two position switching valve where it is directed to one of two static mixing devices. The first is designed for flow rates of less than 0.5 ml/min., and the other for flow rates between 0.5 and 5 ml/min. From the mixer, the solvent is directed to the U6K injector. Standard configuration of this variable loop injector allows sample injection volumes between 1ul and 125ul. From the injector, the fluid is directed to a column, and finally to the photodiode array detector. A unique flow control software package is incorporated into the solvent delivery system to assure smooth, accurate fluid delivery at low flow rates.

The Peptide Analyzer fluid path is designed with three important criteria in mind. The first is a minimum gradient delay volume that will give good performance. Delay volume is defined as the volume from the point in the system where the solvents are blended to the detector flow cell. The system is equipped with two fluid paths in order to minimize system volume for microbore operation without sacrificing standard analytical scale performance. System volume is 950ul and 1300ul for the low flow rate and analytical flow rate fluid paths respectively.

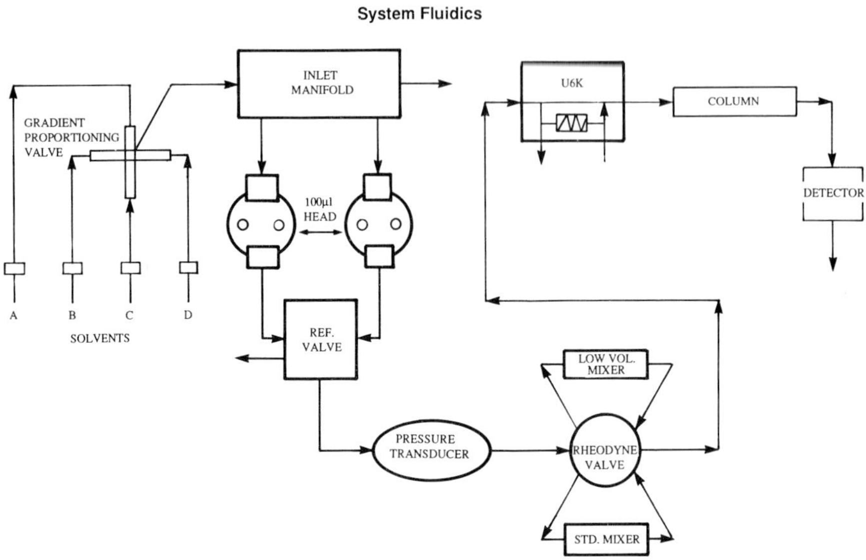

Fig. 1 System Fluidics

The second important design criterion is high resolution of complex peptide mixtures generated through specific proteolytic cleavage of large purified proteins. In order to examine this, we chose a tryptic digest of transferrin as a test sample. We first examined the separation of 1200 pmoles of digest at 1ml/min using a 3.9mm x 15cm Delta-Pak 5um, 300A C18 column. The conditions used and the resulting chromatogram can be seen in Figure 2A. The same separation using 300 pmoles of digest at 250ul/min. using a 2.0mm x 15cm Delta-Pak 5um, 300A C18 column is shown in Figure 2B. Finally, the separation of 25 pmoles of digest at 50 ul/min. using a 1mm x 25cm Aquapore RP-300, C8 column is shown in Figure 2C. In all cases, high resolution separations were obtained.

The final design requirement, very important where high sensitivity is vital, is to achieve smoothness of flow and good compositional mixing. Figures 3A and 3B illustrate high sensi- tivity chromatography using both 1 and 2 mm internal diameter columns respectively. A baseline rise of 0.02-0.03 absorbance units at 214nm is typical for gradients with TFA-containing eluents (9). In addition, small amounts of buffer contaminants can be seen. However, a close examination of the baseline reveals no discernable compositional ripple, and very little other associated noise. Chromatography of a mixture of ACTH 11-24 and

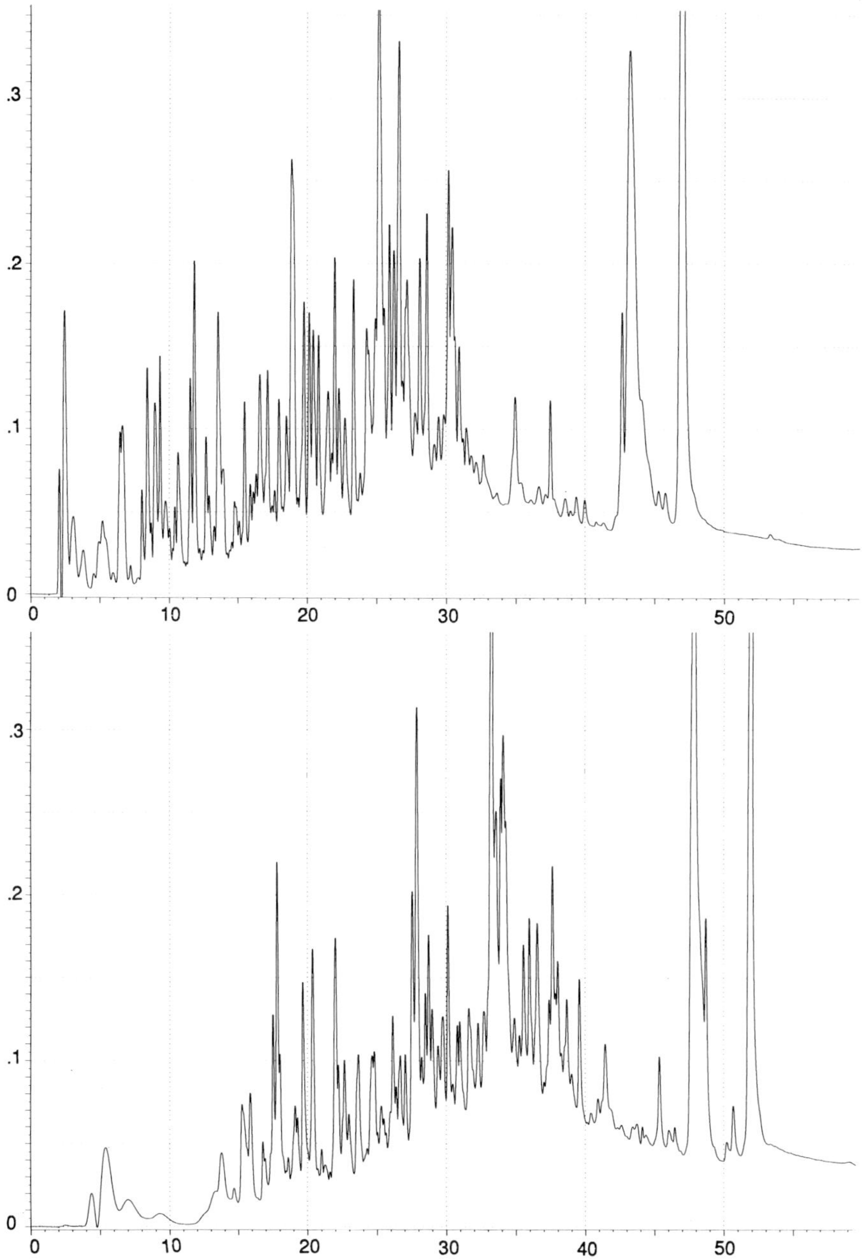

Fig. 2. Resolution of a Tryptic Digest of Transferrin. Buffer A: 0.1% TFA/H_2O; Buffer B: 0.1% TFA/CH_3CN. Peptides were monitored at 214nm.; A & B) 0-50%B, 50 min., Curve 6. C) 0-40%B, 125 min., Curve 6.

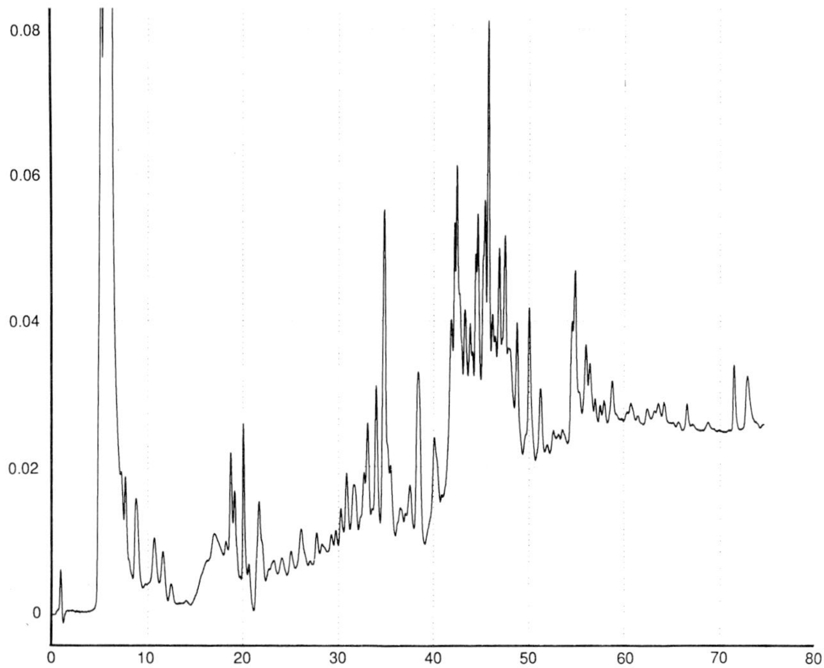

Fig. 2c

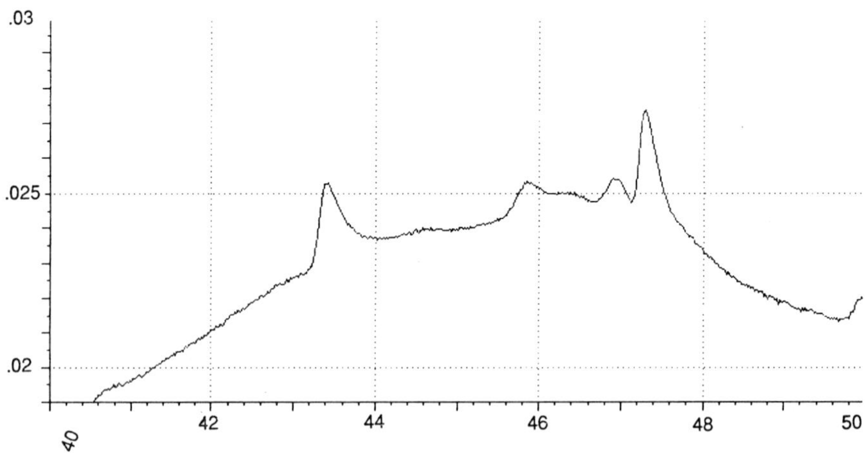

Fig. 3. High Sensitivity Analysis Using the Peptide Analyzer. Buffer A: 0.1% TFA/H_2O; Buffer B: 0.1% TFA/CH_3CN. Peptides were monitored at 214nm. A) 1 pmole each ACTH 11-24 and ACTH 4-10; Aquapore RP-300, 25cm; 1mm x 50ul/min.; Gradient: 0-100%B, 60 mins., Curve 6. B) 4 pmoles each ACTH 11-24 and ACTH 4-10; Delta-Pak C18 2mm x 15cm column; 250ul/min; Gradient: 0-100%B, 30 mins., Curve 6.

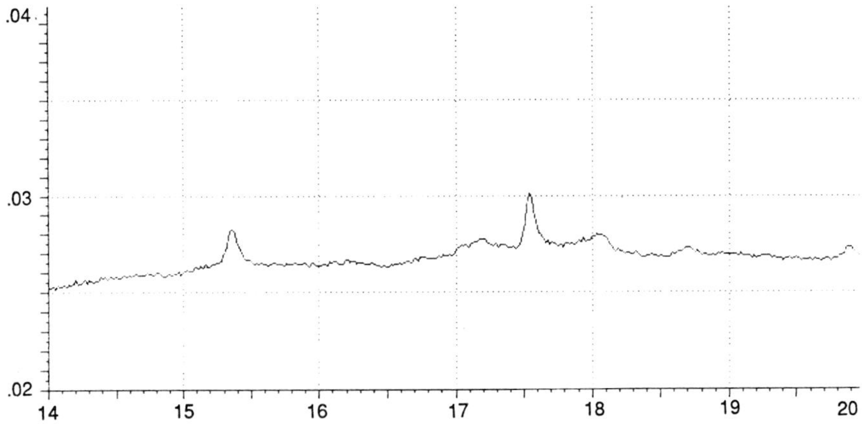

Fig. 3B

ACTH 4-10 demonstrates that 1 picomole sensitivity is achievable when using 1.0 mm I.D. columns and 4 picomole sensi- tivity when using 2.0 mm I.D. columns.

B. Flow Reproducibility

In order to examine the integrity of flow performance throughout the flow rate range of the instrument, we examined the retention time reproducibility for the separation of a four-component ACTH-peptide mixture. Nine identical injections of the peptide mixture were made using conditions described previously. The results can be seen in Table 1. Throughout the flow rate range tested, retention time reproducibility exhibited an average coefficient of variation of less than 0.3%.

TABLE I. REPRODUCIBILITY OF RETENTION TIME

Flow	Average Standard Deviation (Min.)	Average Coefficient of Variation (%)
1 ml/min	0.012	0.012
250ul/min	0.047	0.145
100ul/min	0.112	0.217
50ul/min	0.225	0.260

C. Spectral Analysis of ACTH Peptides

It has been reported previously that the second derivative of peptide spectra can reveal information about amino acid content (10). Specifically, it has been shown that the aromatic amino acid tryptophan can be distinguished from other aromatic amino acids using this technique. A model second derivative spectrum of N-acetyl tryptophanamide can be seen in Figure 4. Two minima at 281.0nm and 289.5nm are seen, with one maximum at 285.5nm (8). We chose ACTH 4-10, which contains a single tryptophan residue, as a test compound for this analysis. The second derivative spectra for 96 pmoles ACTH 4-10 is illustrated in Figure 5. Second derivative spectra generated at the 48 and 24 pmole level yielded equivalent results.

The pattern characteristic of tryptophan can be seen clearly at each concentration. However, at lower concentrations the absolute wavelengths of the minima and the maxima have shifted slightly. In all cases, the spectrum of a blank gradient was first subtracted from that of the peptide prior to differentiating. This data demonstrates that useful spectral information indicative of peptide structure is obtainable at the peptide concentrations targeted by micro-scale peptide mapping methodology.

D. Amino Acid Analysis of Collected Peptides

To examine our ability to collect and analyze small amounts of peptide, a mixture of oxytocin and neurotensin was separated using the 2mm x 15cm Delta-Pak column. The separation of 100 pmoles (A) 25 moles (B) of the peptide mixture are illustrated in Figure 6. The peptide peaks were collected manually in a volume of 50-75ul. Amino acid analysis was conducted using the Pico-Tag method as described previously. Blank gradients were also run, 50-75ul of solvent collected at the retention times of the previously separated peptides, and used as background measurements. The results of the amino acid analysis can be seen in Table 2. At both the 100 and 25 pmole level the analyses show good agreement with the known compositions. This data illustrates the feasability of using the Peptide Analyzer in conjunction with the Pico-Tag Amino Acid Analysis System to purify and analyze peptide composition on tens of picomoles of original sample digest.

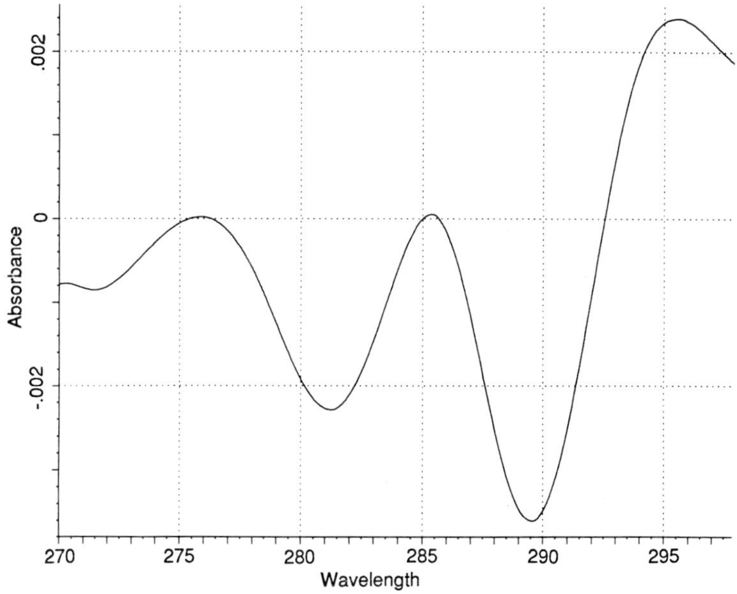

Fig. 4. Second Derivative Spectrum of N-Acetyl Tryptophanamide. A 25mM solution of N-acetyl tryptophanamide was injected into the flow cell of the photodiode array detector.

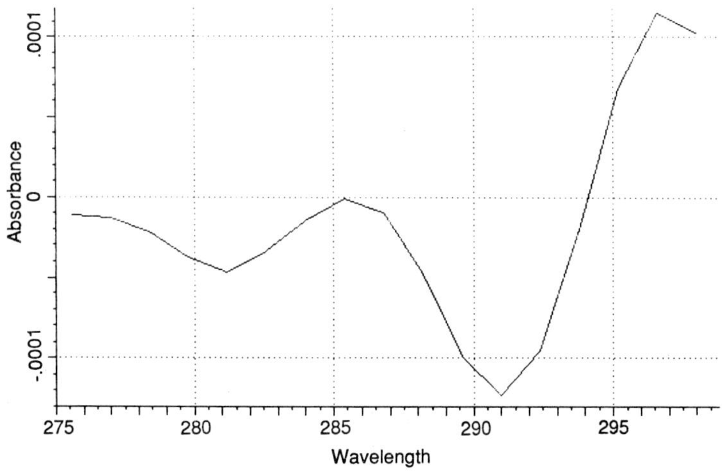

Fig. 5. Second Derivative Spectra of 96 pmoles ACTH 4-10. Elution conditions as in Fig. 3B.

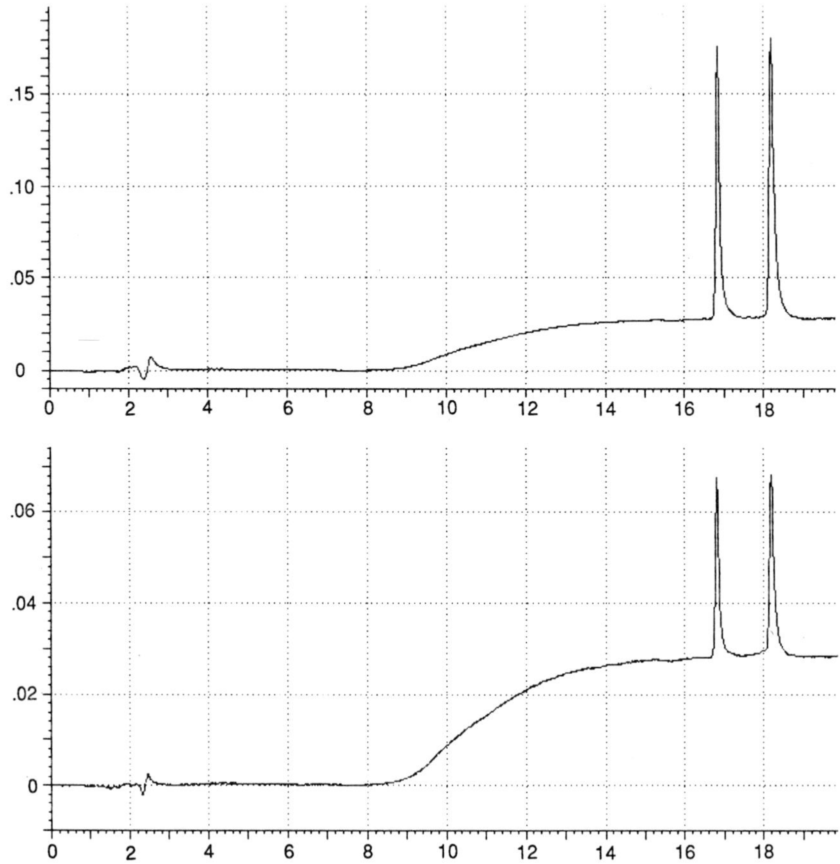

Fig. 6. Separation of Oxytocin and Neurotensin. A mixture of oxytocin (1) and neurotensin (2) was separated using conditions shown in Fig. 3B. A) 100 pmoles of each peptide. B) 25 pmoles of each peptide.

TABLE II. Pico-Tag Amino Acid Analysis of collected fractions

	OXYTOCIN			NEUROTENSIN		
Amino Acid	25pmol	100pmol	Actual	25pmol	100pmol	Actual
ASP	1.01	0.95	1	0.60	1.03	1
GLU	0.97	0.92	1	1.73	1.73	2
SER						
GLY	0.96	0.81	1			
HIS						
ARG				2.20	2.17	2
THR						
ALA						
PRO	1.07	1.19	1	2.19	2.27	2
TYR	0.80	0.83	1	1.25	1.68	2
VAL						
MET						
CYS	*	*	2			
ILEU	0.96	1.10	1	1.01	0.97	1
LEU	1.02	1.19	1	2.17	2.07	2
PHE						
LYS				0.70	0.80	1

IV. DISCUSSION

The structural analysis of proteins is required for a basic understanding of protein biochemistry, as well as in the growing biotechnology industry. A key element in protein sequencing is the process called peptide mapping, the chromatographic separation of the peptides formed by digestion of a protein. Individual peptides purified in this manner are subjected to amino acid compositional determination or sequence analysis. In this paper we present the Waters Peptide Analyzer, a special purpose HPLC designed specifically to optimize both high resolution and high sensitivity separations of complex peptide mixtures.

There are several criteria used to judge the performance of the Peptide Analyzer. The first is the gradient delay volume in the system, a crucial factor when microbore columns are used at very low flow rates. The 950ul system volume on the low flow rate fluid path of the Peptide Analyzer yields a gradient delay which is small when viewed in the perspective of 2-4 hour gradient times typically used in peptide mapping (3). The second criterion used is flow performance, a quality evaluated using retention time reproducibility, baseline smoothness, and compositional ripple. Throughout the flow rate range of the instrument, retention time coefficients of variation are less than 0.3%. The ability to achieve 1 pmole sensitivity of a single peptide with no observable compositional ripple indicates that flow smoothness and mixing are acceptable for high sensitivity analysis.

Next, a tryptic digest of transferrin was used to produce high resolution peptide maps in a variety of flow rate and column size combinations. Microbore chromatography produced peptide maps of equal quality to those produced using standard analytical conditions, with higher sensitivity, and smaller fraction volumes. Furthermore, the high reproducibility of these complex peptide maps permits their use in quality control "fingerprinting" of proteins.

Finally, because the amount of starting material is often limited, it is important to maximize the information gained from a single separation. We have demonstrated the detectable limits of peptides at 214nm to be lower than 1 pmole. In addition, post-run spectral analysis clearly showed the presence of 1 residue of tryptophan in as little as 25 pmoles of peptide. Compositional analysis of as little as 25 pmoles of collected peptides has been demonstrated. In conclusion, the Peptide Analyzer exhibits the high resolution, high sensitivity performance required for micro-scale peptide analysis and purification.

V. REFERENCES

1. Bhown, A.S. and Bennett, J.C. (1985). In "Handbook of HPLC For The Separation of Amino Acids, Peptides, and Proteins" (W.S. Hancock, ed.), Vol II, p. 267. CRC Press, Boca Raton.

2. Regnier, F.E. (1988). BioPharm. 1 (1), 19.

3. Stone, K.L. and Williams, K.R. (1987). In "Macromolecular Sequencing and Synthesis" (D.H. Schlesinger, ed.). Alan R. Liss, New York.

4. Bidlingmeyer, B.A., Cohen, S.A. and Tarvin, T.L. (1984). J. Chromatogr. (Biomed. Appl.) 336, 93.

5. Strickler, J.E., Hunkapiller, M.W. and Wilson, K.J. (1984). Anal. Biochem. 140, 553.

6. Aebersold, R.H., Pipes, G.D., Nika, H., Hood, L.E., and Kent, S.B.H. (1988). Biochem., in press.

7. Hancock, W.S. (1987). At 7th Int. Symp. on HPLC of Proteins, Peptides and Polynucleotides. Abstract No. 204.

8. Phoebe, Jr., C.H., Cohen, S.A. and Tarvin, T.L. (1988). At 39th Pittsburgh Conference on Analytical Chemistry and Applied Spectroscopy. Abstract No. 548.

9. Winkler, G. (1987). LC-GC 5 (12), 1044.

10. Grego, B., Nice, E.C., and Simpson, R.J. (1986). J. Chromatogr. 352, 359.

11. Gooderham, K. (1984). In "Methods in Molecular Biology" (J.M. Walker, ed.), Vol. 1, Protein, p. 185. Humana Press, Clifton, N.J.

Section V

Introduction

SECTION V
PROTEIN AND PEPTIDES: GENERAL TECHNIQUES

Albert Light and Robert F. Pfeifer

Department of Chemistry, Purdue University
and
Waters Chromatography Division, Millipore Corporation

The collection of articles in this Section contain an interesting range of topics that complement and supplement other Sections of this monograph. The detection, isolation, and characterization of peptides and proteins will always present a challenge to the experienced investigator and new results will be useful to old and new investigators alike. The contributions in this Section introduce the reader to several novel and useful techniques.

Affinity chromatography is a highly selective process for protein isolation. The protein of interest is retained by a solid support using biospecific ligand interactions. Later, the bound protein is displaced by a competing ligand or a change in pH. In the elucidation of the metabolic pathway for estrogen biosynthesis, Tan purified human placental estrogen synthetase using a two-step procedure with anion-exchange chromatography, followed by affinity chromatography using an immobilized steroid as a substrate-specific ligand.

The coding strand of DNA is the sense strand and the non-coding strand is the anti-sense strand. Recent investigations have been concerned with the control of transcription from interactions of the gene with the anti-sense strand. At the translation level, some peptides generated from the non-coding strand recognize peptides coded by the sense strand. Such recognition peptides were computer designed by Omichinski et al to enhance their binding to the sense peptide sequence. The goal of these studies was the design and synthesis of a recognition peptide that serves as an affinity chromatography support for the purification of the complementary protein.

Early et al described a non-chromatographic separation of peptides. They prepared amino acid specific reagents attached to porous glass beads. Specific reagents are available for methionine, cysteine, arginine, and tryptophan. The authors illustrate the separation of peptides containing methionine and cysteine as the interacting residue.

An important advance in the production and preparation of large amounts of specific proteins is the cloning of genes. The technique of fusion proteins helps in the later separation of the expressed protein. This is illustrated by Glabe et al in the fusion of a eukaryotic protein to a bacterial cytoplasmic protein encoded by pBindin+4, which is rapidly cleaved at the fusion junction by an endogenous bacterial protease. The bacterial protease cleaves between Arg-Arg and Lys-Arg sequences located in the junction but not in the sequence of the desired protein.

The newly developed technique of high performance capillary electrophoresis provides high sensitivity and resolution to the protein chemist. Gutzman and Hernandez used this technique to rapidly separate and quantitate femtomole amounts of a monoclonal antibody directed against prolyl 4-hydrolylase. McManigill and Swedberg studied three proteins to derive the equations for estimating the mobility of proteins and examined factors affecting plate height in the separation. These two studies will pave the way for continued development of the method and new applications will surely follow.

HPLC was used to identify a group of proteins that appear in lung fluids after exposure to environmental insults. The methods for rapid protein screening presented by Gurley et al will be useful in clinical studies to recognize pulmonary disorders. HPLC has been employed by Janska and Light for a new assay of the proteolytic enzyme enterokinase, which is the physiological activator of trypsinogen. The activation of trypsinogen forms trypsin which is immediately inactivated in a complex with ovomucoid. The HPLC method for separation of the substrate and product is based on the increased size of the complex relative to the substrate. The methodology should be suitable for the detection of a congenital deficiency of the enzyme in newborns.

In protein sequence studies, the generation of peptide fragments and their subsequent purification was discussed by Schlabach et al. Peptides of carboxypeptidase P were prepared by digestion of the substrate with an asparagine-specific endoprotease. The peptide mixture was separated by reverse phase and ion exchange HPLC. Capillary zone electrophoresis was used to determine peptide purity. These studies represent preliminary data with both a new cleavage enzyme and the assessment of purity with capillary zone electrophoresis.

In the chemical synthesis of peptides by solid-phase methods, amino-blocking groups are necessary. Reaction conditions were explored by Harrison et al for protecting groups used in the Fmoc procedure for peptide synthesis. A different type of chemical modification was developed by Day et al in the formation of amide bonds between cyanogen (C_2N_2) and the carboxyl groups of proteins. The reactive derivative formed intramolecular covalent crosslinks with neighboring lysyl or histidyl residues. The authors describe the preparation of derivatives of three proteins. Further work is contemplated by the authors to identify the sites of reaction and the altered properties of the proteins.

Electrostatic properties of proteins were investigated by Jenkins who estimated the stoichiometry of the interaction with ligands (salts, pH, small molecules, etc.) by a "thermodynamic bookkeeping." The calculation of the overall free energies for the displacement of bound charged ligands will be useful in the design of protein chromatographic procedures and in understanding metabolic controls by bound salts and metabolites. In elucidating the role of helix regions in the folding of polypeptides, Bradley et al used circular dichroism and NMR to characterize small synthetic peptides. The variation of temperature, pH, salt, and amino-terminal blocking on the helical content of the peptide was related to secondary structural interactions.

A limited number of potential contributions were included in this Section. However, the size of the Section bears little relationship to the important role that new techniques play in exploring peptide and protein structure and function. Today, a large number of powerful separation methods, physical measurements, synthetic procedures, and chemical modifications make protein chemistry a little easier to study and clearly is more informative. It is also obvious that the more we have, the more we need. We are far from that point in time where the miracle of a protein will be understood. We expect that "New Techniques" will be a continuing subject for many more years to come.

40

RAPID TWO-STEP PURIFICATION OF HUMAN PLACENTAL ESTROGEN SYNTHETASE BY SUBSTRATE-SPECIFIC AFFINITY CHROMATOGRAPHY

L. Tan[1]

Biochemistry Department, University of Sherbrooke
Medical Faculty, Fleurimont, Que., Canada, J1H 5N4

I. INTRODUCTION

In the reproductive system of mammals, the bioconversion of the male steroid hormones into estrogens is catalysed by a membrane-bound and NADPH-dependent cytochrome P-450-type monooxygenase, commonly referred to as estrogen synthetase or aromatase. Whereas the other monooxygenases that are involved in the biogenesis of steroid hormones, namely those that introduce a hydroxyl group at positions 11β, 17α, 18, and 21, have been purified already some time ago (1-6), the microsomal aromatase enzyme has resisted until recently all attempts at its purification.

In 1985, we succeeded in developing a conventional, four-step chromatographic purification procedure which led to a homogeneous aromatase preparation of high catalytic activity (7,8). However, the procedure is laborious, slow, and clearly needed improvement.

In the following, we describe a much faster procedure which requires only two chromatographic stages. Initial prepurification of the crude solubilised microsomes by anion-exchange chromatography was directly followed by substrate-specific liquid affinity chromatography (SLAC) in the second and final step.

Starting with the solubilisation of crude microsomes, we can now purify the aromatase within five days.

---

[1]Supported by the Medical Research Council of Canada

II. MATERIALS AND METHODS

[1β,2β-^{3}H]4-Androstene-3,17-dione (s.a. 40.0 Ci/mmol) was obtained from Du Pont, Boston, MA. QMA-Accell was from Waters, Milford, MA. Diaminodipropylamino-agarose (containing 16-20 μmoles amino groups per ml of gel) was from Pierce, Rockford, Illinois. 1-Ethyl-3-(3-dimethylaminopropyl)carbodiimide (EDC), androstenedione, 19-nortestosterone, glycerol, dithiothreitol (DTT), glucose-6-phosphate, glucose-6-phosphate dehydrogenase and $NADP^+$ were from Sigma, St. Louis, MO. The new competitive aromatase inhibitor, 6-methylenetestosterone, was synthesised by Dr Holland's group (9) at Brock University, St. Catharines, Ontario.

A. Synthesis of 19-Nortestosteronehemisuccinate (NTHS)

To a solution of 2.0 g (7.3 mmol) 19-nortestosterone in 6 ml pyridine (freshly distilled over sodium hydroxide) was added 2.0 g succinic anhydride. The reaction flask was filled with Ar, stoppered, and heated at 85°C for 16 hr. The slightly colored solution was cooled, triturated with 5 ml methanol, let stand for 1 hr, then poured into 400 ml icewater. The precipitate was filtered, washed, and dissolved in 10 ml of hot acetone. Water was then added dropwise (circa 5 ml) until the first appearance of turbidity. After cooling at room temperature, then at 4°C, the crystalls were filtered. Yield: 1.25 g colorless needles; m.p. 168-169°C. It showed a single spot on TLC with R_f 0.25 and chloroform/ethylacetate/ethanol (100:20:7) as the mobile phase. Its infrared spectrum was identical with that of reference steroid.

B. Preparation of Biospecific Affinity Matrix

Diaminodipropylamino-agarose (25 ml) was first washed with water, filtered, then suspended in 50 ml of 50% aqueous dimethylformamide (DMF). After adjusting the pH to about 4.6 with a few drops of constant boiling HCL, NTHS (400mg, dissolved in 6 ml DMF) was added dropwise. The supernatant should remain clear, otherwise more DMF should be added and the suspension heated to 45°C. Solid EDC (900 mg) was then added and the pH readjusted to 4.6. The gel was shaken at 25°C for 16 hr. After adjusting the pH again to 4.6, a second batch of 600 mg solid EDC was added and shaking continued for a further 24 hr. To remove excess steroid and the urea formed during the coupling reaction, the gel was washed successively with DMF, acetone and water until the absorbance of the wash at 240 nm and 280 nm became negligible. The affinity gel remained colorless after these manipulations. It gave a strong, magenta Zimmermann-reaction chromogen with alkaline m-dinitrobenzene, indicating the presence of covalent, matrix-bound steroid with a methylene group at the C-2 position, activated by the adjacent C-3 ketone. Colorimetric estimation of the remaining free amino groups with trinitrobenzenesulfonic acid gave a value of 6-8 µmoles of unreacted amine per ml of gel. About 60% of the initially present free amino groups had thus reacted with the steroid [cf. (10)].

C. Enzymatic Assays

Aromatase activity of the chromatographic fractions was measured by radiometric analysis with [1β,2β-^{3}H]androstenedione as substrate and counting of an aliquot of the tritiated water formed, as described earlier (11,12). Cytochrome P-450 content was determined from the ferrous-carbon monoxide difference spectra with the value of 91/mM/cm as extinction coefficient (13). Microsomal NADPH-cytochrome P-450 reductase was purified from bovine liver (14). Protein concentration was estimated according to Bradford (15) with bovine serum albumin as standard and correction for the presence of detergent.

D. First-step Anion-Exchange Chromatography

All chromatographic elution and fraction collection took place in a modified refrigerator maintained at 6-8°C. The first step was done in a 1.5 x 45 cm medium pressure glass column (Bio-Rad, Richmond, CA), equipped with an adjustable flow adaptor. It was packed with QMA-Accell for a total bed volume of 68 ml. The column was first equilibrated with 20 mM potassium phosphate buffer (Buffer-A) of pH 7.4, containing 20% glycerol, 0.5% cholate, 1 mM DTT, 1 mM EDTA, and 10 µM androstenedione. Frozen microsomes (539 mg protein) were solubilised (50 mM potassium phosphate, 1% cholate; otherwise same as Buffer-A). Insoluble material was removed by centrifugation at 105,000 g for 1 hr. The supernatant (44 ml, 397 mg protein) was poured into an identical, empty precolumn which served as the loading reservoir. With a Pasteur pipette, Buffer-A was carefully layered on top of the denser protein solution in such a way as to maintain separation of the two liquid phases. The two columns were then connected in tandem and loading of the protein solution into the second, anion-exchange column was started by delivering Buffer-A into the precolumn with a titanium HPLC pump (LKB-GTi system, Bromma, Sweden) at a flowrate of 1.0 ml/min. After all of the protein solution had entered the Accell-column, the precolumn was disconnected. A linear gradient was then applied from 0-0.8M KCl. The aromatase-active fractions eluted between 0.5-0.7M KCl. They were pooled (43 ml) and dialysed twice against 1L of Buffer-A which, however, contained neither androstenedione, nor KCl. The dialysed protein solution was then immediately used in the next step.

E. Second-step Substrate-specific Liquid Affinity Chromatography (SLAC)

The NTHS-agarose column (total bed volume 24 ml) was first equilibrated with SLAC-A buffer (Buffer-A minus androstenedione). The dialysed solution from the previous step was then recirculated through the column at a flowrate of 0.3 ml/min for 18 hr with a peristaltic pump. The affinity matrix slowly assumed the faint yellow-orange color of trapped hemoprotein. The column was then connected to the LKB-GTi system. After collecting 50 ml of eluate with SLAC-A buffer, 30 ml of SLAC-B buffer (SLAC-A buffer plus 0.5 mM androstenedione) was pumped through at a flowrate of 0.5 ml/min. The flow was then interrupted for 16 hr to allow displacement of the enzyme from the matrix-bound ligand by competition with the free substrate present in saturating concentration. A two-step linear gradient from 0-1M KCl was then applied by introducing, via a mixing valve, SLAC-C buffer (SLAC-B buffer plus 1M KCl) at a combined flowrate of 0.3 ml/min and with the

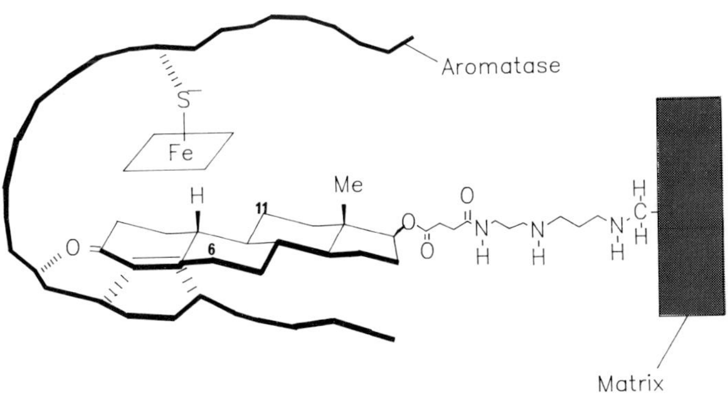

Fig.1. Structure of the substrate-specific NTHS-affinity matrix and 3-dimensional model of its possible interaction with the aromatase active site. The steroid molecule is drawn according to the Fieser convention with the front face shown in heavy lines; the heme plane is schematised as a square.

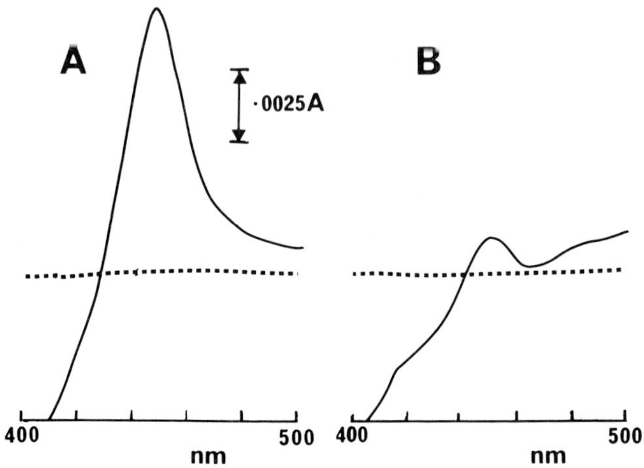

Fig. 2. Efficiency of NTHS-affinity matrix. Difference spectra of two ml each of the dithionite reduced-carbon monoxide complex of the pooled and dialysed aromatase-active fractions from QMA-Accell chromatography. (A) **Before,** and (B) **after** recirculation for 18 hr through the NTHS-column.

detector set at 405 nm. The aromatase-active fractions eluted between 0.6-0.9M KCl. They were pooled (72 ml), dialysed twice, concentrated in Amicon Centriprep-10 tubes, and stored at -80°C in 1.5 ml Eppendorf centrifuge tubes.

III. RESULTS AND DISCUSSION

The ideal biospecific affinity support for aromatase purification should meet the following criteria: a) the matrix itself must be hydrophilic, have good flow characteristics, and be able to withstand a minimum back pressure of 2 bar; b) it must have a spacer arm that is sufficiently long but without causing unwanted nonspecific hydrophobic interactions; c) the arm must be attached to the matrix by a stable N-C- or O-C-type bond so that the support can be used many times; d) the steroid ligand must not only be recognised by the enzyme, but must also bind to the enzyme with a lower affinity than either androstenedione, or testosterone, so that it can later be displaced by these endogenous aromatase substrates.

Our affinity support met several, but not all of these criteria. Agarose turned out to be the gel that gave the best results although it tends to collapse at higher column back-

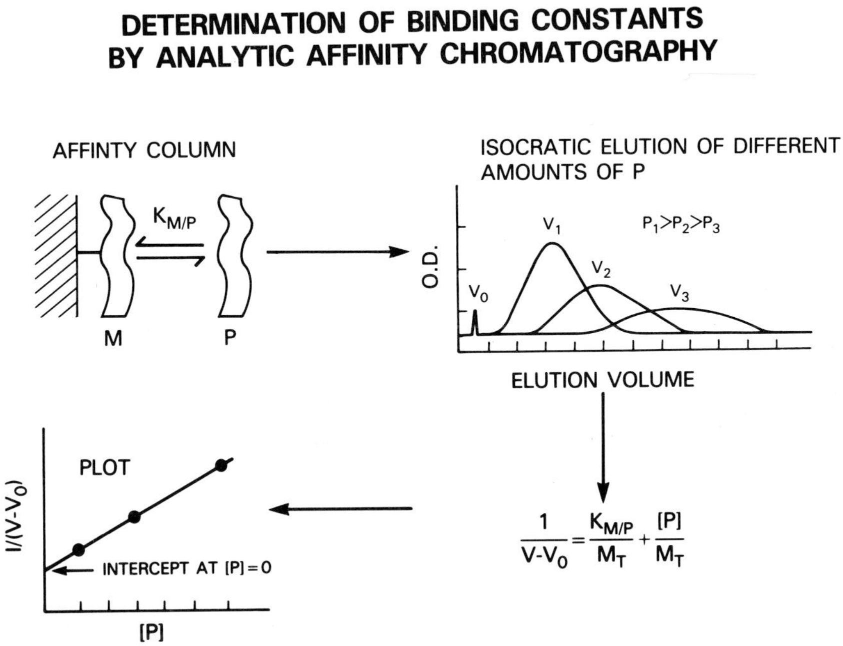

Figure III: Schematic representation of the use of HPLAC techniques to quantitatively determine sense anti-sense equilibrium binding constants.

Using this system, the $K_{M/P}$ values for the recognition peptide were determined (see Table II). Both Codon 1 and Codon 2 gave binding constants in the millimolar range. In contrast, the computer generated recognition peptides gave equilibrium constants ranging from 1×10^{-4} M to 2×10^{-7} M. Within this range, an interesting pattern seems to exist between affinity and the hydropathic complementary score. It seems that when the score is lowered the degree of recognition increases, and that in this particular system the recognition can be enhanced between three and four orders of magnitude relative to the codon frequency pattern.

IV. COMPARISON WITH DNA DERIVED ANTI-SENSE PEPTIDE

In order to evaluate the general applicability of the complementary scoring program, a comparison was made between a DNA derived anti-sense peptide and a computer generated recognition-peptide. The model target peptide chain was the putative ATP binding region of the *raf* oncogene (6). This

Table 2: Equilibrium binding constansts ($K_{M/P}$) for anti-sense recognition peptides. Values were obtained by zonal elution on a column with immobilized target peptide (GP-1) equilibrated with 100mM NH_4OAc, pH= 5.7 with column flow rate of 1.0 ml/min.

PEPTIDE	SEQUENCE	$K_{M/P}$ (μM)
CODON-1	AAQHGGGLGVQVE	920
CODON-2	AAQHGRGFGIQIE	460
AVG-1	TAKDSGSLSVKVE	460
AVG-3	AGKDSGGFSVKVE	160
AVG-5	PAKDSGSFSVKVK	59
AVG-7	TAKDSWSFSVKVK	22
AVG-9	AAKDSRGLGIEIE	40
AVG-11	AGKYSWRLRIKIK	0.7

region is 20 amino acid-residues long (*raf* 357-376) and the DNA sequence information is published (16). To compare the relative affinities of the DNA derived anti-sense peptide (FAR-LYSNISMPLALVHSAKGARA) and the computer generated recognition peptide (FARCG--FHGHISMDFAFVYSSKAAAA), the *raf* ATP binding region (GSGSFGTVYKGKWHGDVAVK) was immobilized on an affinity support and the two anti-sense peptides evaluated by HPLAC techniques. In this system, the computer generated peptide demonstrated almost a 50-fold higher affinity for the target peptide relative to the DNA derived anti-sense peptide (Table III).

Table III: Equilibrium binding constants ($K_{M/P}$) for DNA derived (FAR) and computer generated (FAR-CG) anti-sense recognition peptides to the putative ATP binding site of the *raf* oncogene (16).

PEPTIDE	SEQUENCE	$K_{M/P}$ (μM)
FAR	LYSNISMPLALVHSAKGARA	377
FAR-CG	FHGHISMPFAFVYSSKAAAA	8

V. OVERVIEW

Based on these experiments, it is possible to derive recognition peptides with relatively high binding affinities to a target sequence given only the amino acid sequence of the target peptide. More importantly, it is possible by optimizing the hydropathic complementary score to choose sequences that have higher affinities for the target peptide than the peptide derived from DNA sequence information. This information enables one to apply sense:anti-sense interactions much more generally to many biochemical systems. However, future experiments are needed to define the exact physical chemical nature of the affinities and to further test their specificity.

ACKNOWLEDGMENTS

The authors thank Peter Roller and Dolph Hatfield for their helpful discussions.

VI. REFERENCES

1.K.L. Bost, E.M. Smith and J.E. Blalock, Proc. Natl. Acad. Sci. USA, 82, 1372 (1985).

2.R.R. Brentani, S.F. Ribeiro, P. Potacnjak, J.D. Lopes, and C.R. Nakaie, Proc. Natl. Acad. Sci. USA, 85, 364 (1988).

3.T.S. Elton, L.D. Dion, K.L. Bost, S. Oparil and J.E. Blalock, Proc. Natl. Acad. Sci. USA, 85, 2518 (1988).

4.Y. Shai, M. Flashner and I.M. Chaiken, Biochem., 26, 669 (1987).

5.Y. Shai, M. Flashner and I.M. Chaiken in "Protein Structure, Folding and Design 2", D. Oxender, ed., Alan R. Liss: New York, 1987.

6.J.E. Blalock and E.M. Smith, Biochem. Biophys. Res. Comm., 121, 202 (1984).

7.J. Kyte and R.F. Doolittle, J. Mol. Biol., 157, 105 (1982).

8.G. Fassina, M. Zamai, and I.M. Chaiken, J. Cell Biochem., 35, 9 (1987).

9.H. Raunio, R. Konno, K. Linnainmaa, P.J. Wirth and S.S. Thorgeirsson, Exp. Cell Res., 173, 596 (1987).

10.T. Benjamin, C.H. Niu, D.C. Parmelee, A.C. Huggett, B. Yu, P.P. Roller and S.S. Thorgeirsson, Anal. Biochem. (submitted).

11.S. Aota, T.Gojobori, F. Ishibashi, T. Marvyama and T. Ikamvea, Nucl. Acid Res., 16 Suppl, r. 315 (1988).

12.D. Hatfield and M. Rice, Biochem. Int., 13, 835 (1986).

13.D. Hatfield, C.R. Mathews and M. Rice, Biochim. Biophys. Acta, 564, 414 (1979).

14.I.M. Chaiken, Ed., "Analytical Affinity Chromatography," CRC Press, Boca Raton, 1987.

15.M.E. Swaisgood and I.M. Chaiken, Biochem., 25, 4148 (1986).

16.F. Ishikawa, F. Takaku, M. Nagao and T. Sugimaea, Mol. Cell Biol., 7, 1226 (1987).

42

NON—CHROMATOGRAPHIC SEPARATION OF PEPTIDES

Sherrell L. Early, Sheila G. Magil
Curt Novak, and Marcus J. Horn

Applied Protein Technologies
Cambridge, MA 02139

I. INTRODUCTION

Separations of complex mixtures of peptides present one of the most challenging problems facing researchers today. Applied Protein Technologies, Inc. (APT, Cambridge, MA) has developed a non-chromatographic approach to aid in the separation of peptide mixtures. The separations are based on the development of reversible, amino acid specific reagents, Pep-Seps™.

The Pep-Seps reagents are insoluble reagents which are permanently attached to porous glass beads. Unique chemistries have been developed for each individual Pep-Seps reagents, which allow the peptide to attach to the Pep-Seps reagent through a specific amino acid side chain. The attachment is through a reversible covalent bond formed between the Pep-Seps and the specific amino acid side chain and the peptide.

Pep-Seps reagents require no instrumentation. All reactions can be carried out in a test tube. The immobilization and release procedures for each reagent are very simple and straightforward. To use Pep-Seps, first they are washed. Next the peptide mixture is mixed with the Pep-Seps reagents and allowed to react under mild conditions. During the immobilization procedure a covalent bond is formed between the Pep-Seps and the side chain of the specific amino acid. The mixture is spun down. The Pep-Seps containing the specific amino acids remain immobilized to the glass beads. Non-immobilized peptides remain in the supernatant. The final step is to cleave the bound peptides from the glass beads using mild releasing conditions.

Pep-Seps reagents are currently available that react specifically with cysteine, methionine, arginine, or tryptophan residues. Cysteine, methionine, and arginine peptides are released unmodified. The tryptophan amino acid side chain is modified by use with the Pep-Seps. This amino acid side chain is released as a thiotryptophan, which can still be identified by HPLC from Edman degradation of the thiotryptophan-containing peptide.

II. UPTAKE AND RELEASE OF PEPTIDES BY PEP—SEPS

In order for Pep-Seps to be of value to researchers, the reagents should be able to immobilize the desired peptides nearly quantitatively. In order to achieve this quantitative immobilization, the immobilization conditions have been optimized for each individual Pep-Seps reagent. However, just as one observes different rates of reactivity in a chemical reaction due to differences in nucleophilicity and steric factors of the nucleophile, one would expect <u>a priori</u> different rates of coupling between different reactive peptides and Pep-Seps.

Using methionine Pep-Seps and several different methionine-containing peptides, a difference in the rate of immobilization between the various peptides was observed (Figure 1, Table 1). The peptides most quickly immobilized were those that contained C-terminal methionine, e.g. dansyl-<u>dl</u>-methionine, LWM, and met-enkephalin (YGGFM). Within this subset of methionine-containing peptides, the rate of immobilization was not proportional to the size of the peptide.

The incubation time used for the methionine-containing peptides in Figure 1 and Table 1 was the recommended minimum time of four (4) hours at room temperature. Further studies using incubation times up to 20 hours showed that all peptides could be immobilized in yields of 90% or more (Table 2). The effects of temperature on the rate of peptide immobilization to the Pep-Seps has not yet been studied.

Once the Pep-Seps and peptides have gone through the immobilization and wash steps, the immobilized peptides are ready to be released. Using the conditions described for methionine Pep-Seps in Table 1, most peptides are released in a yield of 60% of the amount taken up by the Pep-Seps (Table 1). Extended incubation times of up to 20 hours can increase the yields to 70-75%. The effects of temperature on peptide release are currently being investigated.

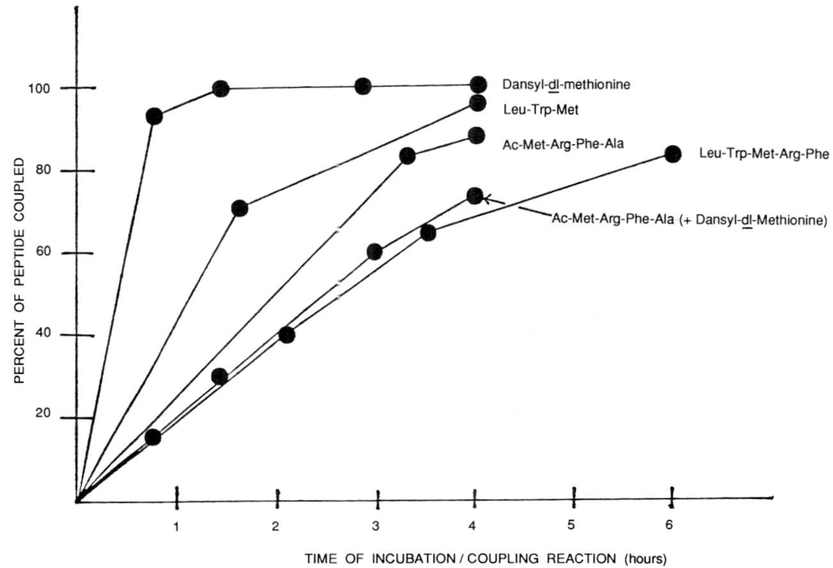

Figure 1

Kinetics of Peptide Coupling to Methionine Pep-Seps

TABLE I. Uptake and Release of Peptides by Methionine Pep-Seps

Peptide	% Uptake [a]	% Release [b]
Dansyl-*dl*-methionine	100.0	59.6
LWM	95.9	56.4
YGGFM	98.6	---
Acetyl-MRFA	87.5	---
LWMRF	72.1	60.0
RF	0.0	0.0

[a] To 100 mg of methionine Pep-Seps was added 1 ml of 0.2 M aqueous acetic acid containing 100 nmoles of the desired peptide. This mixture was mixed gently at room temperature for 4 hours.

[b] To 100 mg of peptide-treated methionine Pep-Seps was added 1 ml 1 M sodium bicarbonate, pH 8.75/100 mM 2-mercaptoethanol. This mixture was mixed gently at room temperature for 3 hours.

TABLE II. Effect of Incubation Time on Uptake of Peptides by Methionine Pep-Seps

Peptide [a]	% Uptake		
	2 hrs	6 hrs	20 hrs
LWM	89	98	98
Acetyl-MRFA	67	92	90
LWMRF	75	94	100

[a] See footnotes *a* and *b* in Table 1 for experimental protocol.

III. SELECTIVITY AND NON—SPECIFIC ADSORPTION

The premise on which the Pep-Seps are based is that a solid-phase chemistry can be used that is specific for a single, amino acid side chain, i.e., selectivity. Non-specific adsorption of peptides is known to occur with peptides and glass surfaces, possibly through processes involving electrostatic interactions or through chelation of the peptides by metals found in the glass. Since non-specific adsorption of peptides would result in the lack of selectivity by Pep-Seps, the significance that non-specific adsorption has in the interaction of Pep-Seps with peptides which should be chemically unreactive with the Pep-Seps must be explored.

This issue was addressed by incubating methionine Pep-Seps with a methionine-containing peptide (met-enkephalin, YGGFM) and with two methionine-lacking peptides (leu-enkephalin, YGGFL, and bradykinin, RPPGFSPFR). The supernatant of the incubation mixture was analyzed by HPLC. After four hours of incubation at room temperature in 0.2M acetic acid (Figure 2b), met-enkephalin was immobilized in a 98.6% yield. The methionine-lacking peptides, leu-enkephalin and bradykinin, showed a loss of only 2.3% and 6.3%, respectively. This indicates that non-specific adsorption does not play a major role with small peptides, thereby allowing the methionine Pep-Seps to retain their selectivity.

The selectivity observed with methionine Pep-Seps has also been observed in initial studies with both cysteine Pep-Seps and tryptophan Pep-Seps. Figure 3 shows that a cysteine-containing peptide, Cys-17 (SCDIFTTSRGKRASKGY), is quantitatively immobilized to cysteine Pep-Seps after a 3 hour incubation at room temperature in 0.5 M Tris-HCl, pH 7.0. Washing of the Pep-Seps with 3.2 M guanidine did not show any non-specific adsorption of Cys-17 to the Pep-Seps. Incubation of the washed Pep-Seps under releasing conditions caused an 80-85% release of Cys-17. Quantitative release of Cys-17 was obtained with a longer incubation time (data not shown).

The cysteine-lacking peptides, MRFA and LWMRF, were used to study the extent of non-specific adsorption with cysteine Pep-Seps. When the two peptides were co-incubated using the conditions described above for the immobilization of Cys-17, the more hydrophobic LWMRF was "immobilized" in a 31.5% yield, whereas only 2% of MRFA was "immobilized". The "immobilized" LWMRF was shown to be non-covalently bound simply by washing the Pep-Seps two times with 3.2 M guanidine. The majority of the non-covalently bound (i.e., non-specifically adsorbed) LWMRF was released in the first guanidine wash.

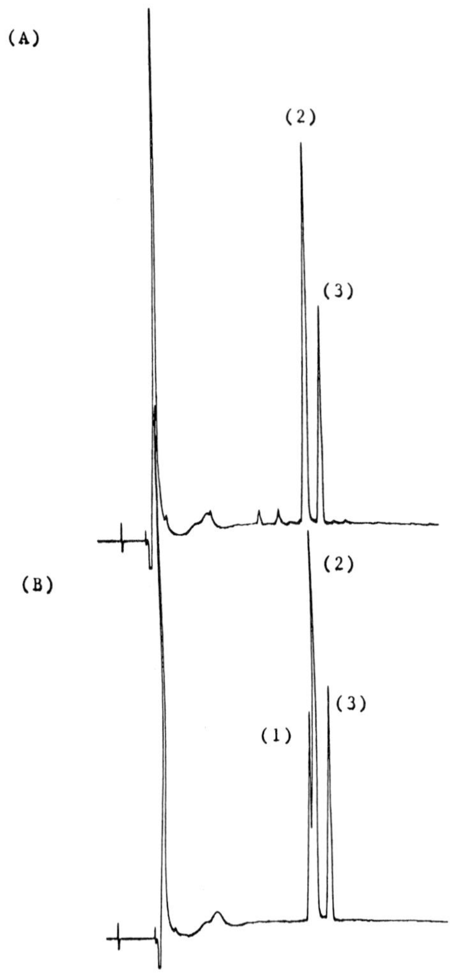

Figure 2

(A) Supernatant of incubation mixture containing methionine pep-seps and (1) met-enkephalin, (2) leu-enkephalin, and (3) bradykinin

(B) Standard solution containing (1) met-enkephalin, (2) leu-enkephalin, and (3) bradykinin

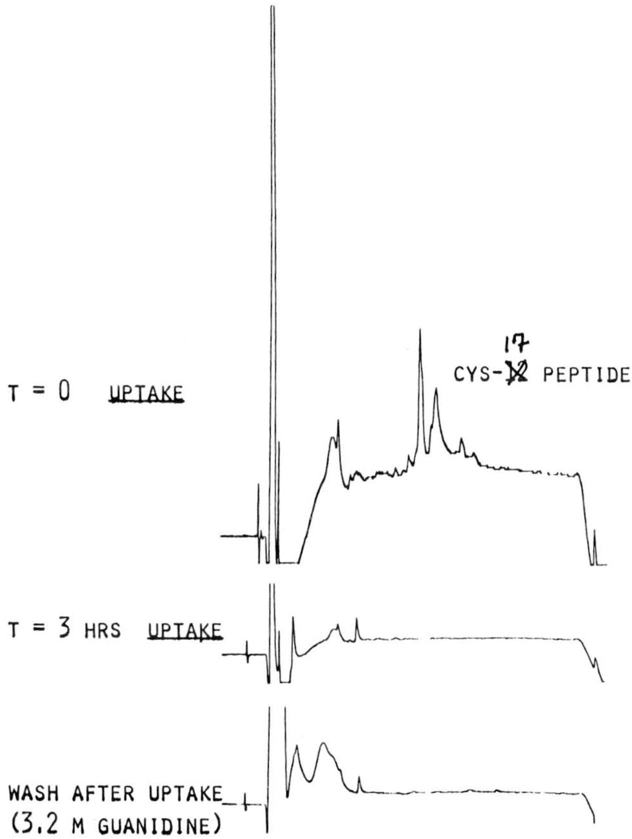

Figure 3

Selectivity and non-specific adsorption of cysteine pep-seps
-HPLC Analysis-

A similar study was conducted with tryptophan Pep-Seps. A mixture of tryptophan-containing peptides (LWM, LWMRF) and tryptophan-lacking peptides (RF, acetyl-MRFA) was incubated for a period of 2 or 3 hours with tryptophan Pep-Seps under immobilization conditions (20% aq. acetic acid) in the absence or presence of 2.5 M guanidine. From the data presented in Table III, a significant amount of non-specific adsorption is taking place, as evident by the difference in the amounts of RF and acetyl-MRFA immobilized in the absence and presence of 2.5 M guanidine (RF, 40% vs 17%; acetyl-MRFA, 35% vs 0%; t= 2 hours). While the tryptophan-containing peptides also were immobilized to a lesser extent in the presence of 2.5 M guanidine than in its absence, the amount of LWM and LWMRF immobilized to the Pep-Seps increased greatly between 2 and 3 hours of incubation. This was not true for RF and acetyl-MRFA. This indicated that the chemical reaction occurring between the tryptophan Pep-Seps and the tryptophan in LWM and LWMRF was still occurring in the presence of 2.5 M guanidine but at a slower rate than in the absence of guanidine.

TABLE III. Selectivity and Non-Specific Adsorption of Tryptophan Pep-Seps

Peptide	% Immobilized (-)[a]	% Immobilized (+)[a]
t = 2 hours		
RF	40	17
Acetyl-MRFA	35	0
LWM	59	30
LWMRF	91	42
t = 3 hours		
RF	38	22
Acetyl-MRFA	41	0
LWM	66	45
LWMRF	93	58

[a] (-) without 2.5 M guanidine
(+) with 2.5 M guanidine

IV. SUMMARY

The ideal interaction between a peptide mixture and a Pep-Seps reagent is one that involves a specific chemical reaction between a Pep-Seps reagent and a single type of amino acid side chain found in a peptide, without any non-specific adsorption to the Pep-Seps by peptides that do not contain the amino acid of interest. An example of this is the reaction of methionine Pep-Seps with a peptide mixture containing met-enkephalin, leu-enkephalin, and bradykinin. However, the data presented here also show that an ideal interaction between a peptide mixture and Pep-Seps is not always possible. Initial data indicates that the reaction conditions can be made to approximate ideal conditions if compatible denaturants and/or detergents are included in the incubation solution.

The success of protein chemistry techniques, such as HPLC and gel electrophoresis, has been shown to be affected by the peptides and proteins analyzed. The above data show the Pep-Seps reagents to be similarly affected. Expanding on these findings, the goal of ongoing research is to define a range of conditions that is optimal for each Pep-Seps reagent.

43

AN ENDOGENOUS BACTERIAL PROTEASE SPECIFICALLY CLEAVES A FUSION PROTEIN JUNCTION CONSISTING OF CONSECUTIVE BASIC AMINO ACIDS TO GENERATE A BIOLOGICALLY-ACTIVE EUKARYOTIC ADHESIVE PROTEIN, BINDIN.[1]

Charles G. Glabe
Sandra Brockman
Angelika Lopez
Ken Kimura
Laura Kennedy
Paul L. DeAngelis

Department of Molecular Biology and Biochemistry
University of California, Irvine
Irvine, CA 92717

I. INTRODUCTION

The ability to produce large amounts of biologically-active proteins in microorganisms is an important tool for analyzing the structure and function relationships of proteins. However, expression of many eukaryotic genes in *Escherichia coli*, at a level sufficient to analyze the activity of the purified product, has proved to be a difficult challenge even in the presence of strong bacterial promoters and ribosome binding sites (1,2,3,4). An alternative strategy for high level expression of many eukaryotic proteins has been through production as fusion proteins (5). Although some fusion proteins retain biological activity, the presence of bacterially-derived polypeptide segments may interfere with the folding and biological activity of the protein product.

In response to this problem, several strategies have been employed to separate the desired eukaryotic protein product from the bacterially-derived segments of the fusion protein. Chemical cleavage of the fusion protein junction at

[1]This work was supported in part by grants from the California Cancer Research Coordinating Committee and the National Institutes of Health (HD 21379).

methionine residues with cyanogen bromide (1) or at aspartyl-glycine bonds by mild acid hydrolysis (4) has been used to release the eukaryotic product. However, the relatively low specificity of the chemical cleavage strategies preclude their use with protein products that contain internal methionine or acid sensitive residues. Alternatively, more specific enzymatic hydrolysis using collagenase (6), endoproteinase LysC (7) and factor Xa (3) has been employed to cleave the fusion junction. Endogenous proteolytic enzymes of the secretory pathway have also been used to process the fusion junction (8,9,10). In this chapter, we describe the use of an endogenous bacterial protease to efficiently cleave between Arg-Arg and Lys-Arg sequences that are located at the junction between the cII and mature bindin sequences. Specific cleavage of the cytoplasmic fusion protein only occurs after the cells are lysed, suggesting that the protease is localized in the periplasm or outer membrane. Since Arg-Arg and Lys-Arg sequences commonly occur at the proteolytic processing site of proteins that are synthesized as larger precursors (11), this observation may be especially useful for the facile production of this class of eukaryotic proteins as fusion proteins in *E. coli*.

II. EXPRESSION OF BINDIN

Bindin is a protein from sea urchin sperm that mediates the species-specific attachment of sperm to the egg surface during fertilization (12). Mature bindin (24,000 MW) is derived from the carboxyl half of a larger precursor (51,000 MW) by proteolysis. (13). In order to understand the structure and function relationships of bindin, we would like to produce altered bindin polypeptides by *in vitro* mutagenesis. High level expression of beta globin (3) and myoglobin (14) has been reported using a novel expression vector described by Nagai and Thogersen, which directs the synthesis of a fusion protein containing the N-terminal 31 amino acids of the lambda cII gene separated from the eukaryotic coding sequence by a linker that codes for the specific tetrapeptide recognition sequence for the protease factor Xa: Ile Glu Gly Arg. Factor Xa cleaves on the C-terminal side of arginine, allowing for the recovery of the authentic eukaryotic protein sequence after proteolysis. We cloned cDNA segments of bindin, corresponding to the mature bindin sequence +4 amino acids of the probindin sequence and -18 amino acids of the mature sequence, into pUC12FX, taking advantage of conveniently located restriction sites which produce blunt ends that are in reading frame with the Stu I site of the factor Xa linker (Fig. 1).

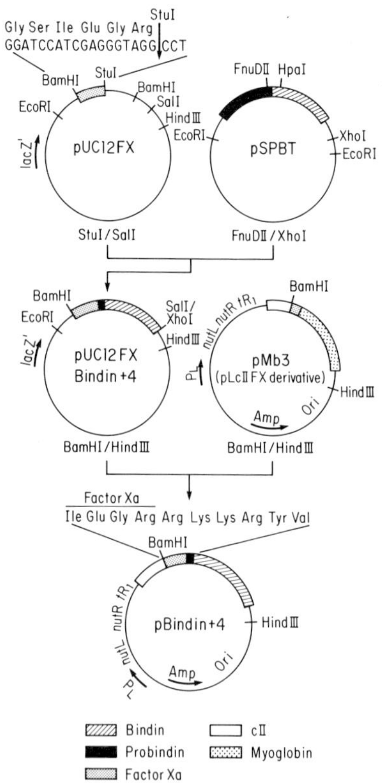

Fig. 1. Construction of pBindin+4 and pBindin-18. The factor Xa-encoding linker from M13mp11FX (3) was excised with Bam H1 and cloned into Bam H1 cut pUC12 to yield pUC12FX. pSPBT (13) was cut either partially with Fnu DII or completely with Hpa I. Both samples were cleaved with Xho I and the 768 bp Fnu DII-Xho I and 702 bp Hpa I-Xho I fragments (12) were isolated. These fragments were ligated into Stu I-Sal I cut pUC12FX. The bindin-factor Xa cleavage site coding regions of pUC12FX-bindin+4 and pUC12FX-bindin-18 were excised by partial digestion with Bam H1 followed by Hind III digestion and purified by agarose gel electrophoresis. The bindin-factor Xa cleavage site coding fragment was used to replace the myoglobin-coding segment of pMb3 by ligating it into Bam H1-Hind III cut pMb3, yielding plasmids pBindin+4 and pBindin-18 after transformation into MZ1 cells. These plasmids direct the synthesis of a fusion protein consisting of the first 31 amino acids of the cII gene product, separated from the bindin-coding region by the factor Xa linker (3).

In the construction of pBindin+4 and pBindin-18, the myoglobin-coding sequence of pMb3 was replaced with the Bam H1-Hind III fragment from pUC12FX, containing the factor Xa linker and bindin coding sequences. After transformation into MZ1 cells, these plasmids direct the expression of high levels of bindin fusion proteins as revealed by the presence of a new, intensely-staining band on SDS polyacrylamide gels with apparent molecular masses of 35K and 33K Da respectively (Fig. 2A, lanes 1 and 4).

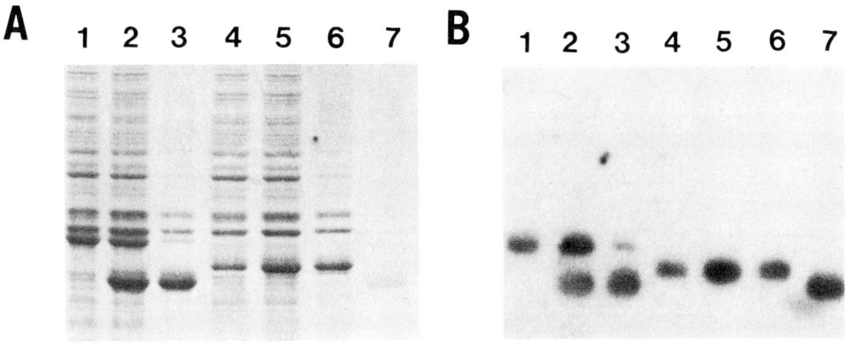

Fig. 2. Expression and processing of bindin fusion proteins in MZ1 cells. A. Coomassie blue-stained SDS-polyacrylamide gel (15%) and B. Western blot of an identical gel stained with antibindin antibodies and ^{125}I-protein A. Lane 1, whole cells containing pBindin+4 after induction for 2 hr. The arrow indicates the position of the fusion protein at an apparent molecular mass of 35 KDa which is stained with antibindin immunoglobulin. Lane 2, a mixture of equal amounts of protein from lane 1 and lane 3. Lane 3, recombinant bindin purified from pBindin+4 containing cells after cell lysis, solubilization in urea and ammonium sulfate precipitation. The major bindin product runs at an apparent molecular mass of 30.5 KDa. Lane 4, whole cells containing pBindin-18. The bindin fusion protein has an apparent molecular mass of 33 KDa. Lane 5, a mixture of equal amounts of protein from lane 4 and lane 6. Lane 6, bindin fusion protein purified from pBindin-18 containing cells. Lane 7, Bindin purified from sperm.

The molecular masses of the fusion proteins are approximately correct for the expected products since bindin migrates anomalously slowly on SDS polyacrylamide gels with an apparent molecular mass of 30,500 (Fig. 2A, lane 7). These bands are also intensely stained with antibindin antibody on Western blots, confirming their identity as bindin fusion proteins (Fig. 2B). From densitometry of the Coomassie blue stained gel, we estimate that the fusion protein for pBindin+4 and pBindin-18 represents approximately 35% and 17% respectively of the total cellular protein.

III. CLEAVAGE OF THE FUSION JUNCTION AFTER CELL LYSIS

We isolated "inclusion bodies" from induced MZ1 cells as previously described (14) and found that approximately 50% of the total fusion protein was recovered in the particulate fraction for both pBindin+4 and pBindin-18. After dissolving the lysate in 6.7 M urea and ammonium sulfate precipitation of the recombinant bindin product, SDS polyacrylamide gel electrophoresis and Western blotting of the purified protein revealed that the bindin product of pBindin+4 ran at a lower apparent molecular mass of 30.5 KDa, while the bindin product of pBindin-18 showed no detectable change in molecular mass (Fig. 2). This suggested that the product of pBindin+4 is cleaved during the isolation procedure. The proteolysis is remarkably efficient since only small amounts of the unprocessed form remain (Fig. 2A & 2B, lane 3). We purified the cleaved bindin product to homogeneity by reverse phase HPLC and sequenced it to determine the nature of the molecular weight change (Table I). The predominant product is the mature bindin polypeptide containing an additional 4 amino acids of probindin and a minor product that corresponds to bindin with a single additional residue. Both products contain arginine as the N-terminal amino acid. These results indicate that the fusion protein is cleaved at two sites: the Arg-Arg junction between the factor Xa linker and the probindin coding sequence and within the probindin segment at a Lys-Arg junction.

The bacterial protease that cleaves the pBindin+4 fusion protein appears to be localized in the external compartments of the cell. The fusion

TABLE I
Sequence of the proteolytic cleavage products of pBindin+4.

Cycle number	Amino acid (picomoles) Major sequence	Minor sequence
1	Arg (62)*	none
2	Lys (512)	Tyr (88)
3	Lys (504)	Val (91)
4	Arg (54)	Asn (nq)
5	Tyr (276)	Thr (40)
6	Val (279)	Met (67)
7	Asn (nq)	Gly (120)
8	Thr (137)	Tyr (64)
9	Met (321)	Pro (95)
10	Gly (285)	Gln (51)

* The conversion of arginine was low on this instrument. The values reported are not corrected. nq - not quantified. Asparagine eluted as a shoulder on a major artifact peak, precluding its accurate quantification.

protein of pBindin+4 is very stable in the intact cell. No smaller processed form of the bindin fusion protein is detected on Western blots of intact cells incubated for 16 hr at 0° C. After lysis of the cells by the lysozyme/freeze thaw procedure, the fusion protein is substantially cleaved by 5 min of digestion (data not shown). The fusion protein is also cleaved in cells lysed by passage of the cell suspension through a French press (data not shown). We have also investigated the effect of a variety of protease inhibitors on the proteolytic cleavage of the bindin fusion protein. None of the inhibitors tested, including antipain, aprotinin, chymostatin, EDTA, leupeptin, p-chloromercuribenzoate, pepstatin, phenylmethylsulfonylfluoride, or soybean trypsin inhibitor significantly slowed the rate of cleavage of the pBindin+4 fusion protein (data not shown).

IV. BIOLOGICAL ACTIVITY OF RECOMBINANT BINDIN

We have analyzed the biological activity of the proteolytically-processed recombinant bindin isolated from pBindin+4 containing MZ1 cells. The purified ammonium sulfate-precipitated bindin product was solubilized in urea and dialyzed against sea water. Bindin isolated from sperm is known to renature to a biologically-active form after solubilization in urea and dialysis into sea water (15). The precipitated recombinant bindin was dispersed in sea water by sonication and its ability to agglutinate unfertilized eggs was determined. The recombinant bindin mediates the agglutination of eggs (data not shown). Ammonium sulfate precipitated control protein prepared from MZ1 cells lacking pBindin+4 does not agglutinate eggs at the same concentration. Approximately 15 ug/ml of the recombinant bindin is required to agglutinate 50% of the eggs as compared to 5 ug/ml for bindin isolated from sperm which has not been subjected to urea solubilization (16).

V. DISCUSSION

We have found that the cytoplasmic fusion protein encoded by pBindin+4 is rapidly and efficiently cleaved by an endogenous bacterial protease after the cells are lysed. The processing of the bindin fusion protein results in the production of a biologically-active, mature bindin sequence containing 1 or 4 additional amino acids at the amino terminus. The amino terminal amino acid is arginine in both cases, suggesting that the protease specifically cleaves on the amino side of arginine residues when the preceeding amino acid is either lysine or arginine. The protease which is responsible for the processing of the bindin-cII fusion protein appears to be located in one of the external compartments of the cell, since no proteolysis is observed until the cell is lysed. One potential drawback of this method is that the protease cleavage specificity produces proteins that contain arginine residues at amino terminus. However, the presence of an additional arginine residue may not affect the biological activity of the protein product as we have observed for the processed form of recombinant bindin.

ACKNOWLEGEMENT

We are grateful to Drs. K. Nagai and S. Boxer for providing components of the expression vector system, pLcIIFX and Drs. R. Britten and E. Davidson for providing the bindin cDNA. We thank Drs. M. Cumsky, L. Villarreal and D. Knauer for helpful discussions.

VI. REFERENCES

1. Itakura, K., Hirose, T., Drea, R., Riggs, A. D., Heyneker, H. L., Bolivar, F., and Boyer, H. W. (1977) Science **198**, 1056-1063.
2. Germino, J., Gray, J., Charbonneau, H., Vanaman, T., and Bastia, D. (1983) Proc. Natl. Acad. Sci. U.S.A. **80**, 6848-6852.
3. Nagai, K., and Thogersen, H. C. (1984) Nature (London) **309**, 810-812.
4. Szoka, P. R., Schreiber, A. B., Chan, H., and Murthy, J. (1986) DNA **5**, 11-20.
5. Harris, J. T. R. (1983) in Genetic Engineering, Vol. 4. (R. Williamson, ed.) pp127-185. (Academic Press, New York, N.Y.)
6. Germino, J., and Bastia, D. (1984) Proc. Natl. Acad. Sci. USA **81**, 4692-4696.
7. Allen, G., Paynter, C. A., and Winther, M. D. (1985) J. Cell. Sci (Suppl.) 3, 29-38.
8. Talmadge, K., Kaufman, J., and Gilbert, W. (1980) Proc. Natl. Acad. Sci. USA **77**, 3988-3992.
9. Dodt, J., Schmitz, T., Schafer, T., and Bergmann, C. (1986) FEBS Lett. **202**, 373-377.
10. Brake, A. J. et al., (1984) Proc. Natl. Acad. Sci. USA. **81**, 4642-4646.
11. Steiner, D. F., Docherty, K., and Carroll, R. (1984) J. Cell. Biochem. **24**, 121-130.
12. Vacquier, V. D., and Moy, G. W. (1977) Proc. Natl. Acad. Sci. USA. **74**, 2456-2460.
13. Gao, B., Klein, L. E., Britten, R. J., and Davidson, E. H. (1986) Proc. Natl. Acad. Sci. USA. **83**, 8634-8638.
14. Varadarajan, R., Szabo, A., and Boxer, S. G. (1985) Proc. Natl. Acad. Sci. USA **82**, 5681-5684.
15. Vacquier, V. D. (1983) Anal. Biochem. **129**, 497-501.
16. Glabe, C. G., and Lennarz, W. J. (1979) J. Cell Biol. **83**, 595-604.

44

A RAPID PROCEDURE FOR THE QUANTITATIVE ANALYSIS OF MONOCLONAL ANTIBODIES BY HIGH PERFORMANCE CAPILARY ELECTOPHORESIS

Norberto A. Guzman* and Luis Hernandez, Protein Research Unit, Princeton Biochemicals Inc., Princeton, NJ 08543

I. INTRODUCTION

The present day expansion of biochemistry-based technologies has increased the need for analyzing and characterizing a great number of macromolecules, in particular immunoglobulins. The uses of immunoglobulins or immunoglobulin fragments are, among others: as immunoadsorbent in purification of a wide variety of substances, in immunodiagnostics for identification and quantitation purposes, and in various aspects of immunotherapy for human and veterinary health care. For many years, conventional chromatography was the common method for purification of immunoglobulins. Today, various techniques including affinity techniques using analytical and preparative high-performance liquid chromatography are the choice for separation and characterization of immunoglobulins. More recently, however, new methods for large-scale processes which can handle up to 1,000-liter volumes are being used for monoclonal antibody production. The use of monoclonal antibodies has increased significantly in the last decade, with even greater demands anticipated in the years to come.

Purification and characterization of immunoglobulins, using the various different methods available today, are often time-consuming and quantitatively poor. We now present a rapid procedure for the quantitative analysis of monoclonal antibodies by high-resolution open tubular capillary electrophoresis or high performance capillary electrophoresis (HPCE)†. HPCE is a powerful nanotechnique characterized by high speed, high efficiency, excellent mass sensitivity, low sample consumption, and high resolution. HPCE has been one of the most practical advances in the purification and analysis of substances using a high voltage electrical field as the means of separation (1-5). Because of the combined action of electroosmotic flow with electrophoretic separation, all compounds in the sample normally travel in one direction, allowing detection of positively charged, neutral, and negatively charged species at one single location along the capillary tubing. As a model of study we have used mouse monoclonal antibodies directed against human placental prolyl 4-hydroxylase. Prolyl 4-hydroxylase catalyzes the cotranslational and post-translational conversion of selected prolyl residues to *trans*-4-hydroxy-*L*-proline. This reaction is carried out in nascent or completed pro-alpha chains of fibrillar procollagen, basement membrane procollagen, and in collagen-like domains of a variety of peptides and proteins (6,7).

II. MATERIALS AND METHODS

Purification of Prolyl 4-Hydroxylase from Human Placentae. Full-term human placentae obtained from normal mothers were collected 30 min after delivery. The umbilical cords were removed and the placentae were immediately stored at -70°C. The frozen placentae were thawed at 4°C for one day. Prolyl 4-hydroxylase was then purified to homogeneity from tissue homogenates by successive application of ion-exchange, affinity and high-performance liquid chromatography as described previously (8,9). Prolyl 4-hydroxylase activity was assayed by the tritium release method of Hutton *et al* (10) as modified by Cutroneo *et al* (11).

Preparation of Antibodies. Monoclonal antibodies directed against native human placental prolyl 4-hydroxylase were prepared as described by the method of Galfre and Milstein (12) with minor modifications. Six-to-8-week old BALB/c mice were immunized with 60 μg of the purified enzyme subcutaneously at multiple sites and intraperitoneally targeted at the spleen area. The mice were immunized 3 times at weekly intervals and test bled after the last immunization. Following antibody titer measurements, the animals were killed 3 days after a final subcutaneous booster injection. Spleen cells were fused with the SP2/O Ag nonsecreting murine myeloma cell line (13) in the presence of poly(ethylene glycol) as described previously (12). The cells producing antibodies were cloned by the dilution method (12,14) and the animals were then injected intraperitoneally with the cloned hybrids.

Purification of Antibodies from Ascites Fluid. Approximately 10 ml of mouse ascites fluid were incubated at 25°C for 2 hr and then at 4°C overnight. The particulate matter present in the sample was removed first by centrifugation and then by filtration using a 0.22 μm membrane filter after the sample had been diluted 3 times with binding buffer (1 M glycine and 3 M NaCl, pH 9.0). Aliquots of about 25 mg of immunoglobulin present in the clarified ascites fluid solution were applied to a HPLC Affi-Prep Protein A MAPS preparative cartridge (BIO-RAD, Richmond, CA). The HPLC column was previously equilibrated with binding buffer. After application of the sample, the column was extensively washed with the same binding buffer. The protein bound to the lectin affinity column was then eluted with sodium citrate buffer pH 6.0, 4.5, and 3.5 respectively in a stepwise manner. The samples eluted from the column were monitored at 280 nm.

Polyacrylamide Slab Gel Electrophoresis. Polyacrylamide slab gel electrophoreses of native and denatured proteins were performed. Slab gel electrophoresis of the native proteins were carried out similarly to the conditions described previously (15). Slab gel electrophoresis of denatured proteins was carried out in the presence of sodium dodecyl sulfate by the method of King and Laemmli (16). All denatured samples were reduced in 2-mercaptoethanol and boiled for 3 min. The proteins in either native or denatured gels were stained with Coomassie brilliant blue R.

High Performance Capillary Electrophoresis. A standard solution of Protein A-purified major component was used to generate a calibration curve. The IgG sample eluted from the lectin affinity chromatography was dialyzed against 0.05 M sodium tetraborate, pH 8.3, and an accurate concentration of the sample then was measured by amino acid analyzer. Several dilutions were obtained using the same borate buffer. Injections of IgG samples into the capillary column were carried out starting from the more diluted sample first. Untreated fused-silica capillaries (Scientific Glass Engineering, Inc., Austin, TX) were used. The capillary column (75 μm I.D. x 100 cm) was filled with 0.05 M sodium tetraborate buffer and the measurements were obtained using an automated capillary electrophoresis instrument (Princeton Biochemicals, Inc., Princeton, NJ). Approximately 3.3 nanoliters of IgG solution were electrokinetically introduced into the capillary by applying 10,000 volts for 15 sec. Samples were then separated at 22,000 volts for 35 min and detected at 210 nm using a modified UV detector.

III. RESULTS

Electrophoretic Migration of Prolyl 4-Hydroxylase Under Native and Denatured Conditions. Electrophoretic studies revealed that human prolyl 4-hydroxylase is a tetrameric enzyme composed of two nonidentical subunits of about 64 and 62 kDa, in a ratio 1 to 2, designated alpha and beta, respectively (data not shown).

Chromatographic Profile of Monoclonal Antibodies. Antibodies were produced against enzymatically active purified human placental prolyl 4-hydroxylase in mice. The specific antibodies were then purified from mouse ascites fluid by Protein A affinity chromatography. As seen in Figure 1, the major IgG component was eluted from the lectin affinity column at pH 6.0. Electrophoresis carried out under native conditions showed a single band with a molecular weight of approximated 160 kDa. Electrophoresis performed under denatured conditions, on the other hand, showed two bands with molecular weights of 50 and 25 kDa respectively (Figure 1, insert).

High Performance Capillary Electrophoresis Studies. The emergence of HPCE has been an important advance in the purification and analysis of substances. After the samples are introduced into the capillary column, and they are separated into pure compounds, the sample zones are detected before leaving the capillary (on-column detection) or immediately after leaving the capillary (off-column detection) (see Figure 2). Many methods are now available to detect the separated compounds, though in these studies we limited ourselves to the use of on-column ultraviolet detection.

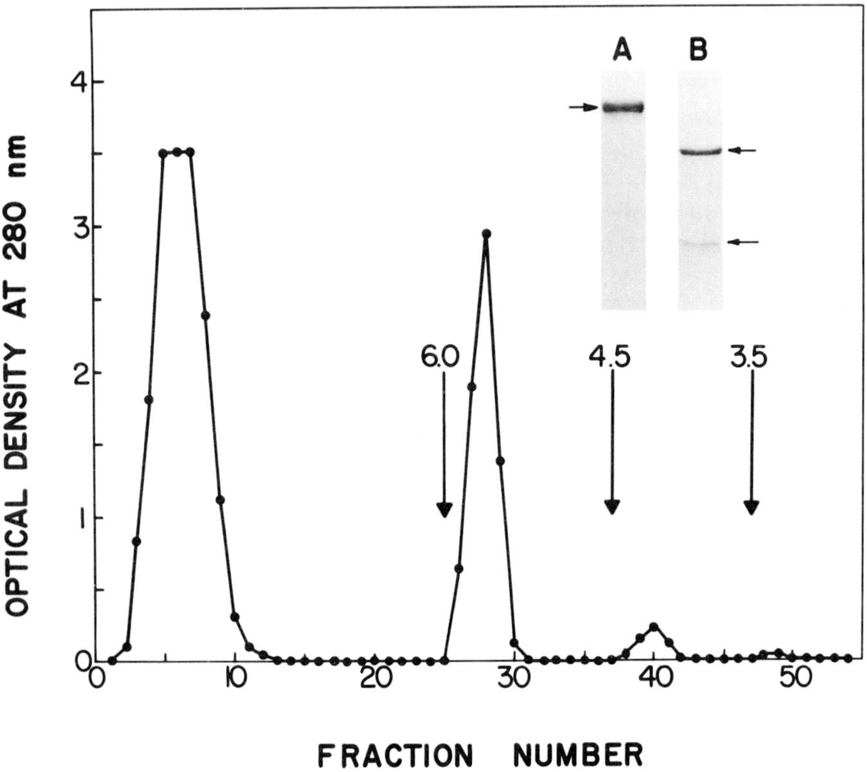

Figure 1. Affinity HPLC Profile of Ascites Fluid Monoclonal Antibodies. The ascites fluid was injected into a Protein A-affinity HPLC column as described in the text. The first peak consisted of material that did not bind to the column. The second peak eluted at pH 6.0 contained most of the IgG present in the ascites fluid. The arrows indicate the beginning of the elution at pH 6.0, 4.5, and 3.5 respectively. **Insert. (A)** Slab gel electrophoresis profile of IgG under native conditions. **(B)** Slab gel electrophoresis profile of IgG under denatured conditions. The arrows indicate the position of the 160 kDa native IgG, the 50 kDa heavy chain and the 25 kDa light chain respectively.

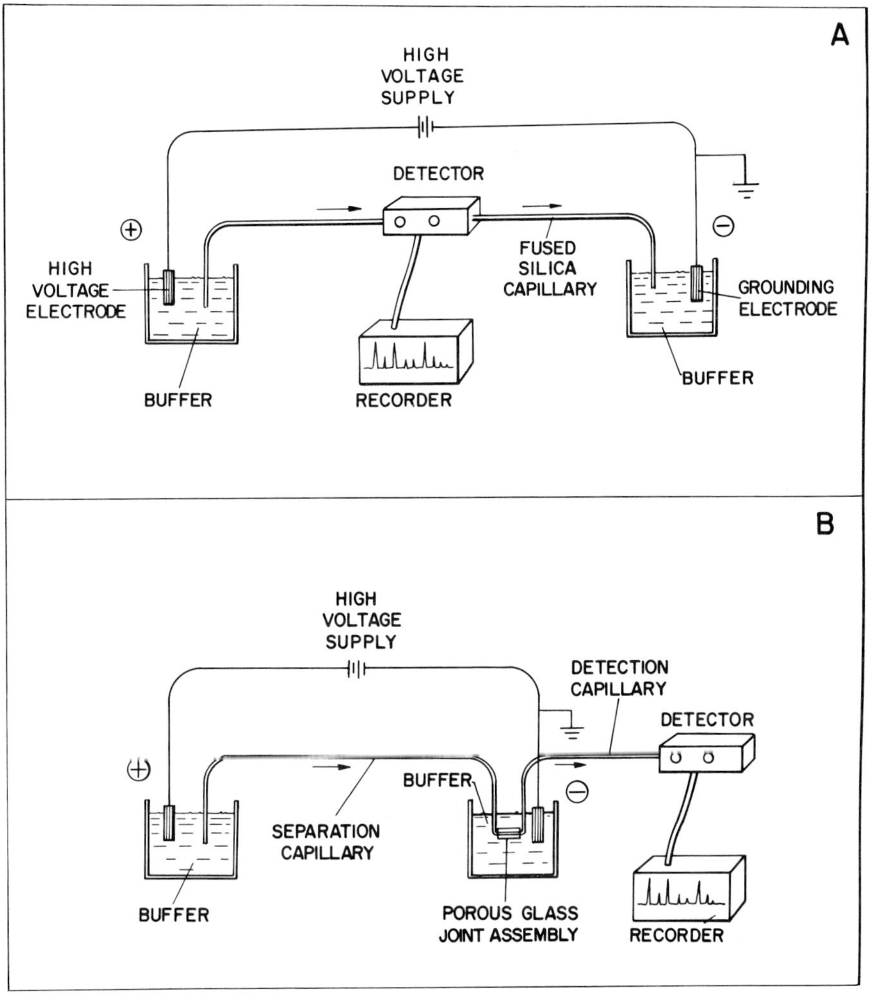

Figure 2. A Schematic Diagram of the Principle of High-Performance Capillary Electrophoresis. **(A)** System with an on-column detection. **(B)** System with an off-column detection.

In order to obtain an accurate quantitation of the immunoglobulins under study and a very reproducible electropherographic profile, it was necessary to use an automated capillary electrophoresis apparatus. In our lab-

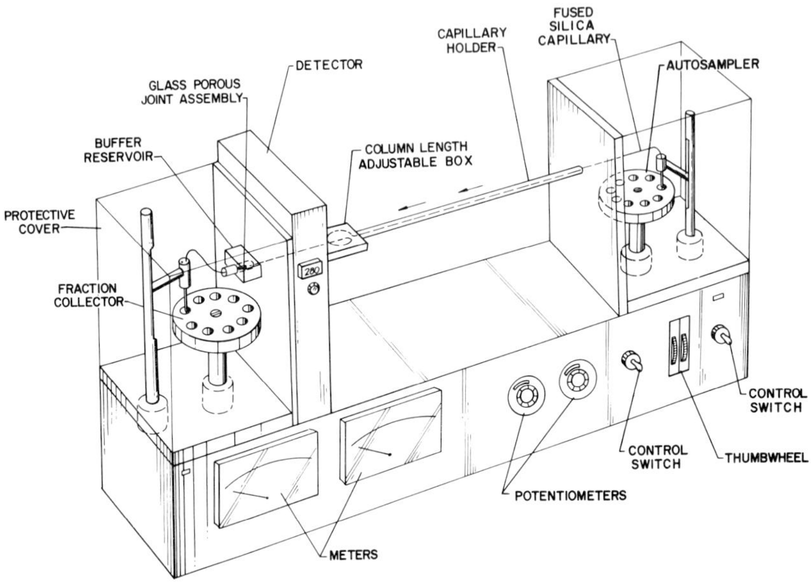

Figure 3. A Schematic Diagram of an Automated Capillary Electrophoresis Apparatus.

oratory we have undertaken the task of developing a unique instrument containing many features of practicality to the user and to the high performance of this technique. As seen in Figure 3, the automated capillary electrophoresis apparatus contained an autosampler to introduce samples automatically into the capillary column, a fraction collector to collect purified components, an adjustable column length cartridge-like box to fit different sizes of capillaries, and several switches to control the entire operation in a semi-automated or complete automated fashion.

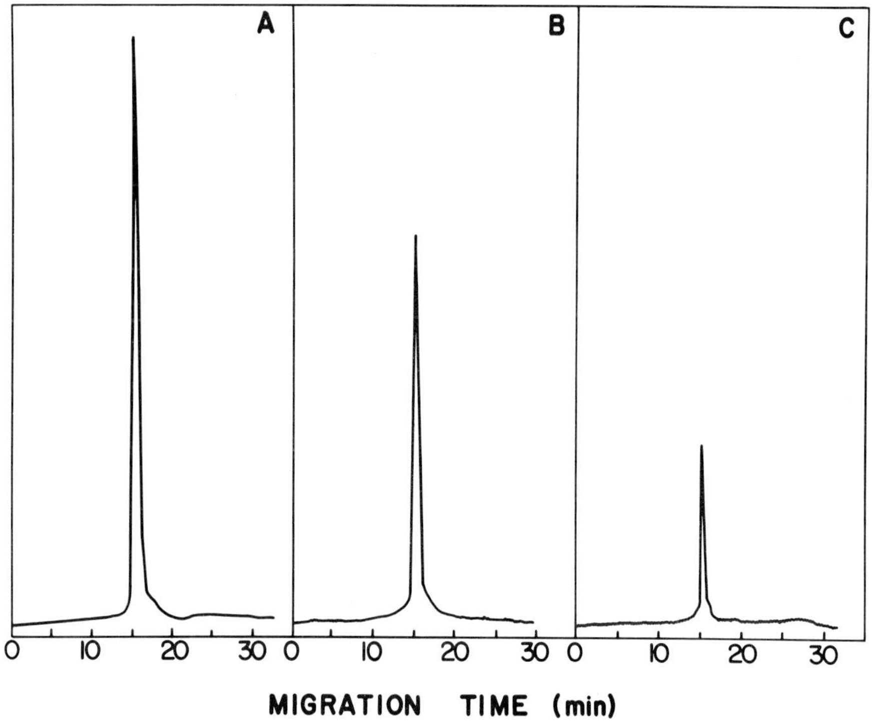

Figure 4. HPCE Profile of Affinity-Purified Monoclonal Antibodies. Electropherograms of affinity-purified monoclonal antibodies were obtained. IgG samples eluted at pH 6.0 from the Protein A HPLC column were electrokinetically injected into the capillary column as described in the text. Samples were monitored at (**A**) 210 nm (0.2 AUFS); (**B**) 230 nm (0.1 AUFS); and (**C**) 280 nm (0.05 AUFS) respectively.

As seen in Figure 4, Protein A affinity purified IgG was separated as a single peak in HPCE when examined at various wavelengths. The highest signal to noise ratio response was obtained at 210 nm, a convenient wavelength to generate a calibration curve (Figure 5). Reproducibility tests showed less than 2% error in peak height and retention time.

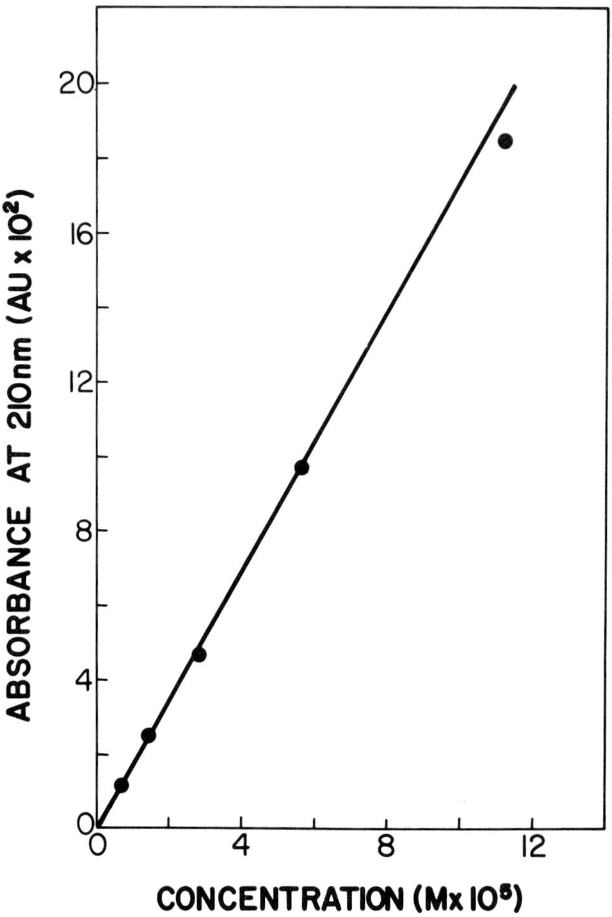

Figure 5. Calibration Curve of Monoclonal Antibody by HPCE. Different dilutions of IgG were carried out as described in the text. Samples were electrokinetically injected starting from the most diluted concentration.

Attempts were carried out to inject samples immediately after being eluted from the Protein A affinity column and without dialysis against the HPCE buffer. No major changes were observed either in the electrophoretic profile or in the height of the peak (data not shown). However, when samples derived from the clarified ascites fluids were immediately injected into the capillary column, without previous separation of the components, a few differences were observed (data not shown). These differences are probably reflected as the ability of some proteins present in the ascites fluid, i.e. albumin, to bind to the walls of the untreated fused-silica capillary and change the properties of the electropherographic profile. In particular, large concentrations of proteins will change the properties of the column and eventually may plug the column. Studies are now underway in our laboratory in order to attach covalently to the column various compounds which may block the sticky properties of some proteins, and improve the quality and consistency of the HPCE method.

IV. DISCUSSION

A significant progress in the availability of modern techniques and instruments for the separation and characterization of macromolecules from complex mixtures has taken place in the past years. Currently, chromatographic and electrophoretical techniques are among the most commonly employed forms of separation and analysis, in particular for immunoglobulins. However, many limitations need to be overcome.

The emergence of HPCE has been an important advance in the purification and analysis of substances of medical and scientific interest (1-5). HPCE is currently under rapid development and has been successfully applied to the separation and analysis of a variety of macromolecules. However, this is the first report about the use of HPCE for separation and analysis of immunoglobulins. We have been able to introduce electrokinetically into the capillary column 3.3 nanoliters of a detectable diluted IgG solution. This is about femtomole range of IgG loaded into the capillary column, making HPCE the most sensitive technique for analyzing monoclonal antibodies by ultraviolet detection. In addition, the number of theoretical plates for IgG using HPCE is about a few orders of magnitude larger than HPLC, which also makes HPCE an analytical technique with the most resolution to separate monoclonal antibodies.

Since HPCE is a technique of high resolution it could have great utility for measuring the productivity of hybridoma clones in producing monoclonal antibodies as well as for determining the homogeneity of the immunoglobulin product.

V. FOOTNOTES

*Dr. Guzman is presently a Senior Scientist at Roche Diagnostic Systems, a subsidiary of Hoffmann-La Roche Inc., Nutley, New Jersey 07110.

†Abbreviations used here: HPCE, high-performance capillary electrophoresis; IgG, immunoglobulin G; kDa, kilodalton; nm, nanometer; AUFS, absorbance units full scale.

VI. REFERENCES

1. J.W. Jorgenson, and K.D. Lukacs, Science **222**, 266-272 (1983).
2. J.W. Jorgenson, in **New Directions in Electrophoresis Methods** (J.W. Jorgenson, and M. Phillips, Eds.), American Chemical Society, Washington, D.C., Chapter 13, pp. 182-198 (1987).
3. A.S. Cohen, A. Paulus, and B.L. Karger, Chromatographia **24**, 15-24 (1987).
4. S.W. Compton, and R.G. Brownlee, BioTechniques **6**, 432-439 (1988).
5. N.A. Guzman, L. Hernandez, and B.G. Hoebel, Biopharm, January (1989).
6. D.J. Prockop, R.A. Berg, K.I. Kivirikko, and J. Uitto, in **Biochemistry of Collagen** (G. Ramachandran, and A.J. Reddi, Eds.), Plenum, New York, pp. 163-273 (1976).
7. N.A. Guzman, G.C. Fuller, and J.E. Dixon, in **Biology of Extracellular Matrix** (R.P. Mecham, and S.W. Adair, Eds.), Academic Press, New York, in press.
8. N.A. Guzman, A.L. Oronsky, G. Suarez, L. Meyerson, K.R. Cutroneo, B.R. Olsen, and D.J. Prockop, Collagen Rel. Res. **2**, 381-396 (1982).
9. N.A. Guzman, W.Q. Ascari, K.R. Cutroneo, and R.J. Desnick, J. Biol. Chem., submitted.
10. J.J. Hutton, Jr., A.L. Tappel, and S. Udenfriend, Anal. Biochem. **16**, 384-394 (1966).

11. K.R. Cutroneo, N.A. Guzman, and M.M. Sharawy, J. Biol. Chem. **249**, 5958-5994 (1974).
12. G. Galfre, and C. Milstein, Meth. Enzymol. **73B**, 3-46 (1981).
13. M. Shulman, C.D. Wilder, and G. Köhler, Nature **276**, 269-270 (1978).
14. V.T. Oi, and L.A. Herzenberg, in **Selected Methods in Cellular Immunology** (B.B. Mishell, and S.M. Shiigi, Eds)., Freeman, San Francisco, pp. 351-372 (1980).
15. N.A. Guzman, P.N. Graves, and D.J. Prockop, Biochem. Biophys. Res. Commun. **84**, 691- 698 (1978).
16. J. King, and U.K. Laemmli, J. Mol. Biol. **62**, 465-477 (1971).

45

FACTORS AFFECTING PLATE HEIGHT IN HIGH PERFORMANCE ZONAL CAPILLARY ELECTROPHORESES (HPZCE)

D. McManigill
S. A. Swedberg*

Hewlett-Packard Laboratories
Palo Alto, CA 94303-0969

I. INTRODUCTION

Electrophoresis of macromolecules in capillary tubing has attracted increasing attention over the past few years since the publication by Mikkers in 1979 [1] followed by the publication by Jorgensen and Lukacs in 1981 [2]. Since these reports, a steadily increasing number of articles have appeared relating to various aspects of the type of experiment performed (i.e., zonal [3-6], micellar [7-9] or gel [10]), the type of detection employed (i.e., on-column [3], post column [11], electrochemical [12], spectrophotometric [3, 13], mass spec [14], etc.) and various aspects of equipment [15-16].

For the zonal experiment, more than a million theoretical plate efficiency can be predicted in considering axial diffusion to be the only contributing factor to band broadening. This makes capillary electrophoresis for macromolecules theoretically comparable to capillary gas chromatography for small solutes. Clearly, no other technique available holds the promise for such high efficiencies for macromolecules. Factors other than the unprecedented efficiencies for this class of solutes which make the technique attractive include rapid time to analysis (1-2 hr vs 1-2 days for conventional electrophoresis), small sample size (nL quantities), little or no sample pretreatment, recovery of biologically active samples, the capability for quantification, and the potential to fully automate the instrumentation.

This paper will provide an analysis of factors influencing plate height in zonal capillary electrophoresis. The behavior of three proteins with significantly different efficiencies under the stated electrophoretic conditions will be inspected with respect to the terms which contribute to the overall height equivalent to a theoretical plate (HETP) term.

II. HIGH PERFORMANCE ZONAL CAPILLARY ELECTROPHORESIS

Description of HPZCE: Equipment and Process

With a paucity of commercially available instrumentation, the majority of reports on high performance zonal capillary electrophoresis to date have been done with systems similar to that shown schematically in Figure 1. For the basic experiment, the components are fairly simple as shown with critical components consisting of

(1) Capillary: generally a fused silica capillary with internal diameter typically 20-75 μ and outer diameters ranging from 150-375 μ. Capillary lengths vary in the range from 30-100 cm.

(2) Detector: while many detection schemes have been reported, as previously mentioned, one that serves well as a universal detector is a UV detector with a cell holder modified to support and place the capillary in the source path.

(3) Power source: since HPZCE is commonly done at 250-300 V/cm, at lengths stated above a power source capable of delivering 30 kilovolts is required.

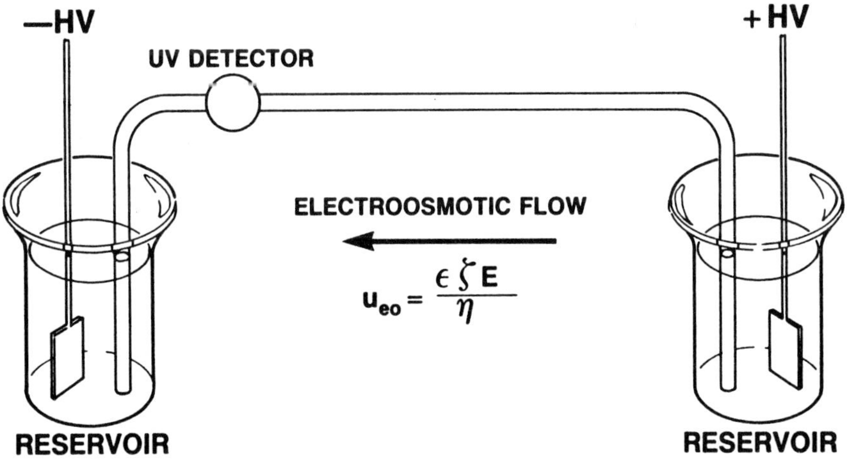

Fig. 1: Schematic showing some representative basic components of a capillary electrophoresis experiment.

A process which occurs during electrophoresis which is noteworthy is that of electroosmosis. At any solid-aqueous interface, a double-ion layer develops (see Fig. 2). The components of this double layer consist of the compact or Stern layer, which is tightly bound and may be considered as immobilized point charges. In the adjacent layer, the diffuse layer, counter charges to the compact layer exist which are more highly organized than like species in the bulk but have greater mobility than the compact layer. The total organi-

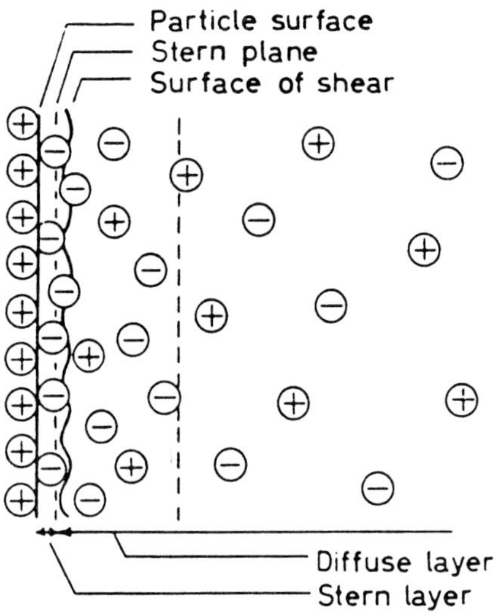

Fig. 2: The organization of the double ion layer.

zation of the double layer occurs in a region approximately 1-100 nm from the wall [17]. In the presence of an applied field, the diffuse layer moves towards its respective electrode and as a result, osmotically drags water in that layer with it. Bulk water in the tube is then moved by capillary action. It is important to note that a plug flow profile results from this process [18] and not a parabolic flow profile as is the case for nonturbulent flowing liquids which occur in a chromatography experiment (see Figure 3). The differences in these flow profiles is very important in accounting for the low efficiencies of macromolecules in chromatography processes vs the high efficiencies obtained for these large solutes in capillary electrophoresis, as will be discussed in some detail below.

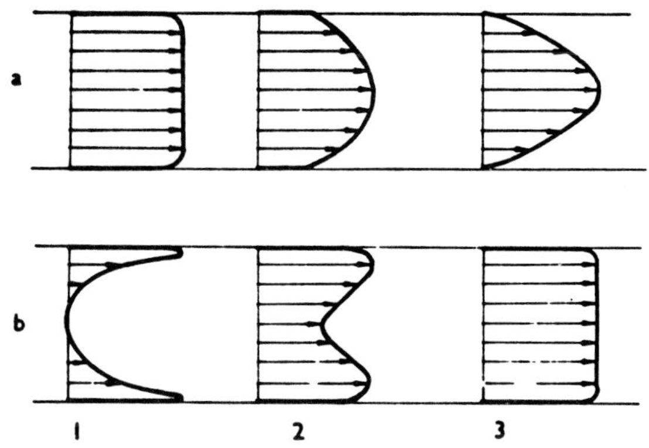

Fig. 3: Parabolic (a) vs plug (b) flow. The organization of flow profiles from left to right represents a time sequence from initial to final flow profile under ideal conditions.

The absence of a gel matrix in this experiment eliminates any sieving effect and hence separation occurs according to net charge. The mobility of proteins in a free zone is described as

$$U = \frac{Ze}{6\pi\eta R} \tag{1}$$

where Z = number of units of electron charge; e = fundamental electrostatic unit of charge; η = viscosity; R = radius of molecule. The net mobility of species is then a net summation of their intrinsic mobility and the electroosmotic flow:

$$\vec{M}_{total} = \vec{M}_{EP} + \vec{M}_{EO} \tag{2}$$

Terms which will contribute to band broadening in HPZCE

The equation used for the purpose of this analysis for HETP as a function of flow rate was adapted after Giddings [19] and is described as:

$$H_{ADS} = \underbrace{\frac{l^2}{12L}}_{\substack{\text{injection}\\ \text{plug}}} + \underbrace{\frac{2D_{soln,T}}{u}}_{\substack{\text{axial}\\ \text{diffusion}}} + \frac{k'\mu}{(1+k')^2}\left[\underbrace{\frac{r_0^2\mu}{4D_{soln,T}}}_{\substack{\text{resistance}\\ \text{to mass}\\ \text{transfer}}} + \underbrace{\frac{2}{k_d}}_{\substack{\text{adsorption-}\\ \text{desorption}\\ \text{kinetics}}}\right] \tag{3}$$

(mechanical: injection plug; physicochemical: remaining terms)

where l = length of the injected plug, u = flow rate, k' = capacity factor, L = length of the column, r_0 = radius of capillary, $D_{soln,T}$ = diffusion coefficient in designated solvent at a specific temperature (for this study water at 20°C), k_d = first order dissociation constant.

While most of these terms require little comment it is worth expanding the explanation of the capacity factor, k'. The simplest definition of k' is as a measure of retention:

$$k' = \frac{t_r - t_0}{t_0} \tag{4}$$

where t_r = elution time of a retained species and t_0 = elution time of an unretained species. While this is appealing from a pragmatic experimental approach, at a more physically significant level, it can be shown that given a thermodynamic distribution constant, K_D, describing the equilibrium of a species between two phases that k' is related to this equilibrium constant through a term known as the phase ratio [20]:

(1) $K_D = M_S/M_M$

(2) $-\Delta\mu_0 = -RT \ln K_D$

(3) $-K_D = \beta k'$

(4) then $-\Delta\mu_0 = RT\beta k'$

given M_M = concentration solute, mobile

M_S = concentration solute, stat. phase

V_m = volume of a mobile phase (5)

V_S = volume of a stationary phase

then $\beta = V_m/V_s$

It can be seen by inspection of Eq. 3 that as k' increases, the radial terms increase. The two discrete terms comprising the radial term consist of:

(1) The uniform flow profile term which can be thought of as a relaxation or broadening of the solute band based on radial transit time of a solute.

(2) The adsorption desorption kinetic term which is inversely proportional to the first order dissociation constant. It is intuitive that as the half-life for desorption increases (k_d decreases) that band broadening would occur in a dynamic separation system.

It is the goal of the experimentalist to control separation conditions to minimize HETP, since this is maximizing the efficiency as can be seen through the relationship:

$$N = \frac{L}{H} \tag{6}$$

where L = length of column, N = theoretical plates.

Previously, it was demonstrated by curve fitting to Eq. 3 that when k' is very small, HETP becomes small and insensitive to flow rate [21] (see Figure 4).

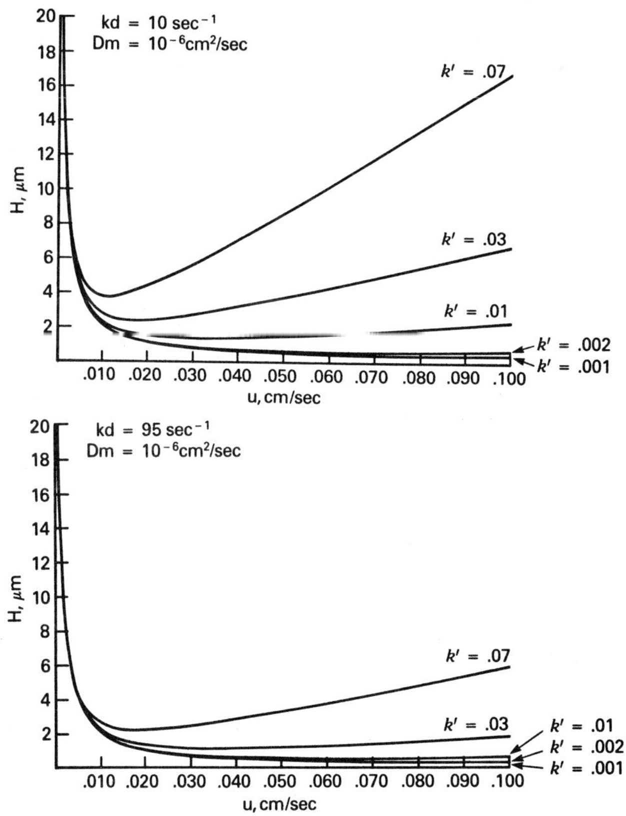

Fig. 4: Curves generated from the HETP equation (3) vs flow rate for various values of kd, Dm and k'.

Originally, contributions of protein interactions at the capillary surface were ignored when high efficiencies were generated in consideration of axial diffusion alone. Experience has demonstrated that significant interactions of proteins at the capillary surface can result, even after surface deactivation of fused silica capillaries with silylating reagents. These interactions may severely degrade the high efficiencies desired for capillary electrophoresis.

Breakdown of the Contribution to HETP

In Eq. 3, all terms are known or easily determined except k' and k_d. Independent measurement of either would result in the determination of the other through computation. For our purposes, a carefully controlled experiment in which k' was determined was done. An apparatus previously described [3] and quite similar to Figure 1 was used except the capillary and buffer reservoirs were jacketed and maintained at 20°C throughout the experiment. Flow was created with a vacuum pump which maintained a constant pressure difference between reservoirs and hence constant flow rate. With this information, the impact of each term could be inspected for each protein studied.

Lysozyme (hen egg white), myoglobin (horse heart) and ribonuclease (bovine pancreatic) were chosen since under defined electrophoresis conditions, their behavior is quite different. Table 1 summarizes important information concerning these proteins.

Protein	MW (KDaltons)	pI	$D_{w,20}$ (cm^2/sec)	(k' chrom.)	k_d(Cal'c) (sec^{-1})	Efficiency N(electrophoresis)
Lysozyme	13.8	10.5–11.0	10.6×10^{-7}	0.071	94	206K
Ribonuclease	13.6	9.6	10.7×10^{-7}	0.013	17	505K
Myoglobin	16.6	7.5	11.3×10^{-7}	0.012	10	373K

Table 1: Summary of pertinent parameters for the three proteins studied.

Table 2 shows a breakdown of the contribution to the overall HETP term by the individual terms.

	Injection Plug		Axial		Radial (Relaxation)		Radial (Adsorp.-Desorp. Kinetics)		Total*	
Solute	E.**	C.†	E.	C.	E.	C.	E.	C.	E.	C.
Lysozyme ($k' = 0.071$)	0.21 (6%)	1.03 (0.9%)	0.37 (10%)	0.21 (0.1%)	2.35 (64%)	114 (99%)	0.75 (20%)	—	3.68	115.24
Ribonuclease ($k' = 0.013$)	0.21 (14%)	1.03 (0.9%)	0.38 (25%)	0.22 (0.1%)	0.09 (6%)	112 (99%)	0.84 (55%)	—	1.52	113.25
Horse Myoglobin ($k' = 0.012$)	0.21 (10%)	1.03 (0.9%)	0.42 (21%)	0.24 (0.2%)	0.07 (3%)	112 (99%)	1.34 (66%)	—	2.04	113.27
KNO_3 ($k' = 0.0$)	0.21 (4%)	1.03 (9%)	5.26 (96%)	3.83 (34%)	—	6.23 (56%)	—	—	5.47	11.09

Table 2: Summary of contribution of terms (see eq. 3) for the three proteins studied vs a small solute.

Several aspects of this table are noteworthy:

(1) Comparing the total HETP term for electrophoresis vs chromatography shows dramatically a near two order of magnitude improvement of efficiency for macromolecules.

(2) For electrophoresis, comparison of the total HETP term for macromolecules vs small solutes (here potassium nitrate) shows better efficiency for even the worse case protein vs the small solute. In comparing the axial term contribution for the different size molecules, clearly the impact of the difference in the diffusion constants can be seen for this term.

(3) While the impact of the flow profile (i.e., plug vs parabolic for electrophoresis vs chromatography) can be seen for both large and small solutes, the contribution of the relaxation term for macromolecules in the chromatography experiment is evident, since this term is inversely proportional to the diffusion coefficient.

(4) In general, the major contribution to band broadening during electrophoresis for the proteins is from the adsorption-desorption term, as long as k' is small. As k' becomes greater, the relaxation of the solute profile becomes a more dominant term (see lysozyme vs ribonuclease and myoglobin).

(5) When the k' values are sufficiently low, the adsorption-desorption kinetics term may have a dramatic impact on efficiency. Myoglobin with a greater τ (sec) than a ribonuclease has a lower efficiency due to this term.

The impact of the contribution of terms can be readily demonstrated by comparing electropherograms to chromatograms (see Fig. 5).

In the electropherograms, the peak with the greatest width at half height has the lowest efficiency. This is a small solute, dimethylsulfoxide (DMSO), used as a neutral marker in electrophoresis. In the chromatograms, the macromolecule (horse heart myoglobin) is a distinctly broader peak than the small solute (potassium nitrate) peak. In electrophoresis, where the k' values are low (i.e., minimal chromatography at the capillary wall) and the

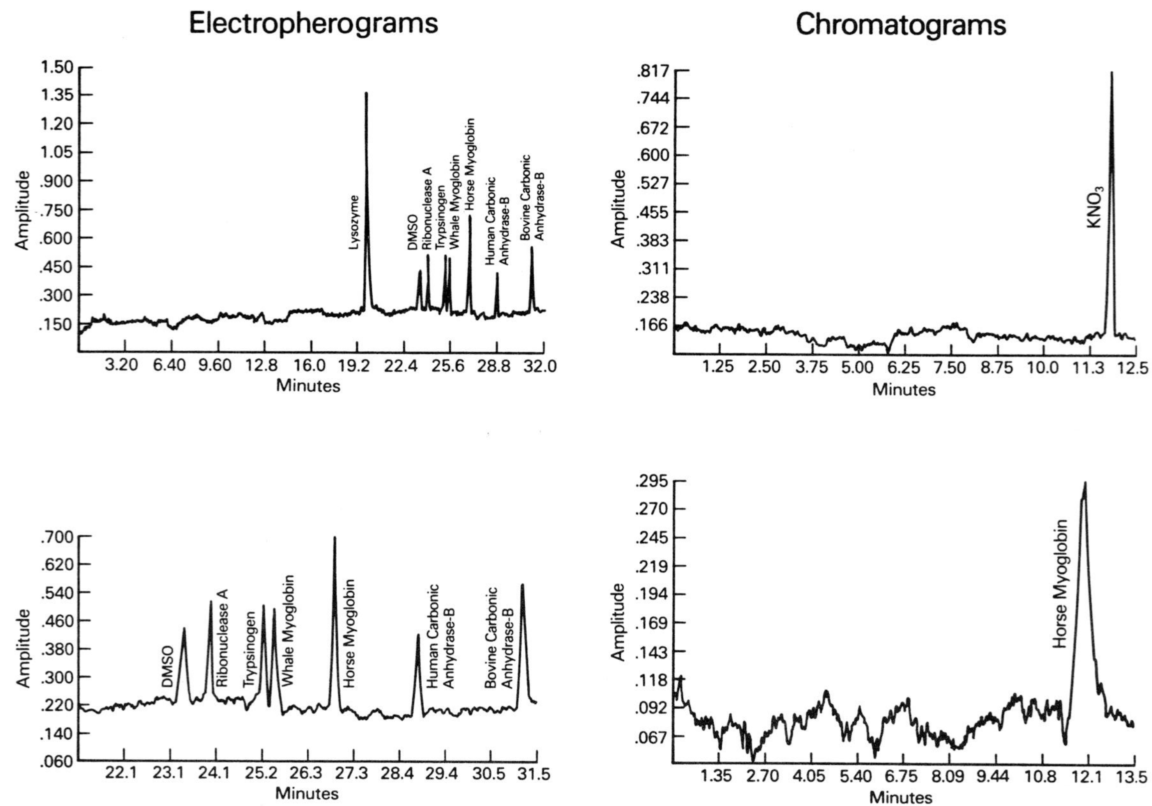

Fig. 5: The visual impact of contributions of terms for macromolecules vs small solutes can be seen by the inspection of electropherograms (left) vs chromatograms (right).

diffusion coefficients are small compared to that of the small solute, the greatest contribution for macromolecules is the adsorption-desorption kinetics term. In chromatography, the relaxation term becomes very dominant and efficiency limiting for macromolecules. Figure 6 demonstrates the HETP vs flow rate relationship for each protein studied.

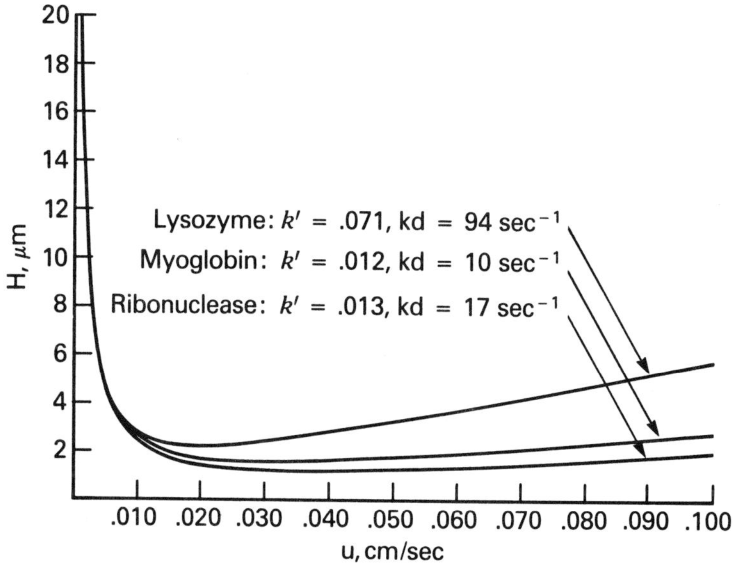

Fig 6: Curves for HETP vs flow rates generated from data for the three proteins studied.

III. SUMMARY

The contributions of the axial and radial diffusion terms to the overall HETP term for three proteins with efficiencies from 206k to 505k theoretical plates was compared.

Very high efficiencies are predicted for capillary electrophoresis in considering the axial diffusion term alone in a system with plug flow. In reality, there is little likelihood that no protein/surface interaction will occur due to the chemical heterogeneity and complexity of the protein molecule.

In this case, minimization of interaction (i.e., low k' values) may result in low HETP (i.e., high efficiencies) as depicted by Figure 4.

The impact of k' and first order dissociation constant on the radial diffusion terms was demonstrated by analysis of data for lysozyme, ribonuclease and myoglobin. For lysozyme with a k' significantly greater than that of the other two proteins studied, a significant increase in the relaxation term was seen to contribute to lower efficiencies. For myoglobin, with a significantly smaller dissociation constant than ribonuclease, the lower efficiency observed for this macromolecule may be attributed to the increase in the adsorption-desorption term.

IV. REFERENCES

1. Mikker, F. E. P.; Everaerts, F. M.; Verheggen, Th. P. E. M. *J. Chrom.* 1979, *169*, 11.
2. Jorgenson, J. W.; Lukacs, K. D. *Anal. Chyem.* 1981, *53*, 1298.
3. Lauer, H. H.; McManigill, D. *Anal. Chem.* 1986, *58*, 166.
4. Fujiwara, S.; Honda, S. *Anal. Chem.* 1986, *58*, 1811.
5. Tsuda, T.; Nomura, K.; Nakagawa, G. *J. Chrom.* 1983, *264*, 385.
6. Hjerten, S. *J. Chrom* 1985, *347*, 191.
7. Otsuka, K.; Terabe, S.; Ando, T. *J. Chrom.* 1985, *332*, 219.
8. Fujiwara, S.; Honda, S. *Anal. Chem.* 1987, *59*, 2773.
9. Cohen, A. S.; Terabe, S.; Smith, J. A.; Karger, B. L. *Anal. Chem.* 1987, *59*, 1021.
10. Cohen, A. S.; Karger, B. L. *J. Chrom.* 1987, *397*, 409.
11. Rose, D. J.; Jorgensen, J. W. *J. Chrom.* in press.
12. Wallingford, R. A.; Ewing, A. G. *Anal. Chem.* 1988, *60*, 258.
13. Gassmann, E.; Kuo, J. E.; Zare, R. N. *Science* 1985, *230*, 813.
14. Olivares, J. A.; Nguyen, N. T.; Yonker, C. R.; Smith, R. D. *Anal. Chem.* 1987, *59*, 1230.
15. Rose, D. J.; Jorgensen, J. W. *J. Chrom.* 1988, *438*, 23.
16. Rose, D. J.; Jorgenson, J. W. *Anal. Chem.* 1988, *60*, 642.
17. Hunter, R. J. *Zeta Potential in Colloid Science*; Academic Press: New York, 1981.
18. Whitehead, R.; Rice, C. L. *J. Phys. Chem.* 1965, *69*, 4017.
19. Giddings, J. C. *Dynamics of Chromatography*; Marcel Dekker: New York, 1965.
20. Karger, B. L.; Snyder, L. R.; Horvath, C. *An Introduction to Separation Science*; John Wiley & Sons: New York, 1973.
21. Lauer, H. H.; McManigill, D.; Sixth International Symposium on HPLC of Proteins, Peptides and Polynucleotides; Baden-Baden, FRG; Oct. 22-26, 1986.

46

ANALYSIS OF PROTEINS IN BRONCHOALVEOLAR LAVAGE FLUIDS DURING PULMONARY EDEMA RESULTING FROM NITROGEN DIOXIDE AND CADMIUM EXPOSURE[1]

L.R. Gurley, J.E. London, L.A. Dethloff, D.M. Stavert, and B.E. Lehnert

Life Sciences Division, Los Alamos National Laboratory
Los Alamos, New Mexico 87545

I. INTRODUCTION

An extracellular lining fluid (ELF) covers the epithelium of the alveoli and conducting airways of the lung. This ELF contains a complex mixture of lipids and proteins, inter alia (1). While extensive work has focused on the phospholipid components in ELF, the protein constituents are much less characterized in terms of their origin, biochemical characteristics, and their significance in physiologic and pathophysiologic processes. In order to study these proteins, our laboratory has developed a high-performance liquid chromatography (HPLC) method for fractionating ELF proteins (2,3), that are obtained by a relatively non-invasive bronchoalveolar lavage procedure (4). In this report, we demonstrate applications of the HPLC method for detecting and quantifying protein alterations in ELF that occur during acute lung injury in which permeability pulmonary edema is a prominent pathologic component.

II. QUALITATIVE ANALYSIS OF THE HPLC FRACTIONS

Since lung lavage is a method by which ELF can be sampled, we based our HPLC method on the adsorption of proteins from bronchoalveolar lavage fluid (BALF) (5) onto a C18 reversed phase chromatography column. Six fractions were eluted from the column using a series of gradients and isocratic steps that progress from H_2O containing 0.2% trifluoroacetic acid (TFA) to 65% acetonitrile/0.2% TFA. Following this, five additional fractions were eluted with methanol (Fig. 1). Constituents in these fractions can be recovered by lyophilization, since all of the HPLC solvents are volatile (6). All eleven fractions can be detected by monitoring the column effluent at 206 nm (Fig. 1A), but only nine can be detected at 280 nm. Details of the development of this procedure have been documented elsewhere (2,3).

---

[1] This work was supported by the Respiratory Branch of the U.S. Army Medical Research and Development Command, Project No. 85pp5855, and was conducted under the auspices of the U.S. Department of Energy.

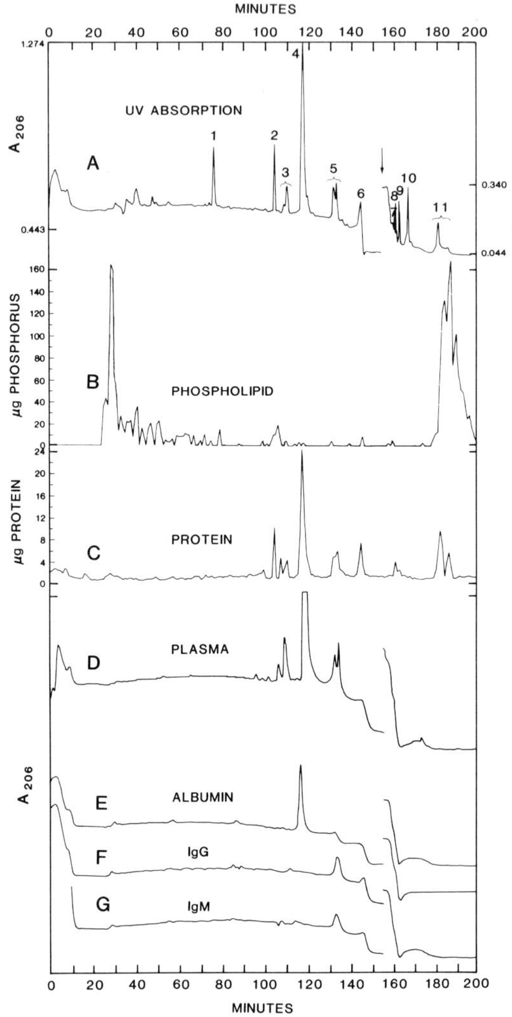

Figure 1. Fractionation of BALF by HPLC. Rat lungs were lavaged with 6 sequential washes of 8 ml phosphate buffered saline recovering 97% of the volume. The lung free cells were removed from the lavage fluid by centrifugation at 300 g and any remaining acellular solid material was removed by centrifugation at 2300 g for 10 min. Samples were stored at -70°C. Prior to HPLC, samples were made 0.2% with respect to TFA, which normally clarified any cloudiness. These samples were then loaded directly on a reversed-phase µBondapak C18 Radial-PAK cartridge without further treatment. HPLC elution was accomplished as described previously. The break in the chromatogram at the arrow results from a recorder baseline shift made at 155 min in order to keep the chromatogram on scale during the methanol elution step. (A) UV absorption profile of rat BALF monitored at 206 nm. (B) Phospholipid profile of rat BALF; (C) protein profile of rat BALF. (D) UV absorption profile of rat plasma. (E) UV absorption profile of rat albumin. (F) UV absorption profile of rabbit polyclonal IgG. (G) UV absorption profile of rabbit polyclonal IgM.

Phospholipid analysis of the HPLC effluent was accomplished by collecting samples in a fraction collector and extracting them with chloroform-methanol (7) as described by the Fiske-Subbarow procedure (8) and modified by Shin (9) (Fig. 1B). Protein analysis was performed in similarly collected samples by the Folin-Lowry procedure (10) (Fig. 1C). From these analyses, we found that phospholipids eluted before fraction 1 and also in the trailing shoulder of fraction 11. Therefore, by monitoring the HPLC at 206 nm, we are not measuring phospholipids. Proteins eluted in fractions 2-9 and 11. No protein was detected in fractions 1 and 10. These experiments indicated that ELF protein fractionation and analysis can be obtained from whole BALF without the need for intervening steps to remove phospholipids.

The proteins of ELF are derived from two sources with the larger quantity coming from the blood and the lesser amount from the cells of the lung itself. To determine which HPLC fractions might be blood proteins, comparisons were made between the elution profiles of rat lung lavage fluid, rat plasma, and certain protein standards. It was found that plasma proteins eluted only in HPLC fractions 3, 4 and 5 (Fig. 1D). Albumin eluted only in fraction 4 (Fig. 1E). Immunoglobulins G (Fig. 1F) and M (Fig. 1G) eluted primarily in fraction 5 with a small amount also occurring in fraction 6. The identification of fraction 4 as albumin was confirmed by treating BALF with anti-rat albumin IgG conjugated to Sepharose. Fraction 4 was essentially all removed from the BALF by this treatment. These results suggested that while fractions 3, 4 and 5 were derived from the blood compartment, fractions 1, 2, and 6-11 probably were derived from the lung cells themselves. Therefore, this fractionation procedure should be useful for distinguishing between injurious effects on the lung's cells and the resulting changes in the permeability status of the air-blood barrier resulting from environmental insults.

III. QUANTITATIVE ANALYSIS OF THE HPLC REACTIONS

For this HPLC method to be useful for investigating protein changes during lung injury, it is most important that the quantitative response of each fraction be linear with respect to the amount of lavage loaded on the column. In Fig. 2, we demonstrate that the total amount of material eluted is proportional to the volume of BALF loaded on the column (Fig. 2A) and that the relationship between sample load and quantity of each fraction is linear (Fig. 2B). The regression coefficients (r) are >0.85 for all fractions, indicating this method can be used to detect quantitative changes in BALF components over a wide range of concentrations.

IV. CADMIUM MODEL OF PULMONARY EDEMA

The heavy metal, cadmium (Cd), can cause an acute inflammatory response in the lung (11) and bring about changes in the lung's

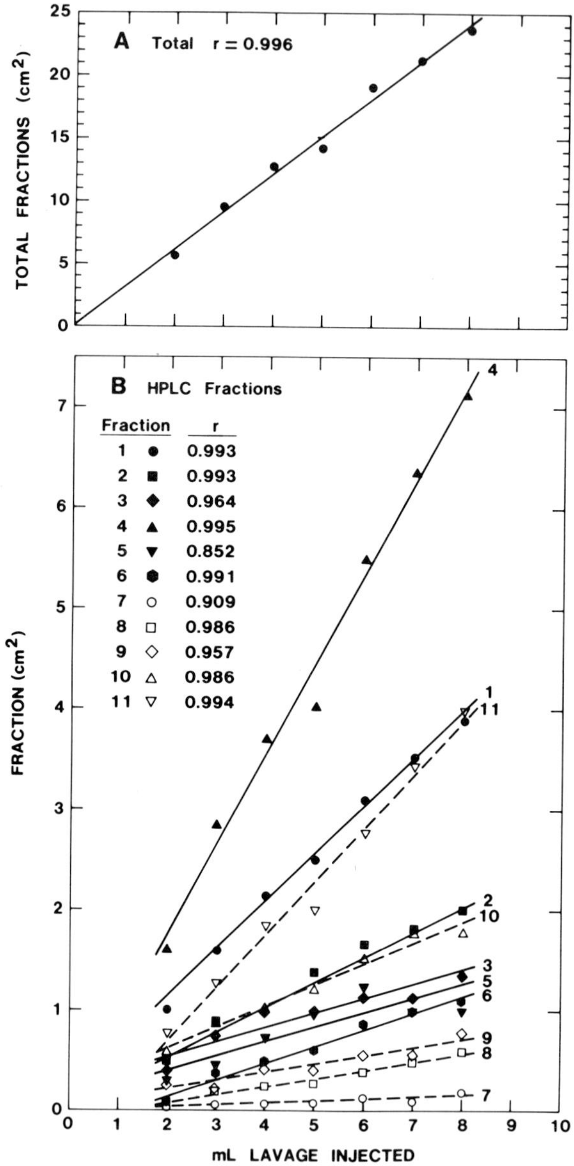

Figure 2. Linear relationship between sample load and quantity of each fraction recovered. Volumes of rat lung lavage ranging from 2 ml to 8 ml were made 0.2% with respect to TFA and pumped through the column at 1 ml/min. Following elution, each fraction was quantified by measuring the area of the eluted peaks detected by UV absorption at 206 nm. (A) The total amount of all eleven fractions recovered from the column. (B) The amount of each of the eleven fractions recovered from the column. The correlation coefficients (r) between volume loads and the areas of the peaks are shown on the figure. All "r" values are significant, $p < 0.05$.

air-blood barrier that result in permeability pulmonary edema (12). During such injury, an abnormal abundance of blood proteins in BALF would be expected due to protein translocations from the lung's vasculature into the air space. In order to confirm that the HPLC fractions 3, 4, and 5 represent protein constituents common to blood, rats were intratracheally instilled with 7 μg of cadmium as $CdCl_2$ and lavaged (5) 24 h later. The BALF was subjected to HPLC and the fractions quantified (Fig. 3). Fractions 3, 4, and 5 comparably increased 25-, 27-, and 21-fold, respectively. This rapid and marked increase in these three fractions pointed to the blood compartment as their primary source, as was suggested by the experiments with standards in Fig. 1D).

Fractions 7 and 8 increased 8-fold (Fig. 3), making it impossible to resolve the two (Fig. 4C). Fractions 1, 6, 9, 10, and 11 increased about 2-fold (Fig. 3). The much smaller increases in these fractions (compared to the large increases in fractions 3, 4, and 5) suggest they are derived from an extra-vascular source, such as the lung tissue cells and/or lung matrix constituents. Concurrent with these general increases in BALF consitutents, fraction 2 was decreased by Cd treatment. Additionally, two new fractions (A and B) were observed in BALF during the inflammatory response (Fig. 3 and 4C). The source of A and B may be blood; there were low levels of fractions eluting similarly to A and B in the plasma standard (Fig. 1D). However, this remains to be confirmed.

V. SOLUBILIZATION OF BALF

The treatment of lungs with agents that cause an inflammatory response can result in lavage fluid that is cloudy. This cloudiness may be caused by the proteinaceous fluid and fibrin commonly observed histologically in the alveolar space following acute lung injury, like that produced by Cd. This proteinaceous material is no longer visible in the lung following lavage, indicating it has been successfully transferred to the BALF. The addition of TFA (which is a protein solubilizing agent in the HPLC system) to the lavage fluid clarified the BALF from Cd treated rats. Therefore, there were no particulate problems with insoluble samples in the HPLC analyses in Fig. 3. However, TFA alone was not sufficient to clarify the BALF from rats exposed to the oxidant gas, nitrogen dioxide (to be discussed). From previous experience, we knew that the use of 6 M guanidine hydrochloride to dissolve insoluble protein complex samples did not interfere with reversed-phase HPLC systems (13). Similarly, we found that BALF from NO_2-treated rats could be clarified by making the sample 6 M in guanidine·HCl.

To determine whether guanidine solubilization would cause any changes in HPLC profiles, comparisons were made between the profiles of untreated BALF and guanidine-treated BALF from control rats and similar BALF samples from rats exposed to Cd (Fig. 4). We found that addition of guanidine to BALF did not quantitatively change the HPLC profile in control rats (Fig. 4A and 4B), nor did it change the HPLC profile of BALF from Cd injured lungs (Fig. 4C). Thus, it was concluded that the clarification of BALF samples with 6 M guanidine·HCl could be used to solubilize constituents lavaged from extensively injured lungs in which alveolar flooding due to permeability pulmonary edema is substantial.

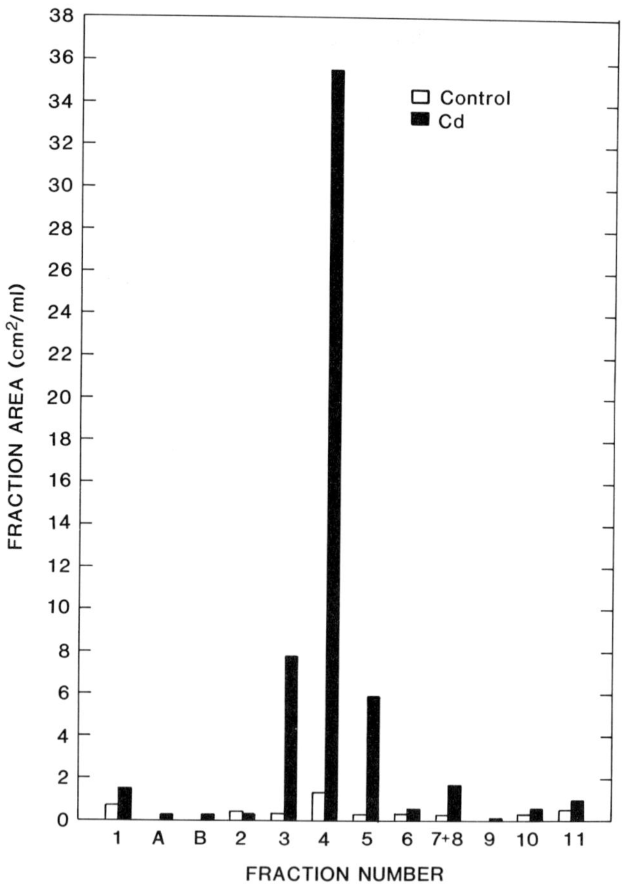

Figure 3. Quantification of the HPLC profile of rat BALF following instillation of $CdCl_2$. Each fraction was quantified by measuring the area under the peak and normalized to the volume of lavage loaded on the column. Due to the large differences in the quantities of the various fractions in cadmium treated rats, HPLC peaks 3, 4, and 5 were measured from small lavage loads (0.2 ml) while peaks 1, 2, 6-11, and A and B were measured from larger lavage loads (1.0 ml). Measurements of peaks from untreated control animals were made on 5 ml lavage loads. Each fraction represents the mean value of four animals.

VI. NITROGEN DIOXIDE PRODUCED PULMONARY EDEMA

NO_2 is a potential health hazard in occupational settings where thermal combustion processes occur. For this reason, our laboratory has pursued a program that explores the physiological and biochemical results of lung exposure to this agent. In the course of these

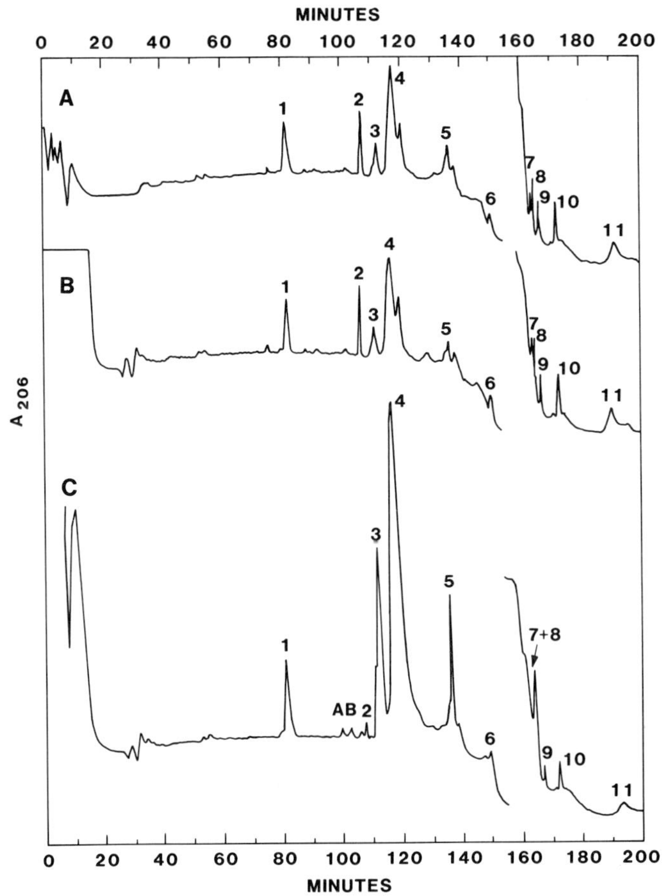

Figure 4. HPLC of BALF solubilized with guanidine. BALF samples were prepared for HPLC with 0.2% TFA as described in Fig. 1. Solid guanidine hydrochloride was added to these samples to a concentration of 6 M. These clarified samples were diluted 1:1 with PBS containing 0.2% TFA thus reducing the guanidine concentration to 3 M. The samples were then loaded on the column. The guanidine was eluted in the void volume. The break in the chromatogram at 155 min results from a recorder baseline shift that was made to keep the chromatogram on scale during the methanol elution step. (A) BALF not treated with guanidine from control rats. (B) BALF treated with guanidine from control rats. (C) BALF treated with guanidine from rats with cadmium-induced lung injury.

experiments, we have described increases in lung water and histopathologic changes; i.e., fibrin accumulation in peri-terminal bronchiolar-alveolar region, following NO_2 exposure (14,15) that are hallmarks of pulmonary edema. To determine the effects of NO_2-induced injury on the protein constituents in BALF, we quantified the HPLC fractions in BALF taken from rats 24 h after exposure to 100 ppm NO_2 for 15 min. BALF was clarified by adding guanidine·HCl to a 6 M concentration. The clarified samples were then diluted to 3 M guanidine with PBS/0.2% TFA and injected into the HPLC (Fig. 5A). As expected, NO_2 exposure produced a major translocation of blood proteins to the lung's bronchoalveolar compartment. The amounts of these proteins (fractions 3, 4, and 5) were too high to quantify when normal control-level loads were injected (Fig. 5A). Therefore, it was necessary to inject very small loads in order to measure the blood proteins (Fig. 5B).

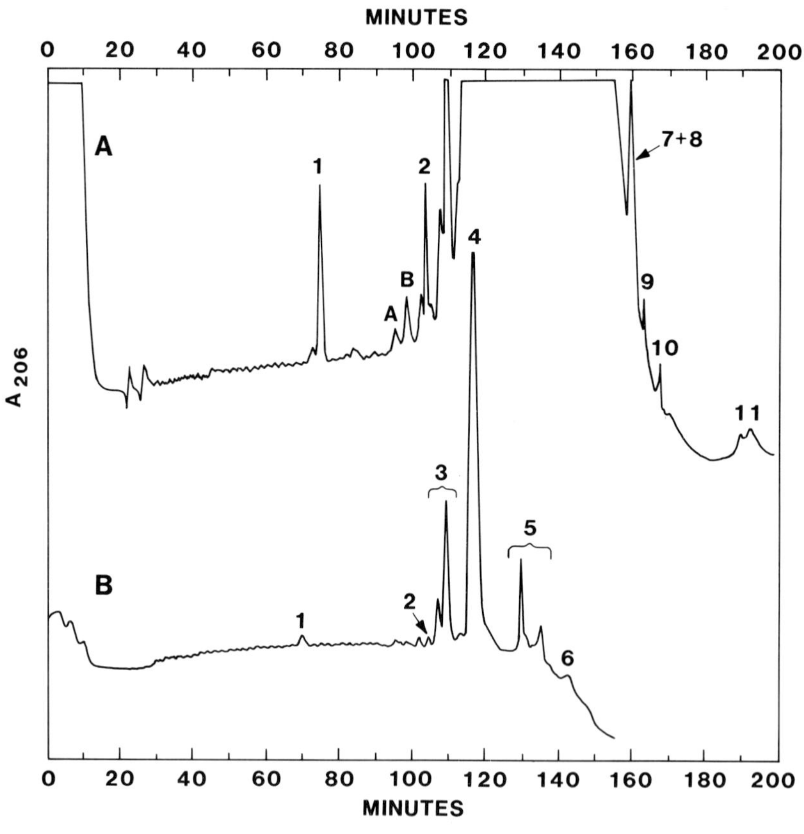

Figure 5. HPLC of BALF from rats treated with NO_2. Rats were exposed to 100 ppm NO_2 for 15 min and lavaged 24 h later. (A) HPLC was run on normal loads of lavage fluid (3.5 ml) that were clarified by adding 0.2% TFA and 6 M guanidine·HCl and then diluted 1:1 with PBS/0.2% TFA. (B) HPLC was run on small loads of lavage fluid (0.2 ml) that were clarified by diluting the fluid 1:35 with PBS and adding 0.2% TFA.

This quantification indicated that the blood-derived proteins, fractions 3, 4, and 5 increased 72-fold, 58-fold, and 35-fold, respectively (Fig. 6). Fractions 7 and 8 increased 8-fold, (as it did following Cd exposure). Fraction 2 increased 6-fold (unlike its loss following Cd exposure). Fractions 2, 10, and 11 increased ~1.5-fold, but fraction 9 decreased to 35% of control (unlike its increase following Cd exposure). The fractions A and B were also observed following NO_2 exposure, as they were in the Cd-treated lungs, except the amount was greater after NO_2 exposure. This again suggests fractions A and B may be translocated from the lung's vascular compartment.

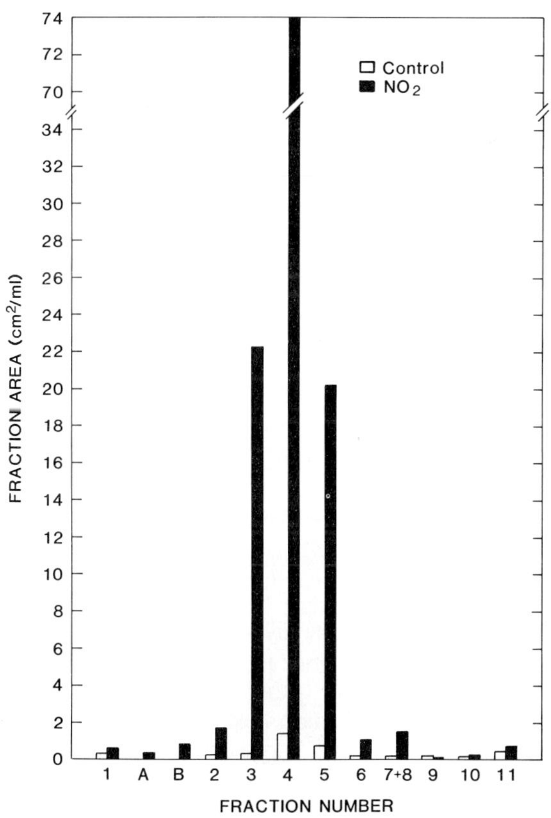

Figure 6. Quantification of the HPLC profile of rat BALF following exposure to NO_2. Each fraction was quantified as described in Fig. 3. Peaks 3, 4, 5, and 6 were measured from small loads (0.2 ml) and peaks 1, 2, 7-11, and A and B were measured from larger loads (3.5 ml). Measurements of the peaks from untreated control animals were made on 4 ml lavage loads. Each fraction represents the mean value of three animals for the NO_2 exposed rats and four animals for the control rats.

While most of the components of BALF responded to NO_2 exposure as they did in the Cd model, there was one outstanding difference between the two models of injury that may be of importance. Whereas the relative increases in the three major blood fractions were virtually identical in the Cd model, the relative increase in fraction 3 was approximately two times greater than the increases in the other two major blood fractions in the NO_2 model. Such findings suggest that the underlying permeability disturbance(s) involved in NO_2-induced pulmonary edema fundamentally differs from that produced by Cd. Further interpretations regarding this difference obviously require the future identification of fraction 3. The comparison of these two models illustrate, nevertheless, that the employment of HPLC to analyze protein constituents in BALF will offer significant utility in elucidating mechanisms at play in a variety of lung disorders in addition to pulmonary edema.

VII. SUMMARY

We have developed a new HPLC method by which quantitative measurements can be made on the biochemical constituents of the extracellular fluid lining of the lung as sampled by bronchoalveolar lavage. Nine of the fractions are proteins, two are phospholipids, and two fractions remained unidentified. Rats were subjected to the intrapulmonary deposition of cadmium, a treatment model known to induce pulmonary edema and cause a translocation of blood compartment proteins into the lung's alveolar space compartment. Resulting pulmonary edema was hallmarked by ~25-fold increases in three major blood compartment-derived HPLC protein fractions, two of which have been identified as albumin and immunoglobulin(s). Analysis of lavage fluid from rats exposed to 100 ppm NO_2 for 15 min, an exposure regimen which also produces pulmonary edema, indicated that the three blood compartment proteins in the lavage fluids were elevated 35- to 72-fold over controls 24 h after exposure.

These results demonstrate that HPLC can be used as a highly sensitive method for quantifying the translocation of blood compartment constituents into the lung's airspaces, as observed in permeability pulmonary edema. Qualitative and quantitative differences in the profiles of the various blood proteins detected by the HPLC method following lung injury may serve to provide fundamental information on the underlying nature of the injury; e.g., changes in the relative conductances of the different proteins and changes in the "effective" air-blood barrier pore sizes. The use of this HPLC method for the sensitive detection and quantification of lavage fluid protein components that are not of blood origin may ultimately yield a relatively noninvasive assessment of the status of the lung's cellular populations and/or extracellular matrix during pulmonary disorders.

VIII. REFERENCES

1. G. George and E.R. Hook (1984) Environ. Health Perspect. 55:227-237.

2. L.R. Gurley, J.G. Valdez, J.E. London, W.D. Spall, L.A. Dethloff, B.E. Lehnert (1988) Los Alamos National Laboratory Report LA-11309-MS, National Technical Information Service, Springfield, VA, pp. 1-36.

3. L.R. Gurely, W.D. Spall, J.G. Valdez, J.E. London, L.A. Dethloff, and B.E. Lehnert (1988) Analyt. Biochem. 172: 465-478.

4. R.G. Crystal, H.Y. Reynolds, A.R. Kalica (1986) Chest 84:122-131.

5. B.E. Lehnert, L.A. Dethloff, and Y.E. Valdez (1988) J. Toxicol. Environ. Health 24:41-66.

6. L.R. Gurley, D.A. Prentice, J.G. Valdez, and W.D. Spall (1983) J. Chromatography 266:609-627.

7. J. Folch, M. Lees, and G.H. Sloane-Stanley (1957) J. Biol. Chem. 226:497-509.

8. C.H. Fiske and Y. Subbarow (1925) J. Biol. Chem. 66:375-400.

9. Y.S. Shin, Anal. Chem. (1962) 34:1164-1166.

10. O.H. Lowry, N.J. Rosabrough, A.L. Farr, and R.J. Randall (1951) J. Biol. Chem. 193:265-275.

11. B.E. Lehnert, Y.E. Valdez, J.E. London, and D.M. Smith (1985) The Toxicologist 5:713.

12. J.S. Bus, A. Vinegar, and S.M. Brooks (1978) Amer. Rev. Respir. Dis. 118:573-580.

13. P.S. Jackson and L.R. Gurley (1985) J. Chromatography 326:199-216.

14. D.M. Stavert and B.E. Lehnert (1988) The Toxicologist 8:556.

15. D.M. Stavert and B.E. Lehnert (1988) Environ. Res. in press.

47

A NEW ASSAY FOR ENTEROKINASE

Hanna Janska[1]
Albert Light[2]

Department of Chemistry
Purdue University
West Lafayette, Indiana, U.S.A.

I. INTRODUCTION

Enterokinase, also called enteropeptidase, is a proteolytic enzyme that is the physiological activator of trypsinogen (1-4). Enterokinase is located in the brush border membrane of the duodenal section of the small intestine, and it converts pancreatic trypsinogen, entering the intestine via the pancreatic duct, into the enzyme trypsin (5). Trypsin is the activator of chymotrypsinogen, proelastase, and other zymogens. Since enterokinase initiates this cascade process, the enzyme is essential for normal digestion.

Enterokinase was partially purified by Kunitz in the 1930's, and he established the physiological function of the enzyme (1). Kunitz developed a two-step procedure for the determination of the enzymatic activity. In the first step, a catalytic amount of enterokinase was used to activate trypsinogen (0.1 mg per ml). The pH of the activation was 5.0, which is a pH where autoactivation by trypsin was negligible. The trypsin activity was then determined by any one of several methods. In this article, the Kunitz method of assay will be referred to as the Classical Assay procedure.

---

[1]Visiting scientist from Institute of Biochemistry, University of Wroclaw, Wroclaw, Poland

[2]Research supported in part by Army Research Office and the Immunex Corp., Seattle Washington

Although the Classical Assay is highly specific and utilizes precise trypsin assays, the method has the following limitations. The activation is at a pH much lower than the optimal pH of 8 to 9. The substrate concentration is below the K_m value, which severely limits the linear range of velocity-enzyme concentrations (see below). Finally, the trypsin formed in the activation step slowly loses activity with time, and the activity may also be inhibited by trypsin inhibitors found in cells and the pancreatic juice.

We developed a direct assay for enterokinase utilizing the strategy outlined in Fig. 1. Trypsinogen is activated to trypsin at the optimum pH of enterokinase, in the presence of an excess of chicken ovomucoid, a known trypsin inhibitor (6). As trypsin is produced in the activation process, it immediately forms a stable inactive complex with ovomucoid. Therefore, trypsin, which is highly active at pH 8, cannot cause autoactivation of trypsinogen. The molecular weight of trypsin, (24,000 daltons), is increased two-fold on forming the ovomucoid-trypsin complex (52,000 daltons). Therefore, we use size-exclusion chromatography in an HPLC system to separate the two components. Quantitative analysis of the rate of disappearance of the substrate trypsinogen and the rate of formation of the product ovomucoid-trypsin complex is easily performed by coupling size-exclusion HPLC with a computer-assisted analysis of the data.

ZYMOGEN ACTIVATION

$$\text{Trypsinogen} \xrightarrow{\text{EK}} \text{Trypsin} + \text{Hexapeptide}$$

$$\text{Val.Asp.Asp.Asp.Asp.Lys.Ile.Val...} \longrightarrow \text{Ile.Val...} + \text{Val(Asp)}_4\text{Lys}$$

Formation of Stable Trypsin-Ovomucoid Complex
and Molecular Weight Changes

$$\underset{\text{Mr } 24{,}000}{\text{Trypsin}} + \underset{28{,}000}{\text{Ovomucoid}} \xrightarrow{\text{pH 8}} \underset{52{,}000}{\text{Trypsin-Ovomucoid Complex}}$$

Figure 1. Activation of trypsinogen and the formation of a high molecular weight trypsin-ovomucoid complex.

II. EXPERIMENTAL

Size-exclusion chromatography employed a Varian model 5020 HPLC system and a Toya Soda G2000SW column (0.75 x 50 cm) with the output recorded at 220 nm using an external Gilson model HM Holochrome detector. The wavelength 220 nm has a high extinction coefficient for proteins compared with measurements at 280 nm. It measures the absorbancy of peptide bonds, and the response was related to the total amount and not the composition of the protein molecules. The solvent system was 0.1M Tris, pH 7, containing 0.3M sodium chloride and 0.02M calcium chloride. The retention time of trypsinogen was 17.7 min with a flow rate of 1 ml per min and that of the trypsin-chicken ovomucoid complex was 14.1 min. The two components were easily separated and the areas of the peaks estimated from computer integration of the absorbancy values. The chromatographic separations were analyzed with an IBM 9000 computer and an IBM computer software program.

The enterokinase-catalyzed activation of trypsinogen was performed by incubation of bovine trypsinogen (1 mg per ml) and chicken ovomucoid (1 mg per ml) with an appropriate amount of enterokinase (see below), in 0.01M Tris buffer, pH 8, containing 0.01M $CaCl_2$, at 30^oC. Samples were removed as a function of time, or after a fixed time of 30 min., and the activation was quenched by lowering the pH to approximately 2.0. Enterokinase was irreversibly inactivated at this pH while trypsinogen, trypsin, and ovomucoid were stable. Samples were stored in the cold prior to analysis. The pH was then adjusted to pH 8.0 and after 10 min., injected onto the size-exclusion column.

The zero-time sample was used for calibration of the separation process and the area of the peak was used for calculating changes during the activation. Highly purified bovine enterokinase was prepared as described by Liepnieks and Light (7).

III. RESULTS

A. Activation and Size-exclusion HPLC

The activation of enterokinase is highly specific for trypsinogen, and the enzyme does not alter the structure of trypsin or ovomucoid. The reaction of trypsin and ovomucoid occurs at pH 8 with the active center of trypsin complexing with the reactive center of ovomucoid in a one to one ratio (8). As mentioned in Section II, the separation on a column of TSK G2000SW has 3 min. between the peak maxima, making it relatively easy to resolve the trypsin-ovomucoid complex and trypsinogen. Since ovomucoid is a glycoprotein, it has a greater Stokes radius and elutes earlier than other molecules of its molecular weight.

B. Rate Measurements

In a typical enzyme assay, one usually follows the disappearance of substrate or the appearance of product. The choice usually rests on the ease of analysis of one of the components. In the assay of enterokinase by size-exclusion chromatography, both the rate of disappearance of substrate (trypsinogen) and the rate of formation of product (ovomucoid-trypsin) are simultaneously measured (Fig. 2). This of course has the obvious advantage that the two rates must be equal and the quality of the data is immediately obvious. In the activation of trypsinogen with samples of enterokinase, changes in the amount of trypsin because of autolysis will be apparent since the rate of trypsinogen disappearance will then be greater than the rate of formation of the ovomucoid-trypsin complex. On the other hand, with crude samples of enterokinase (see later) contaminating proteins may appear at the substrate or product positions and influence their amounts. These possible problems will also be obvious from the changes of substrate and product concentrations.

Figure 2 shows that we found approximately equal rates for the changes in substrate and product concentration for up to 60 min. Clearly, the data fit the scheme presented in Fig. 1 and demonstrate the usefulness of the procedure. If low amounts of sample must be detected, an activation time of 1 hour or longer would give a sufficient change in peak areas to ensure a high precision of measurement. On the other hand, if high amounts are assayed, the activation time could be decreased to 10 or 5 min.

C. Linear Response with Increasing Enzyme Concentration

Figure 3 is the response after a 30 min. activation of trypsinogen as a function of the concentration of enterokinase. The size-exclusion HPLC analysis is linear over a much greater range of concentrations then the Classical

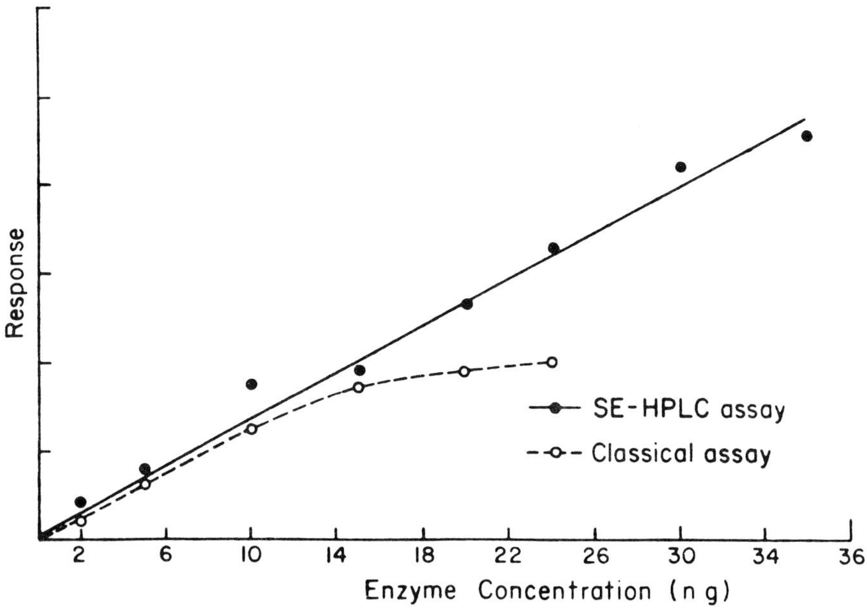

Figure 2. The rate of disappearance of substrate (trypsinogen) and the rate of formation of product (trypsin) shown as the ovomucoid-trypsin complex.

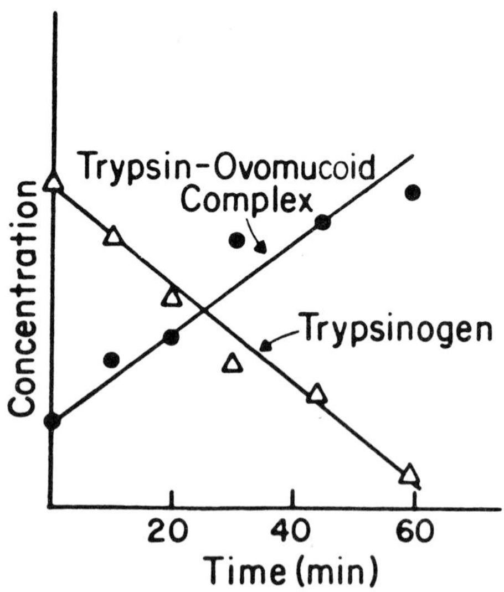

Figure 3. The relationship of the velocity of the activation process as a function of the concentration of enterokinase for the classical assay and the size exclusion HPLC analysis.

Assay. The data fit a straight line relationship in bothcases showing the high precision of the analysis. In a practical sense, the size-exclusion HPLC gives the experimenter a greater opportunity for a sample size to be within the linear region and therefore correct and reliable.

D. Enterokinase Activity at Different Levels of Purity

We were concerned that the separation of trypsinogen and the ovomucoid-trypsin complex on size-exclusion chromatography may be influenced by other protein molecules present in crude samples of the enzyme. A sample of enterokinase, purified by a) extraction of the brush border membrane with detergent, b) a pH 5 precipitation, and c) ammonium sulfate precipitation, was used in the activation of trypsinogen with the same number of enzyme units used with highly purified enzyme (6). The preparation was at a low level of purity and the bulk of the sample represented protein impurities. After 30 min. activation, size-exclusion chromatography showed an equal but opposite change in the amounts of trypsinogen and the ovomucoid-trypsin complex. The changes in the areas of the peaks were not at all influenced by protein contaminants. Indeed, the large amount of trypsinogen and ovomucoid present in the activation allowed us to use a less sensitive range in the integration of peak areas and helped to eliminate the presence of small peaks. Size-exclusion HPLC should be capable of detecting enterokinase activity in the luminal contents of the intestine and after fractionation of the epithelial cell contents.

IV. SUMMARY

The enzymatic activity of enterokinase is easily determined by size-exclusion HPLC. After activation of trypsinogen to trypsin and trapping the trypsin in a ovomucoid-trypsin complex, size-exclusion chromatography separates the product and substrate. Quantitative analysis of the areas of the separated peaks gives the rate of disappearance of substrate and the appearance of product. The rates are equal but of opposite sign. The assay is linear with enzyme concentration and has a greater range than with the Classical Enzyme Assay. Finally, both highly purified and crude preparations of the enzyme are easily determined rapidly and with high precision.

ACKNOWLEDGMENTS

We thank Dr. Jon Amy and David Henady of the instrument shop of the Chemistry Department for help and technical advice in the use of the computer system. Professor M. Laskowski and Dr. J. Schrode of Purdue University supplied samples of purified chicken ovomucoid. Urvashi Roongta gave valuable technical assistance. Ms. Judith Heine has been helpful in the preparation and the typing of the manuscript.

V. REFERENCES

1. Kunitz, M. (1939) J. Gen. Physiol. 22, 429-446.
2. Yamashima, I (1956) Acta Chem. Scan. 10, 739-743.
3. Maroux, S., Baratti, J., and Desnuelle, P. (1971) J. Biol. Chem. 246, 5031-5039.
4. Baratti, J., Maroux, S., Louvard, D., and Desnuelle, P. (1973) Biochim. Biophys. Acta 315, 147-161.
5. Stryer, L. (1988) Biochemistry, 3rd edn., p. 246, W. H. Freeman and Co., NY
6. Janska, H., and Light, A. (1988). Anal. Biochem., manuscript in press.
7. Liepnieks, J. J. and Light, A. (1979) J. Biol. Chem. 254, 1677-1683.
8. Hadorn, B., Tarlow, M. J., Lloyd, J. K., and Wolff, O. H. (1969) Lancet 1, 812.

48

MEETING THE CHALLENGE IN PEPTIDE FRAGMENT PURIFICATION FOR PROTEIN SEQUENCING

TIMOTHY D. SCHLABACH
JOEL C.COLBURN
ROBERT J. MATTALIANO
AND SYLVIA YUEN

APPLIED BIOSYSTEMS, INC.
FOSTER CITY, CA 94404

I. INTRODUCTION

The determination of primary structural data often requires that the protein of interest be fragmented into peptides suitable for sequence analysis. This information can then be used to directly or indirectly (DNA-cloning) construct the protein's amino acid sequence. Production of peptide fragments by enzymatic digestion often results in complex, peptide mixtures. Individual peptides must be separated from all other peptides to generate reliable sequence data. Reversed-phase chromatography (RPC) is the primary technique for peptide purification (1,2), and is widely used for separating tryptic peptides (3,4). However, complex digests with thirty or more species can overtax the resolving power of RPC. This results in multi-component peaks that contain two or more peptide species. Such peaks are not suitable for protein sequencing (5).

Improved purification of complex, protein digests was reported for a chromatographic system that utilized two dimensions (6). The entire protein digest was separated by RPC at a neutral pH and fractions were collected. Each individual fraction was further purified by RPC at an acidic pH. This technique utilizes changes in side change ionization that produce retention time shifts for peptides separated by RPC (7). An alternative system is to employ ion-exchange chromatography (IEC) in the first dimension, and follow with RPC in the second dimension (8). In general peptides that are strongly retained by IEC are weakly retained by RPC and vice versa. This complementarity enhances the resolving power of the two-dimensional, IEC-RPC system.

Peptide separations historically required an anion-exchange column for acidic peptides (9) and a cation-exchanger for basic peptides (10), because some peptides would have the same charge as the column packing and would not be retained. A newly developed, strong, cation-exchange column overcomes this deficiency by operating at an acidic pH where most peptides have a positive charge (11). Peptide retention at pH 3.0 is directly proportional to nominal peptide charge for this IEC column (12).

We report on the advantages of a two-dimensional chromatography system, combining RPC with IEC, for the purification of enzymatically cleaved peptides derived from a 44 kd protein, carboxypeptidase P. Our ultimate objective is to determine the primary structure of this protein. The focus of this study is to report on a strategy for the actual purification of the requisite peptide fragments. We found that a single dimension of RPC resulted in many heterogeneous peaks, as judged by automated Edman degradation. All fractions purified by two-dimensional chromatography, however, resulted in homogeneous peptides and unambiguous sequencing data.

During the course of this study, a need emerged for a rapid purity check to confirm the apparent chromatographic purity of a peak. On this basis the sample could be submitted for sequencing or further purification. Capillary electrophoresis was reported to rapidly analyze peptides and proteins (13). We found that heterogeneous peptide samples could be detected by capillary electrophoresis in less than thirty minutes.

II. METHODS

A. Digestion of Carboxypeptidase P. with ASP-N

A highly purified preparation of carboxypeptidase P. (CPP) was obtained commercially (Sigma, St. Louis, Mo). Prior to proteolytic cleaveage, the protein was subjected to performic acid oxidation (PAO) according to Hirs (14). PAO-CPP was amidated (15) at aspartic residues with 1-ethyl-3-(3-dimethylaminopropyl) carbodiimide (Pierce, Rockford, IL). and glycinamide (Sigma). The sample was digested for 18 hours at ambient temperature with 2% w/w ASP-N (Boehringer Mannheim, Indianapolis, IN). ASP-N cleaves at the N-terminal side of aspartic and cysteic acid (16).

B. Reversed-Phase Peptide Separations

The peptide digest (ca. 300 picomoles) was directly loaded onto a 220x2.1-mm RP-300 column (Applied Biosystems, Foster City, CA). The column was first equilibrated with solvent A, 0.1% trifluoroacetic acid (TFA). Peptides were eluted at flow of 150 uL/min with a four-step gradient: 0-15 min, 5-18% solvent B (B = 70% acetonitrile with 0.088% TFA); 15-25 min, 18-23% B; 25-60 min, 23-55% B; 60-70 min, 55-85% B, with a column temperature of 40 degrees. Ten microliter aliquots from the IEC-fractions were analyzed under the same conditions except that the first and second steps were 5-20% B and 20-30% B, respectively.

C. Cation-Exchange Peptide Separations

The peptide digest (500 picomoles) was loaded directly on a 200x2.1-mm Sulfoethyl-A column (PolyLC, Columbia, MD)., equilibrated with 10 mM potassium phosphate, pH 3.0 (buffer A). Peptides were eluted with a three-step gradient: 0-20 min, 5-20% buffer B (B= buffer A plus 0.25 M sodium chloride); 20-30 min, 20-25% B; 30-55 min, 25-55% B. The column temperature was 30 degrees.

D. Capillary Electrophoresis

Capillary electrophoresis was performed with a Model 270A electrophoresis systems (Applied Biosystems) using a 50-cm capillary with a 50 micron diameter. About 5 nL of an internal standard, dynorphin (YGGFLRRIRPKLK), was first aspirated into the capillary column with modest vacuum, 5 inches of mercury, applied for 2 seconds. This was followed by sample, ca. 5-nL, using the same injection technique. The capillary was automatically placed in the running buffer, 20 mM sodium citrate, pH 2.5, and the sample was electrophoresed at 20 kV for about 20 min with detection at 200 nm.

E. Protein Sequencing

Peptide fractions were applied in two, 20 uL aliquots to a precycled, Polybrene-treated filter. Samples were sequenced with a Model 477 Sequencer (Applied Biosystems) using standard sequencer cycles.

III. RESULTS

Amino-acid composition data for CPP predicted that cleavage with ASP-N should yield about fifteen large fragments, assuming that fragmentation occcurred only at cysteic acid residues. The peptide map in Figure 1 reveals more than fifty peaks. This result indicates that cleavage occurred at many unmodified aspartic acid residues as well. Although most peaks in the peptide map appeared symmetrical and apparently homogeneous, only peaks c, f, and h (Table 1) proved homogeneous by direct sequence analysis. The other six peaks exhibitted minor peptide signals that ranged from 20 to 40% of the major peptide sequence. The sequencing results in Table 1 confirm that most of the peptide fractions purified by a single dimension of RPC were not sufficiently pure for sequence analysis.

TABLE 1. PEPTIDE SEQUENCES FROM ONE DIMENSIONAL RPC

Peak Label	Peptide Sequence(s)	
b	AIGAQSTYQE	(Major)
	EQFKNK	
c	DFSY_NTYQQLIS	
e	CLPAIQK	(Major)
	APNGSY	
	GF__L	
f	LFAGHAEFSAK	
h	DSV_YSGIEGPISSSGG	
j	DQPIGVGFSYGT	(Major)
	_EAAYSNDPNP	
	_LV	
m	DNPRSFLSTLSSVVK	(Major)
	_LQAFYAQFPEYANA	
n	DFAIFPESYGGAGPEFASYI	(Major)
	_ARNNTQ	
o	VYGAGHEVPYYQPETALQAFYQI	(Major)
	RIFETAYGVNSM	

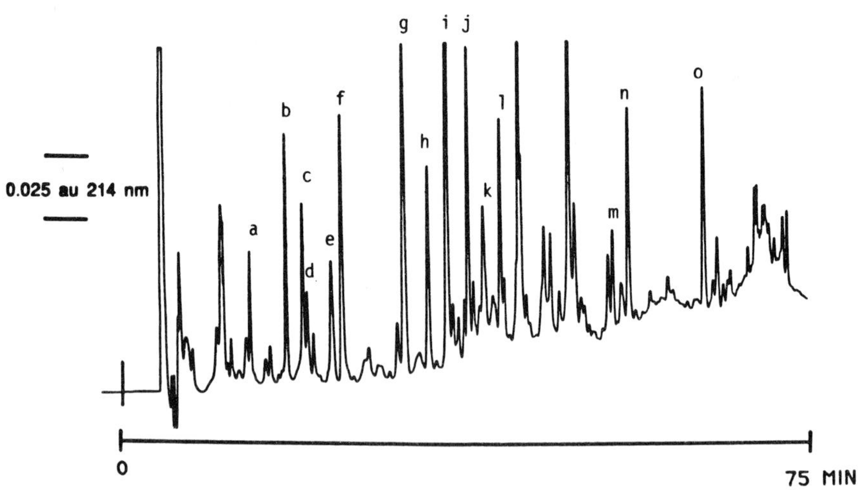

Fig. 1. Peptide map of carboxypeptidase P. About 300 picomoles of CPP-digest was separated on a narrow-bore, reversed-phase column. Peaks a, d, g, and k were first purified by IEC/RPC before direct, sequence analysis.i

The same digest was purified by two-dimensional chromatography, IEC followed by RPC. Figure 2 depicts the separation of the peptide digest on the strong, cation-exchange column. The peaks were broader in the IEC-separation than observed for the reversed-phase column (Figure 1), which suggested that many fractions contained multiple peptide species.

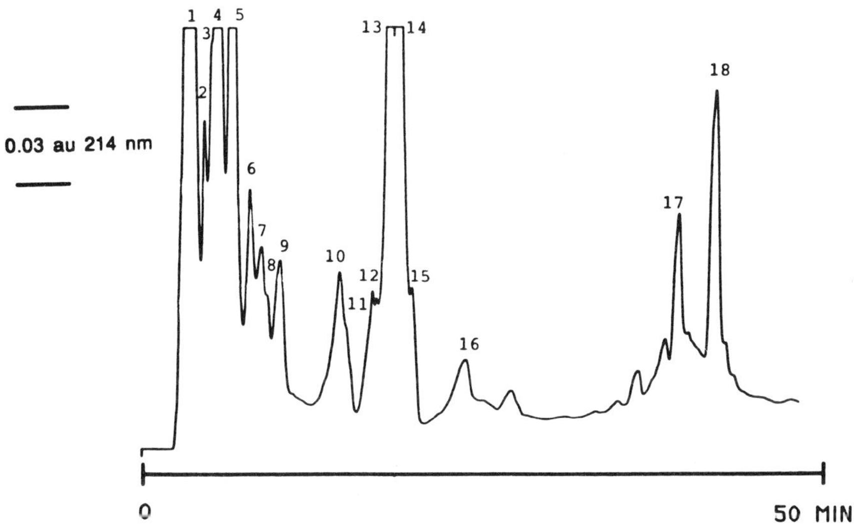

Fig. 2. Fractionation of the CPP-digest on a strong, cation-exchange column. About 500 picomoles of CPP-digest was separated on the narrow-bore, cation-exchange column and a total of 18 fractions were collected

A typical IEC-fraction, number four in Figure 2, was analyzed by RPC in Figure 3. This fraction had three peptides, which corresponded to peaks i, j, and l in Figure 1. All the IEC fractions were first subjected to an analytical separation by RPC prior to preparative RPC under the same conditions but using a larger injection volume to prepare sequence ready peptides. RPC-analysis of the IEC fractions revealed that most fractions contained two or three peptides, but fractions eleven and twelve were devoid of any peptide fragments, and fractions seven and nine were actually homogeneous peptides, DDSQRTSLLSAYNNQ and DLAPYTVNGVEKGQFKN, respectively.

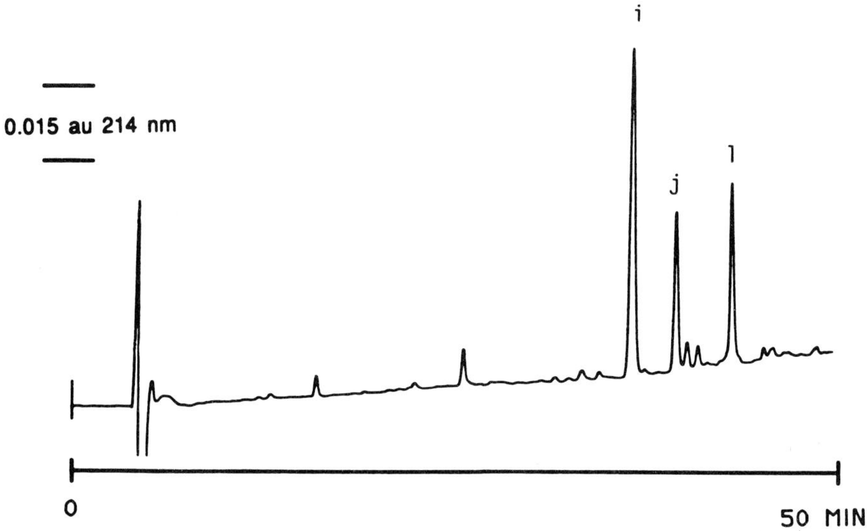

Fig. 3. RPC-analysis of cation-exchange, fraction four in previous figure.. Peaks i, j, and l were identified by direct sequence analysis.

Fractions collected from IEC ranged from 60-80 uL, which allowed 10 uL to be used for the analytical RPC and 50 uL for preparative RPC. Six IEC-fractions had sufficiently intense, well separated peaks that they were selected for preparative RPC. Fractions 1, 4, 7, 9, 14, and 18 (Figure 2) yielded nine peaks that were submitted for direct sequence analysis. The sequencing data in Table 2 reveals that all nine samples proved homogeneous. Most of the sequences began with aspartic acid which proved that amidation of the aspartic residues was not complete.

Five of these sequences have been confirmed by overlaps with other peptide fragments from cyanogen bromide or tryptic digests. All of the recovered peptides were greater than ten residues in length, but none were greater than twenty. The larger, more hydrophobic peptides, such as peaks n and o in Table 1, were not recovered by the two-dimensional purification method. Solubility of these peptides in the aqueous buffers may have been the problem.

TABLE 2. PEPTIDE FRAGMENTS FROM TWO-DIMENSIONAL HPLC

Peak Label	Peptide Sequence
a	CLPAIQK_TQSGSN
d	CPNGPYNKFAST
f	LFAGHAEFSAK
g	DDSQRTSLLSAYNNQ
h	DSV_YSGIEGPISSSGG
i	DPNPPSTYSTYLSNA
j	DQPIGVGFSYGT
k	DLAPYTVNGVEKGQFKN
l	DFSY_NTYQQLI

To verify the purity of the peaks produced by this two-dimensional method, three such peaks (i, j, and l in FIgures 1 and 3) were analyzed by capillary electrophoresis. The approximate peptide concentrations were one picomole per microliter (1 uM). The fraction containing peak "i" produced the electropherogram shown in Figure 4. Only one peak is observed in addition to the internal standard, which served as a marker much like tracking dye in slab-gel electrophoresis. The other two samples (peaks j and l) were also judged homogeneous. Two tryptic peptide fractions, purified by one-dimensional RPC, were heterogeneous by capillary electrophoresis and also by sequence analysis. The capillary technique was fast and consumed almost no sample (ca. 5 nanoliters).

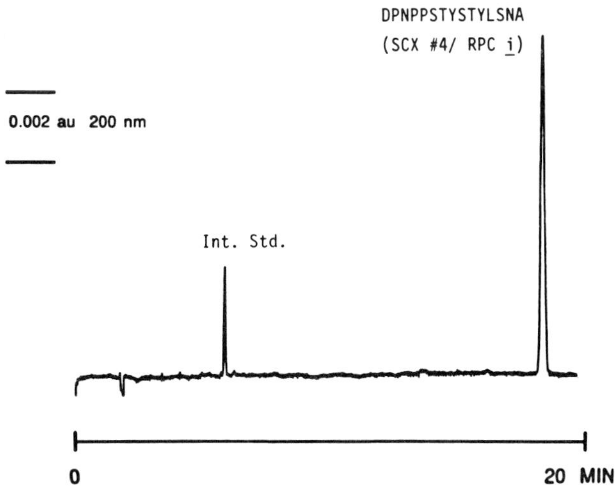

Fig. 4. Analysis of peak "i" by capillary electrophoresis. About 5 nanoliters of fraction i was electrophoresed at 20 kV.

IV. DISCUSSION

Two-dimensional chromatography greatly increases the number of components that can be fully resolved and purified. If the dimensions are truly orthogonal, then there is no cross correlation between retention in one dimension and retention in the other. For orthogonal systems the peak capacity, which is the total number of completely separable components, is the simple product of the individual peak capacities for each chromatographic dimension. For instance, the peak capacity of the strong, anion-exchanger was conservatively estimated to be about thirty components, which was derived from the observation that peak widths were about 1 minute. Thus, a thirty minute separation could maximally generate thirty, fully resolved components. Likewise the peak capacity of the reversed-phase column was about sixty components, which reflects an average peak width of 0.5 minutes and also assumes a thirty minute separation. Based on the estimated peak capacities for IEC and RPC, the combined capacity of the two-dimensional system is eighteen hundred. Excess peak capacity is quite useful because it lowers the statistical probability that any given peak will be heterogeneous.

Comparing the sequencing results for the CPP-digest in Table 1 with the data in Table 2 highlights the advantages of two-dimensional chromatography in purifying peptides from a complex, enzymatic digest. The two-dimensional technique produced highly purified peptides, as judged by sequence analysis, whereas six of the nine fractions purified by a single, RPC-dimension had substantial amounts of contaminating peptides. The chromatographic, electrophoretic and sequencing data suggest that two-dimensional chromatography should become standard practice when homogeneous peptides must be recovered from complex, protein digests. Because it is difficult a priori to predict the complexity of a protein digest, even from an analytical separation by one-dimensional RPC, the two-dimensional method should be routinely employed in preparing peptide fragments for sequencing.

Committing a sample for sequencing inevitably occupies the protein sequencing instrument for several hours or overnight, even when no peptide or protein is present in the sample. Because the project to determine the primary structure for CPP consumed weeks of sequencer time, a quick but sensitive purity check was needed to ensure that the this time would be productive. The new technique of capillary zone electrophoresis meets the speed and sensitivity requirements of such a purity check. It also offers the added benefit of almost no sample loss. This technique proved very useful in our work, because it increased the productive output of our protein sequencers.

V. REFERENCES

1. M.T.W. Hearn and B. Grego (1983) J. Chromatogr. 266:75-87.
2. M.A. Stadalius, M.A. Quarry, and L.R. Snyder (1985) J. Chromatogr. 327:93-133.
3. K.L. Stone and K.R. Williams (1986) J. Chromatogr. 359:203-212.
4. J.J. L'Italien (1986) J. Chromatogr. 359:213-220.
5. C.Y. Yang, E. Pauly, H. Kratzin and N. Hilschmann, Hoppe-Seyler's Z. Physiol. Chem. (1981) 362:1131-1146.
6. B. Grego, I.R. Van Driel, P.A. Stearne, J.W. Goding, E.C. Nice, and R.J. Simpson (1985) Eur. J. Biochem. 148:485-491.
7. W.H. Vensel, V.S. Fujita, G.E. Tarr, E. Margoliash, and H. Kayser (1983) J. Chromatogr. 266:491-500.
8. N.Takahashi, Y. Takahashi and F.W. Putnam (1983) J. Chromatogr. 266:511-522.
9. K.Y. Kumagaye, M. Takai, N. Chino, T. Kimura and S. Sakakibara (1985) J. Chromatogr. 327:327-332.
10. M. Dizdaroglu (1985) J. Chromatogr. 334;49-55.
11. A.J. Alpert and P.C. Andrews (1985) J. Chromatogr. 443:85-96.
12. D.L. Crimmins, J. Gorka, R.S. Thoma and B.D. Schwartz (1988) J. Chromatogr. 443:63-71.
13. H.H. Lauer and D. McManigill (1986) Anal. Chem. 50:166-170.
14. C.H.W. Hirs (1967) Methods Enzymol. 11:197-206.
15. D.G. Hoare and D.E. Koshland (1967) J. Biol. Chem. 242:2447-2452.
16. G.R. Drapeau (1980) J. Biol. Chem. 255:839-840.

49

Fmoc Chemistry: Synthesis, Kinetics, Cleavage, and Deprotection of Arginine-Containing Peptides

Jeanette L. Harrison, Gordon M. Petrie,
Richard L.Noble, Hal S. Beilan and Sarah N. McCurdy

Applied Biosystems, Inc., Foster City, CA 94404.

Alan R. Culwell

Athena Neurosciences, Inc., San Carlos, CA 94070

I. INTRODUCTION

With increasing interest in solid phase peptide synthesis using the 9-fluorenylmethoxycarbonyl (Fmoc) strategy, a considerable amount of work has been done in an effort to understand and optimize the use of this synthetic approach. This paper presents data on synthesis kinetics and techniques for cleavage and deprotection of peptides containing arginine.

Studies concerning the kinetics of chain assembly represent three categories: (a) activation of the Fmoc amino acid and the stability of the resulting active species, (b) deprotection of the Fmoc peptide resin, and (c) coupling of the Fmoc amino acid to the growing peptide chain.

The Fmoc strategy to solid phase peptide synthesis (1) is very desirable for peptides containing acid-sensitive amino acids. However, the benefits of mild synthesis and cleavage conditions are offset by incomplete deprotection of some side chain protecting groups. Arginine derivatives are prime examples of the difficulties encountered when removing side chain protection during cleavage in 95% trifluoroacetic acid (TFA). 4-methoxy-2,3,6-trimethyl-benzenesulfonyl (Mtr) and 2,2,5,7,8-pentamethyl-chroman-6-sulfonyl (Pmc) were incorporated into peptides containing one, two and five arginine residues. Different cleavage and deprotection conditions were investigated.

II. MATERIALS AND METHODS

All reagents, solvents, amino acids and resins were supplied by Applied Biosystems, except Fmoc-Arg(Pmc)-OH, Fmoc-Cys-(Trt), and Fmoc-Cys(Acm)-OH which were purchased from Bachem Biosciences Inc., Philadelphia.

A. Activation Kinetics and Stability Studies

Activations were carried out either manually or on an Applied Biosystems Model 430A Peptide Synthesizer. 1 mmol Fmoc amino acid was activated to either a symmetric anhydride in 72% dichloromethane (DCM)/28% N-methylpyrollidone (NMP) using diisopropylcarbodiimide (DIC) or an HOBt ester in 12% DCM/88% NMP using dicyclohexylcarbodiimide (DCC). The activating amino acid solution was removed from the instrument after all reactants had been added to the reaction, except when DCC was used as the activating agent. In such cases, the reaction was allowed to proceed on the instrument until most of the dicyclohexylurea (DCU) could be removed from the reaction. A small aliquot of the reaction mixture was added to a temperature controlled NaCl Infrared (I.R.) Spectroscopy cell and placed on a Perkin-Elmer Model 1320 I.R. Spectrophotometer. An initial spectrum of the activating solution was taken to determine the optimal wavenumber (wavenumbers 1820-1840) to monitor. The reaction was monitored at the optimal wavenumber for two to three hours to yield a time vs. absorbance plot for the reaction (2).

The time required for the Fmoc amino acid to reach 90%, 95%, and 100% activation was determined from the time versus absorbance plot. The time required for the active species to degrade to 90%, 80%, and 50% was also calculated.

B. Deprotection and Coupling Kinetics

All deprotection studies were performed on a Model 430A Peptide Synthesizer. The 0.1 mmol Fmoc-Gly-HMP polystyrene resin was deprotected for varying amounts of time with 20% piperidine in NMP. The partially deprotected Gly-HMP resin was capped with 30% acetic anhydride in DCM for 15 minutes. The resin was then treated with 20% piperidine in NMP for 30 minutes to deprotect any residual Fmoc-Gly-HMP. Resin deprotected at this step represents the portion not deprotected initially. A final sample was subjected to the quantitative ninhydrin test (3). An estimation of the amount of resin still protected after the first piperidine treatment was determined by comparing the absorbance of the sample taken after the second deprotection against the absorbance of a ninhydrin reagent blank.

The 0.1 mmol Fmoc-Gly-HMP polystyrene resin was deprotected with 20% piperidine in NMP for 20 minutes, rinsed, and dried. The 1.0 mmol of Fmoc-Ile was activated by the previously mentioned protocol until the amino acid reached 90% of maximum activation. The activated solutions were added to individual test tubes and allowed to react with ~10 mg Gly-HMP resin for varying amounts of time. The coupling reactions were stopped by dilution with DCM/methanol (MeOH) (40:60). The resin was immediately rinsed several times with the DCM/MeOH mixture and subjected to ninhydrin analysis (3) to determine the extent of coupling.

C. Cleavage and Deprotection

All syntheses were performed on a Applied Biosystems Model 430A Peptide Synthesizer employing the Fmoc approach with either NMP or dimethylformamide (DMF) as the solvent system. The first amino acid was automatically loaded onto the hydroxymethylphenoxymethyl (HMP) resin on the instrument for one hour using a single couple cycle with 2 equivalents of Fmoc symmetric anhydride and a catalytic amount of dimethylaminopyridine (DMAP). The remaining amino acids were single coupled with 4 equivalents of active HOBt esters in either 100% NMP or 100% DMF for approximately one hour. Coupling yields were determined by a ninhydrin analysis (3) of a small aliquot of peptide resin removed automatically after each amino acid addition. The capping protocol employed was 10% acetic anhydride in NMP for 8 minutes.

All peptide resins were N-alpha deprotected either manually or automatically in 20% piperidine in NMP for 20 minutes. The peptide resin was washed and dried thoroughly prior to cleavage and deprotection. One of the following three cleavage protocols were utilized at room temperature, either 95%TFA:5%thioanisole or 95%TFA:3%anisole:1%1,2-ethanedithiol (EDT):1%ethyl methyl sulfide (EMS) or 93%TFA:3%anisole:3%EDT: 1%EMS. Exact conditions for each experiment are specified in the results section. At the end of the cleavage time, the peptide resin solution was filtered to remove the resin. The solution was then diluted with DCM and rotovaped to remove as much TFA/DCM as possible to simplify dissolution of the peptide. The peptide was dissolved in an appropriate aqueous solution and desalted over either a Sephadex™ G-10 or G-25 disposable column.

A few peptide resin samples cleaved by trifluoromethanesulfonic acid (TFMSA)/TFA or hydrogen fluoride (HF) are described. The TFMSA cleavage mixture consisted of thioanisole:EDT:TFA:TFMSA (1:0.5:10.1). The reaction proceeded for four hours at room temperature. The HF protocol was HF:Anisole:Dimethylsulfide (DMS):p-Thiocresol (10:1:1:0.2) at -5°C to 0°C for 60 minutes.

All crude, cleaved and desalted peptides were analyzed by reverse phase high performance liquid chromatography (HPLC) on a Applied Biosystems Model 130A Separation System using a 0% to 60% Buffer B gradient over 45 minutes (Buffer A: 0.1% TFA and Buffer B: 99.91% Acetonitrile/0.09%

TFA). A Brownlee Labs™ C-8, 300Å, 7μm, 22 cm x 2.1 mm column was used for separartions.

Amino acid analysis was performed on a Applied Biosystems Model 420A Amino Acid Analysis System for either free peptides or resin-bound peptides. Mass spectral analysis was performed by Dr. Terry Lee at Beckman Research Institute of the City of Hope.

III. RESULTS AND DISCUSSION

A. Activation Kinetics and Stability Studies

It was determined that Fmoc symmetric anhydrides are formed faster than the HOBt esters (Table I). However, the Fmoc symmetric anhydrides are less stable than the Fmoc HOBt esters, particularly as the amount of NMP or DMF is increased relative to DCM. It should be noted from subsequent experiments that some of the Fmoc symmetric anhydrides degrade almost instantaneously in solutions containing greater than 90% DMF or NMP in DCM (unpublished observation).

The Fmoc amino acid symmetric anhydride is not soluble in solutions greater than 80% DCM. In such solutions, we observed the appearance of a white Fmoc symmetric anhydride precipitate as the activation reaction proceeded. Therefore, when activating the Fmoc amino acid to the symmetric anhydride, the solvent composition and activation times must be balanced such that the Fmoc symmetric anhydride can form rapidly and is stable throughout the coupling period.

Several Fmoc amino acids Arg(Mtr), Asn, Gln, and His(Trt) are not stable as symmetric anhydrides under any of the conditions tested. Yet, Arg(Mtr) and His(Trt) can be loaded onto the HMP resin when activated with one equivalent of carbodiimide in situ with the HMP resin in the presence of 4-dimethylaminopyridine (DMAP). It has been postulated that the active species in the in situ loading is the O-acyl urea form of the amino acid and not the symmetric anhydride (4,5).

The Fmoc HOBt esters took longer to activate, but were considerably more stable than the Fmoc symmetric anhydrides. The HOBt esters did not display the same solubility constraints exhibited by the Fmoc symmetric anhydrides.

B. Deprotection and Coupling Kinetics

Traditional approaches to the piperidine deprotection of Fmoc peptide resins have utilized solutions of about 20% piperidine in either DMF or NMP for 18-20 minutes. The use of higher concentrations of piperidine have been avoided due to the fear of side reactions generated by the use of a strong, concentrated base.

TABLE I: ACTIVATION AND STABILITIES OF ACTIVATED FMOC AMINO ACIDS IN DCM AND NMP USING DCC ATAMBIENT TEMPERATURE

HOBT ESTERS
ACTIVATED WITH HOBT AND DCC

Fmoc Amino Acid	TIME TO % ACTIVATION 90%	95%	100%	TIME TO % DEGRADATION Trace	10%
Ala	35	69	90	STABLE	
Arg (Mtr)	<40	60	160	STABLE	
Asn	<40	<40	~40	120	?
Asp (OtBu)	<35	<35	69	187	?
Cys (tBu)	<35	<35	87	STABLE	
Glu (OtBu)	<35	47	135	STABLE	
Gln	<40	<40	100	160	?
Gly	<35	35	155	STABLE	
His (Trt)	<41	41	161	STABLE	
Ile	35	83	150	STABLE	
Leu	35	51	155	155	?
Lys (Boc)	30	67	161	STABLE	
Met	30	49	120	STABLE	
Phe	<35	35	140	STABLE	
Pro	<35	<35	115	STABLE	
Ser (tBu)	<35	51	150	STABLE	
Thr (tBu)	<35	35	147	STABLE	
Trp	<41	75	160	STABLE	
Tyr (tBu)	35	67	140	STABLE	
Val	65	97	150	STABLE	

SYMMETRIC ANHYDRIDES
ACTIVATED WITH DIC

Fmoc Amino Acid	TIME TO % ACTIVATION 90%	95%	100%	TIME TO % DEGRADATION Trace	10%	20%	50%
Ala	8	10	13	22	25	32	67
Arg (Mtr)		UNSTABLE					
Asn		UNSTABLE					
Asp (OtBu)	5	10	16	21	33	41	80
Cys (OtBu)	4	8	20	30	44	60	130
Glu (OtBu)	8	11	17	35	52	90	140
Gln		UNSTABLE					
Gly	6	10	14	20	36	41	67
His (Trt)		UNSTABLE	4	8	9	15	18
Ile	10	13	45		STABLE		
Leu	10	12	20	25	43	60	120
Lys (Boc)	6	12	24	60	?	?	?
Met	6	8	10	21	34	43	70
Phe	3	13	23	47	?	?	?
Pro	15	23	40		STABLE		
Ser (tBu)	8	11	18	28	90	?	?
Thr (tBu)	8	15	60		STABLE		
Trp	8	10	23	70	?	?	?
Tyr (tBu)	8	12	25	55	137	?	?
Val	12	17	40		STABLE		

Our study led us to believe that the use of a more concentrated piperidine solution is not necessary for the rapid deprotection of the Fmoc peptide resin. We have demonstrated that the Fmoc peptide resin was completely deprotected within four minutes of exposure to 23% piperidine in NMP, given the scale and conditions explored. The half time of this reaction was estimated to be on the order of 15 seconds for the early stages of the deprotection, and one minute for the last 0.5% of the resin to be deprotected. (Figure 1) In light of this data, shorter deprotection times (as short as 5-10 minutes) may be used with confidence to obtain complete deprotection.

Fmoc symmetric anhydride coupled to the Gly HMP resin at a faster rate compared to the Fmoc HOBt ester (Figure 2), as previously noted (6). The symmetric anhydride coupled to 90% of the available Gly within one minute of coupling, and within four minutes, 99% of the available sites had been coupled. The reaction appeared to be complete (99.9% coupled) within 6 minutes. By contrast, the HOBt ester had coupled to 90%, 99% and completion in two, six, and nine minutes, respectively.

These reaction rates probably represent the upper limit at which these activated species can couple to the unprotected, unhindered peptide HMP resin. However, studies have shown that with increased concentrations of Fmoc HOBt active ester relative to Fmoc symmetric anhydride, coupling with the HOBt ester can proceed at rates similar to those of the symmetric anhydrides. Solvent compositions and other factors may also influence the rate at which these coupling reactions take place (7).

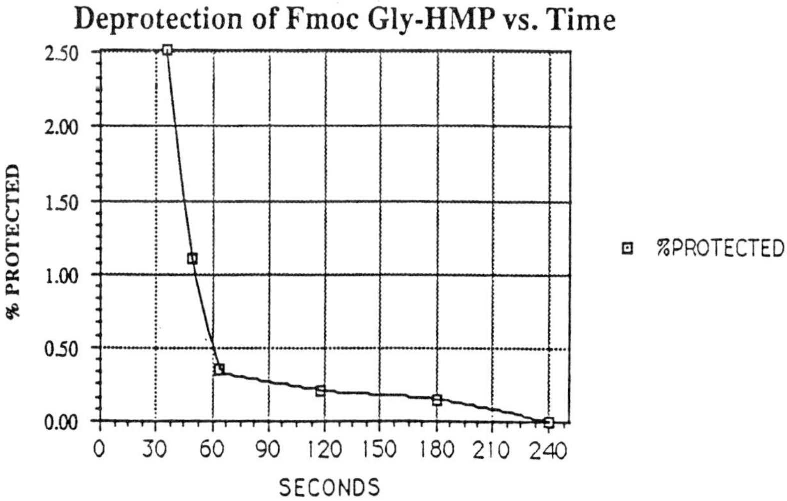

Figure 1
Deprotection was performed in 23% piperidine at room temperature. One treatment was performed four times under 60 seconds and two piperidine treatments were performed for longer times.

Fmoc Ile to Gly-HMP Coupling Yields vs. Time

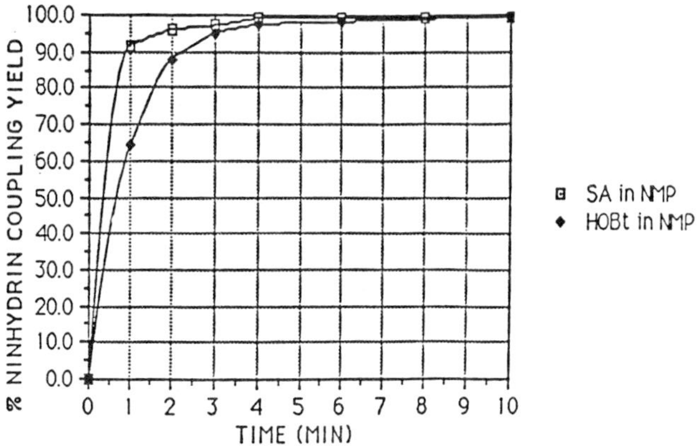

Figure 2
Concentration of the Fmoc Ile active species was 0.15 M coupled at room temperature. The Fmoc Ile symmetric anhydride was formed with 0.5 equivalents DIC in DCM and coupled in 72% DCM in NMP. The Fmoc Ile HOBt ester was formed with 1 equivalent of each DIC in DCM and HOBt in NMP. Coupling of the HOBt ester was in 12% DCM in NMP.

C. Cleavage and Deprotection

The first peptide analyzed was Angiotensin containing one arginine residue. The sequence, H-Asp-Arg-Val-Tyr-Ile-His-Pro-Phe-His-Leu-COOH, is short and straightforward. Average coupling yield for this synthesis was 99.3% which agreed with the amino acid analysis results. The HPLC chromatogram after one hour of cleavage (95% TFA:5% thioanisole) for Angiotensin synthesized with Fmoc-Arg(Mtr)-OH shows the peak of interest, a small unidentified peak, and a peak representing the peptide with the Mtr groups still attached. As the cleavage progresses, the unidentified peak, possibly an intermediate, becomes larger while the Mtr protected peak becomes smaller. Optimal reaction time for complete removal of the Mtr groups was four hours for a 78% yield.

Angiotensin synthesized using Fmoc-Arg(Pmc)-OH achieved an average coupling yield of 99.1%. Approximately 74% of the desired peptide is observed after only one hour of cleavage (95% TFA:5% thioanisole). Extending the reaction time to a total of 3 hours yields an additional 5% of peptide. The Pmc group incorporated into Angiotensin is sufficiently cleaved within one hour. Pmc is a more efficiently removed protecting group than Mtr.

The next peptide example contains two arginine residues and multiple tryptophan residues. The exact sequence is proprietary. The peptide was synthesized twice, once using Fmoc-Arg(Mtr)-OH and the second time with Fmoc-Arg(Pmc)-OH. The syntheses resulted in an average coupling yield of 99.3% for the Fmoc-Arg(Mtr)-OH synthesis and 98.9% for the Fmoc-Arg(Pmc)-OH synthesis. Amino acid analysis verified the integrity of each peptide. The peptides were cleaved in 93%TFA:3%anisole:3%EDT: 1%EMS for one hour at room temperature and the HPLC traces of the two crude products are shown in Figures 3 a and b. The chromatogram of the Fmoc-Arg(Pmc)-OH synthesis has one major peak, the desired peptide verified by Fast Atom Bombardment Mass Spectroscopy (FAB-MS). The doublet peak later in the chromatogram contains two peptides. Each peptide was isolated by preparative HPLC and shown by FAB-MS to have a single Pmc-modified tryptophan residue. The Fmoc-Arg(Mtr)-OH chromatogram contains multiple peaks. The desired peptide (first peak in Figure 3a) was not the major component and, thus, the yield was low. The difficulty in cleavage and deprotection maybe the result of side-reactions occurring simultaneously. Side reactions altering tryptophan during Mtr removal have been previously noted (8). It has been suggested that side product formation can be minimized if EDT and 4-(methylmercapto)phenol are used as scavengers with 95% TFA. The shortest reaction time possible should be implemented to keep dithiolketal formation to a minimum (8). This synthesis incorporating the Pmc protecting group for the guanidino function of arginine resulted in an easier purification and greater overall yields.

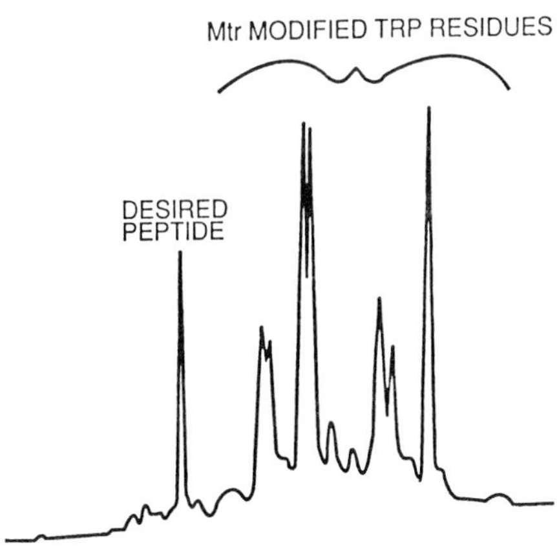

Fig. 3a
(HPLC trace at 220 nm)

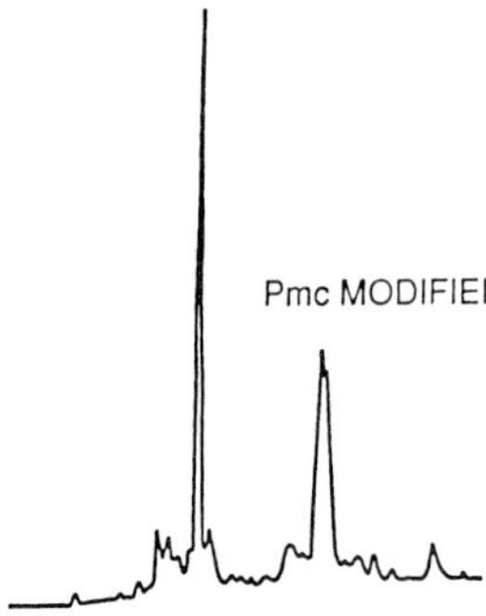

Fig 3b
(HPLC trace at 220 nm)

The final peptide to be investigated was rat Atrial Natriuretic Peptide 1-28, H-Ser-Leu-Arg-Arg-Ser-Ser-Cys-Phe-Gly-Gly-Arg-Ile-Asp-Arg-Ile-Gly-Ala-Gln-Ser-Gly-Leu-Gly-Cys-Asn-Ser-Phe-Arg-Tyr-COOH. This peptide has two difficult sequence regions and, thus, a capping protocol was employed throughout the synthesis. Coupling yields were determined both before and after capping to provide a baseline for the entire synthesis of rANF 1-28 using Fmoc-Arg(Mtr)-OH. The average coupling yield before capping was 98.57% and 99.2% after capping. The peptide resin was treated with 95%TFA:3%anisole:1%EDT:1%EMS for eight hours at room temperature. The HPLC traces (λ=220 nm) from 5, 7, and 8 hours show only 2%, 6%, and 8% (respectively) of the desired peptide is detected in agreement with final yields. A major side product is found later in each of the chromatograms, all with identical profiles. By the eighth hour, the smaller additional contaminant peaks are less prominent. A second sample of the same peptide resin was cleaved by the TFMSA protocol (8%thioanisole:4%EDT:80%TFA:8%TFMSA) for 4 hours at room temperature. An aliquot removed after 30 minutes of cleavage time results in a similar chromatogram as that of the TFA cleavage after 8 hours. Extending the TFMSA reaction time to 4 hours did not increase yields of the desired peptide. A third sample of the peptide resin was exposed to HF cleavage with slightly higher yields of the desired product, approximately 16%, however, the side product is still dominant. This major side product forms quickly and is very stable to extended cleavage conditions.

The same peptide was synthesized using the Fmoc-Arg(Pmc)-OH derivative and a capping protocol was employed. An average coupling yield of 98.33% was obtained for this synthesis and agreed with the amino acid analysis results. The TFA cleavage (95% TFA:3%anisole:1%EDT:1% EMS) proceeded for 4 hours at room temperature. A larger amount of the desired peptide was obtained, approximately 30%, but was not the major component. The chromatogram shows a similar trace to the profiles from rANF 1-28 - Mtr cleavages. The two peaks, the desired peptide and the major side product, were isolated and submitted to FAB-MS analysis. The first peak proved to be the desired peptide, rANF 1-28 with the Acm protecting groups still attached to the cysteine residues. The later peak was determined to be approximately 222 mass units larger than the first peak. The difference in mass units is too small to account for the incomplete removal of Pmc which agrees with the stability of this side product under extended reaction times. Investigation of this side product is continuing. Both rANF 1-28 peptides were synthesized successfully. Nevertheless, the Fmoc-Arg(Pmc)-OH synthesized rANF 1-28 had a higher overall yield with a shorter cleavage time than the same sequence incorporating Fmoc-Arg(Mtr)-OH.

IV. CONCLUSIONS

The kinetic studies demonstrate a faster rate of formation of Fmoc symmetric anhydride over the formation of Fmoc HOBt esters. It was also observed that the Fmoc symmetric anhydrides coupled to the peptide HMP resin at a faster rate than the activated HOBt ester. However, Fmoc symmetric anhydrides are generally less stable than the Fmoc HOBt esters. Some symmetric anhydrides are generally not as soluble as the Fmoc HOBt esters in solutions greater than 80% DCM. For these reasons, Fmoc HOBt esters can be used with a more diverse array of solvent mixtures and coupling times in comparison with Fmoc symmetric anhydrides. Longer activation and coupling times may be required when using Fmoc HOBt esters.

Our investigations revealed that deprotection using 20% piperidine in NMP can, under the right conditions, occur very rapidly (within 4 minutes). In addition, coupling reactions proceed quickly (approximately 9 minutes with tested conditions). Further development of the Fmoc synthetic approach using this data could yield a highly efficient Fmoc synthetic approach with shorter cycle times.

The cleavage and deprotection results reported in this paper favor the Pmc protecting group over the Mtr for ease and efficiency of yielding arginine containing peptides synthesized by Fmoc chemistry. For peptides containing one or two arginine residues, the cleavage is relatively fast and straightforward. If tryptophan residues are present, scavengers should be selected carefully and cleavage times should be minimized. When four or more arginine residues are present in the sequence, a small sample should be cleaved first to determine the optimum reaction conditions for maximum yields with minimum side reactions. Additional research is necessary to find a more acid labile protecting group suitable for the guanidino function of Fmoc arginine when synthesizing peptides containing multiple arginine residues.

REFERENCES

1. Meienhofer, J., Waki, M., Heimer, E.P., Lambros, T.J., Makofske, R.C. and Chang, C.-D. (1979). Int. J. Peptide Protein Res., 13, 35-42.
2. Geiser T., Beilan H., Bergot B.J., and Otteson K.M., (1988). In "Macromolecular Sequencing and Synthesis" (Schlesinger,D.H., ed), Alan R. Liss, NY; (pp 199-218).
3. Sarin, V.K., Kent, S.B.H., Tam, J.P., and Merrifield, R.B. (1981) Analytical Biochem., 117, 147-157.
4. Personal Communication with Hal Beilan and Jeanette Harrison, Applied Biosystems, Inc. ; Foster City, CA 94404.
5. Khorona H.G., (1951) "Chemistry of Carbodiimides", Chem. Revs 53, 145, J. Org. Chem, 53, 617-624.
6. Hudson D., (1988) J. Org. Chem, 53, 617-624.
7. Otteson K.M.O., McCurdy S.N., Harrison J.L., (1988) "N-Methyl pyrrolidone as a Solvent in Solid Phase Peptide Synthesis", Proceedings of the ABI Peptide Synthesis Symposium, ASBC/FASEB, May 1988; Las Vegas, NV.
8. Sieber, P. (1987) Tetrahedron Letters, 28, 1637-1640.

50

SALT BRIDGE CONVERSION TO AMIDE BONDS WITH C_2N_2

Richard A. Day, Jane Kirley, Rebecca Tharp,
David Ficker, Candace Strange and Ghiorgis Ghenbot

Department of Chemistry
University of Cincinnati
Cincinnati, OH 45221-0172

I. INTRODUCTION

Salt bridges currently are detected and identified essentially only by X-ray diffraction studies on crystalline proteins. Cross-linking by bifunctional reagents can be used to provide approximate spacing of appropriately reactive functional groups; most cross-linking involves introduction of a residue of the reagent which itself may significantly alter the properties of the system being measured. Cross-linking methods that do not introduce residues are oxidation of sulfhydryls to disulfides as has been utilized to detect folding intermediates (1), amide bond condensing agents as are used in peptide synthesis (2), and cyanogen (C_2N_2, ethane dinitrile) (3-5). C_2N_2 appears to drive only intra-molecular condensations (5); no inter-molecular condensation to form amide bonds has been detected to date. Thus, amide bonds form only under conditions where salt bridge formation in ω-amino acids is favored (5), or in proteins where the spatially close (< 4Å) carboxylate-ammonium ion pairs are an integral part of the structure. Associated subunits may become cross-linked as seen with hemoglobin and chromatin.

The important physical properties and chemical reactivities of C_2N_2 are summarized by Brotherton and Lynn (6) and by Ciganek et al (7). No C_2N_2 driven amide bond formation has been noted prior to our work. Amide bond formation has been verified with model compounds (5), in carbonic anhydrase (3,8), in β-lactamase I (4), in hens egg white lysozyme (4), hemoglobin (4,9), and chromatin as nucleosomes (R.A. Day, unpublished). When ^{14}C-C_2N_2 was utilized, very little ^{14}C was incorporated into proteins (3,4,8), nor was C_2N_2 incorporated into the model compounds (5). This result appears to exclude the reliance on radioisotopes in labeling C_2N_2 modified sites; nevertheless, with the known organic chemistry and a small amount of incorporation with long reaction times and higher concentrations of C_2N_2, it is important to verify that no C_2N_2 is incorporated in any given study. The newly formed amide bonds in the protein are neither detectable by physical means nor by labeling with ^{14}C-C_2N_2. Accordingly, appropriate strategies are being developed to isolate and identify the amino acid residues in the modified sites.

The methods given below describe the synthesis of labeled C_2N_2, identification of functional groups modified, quantitation, and changes in the properties of the modified proteins.

II. CYANOGEN MANIPULATION, MEASUREMENT AND PREPARATION

The handling of C_2N_2, a toxic gas, requires due care (see Ref. 10). Cyanogen is commercially available (Matheson, East Rutherford, NJ 02073 or A-L Compressed Gases, 11445 Rockfield Court, Cincinnati, OH 45241). Purity of commercial and synthetic C_2N_2 was assessed by mass spectrometry (3,4); no measurable amounts of other components such as HCN have been detected.

A. Measurement of C_2N_2 Concentration

1. Electrochemical Measurement as Cyanide. Cyanogen is quantitatively hydrolyzed to cyanide and cyanate in basic solutions. The cyanide concentration is determined by an ion specific cyanide electrode (Orion Research, Model 94-06). The cyanide concentrations should be in the range of 10^{-2} to 10^{-6} M. The concentration of a cyanogen was determined by comparison with standard solutions.

2. Fluorescence measurement. Cyanogen aliquots giving final concentrations of 10^{-8} to 10^{-4} M were mixed with 3M hexamethylene tetramine (HMTA) and H_2O in capped cuvettes

to give a final HMTA concentration of 2.5 M. Incubate at 35°C for one hour. Measure emission at 450 nm with excitation at 365 nm. Quantitation is achieved by comparison with suitable standards. The electrochemical method will measure C_2N_2 and any cyanide already present. The fluorometric procedure of Mothersole (5,11) avoids cyanide interference.

B. Synthesis of ^{14}C-C_2N_2. A modification of the procedure of Kim and King was used (12). ^{14}C-labeled cyanogen was prepared by the pyrolysis of solid AgCN-$Ag^{14}CN$. Heating AgCN to temperatures of 400°C in a Vycor vessel for 2 hrs forms cyanogen gas which is condensed by liquid nitrogen to the solid cyanogen. The design of the system is as described elsewhere (3,4).

The proteins treated with ^{14}C-C_2N_2 were verified to have been functionally modified. Repeated dialysis to a constant ^{14}C specific activity aliquots was carried out (3,4).

III. PROTEIN MODIFICATION

A. Bovine Carbonic Anhydrase II (BCAII). Cyanogen behaves as an active site directed affinity label and is effective at low concentrations (13). Stock solutions of BCAII were prepared by dissolving 5 mg of the enzyme in 300 µl of water. The BCAII was diluted in water to 5×10^{-5} M and a final pH of 4.25. The cyanogen solution of suitable concentration was added to the BCAII solution by a calibrated syringe yielding final molarities between 5×10^{-4} M and 5×10^{-2} M. The solutions were kept in 500 µl capped vials at room temperature for one and one-half to two hours. After incubation the cyanogen was entrained from the solution by gently passing a nitrogen stream across the top of the solution. Appropriate controls were run with cyanide and cyanate. Both esterase and hydratase activities were measured.

B. β-Lactamase I of Bacillus cereus 569/H. This monomeric protein was treated exhaustively with C_2N_2. Cyanogen was injected into a 1 ml teflon faced septum capped vial containing 0.5 cc of an aqueous unbuffered β-lactamase solution (1-2 mg/ml, pH 4 or 7) via a 5 cc disposable syringe. With the septum slightly loosened, the cyanogen was injected above the surface of the solution. The cap was then immediately tightened and the vial gently inverted several times. This procedure was repeated until 20 cc of cyanogen had been passed through the head space. The protein solution and cyanogen gas were allowed to incubate together for 2 hours, the vial was uncapped, and a stream of nitrogen was passed over the solution for 5 minutes. On occasion, a protein solution (3-4 ml) contained in a 5 ml reacti-vial would be saturated with 30 cc of cyanogen.

C. **Human Hemoglobin**. This tetrameric protein illustrates cross-linking among subunits. Modification was done at 0° and 25°, pH 4 and 7. Typically, a 5 ml syringe was used to take 1 ml of hemoglobin solution in water adjusted to pH 4 or 7 with HCl, NaOH, or in phosphate buffer (310 mOsM) of pH 6.8 and 3 ml cyanogen gas at one atmosphere and the syringe was mixed appropriately. After a while, the gas in the syringe was replaced by the same volume of fresh cyanogen. This was repeated once again. The above procedure was controlled to be finished in one to two minutes. The solution was stored up to 120 minutes. The cyanogen was then expelled at the desired time from the syringe; the dissolved C_2N_2 was entrained in a stream of N_2.

IV. ANALYSIS OF C_2N_2 MODIFIED PROTEINS

The proteins were analyzed by disk or slab PAGE, by isoelectric focusing, and/or by reverse phase liquid chromatography, for changes in functionality and for quantitative changes in the modified amino acid residues.

A. **Electrophoresis**. Essentially standard procedures were used (14). Mercaptoethanol was omitted for the β-lactamase which contains no disulfide links (4).

B. **Isoelectric Focusing**. The analytic procedure of Righetti and Drysdale was used (15).

C. **Reverse Phase HPLC**. Reverse phase chromatography of the modified protein has been dealt with (1) at the level of the modified protein and (2) with enzymatic digests. Essentially the method of Shelton et al. (20) using C-4 reverse phase columns with 0.1% trifluoroacetic acid (TFA) water-acetonitrile gradients has been used with intact or cross-linked polypeptide chains. Separation of peptides in enzymatic digests uses C-18 reverse phase columns and water-acetonitrile gradients with 0.1% TFA. Denaturing with harsh conditions (extremes of pH, heat, neutral nucleophiles) must be avoided in order retain the more labile cross-links such as the GLU-HIS; thus, carbonic anhydrase tryptic digests of the undenatured native and C_2N_2 treated enzyme displayed characteristic differences in their HPLC elution profiles.

D. Measurement of ω,ω'-Links: The Imidazole to γ-Carbonyl Link. Acyl imidazoles are reactive towards nucleophiles and indeed constitute a class of reactive intermediates of considerable use to the synthetic efforts in peptide chemistry (16). The link can be identified and quantitated by the procedure of Kirley et al where the GLU is converted to 2,4-diaminobutanoic acid (DAB) (8). In this procedure the Lossen rearrangement was effected by the procedure of Wilcox (17). The modified protein is hydrolyzed and the DAB is quantitated by amino acid analysis. In single column ion-exchange amino acid analysis DAB elutes immediately after lysine (3).

E. Measurement of ω,ω'-Links: ε-Amino to γ- and β-Carboxyls. Modified protein is anticipated to have some ε-amino groups blocked by cross-links (α-amino groups may also be blocked but require a modification of this analysis). The modified and control protein is treated with an excess of the Sanger reagent, 2,4-dinitrofluorobenzene (4). The cross-linked LYS residues are not dinitrophenylated. Amino acid analysis of the hydrolyzate provides quantitation for the lysine modification.

The number of carboxyls modified are determined by treatment of the modified and control protein with 1-ethyl-3-(3-dimethylaminopropyl)carbodiimide·HCl and glycinamide. Cross-linked GLU and ASP do not react with the reagent. Amino acid analysis provides the quantitation of glycines incorporated at the carboxyls not modified by cyanogen (4).

V. RESULTS AND DISCUSSION

The gaseous reagent C_2N_2 is conveniently manipulated. It is relatively easy to introduce and to remove any unreacted material through entrainment or evacuation of the reaction vessel. Its concentration is measured readily electrochemically as cyanide or fluorometrically by conversion to a fluorophor.

Synthesis of the ^{14}C-C_2N_2 is described. To date we have observed only a small amount of ^{14}C-C_2N_2 incorporation (3,4,8). It approaches one mole/mole of β-lactamase with long reaction times (72 h) at one atmosphere C_2N_2 pH 7 (4). It is important to ascertain the degree of C_2N_2 incorporation in order to correctly assess amide bond formation through the carbodiimide like process.

The cyanogen driven modification of bovine carbonic anhydrase II (BCAII) gives a clean 1:1 stoichiometry of HIS and GLU modification per BCAII (8) perhaps due to the C_2N_2 having two roles: (1) it appears to behave as a CO_2 structural analog and (2) cross-linking of the GLU-γ-$COO^{\ominus}$ and HIS-<u>im</u> known to be in the active site (18). The stoichiometry for the GLU participation was settled by chemical means; high resolution NMR provided direct evidence for participation of a HIS. Chemical modification by the nucleophiles OH^-, NH_2OH and glycine ethyl ester are in accord with the assigned modification (8, and Ghenbot and Day, unpublished). The modified enzyme remained inactive indefinitely at pH 7; at pH 9 activity returned in a few hours. Examination of the C_2N_2 treated carbonic anhydrase by PAGE shows no change in molecular weight. The CD spectra of modified and control enzyme were indistinguishable.

β-Lactamase I modification proceeds with very rapid inactivation of the enzyme with $\tau\frac{1}{2}$ = 7 m at 1 atm C_2N_2 (7). A similar rapid inactivation by a carbodiimide was ascribed to <u>intra</u>-molecular amide bond formation in the active site (19). More protracted C_2N_2 treatment produces almost completely inactivated enzyme It displayed increased thermal stability (ΔT = $+3^{o}$), increased polydispersity, lower pI values by isoelectric focusing, and no di- or oligomerization when examined by SDS-PAGE. Spectroscopic examination by IR and UV showed no significant differences. More critically CD spectra were essentially unaltered showing that the modification did not alter the conformation to a detectable degree as was also noted with BCAII. A higher degree of cross-linking was seen at pH 7 than 4; also measurable amounts of ^{14}C-C_2N_2 were incorporated only at pH 7. The exhaustively C_2N_2 modified enzyme showed 4.6-6.1 LYS residues and 1.7-5.4 GLY+ASP residues modified. No other modifications were noted.

Oxy-, deoxy-, and methemoglobin are rapidly modified by C_2N_2. PAGE reveals that covalently linked subunits corresponding to dimers, trimers and tetramers are formed. HPLC analysis of C_2N_2 treated human oxy-Hb by the method of Shelton et al. (20) shows that the α and β chains are modified and completely disappear from the profile. The α chain is depleted more rapidly by C_2N_2 treatment and a higher molecular weight component elutes with the composition of the α-chain, presumably α_2. Other products with compositions closer to β_n and $(\alpha\beta)_m$ are noted. On the basis of PAGE analysis and retention times they are presumed to be dimers.

The modified oxy-Hb displays the same visible spectrum as the native. However, the modified Hb shows two properties of interest. The first is the reversibility of some of the modification. PAGE always shows a major monomer band even though no α or β remain on the HPLC profile. The susceptibility of carboxylimidizole to nucleophile treatment could mean that PAGE treatment itself cleaves such bonds. Likewise any acyl-guanidino links identified as being formed in the model system ARG in 90% aq. diglyme (Day, unpublished) would likewise be vulnerable. Treatment of the C_2N_2 modified Hb at pH 2.2 at 2° for 3 days restores a substantial fraction of the α and β components to the HPLC profile. Again the same two linkages would be acid labile.

The second phenomenon is the marked increase in the rate of autooxidation in hemoglobin treated in either of two ways. First, if the hemoglobin has been treated with C_2N_2 for a very long time (>30 min.), then the autooxidation proceeds with half time in the tens of minutes. The second way is to physically agitate the hemoglobin solution during the C_2N_2 treatment (product = XL-Hb'); the agitation is mild enough that control Hb agitated for up to ten times as long displays no measurable increase in susceptibility to autooxidation. Thus, longer exposure times or physical agitation produced a lability in the modified Hb. The degree of subunit aggregation as measured by HPLC and by PAGE was somewhat increased. No oligomers of $n>4$ were seen in any case. Dithionite reduction of control methemoglobin produced a stable deoxy Hb whereas reduction of the modified XL-Hb' produced a deoxy Hb that retained the susceptibility to autooxidation. These results are consistent with the "trapping" of (a) salt-bridge(s) in a non-native, distorted form (Hb'). The new covalent bond then locks the polypeptide chain into a permanent conformation inimical to protection of the Fe(II)-heme (XL-Hb') (see Figure 1). The XL-Hb arising from a 2 minute, 1 atm C_2N_2 treatment of Hb does not show the susceptibility to autooxidation. It has been suggested that ion pairs may not exert a crucial influence on the stability of structures or on a pathway of folding (21); this may be correct. However, it appears from our data that we may indeed detect such intermediate forms; it remains to sort out conformations that are "productive" i.e. on their way to the correct conformation as opposed to those that may not. This is a problem common to all such approaches (see Ref. 10).

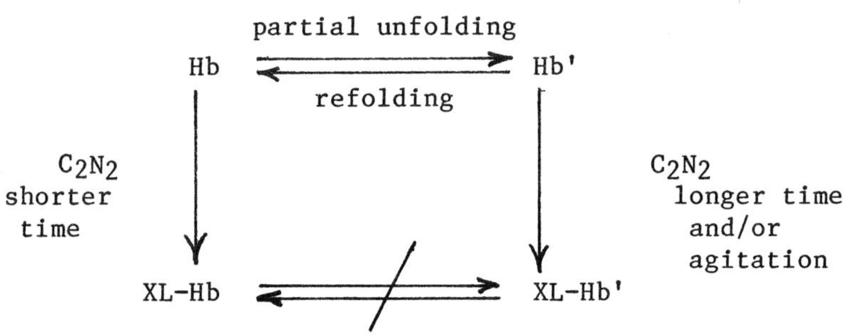

Figure 1: Hemoglobin Modification with C_2N_2. See text for definition of terms.

In conclusion C_2N_2 affords a means of converting some salt bridges into amide links. The reactivity may be a function of hydrophobicity of the salt bridge environment among other factors. The major effort now is focused on developing a HPLC based strategy for identifying the specific amino acids cross-linked.

REFERENCES

1. T.E. Creighton and D.P. Goldenberg (1984) J. Mol. Biol. 179, 497.
2. M. Bodanszky (1984) "Principles of Peptide Synthesis" p. 9, Akademie-Verlag, Berlin.
3. J.W. Kirley (1985) Ph.D. Dissertation, University of Cincinnati.
4. R.M. Tharp (1987) Ph.D. Dissertation, University of Cincinnati.

5. R.A. Day, S. Mothersole, and M.E. Madis (1986), Proc. Am. Soc. Mass Spectrom. 34, 36.
6. T.K. Brotherton and J.W. Lynn (1959) Chem. Revs. 59, 841.
7. E. Ciganek, W.J. Linn, and O.W. Webster (1970) "The Chemistry of the Cyano Group" p. 428, Interscience New York.
8. J.W. Kirley, R.A. Day, and G.P. Kreishman (1985) FEBS Lett. 193, 145.
9. H. Gu (1987) M.S. Thesis, University of Cincinnati.
10. D.W. Fassett, D.D. Irish, and F.A. Patty (1963) in "Industrial Hygiene and Toxicology" (F.A. Patty, ed.) Vol. II, 2nd Ed., p. 2003 Interscience, New York.
11. S.J. Mothersole (1986) M.S. Thesis, University of Cincinnati.
12. K. Kim and W.T. King (1984) J. Chem. Phys. 80, 974.
13. J.W. Kirley and R.A. Day (1985) Biochem. Biophys. Res. Commun. 126, 457; ibid., 130, 1324.
14. J.M. Clark, Jr. and R.L. Switzer (1977) "Experimental Biochemistry" Second Edition, p. 19, Freeman, New York.
15. P.G. Righetti and J.W. Drysdale (1979) "Isoelectric Focusing" p. 419, North Holland, Amsterdam.
16. M. Bodanszky op. cit., pp. 43, 142.
17. P.E. Wilcox (1972) Methods Enzymol. 25, 596.
18. J.E. Coleman (1984) in "Biology and Chemistry of the Carbonic Anhydrases", pp. 26-48, 429. New York Academy of Science.
19. S. Waley (1975) Biochem. J. 149, 547.
20. J.B. Shelton, J.R. Shelton and W.A. Schroeder (1984) J. Liq. Chrom. 7, 1969.
21. R.L. Baldwin and D. Eisenberg (1987) in D.L. Oxender and C.F. Fox, ed. "Protein Engineering" Allan R. Liss, NY p. 140.

51

THERMODYNAMIC BOOKKEEPING

W. Terry Jenkins

Department of Chemistry
Indiana University
Bloomington, Indiana, U.S.A.

I. INTRODUCTION

If a thermodynamic equilibrium is affected by a perturbant ion or molecule (X) then a graph of the Logarithm of the observed equilibrium constant (Log K_{obs}) with respect to Log[X] has a slope (n_x) corresponding to the net number of ions or molecules consumed in the equilibrium (Wyman, J. Jr.,1969, Record, M. T., et al. 1978) since:

$$d.\text{Log } K_{obs} = n_x.d.\text{Log}[X] \qquad (1)$$

Eq.1 has been used to study the effects of protons on binding O_2 to hemoglobin (Wyman, J. Jr.,1969) and substrates/inhibitors to enzymes (Dixon, M.,1953). Eq.1 has also been used to analyze salt effects on the interactions between proteins and nucleic acids(Record, M. T. et al. 1978), resin-nucleotide interactions (Jenkins, W. T., and Bowman, W. S.,1988), and protein-nucleotide interactions (Jenkins, W.T., 1989).

A. pH as a perturbant

When the pH is varied interactions with both protons and hydroxide ions should be considered:-

$$d.\text{Log } K_{obs} = n_{pH}.d.(pH) = (n_{OH} - n_H).d.(pH) \qquad (2)$$

However, since proton consumption is equivalent to a combination of water consumption with hydroxide formation, the individual stoichiometries, n_H and

n_{OH} cannot be determined experimentally from the effects of pH on the equilibrium constant.

Plots of Log K_{obs} versus pH are frequently curved. This is due to titratable groups on either the reactants or products. Dixon (I953) described how the pKa's of such groups could be derived from the curved Log K_{obs}/pH graphs.

B. Salts as perturbants

When the salt concentration is varied interactions with both anions and cations should be considered. It can be shown that if:-

$$d.\text{Log } K_{obs} = n_{salt}.d.\text{Log}[\text{salt}] \quad (3)$$

then $n_{salt} = (n_{cation} + n_{anion})$ regardless of the relative charges on the ions. Interestingly, for reactions involving macromolecules plots of Log K_{obs}/Log[salt] are frequently linear.This is due to the fact that the salt ions are acting as counterions for the macromolecule and hence do not, thermodynamically, dissociate. Consider for example the chloride in Dowex-1-Cl. For the same reasons the ions vary markedly in effectiveness according to their ion exchange selectivities (Norne, J-E., et al. 1975).

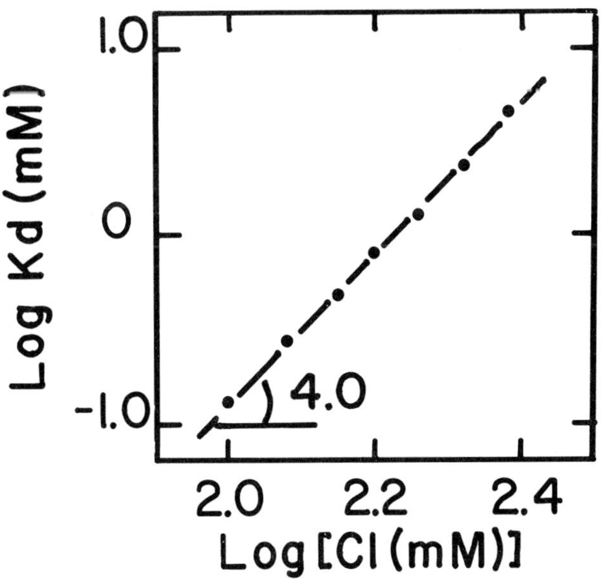

Fig.1 Ion-exchange effect of chloride on binding ATP to Dowex-1-Cl. Data taken from Fig.3 of Jenkins and Bowman,1988.

Fig.1 shows the effect of salt on the dissociation of ATP^{4-} from Dowex-1-Cl, a reaction which involves the uptake of 4 chlorides.:-

$$R.ATP + 4\,Cl^{-} \rightleftharpoons R.Cl_4 + ATP^{4-} \qquad (4)$$

When small charged ligands bind to macromolecules the thermodynamic stoichiometries for cations and anions can be elucidated. Because the ultrafiltrate must be neutral, the summation $\Sigma n_x.v_x = 0$ where v_x is the valency of the species X of which n_x moles are used in the reaction. For example, one can show that for NaCl:-

$$n_{Na} = (n_{NaCl} + n_{pH} + v_L)/2 \qquad (5)$$

and

$$n_{Cl} = (n_{NaCl} - n_{pH} - v_L)/2 \qquad (6)$$

where v_L is the charge on the free ligand. When ATP dissociated from bovine serum albumin at pH 7.7 it was found, experimentally, that $v_L = -3.9$, $n_{pH} = 0.87$ and $n_{NaCl} = 0.97$ from which it was calculated (Eqs.5,6) that $n_{Na} = -1.03$ and $n_{Cl} = +2.0$. The data thus indicated that the dissociation of ATP at high pH could be written:-

$$P.ATP + 2Cl^{-} \rightleftharpoons P.Cl_2 + Na^{+} + H^{+} + ATP^{4-} \qquad (7)$$

and that the bound ATP was, in effect, the dianion $(ATP.Na)^{2-}$.

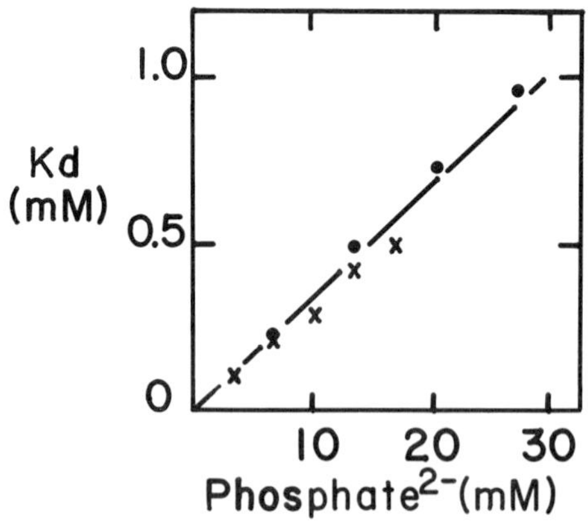

Fig.2 Ion exchange effect of dibasic phosphate on ATP binding to bovine serum albumin. ● – ●, pH 7.2, x - x, pH 6.6. Kd values were determined from Scatchard plots, using 0.15 mM protein, 0.1 mM total ATP and Amicon Centricon-30™ ultrafilters.There was approximately one ATP binding site per albumin in this pH range.

Fig.2 shows the effect of phosphate on this same reaction in the absence of sodium chloride. The data indicate that the dibasic phosphate anion competed (c.f. Velick, S.F. 1949) with the ATP and that no ATP dissociated in the absence of phosphate.These experiments indicate that ATP is binding to two vicinal protein positive charges through its terminal dibasic gamma phosphate moiety. This interpretation was supported by the observation that magnesium ions, which are known to complex with the terminal phosphate(Huang and Tsai, 1982), prevent ATP binding to serum albumin(not shown).

Fig.3 shows the effect of sodium chloride on the chromophoric proton dissociation from the pig heart cytosolic aspartate aminotransferase (Jenkins,1980). The proton is thought to arise from the Schiff base nitrogen through which the phosphopyridoxal is linked to a specific lysine epsilon amino group of the apoenzyme. In order to maintain charge neutrality in the ultrafiltrate this proton dissociation must be associated with the dissociation of an anion and/or the uptake of a cation. The log/log plot has a slope $n_{NaCl} = n_{Na} + n_{Cl} = -0.42$. Since for neutrality $n_{Na} - n_{Cl} = +1$ one can calculate that $n_{Na} = 0.29$ and $n_{Cl} = -0.71$. Both anions and cations were, therefore, involved and the acid dissociation can be written:-

$$0.29\ Na^+ + EH \rightleftharpoons E + H^+ + 0.71\ Cl^- \qquad (8)$$

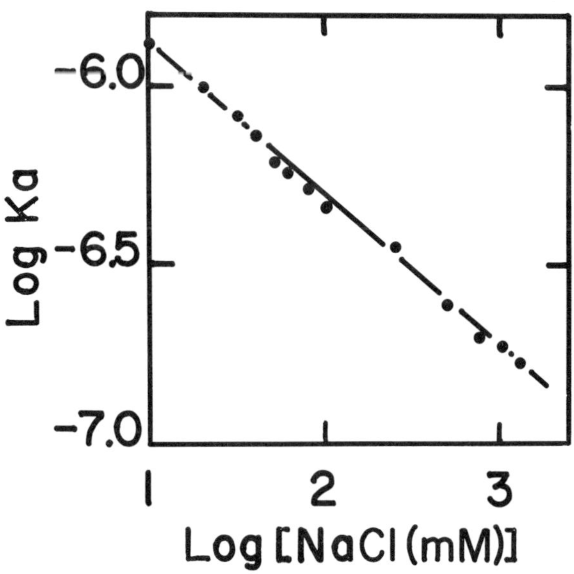

Fig.3. Effect of salt on the chromophoric dissociation of a proton from the phosphopyridoxal prosthetic group of pig heart cytosolic aspartate aminotransferase. The data were taken from Table I of Jenkins(1980).

II. BIOENERGETICS

For competitive ligands the energy contribution to the observed dissociation constant free energy (ΔG_o = -RT.Ln K_{obs}) is :-

$$-RT\ Ln(1 + [X]/K_x) = +RT.Ln(1 - n_x) \qquad (9)$$

When [X] = K_x then $n_x = 0.5$ and the energy contribution is -RT.Ln2= -409 cals (-1.7 kJ). For macromolecular counterions on the other hand, where $K_{obs} = K_L[X]^n x/K_x$, the energy contribution is $-RT\ Ln([X]^n x/K_x)$ where $K_x = [M][X]^n x/[MX_{nx}]$ is the dissociation constant of the perturbant X from the macromolecule(M). Counterions are thus extremely important, bioenergetically, since they contribute much more energy ($[X]^n x/K_x >> 2$) than do competitive ligands.

III. SUMMARY

The dissociations of charged ligands from macromolecules necessarily involve other charged ions as reactants in order to maintain charge neutrality in the ultrafiltrate. The stoichiometries for salt anions and cations can be elucidated from the slopes of Log K_{obs}/Log[salt] and Log K_{obs}/pH plots together with a knowledge of the charge on the free ligand (v_L). Evaluation of these thermodynamic stoichiometries allows one to account for the charges in the reaction and the overall free energies contributed by these reactant protons and salt ions.

IV. REFERENCES

Dixon, M.(1953) *Biochem. J.* **55**, 161-170.

Huang, S. L., and Tsai, M-D. (1982) *Biochemistry* **21**, 951-959.

Jenkins, W. T.(1980) *Arch. Biochem. Biophys.* **205**, 579-586.

Jenkins, W. T.(1989) Manuscript submitted for publication.

Jenkins, W.T., and Bowman, W. S.(1988) *Biochem. Biophys. Acta* **932**, 268-276.

Norne, J-E., Hjalmarsson, S-G., Lindman, B., and Zeppezauer, M. (1975) *Biochemistry* **14**, 3401-3408.

Record, M. T. Jr., Anderson, C. F., and Lohman, T. M. (1978) *Quart. Revs. Biophys.* **11**, 103-178.

Velick, S. F. (1949) *J. Phys. Chem.* **49**, 135-149.

52

INVESTIGATION OF HELIX DIPOLE AND SALT BRIDGE STABILIZATION OF SYNTHETIC HELICAL PEPTIDES USING CD AND NMR

Erin K. Bradley
John F. Thomason
Fred E. Cohen
Irwin D. Kuntz

Department of Pharmaceutical Chemistry
University of California, San Francisco
San Francisco, California

I. INTRODUCTION

The α-helix is a predominant structural motif in proteins, but it was thought for many years that small peptides (<20 residues) could not exist as stable monomeric helices in aqueous solution. Brown and Klee (1971) found, however, that monomeric C-peptide (residues 1-13 of Ribonuclease A) is partially helical under aqueous conditions at low temperatures. Bierzynski et al. (1982) investigated the C-peptide further to determine the forces responsible for helical stability. They initially proposed that the helical conformation was stabilized by a salt bridge between Glu 9 and His 12 because of the pH titration behavior of the helicity (Bierzynski et al., 1982; Kim et al., 1982). The maximum helical content occurred at pH 5 and dropped upon raising or lowering the pH. However, subsequent removal and/or substitution of the proposed salt bridge residues by Shoemaker and co-workers (1985, 1987) did not always reduce helical content or alter the pH titration behavior. Based on their work with C-peptide analogs, Shoemaker et al. (1985, 1987) have suggested that interactions of charged side chains with the helix dipole can both stabilize and destabilize helical structures.

The helix dipole results from the alignment of successive peptide bond dipoles along the helix axis. Hol et al. (1978, 1981) have estimated the net charge at each end of an α-helix to be one half an elementary charge; +1/2 at the N-terminus and -1/2 at the C-terminus. As a result, a positively charged side chain should stabilize helical structure if it occurs near the C-terminus (negative pole of the helix) and destabilize it if it were present at the N-terminus (positive pole).

Marqusee and Baldwin (1987) found that salt bridges may contribute to helix stability in a poly-alanine based peptide with *i ->i+4* spacing of Glu and Lys side chains; whereas, the corresponding peptide with *i->i+3* spacing does not show stabilization by salt bridges. Both salt bridges and the helix dipole have been examined as electrostatic interactions that might stablize helical structure in the system described here.

This investigation started with the *a priori* design of a 17 residue monomeric helix with the premise that salt bridges would stablize helical structure. The first peptide sequence was chosen to allow *i ->i+3* salt bridges between charged side chains that alternate around the face of the helix to avoid an amphipathic structure. The selection of amino acid residues was based on preference for helical conformation (Chou and Fasman, 1974), ease of synthesis and purification, avoiding oligomerization, and of particular interest, ability to perform subsequent NMR assignments. The peptide sequence and possible salt bridges are shown in Figure 1. Four forms of the *(i+3)* peptide were examined: unblocked (free amino and carboxyl termini), N-blocked (acetylated amino termini, free carboxyl), C-blocked (amidated carboxyl termini, free amino), and N,C-blocked (acetylated amino termini, amidated carboxyl termini). An additional sequence was also synthesized to allow *i->i+4* salt bridges n helical conformation. The *(i+4)* peptide has the same amino acid composition as the *(i+3*) peptide, with only minor sequence alterations; also shown in Figure 1. Only two forms of the *(i+4)* peptide have been examined: unblocked and N,C-blocked. The secondary structure of these peptides has been examined with both circular dichroism (CD) and nuclear magnetic resonance (NMR) under various experimental conditions.

FIGURE 1: Peptide Sequences

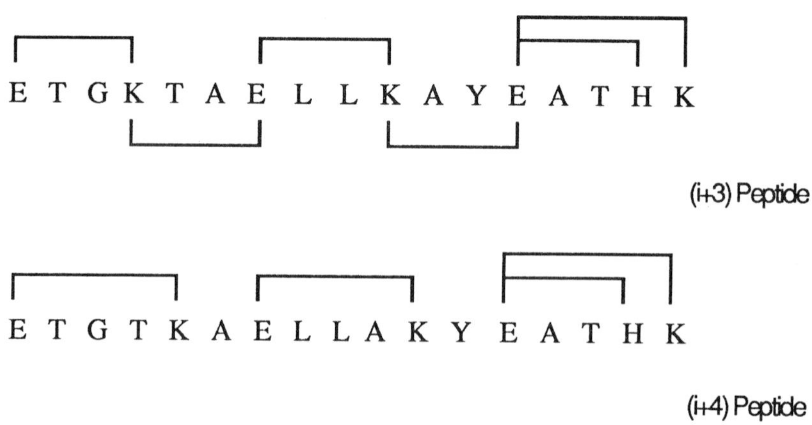

II. MATERIALS AND METHODS

A. Peptide Synthesis and Characterization

The peptides were synthesized with solid phase peptide synthesis techniques (Merrifield, 1963) using standard tBoc protocol on an Applied Biosystems 430 Peptide Synthesizer. HF cleavage was used to remove the peptides from the resin (Sakakibara et al., 1965,1968). Peptides with free carboxyl termini were obtained after HF cleavage from synthesis using 4-(oxymethyl)-phenylaceto-amidomethyl-resin (Mitchell et al., 1976, 1978), and peptides with carboxy-amide termini resulted from use of p-methyl-benzylhydroxylamine resin. The amino termini were acetylated using 10% acetic anhydride in dichloromethane for 10 minutes.

The peptides were purified using HPLC; both reverse phase C-18 with water/acetonitrile gradients in the presence of 0.1% trifluoro acetic acid (TFA) and cation exchange MONO S (Pharmacia) with 30% isopropanol/ 0.1% TFA with NaCl gradient. Amino acid composition was verified using amino acid analysis and mass spectrometry. Subsequent NMR resonance assignments also verified expected sequences.

B. Circular Dichroism

Samples for CD measurements were approximately 150 μM in peptide. These concentrations were obtained by dilution of stock 1.5 μM solutions of peptide in buffer solution of 1mM sodium citrate/ 1mM sodium borate/ 1mM sodium phosphate and the indicated sodium chloride concentrations. The pH was adjusted with HCl and NaOH.

Spectra were recorded at 5°C on a Jasco J500A spectropolarimeter; the instrument was calibrated with (+)-10-camphor sulfonic acid (Chen and Yang, 1977). Cuvettes of 10 mm and 1mm path lengths were used. Ellipticity is reported as [Θ] mean residue ellipticity (deg cm^2/dmol).

C. Nuclear Magnetic Resonance

Samples were prepared for 2 dimensional NMR (2D NMR) as approximately 5 mM solutions of N,C-blocked *(i +4)* peptide in either D_2O or 90% H_2O/ 10% D_2O. The pH was adjusted using 0.1 N HCl or 0.1 N NaOH. The sample for pH titration was prepared as approximately 0.5 mM solution in D_2O and the pH was adjusted with KOD and DCl. Sodium trimethylsilyl propionate (TSP) was used as an internal standard with appropriate pH corrections (Demarco, 1977). pH is reported without correction for deuterium.

Double quantum filtered phase-sensitive COSY (Marion and Wuthrich, 1983; Rance et al., 1983) and HOHAHA (Bax et al.,1985; Davis et al.,1985)) were used for spin system assignments. Phase-sensitive two dimensional

NOESY (Jeener et al.,1977) was used to evaluate secondary structure. Spectra were taken using different mixing times: 200 ms, 150 ms, 100 ms.

All spectra were obtained at 5°C on a General Electric GN-500 NMR; spectrometer was modified for the HOHAHA experiment as described in Basus et al. (1988). A total of 512 t_1 increments were acquired for the NOESY and 1024 for the COSY experiments, while 4096 complex data points were acquired in t_2. Spectral widths of 6024 Hz were used in both dimensions. The carrier frequency was placed on the H_2O resonance, which was irradiated at all times except for evolution and acquisition.

III. RESULTS

A. CD Measurments

The CD measurements indicate that all the peptides have spectra characteristic of α-helical structure at 5°C and pH 2. The spectra all show a minima at 222 nm (helical *n->π** transition) and a minima between 202 - 208 nm (overlapping helical and random coil π->π* transitions at 208 nm and 200 nm respectively). The spectra for the unblocked (*i+4*) peptide are shown in Figure 2. The ellipticity at 222 nm is used as a measure of helicity.

FIGURE 2:
CD Spectra for pH Titration of Unblocked (i+4) Peptide
5°C, 0.1 M NaCl

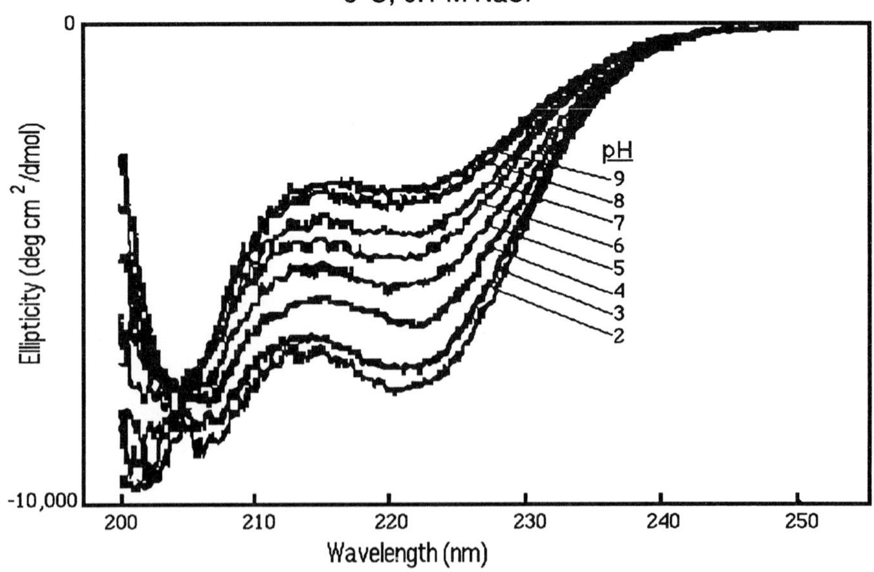

The helical content is affected by the pH, salt concentration, temperature, and state of the N and C-termini. The maximum helical content achieved by both the *(i+3)* and the *(i+4)* N,C-blocked peptides was approximately 30% for a helix of 12 residues; estimated as in Wu et al. (1981) with a constant of -28,400 units for a 12 residue helix. The ellipticity constant for 100% helix depends on the length of the helix involved. For example 100% helix results in a mean residue ellipticity of -37,600 and -28,100 for a poly-Lys helix and a 10 residue protein helix, respectively (Yang et al., 1986); -28,400 for a 12 residue protein helix (Wu et al., 1981). Marqusee and Baldwin (1988) have found -36,000 to be the maximum ellipticity for their 17 residue helix.

All the peptides have maximum helical content at low pH (Figures 3a and 3b). When the charges were removed from the N- and C-termini of the initial peptide (acetylation of the amino-terminus and/or amidation of the carboxyl-terminus), the helical content increases at pHs where the free termini would be charged. For example the C-blocked *(i+3)* peptide at high pH and the N-blocked *(i+3)* peptide at low pH in Figure 3a. When both the termini have been blocked the ellipticity at 222 nm appears to titrate in the pH range expected for the histidine side chain.

Increasing the salt concentration, from 0.1 M to 1.0 M NaCl, decreased the helical content across the entire pH range for the *(i+3)* and *(i+4)* N,C-blocked peptides. However, it did not eliminate the dependence of helix content on pH. The salt concentration effects are shown in Figure 4.

The amount of helix decreases as the temperature increases for both the *(i+3)* and *(i+4)* N,C-blocked peptides. The transition appears to be two state, indicated by an isodichroic point in both sets of spectra. The range of "melting" is from 5°C to 55°C, however, 5°C is an artificial limit imposed by working with aqueous solution. The helical content may increase further with decreasing temperature. The CD spectra for the temperature transition of the N,C-blocked (i+4) peptide is shown in Figure 5.

FIGURE 3a:
pH and Blocked Termini Effects on Helicity

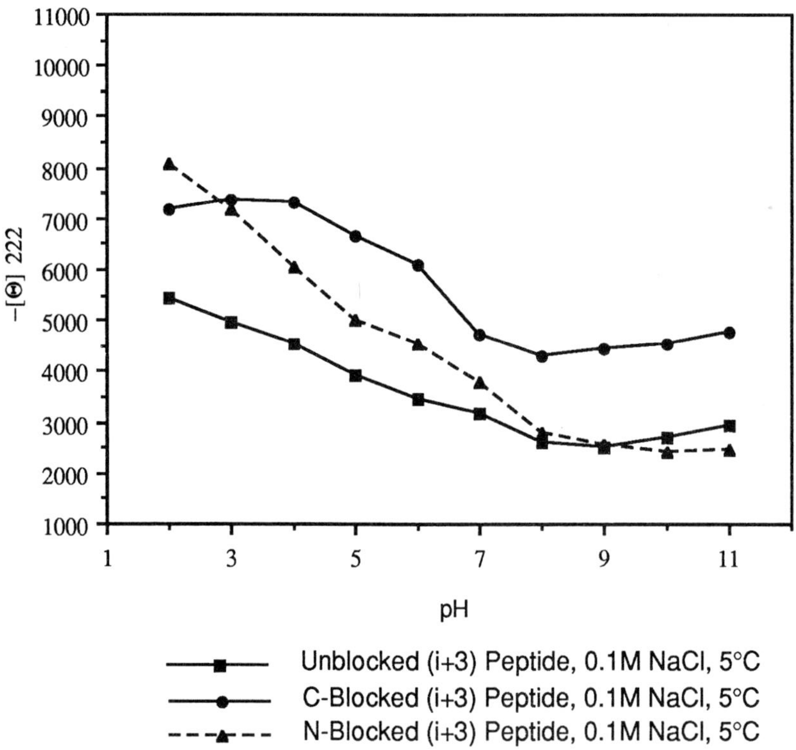

FIGURE 3b:
pH and Blocked Termini Effects on Helicity

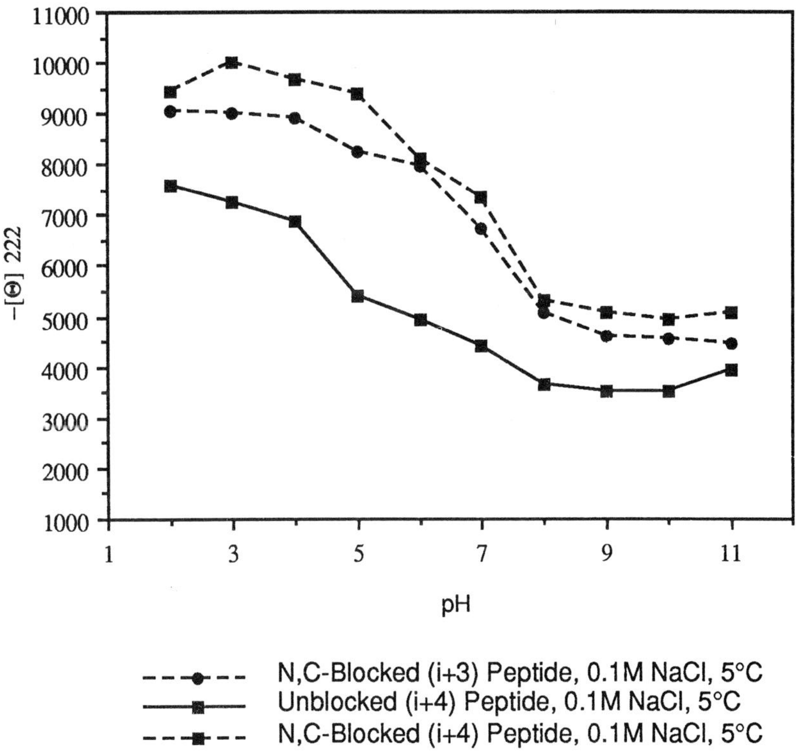

FIGURE 4:
pH and Salt Effects on Helicity

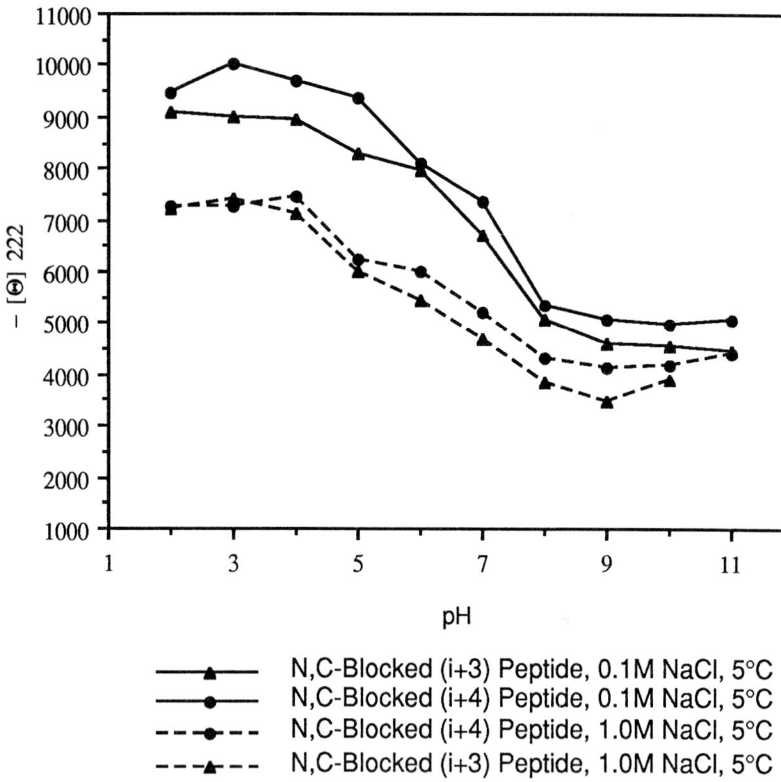

FIGURE 5:
Temperature Transition of the N,C-Blocked (i+4) Peptide
pH 2.0, 0.1 M NaCl

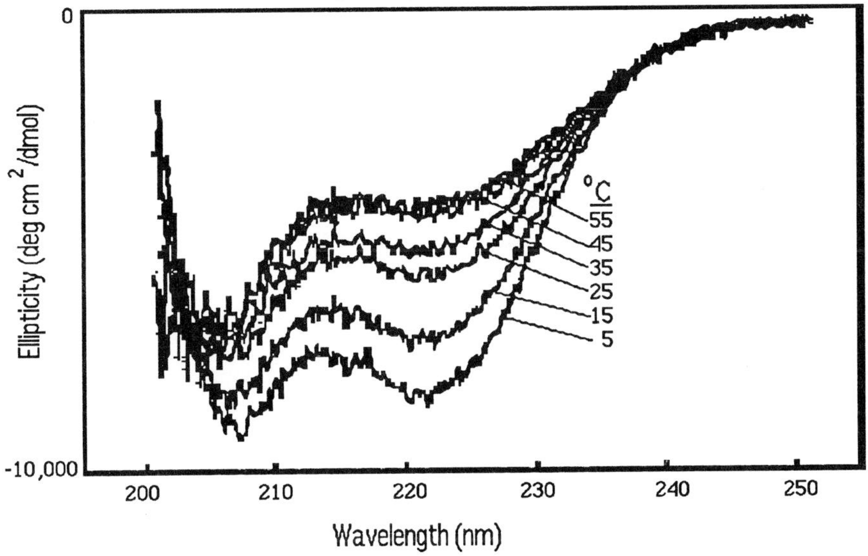

B. NMR Resonance Assignments

Resonance assignments of the side chains were achieved using COSY and HOHAHA spectra. There was some initial difficulty with main chain assignments because Glu 7, Ala 10, and Lys 11 have overlapping Cα–NH crosspeaks in the COSY and NOESY. Ala 10 and Lys 11 actually have identical NH and Cα proton resonances. The HOHAHA spectra made it possible to distinguish which side chains were connected to the overlapping crosspeaks in the finger print region.

All resonance assignments were made in manner of Wuthrich et al. (1982) and are listed in Table 1. Amide proton resonance assignments were made on the basis of NH_i-$C\alpha_i$ crosspeaks in the COSY performed in 90% D_2O. Sequence specific assignments were performed using the $C\alpha_i NH_{i+1}$ crosspeaks in the NOESY spectra performed in 90% D_2O.

TABLE 1:
NMR Resonance Assignments for the N,C-Blocked (i+4)Peptide at 5°C, pH 2.0

Residue	NH	CαH	CβH	CγH	CδH	CεH
Acetyl		-----				
Glu 1	8.55	4.40	2.14 2.05	2.52		
Thr 2	8.38	4.41	4.28	1.23		
Gly 3	8.54	4.07				
Thr 4	8.06	4.31	4.42	1.26		
Lys 5	8.68	4.07	1.86	1.42	1.72	-----
Ala 6	8.43	4.16	1.42			
Glu 7	8.08	4.19	2.18	2.53		
Leu 8	8.36	4.16	1.73 1.83	1.56	0.89	
Leu 9	8.13	4.32	1.74	1.67	0.95 0.89	
Ala 10	8.05	4.19	1.49			
Lys 11	8.05	4.19	1.91	1.44	1.71	-----
Tyr 12	8.23	4.31	3.16 3.24		7.04	6.62
Glu 13	8.73	4.14	2.16	2.69		
Ala 14	8.01	4.24	1.53			
Thr 15	7.73	4.31	4.21	1.19		
His 16	7.96	4.63	2.75 3.17		7.16	8.41
Lys 17	8.32	4.28	1.82	1.46	1.70	-----
Amide	7.78					

C. Evidence of Secondary Structure in NOESY Spectra

We have attempted to both evaluate the position of the helix within the peptide chain and estimate the relative populations of helical and random coil conformations using the intensities of the NOESY crosspeaks. The crosspeak intensities in a NOESY depend on the distance between the two protons. Different secondary structures result in different intervening distances between main chain protons. For example NH_i and NH_{i+1} are approximately 2.8 Å apart in an α-helix ($\phi = -57°$, $\psi = -47°$) and 4.3 Å apart in an extended (β) chain ($\phi = -139°$, $\psi = -135°$)(Wutrich et al., 1984). First, the expected intensity ratio of $NH_{i+1}NH_i$ to $NH_{i+1}C\alpha_i$ crosspeaks were calculated for both an extended chain and an α–helix as shown in Equation 1, using the distances from Wuthrich et al. (1984). The extended chain was used as a model of a random coil.

$$\text{Ratio} = \frac{I_{NN}}{I_{C\alpha N}} = \frac{(d_{NN})^{-6}}{(d_{C\alpha N})^{-6}} \qquad \text{Equation 1}$$

The calculated intensity ratios (3.81 for α–helix and 0.018 for extended chain) were then compared to our experimental intensity ratios for each dipeptide along the chain to determine the relative populations of helical and extended chain conformations; shown in Figure 6.

FIGURE 6:
Crosspeak Intensity Ratios

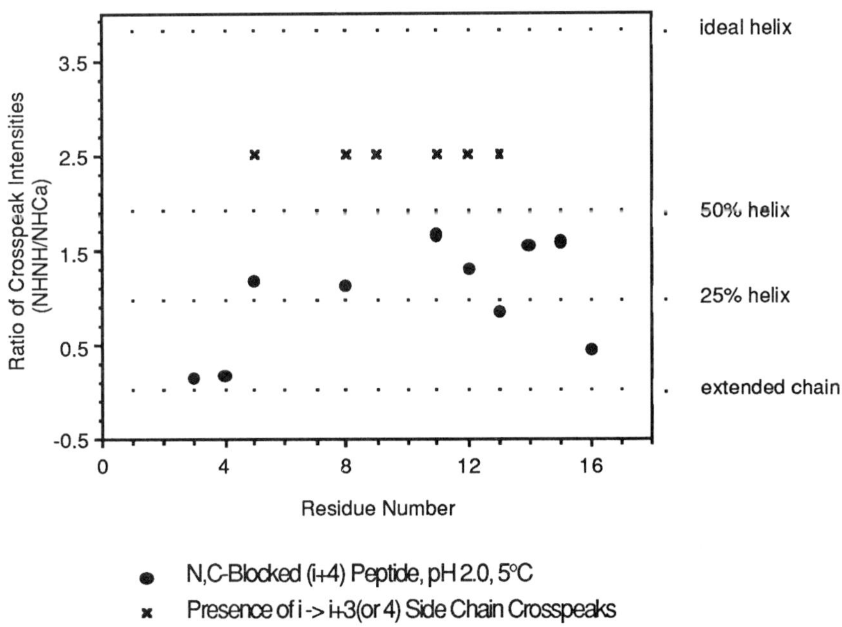

These comparisons indicate that the N,C-blocked (*i+4*) peptide has helical and non-helical regions. Residues 1 through 4 appear to have little or no helical content, whereas residues 5 through 15 appear to be helical in 30 to

40 percent of the time averaged population.This compares with the calculated helical content from the CD measurements at 30 percent for a helix of 12 residues.

In addition to crosspeak intensity ratios there are several long range crosspeaks between side chain protons of residues with spacing *i->i+3* and *i->i+4* along the sequence. This pattern of long range crosspeaks occurs in an α-helix, where the secondary structure places residues close to each other that are three or four apart in the primary sequence. These crosspeaks are shown schematically in Figure 7, and indicated as 'X's in Figure 6.

FIGURE 7: Long Range NOESY Crosspeaks

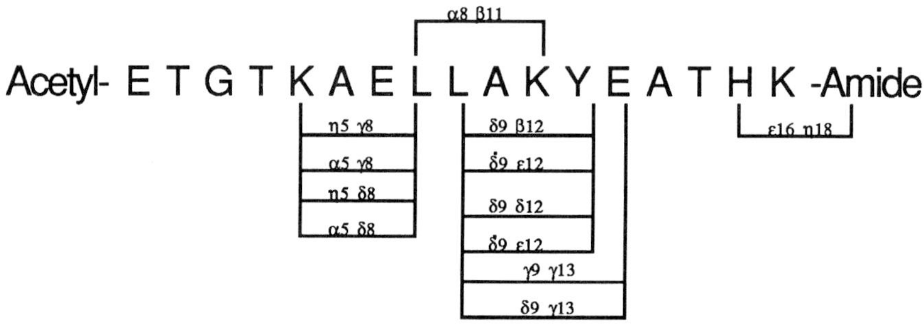

D. NMR Histidine Titration

The His 16 side chain appears to titrate between pH 6 and 8 with a pK_a of approximately 7.2. The CδH and CεH resonance were followed over this pH range and the results are shown in Figure 8.

IV. DISCUSSION

The pH titration data for neither the (*i+3*) peptides nor the (*i+4*) peptides are consistent with the hypothesis that salt bridges play a dominant role in the stabilization of helical conformation. Salt bridges between Glu and Lys would be expected to make the largest contribution when both side chains are charged, somewhere around neutral pH. Whereas all the peptides studied here have maximum helical content at low pH, between 2 and 4. In contrast to the work of Marqusee and Baldwin (1987), the (*i+4*) peptide does not have maximum helicity at neutral pH. Rather, as in the case of the *(i+3)* peptide, low pH yields maximum helicity (Figure 3). Thus the CD pH titration results indicate that salt bridges between the Glu and Lys side chains are not a major force in the stabilization of helical structure in either the (*i+3*) or the (*i+4*) peptides.

FIGURE 8:
pH Titration of His 16 Proton Resonances

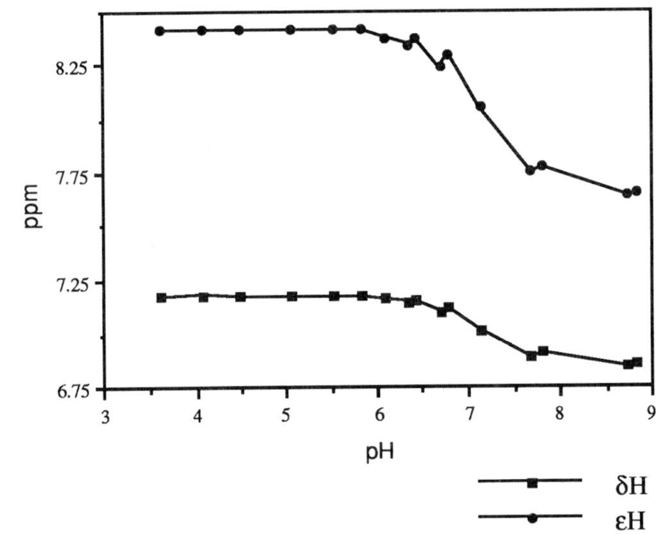

The data are much more consistent with the helix dipole model put forth by Shoemaker and co-workers (1986), where charged groups interact favorably and/or unfavorably with the helix dipole. This explains the increase in overall helical content when the N and C-termini are blocked (unfavorable charge interactions removed). For example, compare the N-blocked (*i+3*) peptide to the unblocked (*i+3)* peptide in Figure 3a. At pH 9, where both peptides have an uncharged N-terminus, they have the same ellipticity at 222nm. However at pH2, the N-blocked (*i+3*) peptide has a greater helical content than its unblocked analog. At this pH the N-blocked peptide has uncharged C- and N-termini, in contrast, the unblocked peptide still has a charged N-terminus which can interact unfavorably with the helix dipole.

The helix dipole model also largely explains the pH titration behavior remaining after both the N and C-termini are blocked. The increases in helicity of both the (*i+3*) and (*i+4)* N,C-blocked peptides is due to the titration of His 16 between pH 8 and 6, as indicated by the CδH resonance shifts in this range (Figure 8). As His 16 is protonated a favorable charge interaction is formed between the positive His side chain and the negative helix pole. The other helicity change, between pH 6 and 4 is probably due to one of the Glu side chains. Glu 14 is a good candidate because it is a negative side chain near the negative pole of the helix, which the NMR data suggest ends at Thr 15 (Figure 6). The unfavorable interaction is eliminated when the side chain is neutralized at low pH.

It is apparent from the CD spectra that all of the peptides mentioned here contain some helical structure in aqueous solution at low temperature and low pH. The question remaining is what portion of the molecules are helical and where is the helical structure located along the peptide chain. In this system 2D NMR holds some promise in answering these questions. Under one set of conditions (pH 2, 5°C) the combination of main chain cross peak intensities and long range side chain crosspeaks has allowed the location of the preferrred helical structure to be identified from residues 5 to 15, and the time averaged population of helical molecues to be estimated at 30 to 40%. These estimates based on NMR agree well with estimates from CD measurements under the same conditions.

Of interest in the comparison of NMR and CD is what each method is measuring. 2D NMR spectra can be used to look at each individual residue. NOESY crosspeaks contain information about distances between protons belonging to different residues; thus exact structural information can be obtained. The intensities of the crosspeaks, however, are a function of the population of molecules as well as distance. NMR in the present system measures an average crosspeak intensity over a population of structures.

The CD also measures a conformational average. Small peptide systems, like those investigated here, are composed of an ensemble of different structures. When this is taken into account, the agreement between NMR and CD is good for the N,C-blocked (*i+4*) peptide. The NMR indicates that residues 5 through 15 are in helical conformation in one of the structural states. Thus for CD calculations the 100% helix constant for 12 residues was utilized (Wu et al., 1981).

In contrast to our peptides, Dyson and co-workers (1988) found no evidence of helical structure in myohemerythrin C-helix peptide (residues 69-87) with CD. However, there were 2D NMR cross peaks (N*i*N*i+1*) which indicated the presence of helical structure. They have suggested that CD measurements of helix require long range ordering of residues in the α-helix, whereas N*i*N*i+1* crosspeaks reflect local order and the presence of helical φ,ψ angles or helical turns in dipeptides. They further conclude that their C-helix peptide exists as an ensemble of structures that are interconverting with different residues in helical turns, while the predominant structure is an extended chain. This local conformation of helical turns they called a 'nascent helix' because it is readily stabilized in an α-helix by the addition of trifluoro ethanol (TFE) or in the native myohemerythrin.

While the (*i+3*) and (*i+4*) peptides mentioned here do not show long range backbone interactions they do show helical CD signals and long range crosspeaks reflective of helical structure in the NMR spectra. The question of what is actually occurring with regards to structure and populations when the CD helix content drops still remains to be answered. Present work is focusing on obtaining this information under different conditions (pH 5, 5°C) for the N,C-blocked (*i+4)* peptide where the helix content as shown by CD, appears to decrease. This will hopefully shed some light on the kinds of transitions that are occurring with pH. There is also the possibility of examining temperature transitions. At pH 2 and 25°C the helical content of the N,C-blocked*(i+4)* peptide is approximately the same as it is at pH 6 and 5°C.

It is not clear at this point whether these peptides will move to the state of nascent helices as the helix content decreases with increasing temperature or pH. But the NMR experiments under these conditions should prove interesting and useful in elucidating some of the different states of folded and unfolded secondary structure, with obvious ramifications to protein folding questions.

ACKNOWLEDGMENTS

We thank V.D. Daggett, Dr. J.T. Yang, and Dr. C-S.C. Wu for helpful discussion. This work was supported in parts by grant GM19267 and RR01695 from the NIH and the Macromolecular Work Bench Project (DARPA) and Instrument Grants from NIH and NSF.

V. REFERENCES

Basus, V.J., Billeter, M., Love, R.A., Stroud, R.M., Kuntz, I.D. (1988) Biochemistry 27, 2763-2771.

Bax, A., Davis, D.G. (1985) J. Magn. Reson. 65, 355-360.

Bierzynski, A., Kim, P.S., Baldwin, R.L. (1982) Proc. Natl. Acad. Sci. USA 79, 2470-2474.

Brown, J.E., Klee, W.A. (1971) Biochemistry 10, 470-476.

Chen, G.C., Yang, J.T. (1977) Anal. Lett. 10, 1195-1207

Chou, P.Y., Fasman, G.D. (1974) Biochemistry 13, 211-221.

Davis, D.G., Bax, A. (1985) J. Am. Chem. Soc. 107, 2820-2821.

Demarco, A. (1977) J. Magn. Reson. 26, 527-528.

Dyson, H.J., Rance, M., Houghten, R.A., Wright, P.E., Lerner, R.A. (1988) J.Mol. Biol. 201, 201-217.

Hol, W.G.J., van Duijnen, P.T., Berendsen, H.J.C. (1978) Nature 273, 443-446.

Hol, W.G.J., Halie, L.M., Sander, C. (1981) Nature 294, 532-536.

Jeener, J., Meier, B.H., Bachman, P., Ernst, R.R. (1979) J. Chem. Phys. 71, 4546-4553.

Kim, P.S., Bierzynski, A., Baldwin, R.L. (1982) J. Mol. Biol. 162, 187-199.

Marion, D., Wuthrich, K. (1983) Biochem. Biophys. Res. Commun. 113, 967.

Marqusee, S., Baldwin, R.L. (1987) Proc. Natl. Acad. Sci. USA 84, 8898-8902.

Merrifield, R.B. (1963) J.Am.Chem.Soc. 85, 2149-2153.

Mitchell, A.R., Erickson, B.W., Ryabtsen, M.N., Hodges, R.S., Merrifield, R.B. (1976) J. Am. Chem. Soc. 98, 7357-7362.

Mitchell, A.R., Kent, S.B., Englhard, M., Merrifield, R.B. (1978) J. Org. Chem. 43, 2845.

Rance, M., Sorensen, O.W., Bodenhausen, G., Wagner, G., Ernst, R.R., Wuthrich, K. (1984) Biochem. Biophys. Res. Commun. 117, 479.
Sakakibara, S., Yasutsugu, S. (1965) Bull. Chem. Soc. Japan 38, 1412.
Sakakibara, S., Nobuhiro, N., Yasiro, K., Shako, Y. (1968) Bull. Chem. Soc. Japan 40, 1477.
Shoemaker, K.R., Kim, P.S., Brems, D.N., Marqusee, S., York,E.J., Chaiken, I.M., Stewart, J.D., Baldwin, R.L. (1985) Proc.Natl. Acad. Sci. USA 82, 2349-2353.
Shoemaker, K.R., Kim, P.S., York, E.J., Stewart, J.D., Baldwin,R.L. (1987) Nature (London) 326, 563-567.
Wuthrich, K., Wider, G., Wagner, G., Braun, W. (1982) J. Mol. Biol. 155, 311-319.
Wuthrich, K., Billeter, M., Braun, W. (1984) J. Mol. Biol. 180,715-740.
Wu, C.C., Keda, K.I., Yang, J.T. (1981) Biochemistry 20, 566.
Yang, J.T., Wu, C-S.C., Martinez, H.M. (1986) Meth. Enzym. 130, 208-256.

Section VI

Introduction

THE STP-3 SAMPLE WORKSHOP

Jean Rivier

Clayton Foundation Laboratories for Peptide Biology
The Salk Institute for Biological Studies
La Jolla, California 92037

The first sample workshop was held on September 30, 1985 at the occasion of the first meeting of the American Protein Chemists. A 300 picomole aliquot of a synthetic 19-residue peptide was submitted to the registrants at a time when microsequencing was still in its infancy. Few participants could report a successful elucidation of the full structure of this peptide.

The primary structure of what was referred to as STP-1 (Symposium Test Peptide 1; reference 1) is reported in the Tablo shown below. Emphasis at the workshop dwelt on the fact that microsequencing could be carried out at very low levels (20 to 50 picomoles) and that reliable sequence information could only be obtained for approximately twenty residues, provided that conditions for the gas phase sequencers were optimized. Recovery of material from test tubes and stability of the samples were addressed. It was also recognized that amino acid analysis yielded little more information than that given by sequence analysis and very few participants reported on that basic technique.

STP-2 was provided to the participants of the first Symposium of the Protein Society and was designed to test not only the capabilities of various analytical instruments, but also the investigators' ingenuity and ability to detect unusual amino acids . The structure of STP-2 is shown in the Table. It is a 56-residue molecule branched at Lysine 19 (reminiscent of ubiquinated proteins) and has two hydroxyprolines as the N-termini of the two identical chains. In addition there are two palindromes within the first and second halves of the molecule: residues 2 to 17 (19 to 36) were palindromes of 37-53 except for Ala-Gly at positions 6-7 (24-25) and 48-49. Except for the fact that there were only two hydroxyprolines in the molecule, the overall composition was a multiple of 19 and could have been that of a l9-, 38-, or possibly 57-residue peptide. A molecular weight of 5446 was obtained by mass spectrometry which corresponded best to the largest of the three possibilities. Approximately three nanomoles of the peptide had been submitted and three groups of investigators obtained the correct sequence on the basis of microsequencing experiments and mass spectrometric analysis combined.

STP-3, whose structure is also shown in the Table was designed as shown in the contributions of this section.It proved to be a challenge best met using several techniques including Edman based sequence analysis, mass spectrometry and extensive use of fragmentation procedures, be they cyanogen bromide cleavage, enzymatic degradation, cysteine derivatization or hydrolysis of an Asn-Gly peptide bond using hydroxylamine. Approximately three nanomoles of STP-3 were provided to the participants attending the second meeting of the Protein Society in 1988.

In all three cases, results were discussed at a workshop on the evening of the first day of the meeting. These worshops have served the purpose of both defining the limitations of presently available techniques used for the characterization of peptides and proteins, and to help define those strategies found to be most successful. Mass spectrometry which appeared a year ago to be an invaluable technique for the determination of peptide size, appeared this year as a strong contender for being the most sensitive and most economical (in term of time) technique for the determination of peptide sequences. This is certainly well illustrated by the contributions of the groups of Drs. R. Paxton and D. Hunt. At a time when technological developments advance at a rapid pace, it is hoped that new developments in Edman chemistry as well as in mass spectrometry will provide a ten fold improvement in sensitivity within the next year or two. It is to be remembered by all, however, that the challenges presented by the analysis of the STPs are often dwarfed by those provided by nature. The contributions that follow in this chapter should give the reader a wider appreciation of the strengths and limitations of the presently available techniques in protein analysis.

---

1. Rivier, J., Galyean, R., Woo, W., Karr, D., Richmond, T., and Spiess, J. (1987), in: "Proteins, Structure and Function" (J. J. L'Italien, ed.), p. 789, Plenum Press, New York.

STP STRUCTURES

STP-1

1 5 10 15
Trp-Asn-His-Ala-Ala-Ala-Thr-Ser-Gln-Ala-Pro-Arg-Val-Ile-Ile-

19
Glu-Glu-Asn-Asp-OH

STP-2

1 5 10 15
(Hyp-Ile-Ile-Ser-Thr-Ala-Gly-Arg-Gly-Gly-Ser-Glu-Val-Pro-Val-

40 45
Ser-His-Lys)$_2$-Lys-His-Ser-Val-Pro-Val-Glu-Ser-Gly-Gly-Arg-

50 56
Ala-Gly-Thr-Ser-Ile-Ile-Gly-Gly-Gly-NH_2

STP-3

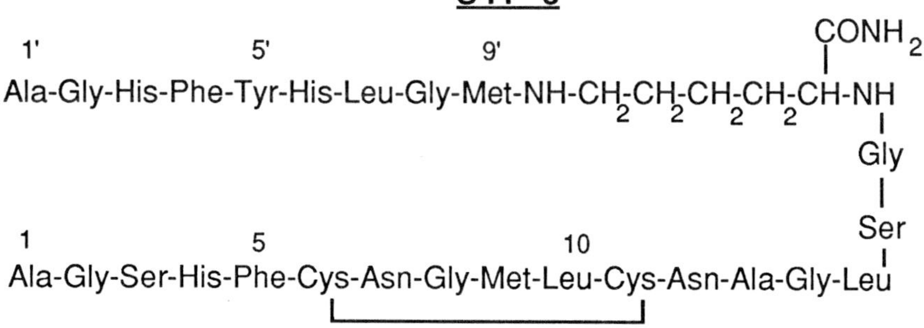

53

DESIGN AND SYNTHESIS OF SYMPOSIUM TEST PEPTIDE-3 (STP-3)

Carl Hoeger, Robert Galyean, Richard McClintock, John Dykert, N.Leo Benoiton[1], and Jean Rivier

Clayton Foundation Laboratories for Peptide Biology
The Salk Institute for Biological Studies
La Jolla, California 92038

I. INTRODUCTION

As a result of technological advances in the separation sciences such as HPLC (1) and electrophoresis (2), the number of new, biologically significant peptides/proteins isolated has increased (3) markedly. In addition, analytical techniques such as gas phase sequencing (4) and mass spectroscopy (5) have become ever more sensitive. Handling and transfer of minute quantities of peptides/proteins has become one of the major stumbling blocks being addressed by biochemists. Under the aegis of the Protein Society, we have, for the third year, designed and prepared a synthetic sample peptide that has come to be known as the Symposium Test Peptide-3 (STP-3). The sample, the structure of which was undisclosed until the evening of the workshop, was provided to approximately 180 registrants upon request. It was designed to provide an opportunity for researchers to utilize isolation and characterization methods applied or developed in his laboratory and to share with others some of the frustrations

[1]Present address: Department of Biochemistry, University of Ottawa, Ottawa, Ont., Canada.

and enlightenment he may have encountered during this process of structure determination.

The structure of STP-3 is given in Figure 1a. An examination of its primary structure reveals that it is modeled after the ubiquinated class of proteins (6) in that both the alpha and epsilon amino groups of lysine (at position 18) are acylated. This produces two N-termini (positions 1 and 1'). STP-3 also contains an intrachain disulfide linkage between cystiene side chains at positions 9 and 14, the first time this feature has been incorporated into an STP. The two chains of different length are referred to as the long (17 residues; 1-17) and short (9 residues; 1'-9') arms of this peptide; they are partially homologous and will give two independent sequences when subjected to Edman degradation . STP-3 was designed to be resistant to the usual enzymes employed in degradative structure determinations (i.e. trypsin), but can be chemically degraded using cyanogen bromide (CNBr) because

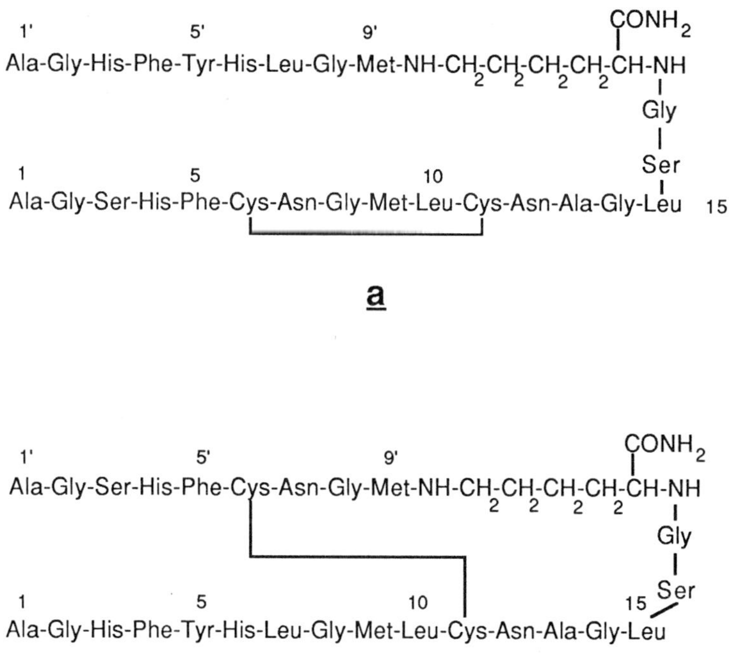

Figure 1. Structure of: a) STP-3 and b) STP-3 alternative (STP-3a)

two methionine residues are present. This reagent generates two fragments from the unreduced peptide and three from the fully reduced and cysteine-modified peptide. Thus, one conceivable sequencing strategy would consist of: 1) amino acid composition; 2) mass spectral analysis to determine the presence of the C-terminal amide of lysine; 3) sequencing of both the reduced (and alkylated) and unreduced forms of STP-3 ; 4) cleavage, isolation, and sequencing of the CNBr-derived fragments (from both the reduced and unreduced forms of STP-3); and 5) duplication by chemical synthesis. This strategy could lead to a correct structural assignment. As shown by the accompanying papers, other strategies taking advantage of partial cyanogen bromide cleavage did generate fragments that could be easily identified by mass spectrometry, fully characterized, and capable of delivering a definition of the final structure. If the first strategy shown above had been followed, a structural ambiguity would have arisen, because the location of the two CNBr fragments (1'-9' and 1-9) is not unequivocally assignable. Thus, two structures could have been proposed, the second one being shown in Figure 1b; we have called it STP-3a (for STP-3 alternative). This structure is identical to that of STP-3 except for the positioning of the aforementioned two fragments. This leads to a structure containing an *inter*chain disulfide bond. In the absence of a further chemical cleavage step utilizing hydroxylamine to cleave the Asn-Gly bond and also allow an unambiguous assignment to be made, duplication by chemical synthesis is necessary to determine the correct structural assignment. Here we report on the synthesis, characterization and aliquoting of STP-3 and STP-3a as it was carried out in our laboratory.

II. SYNTHESIS AND PARTIAL CHARACTERIZATION OF STP-3 AND STP-3a

Both peptides were synthesized using a mixture of standard solid phase techniques (7) on a methylbenzhydrylamine (MBHA) resin. Boc-Lys(Fmoc)-OH was covalently linked to the resin via an amide bond using diisopropylcarbodiimide, the Boc group removed by treatment with TFA, and synthesis of residues 1-17 was completed so that the following protected peptide was obtained: Z-Ala-Gly-Ser(Bzl)-His(Tos)-Phe-Cys(Mob)-Asn-Gly-Met-Leu-Cys(Mob)-Asn-Gly-Ala-Leu-Gly-Ser(Bzl)-Lys(Fmoc)-MBHA. After treatment of the peptide resin twice with 20%

piperidine in DMF (3 min and 7 min), the short arm of STP-3 was built up to give the fully protected STP-3: Z-Ala-Gly-Ser(Bzl)-His(Tos)-Phe-Cys(Mob)-Asn-Gly-Met-Leu-Cys(Mob)-Asn-Gly-Ala-Leu-Gly-Ser(Bzl)-Lys[Met-Gly-Leu-His(Tos)-Tyr(2-Br-Z)-Phe-His(Tos)-Gly-Ala-Z]-MBHA.

The peptide was then deprotected and cleaved from the resin by treatment with anhydrous HF at 0° C for 1 hour. Anisole and ethyl methyl sulfide were added to the cleavage reaction to scavenge the carbocation species produced during the removal of the protecting groups. After distillation of the HF the peptide was washed with anhydrous diethyl ether, filtered, and extracted with dilute acetic acid. This solution was then diluted to a volume of 4 liters with cold, nitrogen-degassed water, the pH adjusted to 6.8 with ammonium hydroxide, and the sample was stirred at ambient temperature in an open beaker until the Ellman test for free sulfhydryls was negative (approximately 16 hours). The pH of the solution was adjusted to 5.0 and the peptide concentrated on a cation exchange column (BioRex-70). Elution of the peptide from the column was accomplished with 50% acetic acid. The eluant was then evaporated to dryness under vacuum and the residue obtained lyophilized from water to yield the crude STP-3 as a white fluffy powder.

The HPLC profile of the crude STP-3 is shown in Figure 2. Purification of this peptide by standard preparative HPLC methods employed in this laboratory gave the purified STP-3 (1); this was then dissolved in 10% acetic acid and aliquoted at a concentration of 1 mg/mL into 15 mg samples which were then dried and shipped for subsequent distribution.

The synthesis and purification of STP-3a was accomplished in identical fashion. It, however, was not aliquoted for distribution.

The analytical HPLC profiles of the purified STP's are shown in Figures 3a and b; the co-elution experiment of these two is shown in Figure 3c. As one can see, the two peptides, while very similar in structure have close but different retention times on HPLC. Both compounds gave similar and satisfactory amino acid compositions as well as sequence information (on both the unreduced and reduced and alkylated peptide). Gel permeation chromatography on HPLC (8) indicated that they were both of the same relative molecular size, with an apparent molecular weight of 3.9 kDa and 3.6 kDa respectively, indicating that the interchain bridged molecule (STP-3a) presents itself as a smaller molecule than STP-3 as one would have intuitively expected.

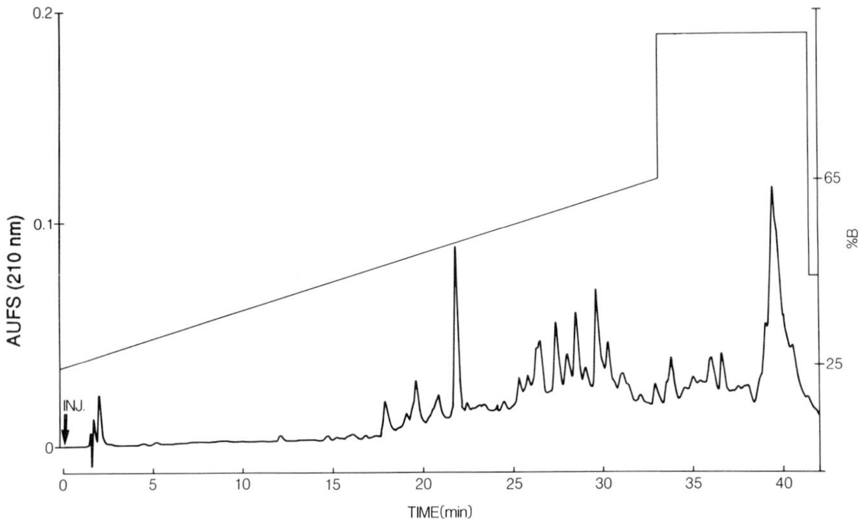

Figure 2. Analytical HPLC of crude STP-3. Column: Vydac 5 μm, C18, 25 x 0.46 cm I.D. Buffers: A=0.1% TFA in H_2O; B=0.1% TFA in acetonitrile-H_20 (60-40); Flow rate: 2 mL/min; Detector 0.2 AUFS (210 nm).

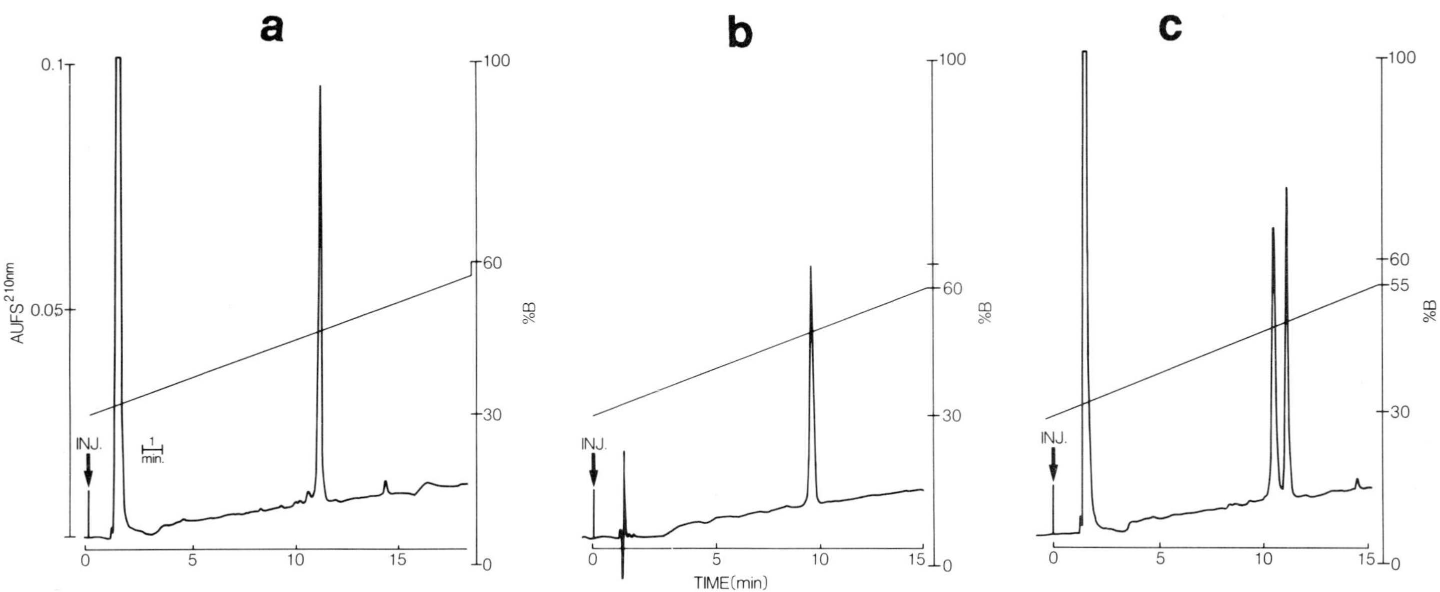

Figure 3. Analytical HPLC of a) purified STP-3; b) purified STP-3a; and c) coelution experiment (STP-3a is the early eluting peak). Column: Vydac 5 µm, C18, 25 x 0.46 cm I.D. Buffers: A=0.1% TFA in H_2O; B=0.1% TFA in acetonitrile-H_20 (60-40); Flow rate: 2 mL/min.

III. REFLECTIONS ON THE NATURE AND FUNCTION OF STP-3

As this report appears in a compendium of papers dealing with techniques in biochemistry, it may be relevant to make the following points: 1) It is remarkable that three groups of investigators could deduce from their data the correct structure of STP-3 using as little as approximately three nanomoles of the peptide received as a dried film which is recognized as a possible hurdle for optimized recovery. Indeed most registrants have experienced better recovery from samples kept in solution rather than from reconstituted samples. 2) We had experienced in the preparation of STP-2 (isolated as the heptafluorobutyrate salt), that it was not amenable to mass spectrometric analysis until converted to the trifluoroacetate suggesting that some salt forms may be less amenable to characterization than others. 3) From the results presented at the workshop and analysis of a sample just prior to the workshop,we can infer that STP-3 was stable under the storage conditions that had been used (i.e dry film at refrigerated or room temperatures) despite the presence of two methionines. Only one sample was reported to have contained significant quantities of methionine sulfoxide upon storage in dilute acetic acid. It was concluded that after reconstitution in mixed aqueous/organic solvents a sample should certainly remain refrigerated but not necessarily frozen. 4) The synthesis and air oxidation of STP-3 and STP-3a, which resulted in the formation of intramolecular bridges of significantly different sizes, occured with an ease that seemed equivalent if evaluated by the final yields of desired peptides. This suggests that both STP's were structurally quite flexible in solution and that thiol capture is indeed a very efficient chemical process. 5) From the comments made during the workshop, and as indicated in the following papers, it is evident that investigators are comfortable when sequencing at the level of 50 picomoles or carrying out derivatization or enzymatic digests at four times that level. This suggests that this level of sophistication can be routinely attained in more than a few isolated laboratories. 6) Total synthesis and coelution experiments appear not to be as critical as in this case for final confirmation of structure considering the new developments brought about in the field of mass spectrometry. However this aspect should not be overlooked in the case of biologically active peptides. There is at least one documented case for the presence of a D-amino acid (DAla) as in dynorphin, an opiate peptide isolated from frog skins.

IV. CONCLUSION

It is our contention that the solvable analytical and intellectual challenges that the STPs have provided participants of the Symposium have made these endeavors worth the time and effort put into them. They defined the strengths and limitations of the analytical tools that are now commonly available to biochemists.

V. ACKNOWLEDGEMENTS

We would like to thank Ron Kaiser and Duane Pantoja for expert technical assistance. This work was supported in part by NIH grants DK26741 and HD13527 and contract NO1HD7-2907.

VI. REFERENCES

1. Rivier, J., McClintock, R., Galyean, R., Anderson, H., (1984), J. Chromatog., 288, 303.

2. Lauer, H.H., and McManigill, D. (1986) Anal. Chem., 58, 166.

3. Rivier, J., and McClintock, R.(1988), in: "The Use of HPLC in Protein Purification and Characterization", (Kerlavage, A.R., ed.), In Press.

6. Shively, J.E., (1986) in: "Methods of Protein Microcharacterization: A Practical Handbook" (J.E. Shively, ed.), p.195, Humana Press, Clifton, N.J.

5. Gibson, B.W., Poulter, L., Williams, D.H., and Maggio, J.E.(1986). J. Biol. Chem., 261, 5341.

6. Hershko, A., and Ciechanover, A. (1982). Ann. Rev. Biochem. 51, 335.

7. Stewart, J.M. and J.D. Young,J.D. (1984) "Solid Phase Peptide Synthesis", 2nd Edition, Pierce Chemical Co., Rockford, IL.

8. Rivier, J., (1980). J. Chromatog., 202, 221.

54

STRUCTURE DETERMINATION OF STP-3

Stanley A. Hefta, Marc J. Besman, Terry D. Lee,
John E. Shively, and Raymond J. Paxton

Division of Immunology
Beckman Research Institute of the City of Hope
Duarte, California 91010

I. INTRODUCTION

The third symposium test peptide (STP-3) was designed with both common and uncommon structural features. Amino terminal sequencing and fast atom bombardment mass spectrometry of native and reduced and alkylated STP-3 suggested the presence of two polypeptide chains linked by an uncommon mechanism. In addition to this linkage, the presence of two cysteine residues and the mass spectral data suggested the presence of an inter- or intrachain disulfide bond. Therefore, the challenge was to properly assign the amino terminal sequences to their respective chains, to determine the disulfide linkage, and to determine the manner by which the chains were linked. In this report we describe our studies with native and pyridylethylated STP-3 and cyanogen bromide fragments derived from the latter that resulted in the determination of the structure of STP-3.

II. EXPERIMENTAL PROCEDURES

Amino terminal sequence analysis was performed on a City of Hope-built gas-phase microsequencer (Hawke et al., 1985) equipped with an on-line HPLC system for PTH-amino acid analysis. Samples were spotted on a 1 x 10 mm strip of polybrene-coated polyvinylidene fluoride membrane (PVDF, Immobilon), which was loaded into a continuous-flow reactor (Shively et al., 1987; Xu and Shively, 1988) for sequence analysis. Positive ion fast atom bombardment mass spectrometry (FAB-MS) was performed on a JEOL HX-100HF high-resolution mass spectrometer operating at 5 kV accelerating potential. Sample ionization was accomplished using a 6 keV Xe atom beam. Samples for FAB-MS were concentrated to dryness in polypropylene microcentrifuge tubes, redissolved in 2 μl of 5% (vol/vol) aqueous acetic acid and added to 2 μl of dithiothreitol/dithioerythritol (1/5, wt/wt) containing 6 mM camphorsulfonic acid on a 1.5 x 6-mm stainless steel sample stage. Reversed-phase high performance liquid chromatography (HPLC) was performed on 2.1 x 250 mm Vydac C18 column using a trifluoroacetic acid (TFA)/acetonitrile solvent system; solvent A = TFA/water (0.1/99.9, vol/vol) and solvent B = TFA/water/acetonitrile (0.1/9.9/90, vol/vol/vol).

STP-3 was dissolved in 5% (vol/vol) aqueous acetic acid and analyzed by reversed-phase HPLC, which showed it to be >95% pure; amino terminal sequence, amino acid compositional, and FAB-MS analyses of STP-3 were performed without further purification of the peptide. The remainder of the sample was concentrated to dryness, dissolved in 0.1 M Tris-HCl, 6 M guanidine hydrochloride, pH 8.2, reduced with 2-mercaptoethanol (70 mM), and S-alkylated with 4-vinylpyridine (350 mM). Each reaction was for 2 hours at 25°C. Pyridylethylated (PE)-STP-3 was desalted by reversed-phase HPLC and collected for further analysis. Cyanogen bromide (CNBr) peptides were prepared by treating PE-STP-3 with an excess of CNBr in 0.1 M HCl for 16 hours at 25°C. The resulting peptides were separated by reversed-phase HPLC using the solvent system described above.

III. RESULTS

The amino terminal sequence obtained for native STP-3 is shown in Table I. The yields and PTH-amino acid assignments on cycles 1-9 were indicative of two amino terminal sequences, while subsequent yields and assignments suggested a single sequence. One equivalent of histidine was observed on cycle 6 and cycle 11 was blank, suggesting the possibility of cysteine at these positions. To determine if cystine/cysteine was present in STP-3, the peptide was reacted with 2-mercaptoethanol and 4-vinylpyridine. Microsequence analysis of the product showed the same sequences as native STP-3, but PTH-pyridylethylcysteine was observed on cycles 6 and 11. FAB-MS showed a mass of 2777.3 for native STP-3 and a mass of 2989.4 for PE-STP-3 (Figure 1), consistent with the addition of two molecules of vinylpyridine to STP-3. Amino acid analysis of native STP-3 (Table I) agreed with the microsequence and FAB-MS analyses.

The fragmentation of PE-STP-3 by cyanogen bromide resulted in 5 major peptides (Figure 2). CNBr-1 and CNBr-2 had identical sequences but differed in mass by 18 mass units due to the presence of either homoserine (CNBr-1) or homoserine lactone (CNBr-2) at the carboxyl terminus (Table II). Similarly, CNBr-6 and CNBr-8 had the same sequences but also differed in mass by 18 mass units. The sequences of

TABLE I. Amino terminal sequence and amino acid composition for STP-3

Amino terminal sequence [a]			Amino acid composition [b]	
Cycle	Residue (pmoles)		Amino acid	Residues (actual)
1	Ala (148)		Cys	1.3 (2)
2	Gly (110)		Asx	2.4 (2)
3	His (7)	Ser (14)	Ser	2.1 (2)
4	His (5)	Phe (32)	Gly	6.4 (6)
5	Phe (40)	Tyr (44)	Ala	3.2 (3)
6	His (8)		Met	1.5 (2)
7	Asn (25)	Leu (34)	Leu	3.1 (3)
8	Gly (58)		Tyr	1.1 (1)
9	Met (68)		Phe	1.8 (2)
10	Leu (23)		His	3.2 (3)
11			Lys	1.2 (1)
12	Asn (21)			
13	Ala (19)			
14	Gly (24)			
15	Leu (14)			
16	Ser (6)			
17	Gly (2)			
18	Lys (1)			

[a]A 10% aliquot of native STP-3 was analyzed as described in Experimental Procedures. Picomoles of PTH-amino acids analyzed (50% of the total) are shown in parentheses. Cysteine residues were later identified on cycles 6 and 11 when pyridylethylated STP-3 was sequenced.

[b]A 5% aliquot of native STP-3 was hydrolyzed at 110°C for 24 hours with 6 N HCl in the vapor phase, and 50% of the hydrolysate was analyzed on a Beckman 6300 amino acid analyzer. For the determination of cystine/cysteine as cysteic acid, a second aliquot of STP-3 was oxidized with performic acid prior to hydrolysis. The residues of each amino acid in STP-3 based on the final structure are shown in parentheses.

CNBr-1/2 and CNBr-6/8 corresponded to those previously determined for the amino terminus of STP-3, thus allowing the assignment of the amino terminal sequences to individual chains. The sequence of the other major peptide, CNBr-3, corresponded to that for the carboxyl terminal portion of STP-3, and FAB-MS of the peptide showed that it terminated in a carboxyl terminal amide rather than a carboxyl terminal acid (Table II).

The remainder of the structure of STP-3 was determined by characterizing two minor peptides from the cyanogen bromide digest. Microsequence analysis of CNBr-5 (Figure 3) revealed a single sequence identical with that of CNBr-1/2 for 9 cycles and continuing with that of

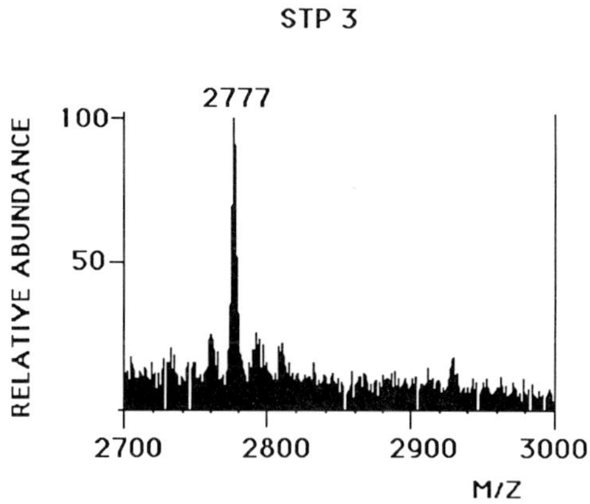

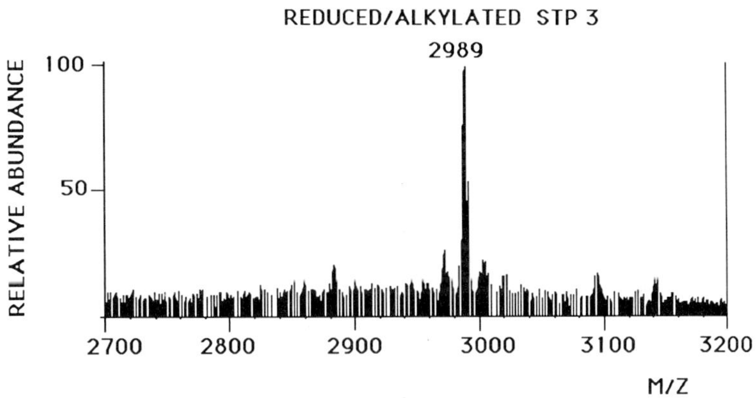

Figure 1. FAB mass spectra in the molecular ion region for native STP-3 (top) and pyridylethylated (PE)-STP-3 (bottom). Spectra were obtained as described in Experimental Procedures, and monoisotopic masses are indicated for each peptide. The 212 mass unit increase for PE-STP-3 is consistent with the presence of cystine in native STP-3.

CNBr-3, ending with lysine on cycle 18. Cycle 9, expected to be methionine, was shown to be homoserine by comparing the retention time with a PTH-homoserine standard. The mass of CNBr-5 (Table II) was consistent with the observed sequence, including the presence of homoserine and a carboxyl terminal amide. Microsequence analysis of CNBr-9 yielded a double sequence corresponding to those determined for CNBr-6/8 (with homoserine on cycle 9) and CNBr-3, although FAB-MS indicated that the sequences arose from a single peptide. A structure for CNBr-9 in which these peptides are linked via the e-amino group of the lysine residue is shown in Table II. The mass of CNBr-9 is consistent with

TABLE II. Structures of the cyanogen bromide peptides of PE-STP-3 [a]

<table>
<tr><th></th><th></th><th colspan="2">Mass [c]</th></tr>
<tr><th>Peptide</th><th>Sequence [b]</th><th>Observed</th><th>Calculated</th></tr>
<tr><td>1</td><td>PE (on C)
|
A-G-S-H-F-C-N-G-X</td><td>998.6</td><td>998.3</td></tr>
<tr><td>2</td><td>PE (on C)
|
A-G-S-H-F-C-N-G-X'</td><td>980.5</td><td>980.3</td></tr>
<tr><td>3</td><td>PE (on C)
|
L-C-N-A-G-L-S-G-K-$CONH_2$</td><td>966.7</td><td>966.5</td></tr>
<tr><td>5</td><td>PE (on C) PE (on C)
| |
A-G-S-H-F-C-N-G-X-L-C-N-A-G-L-S-G-K-$CONH_2$</td><td>1946.0</td><td>1945.8</td></tr>
<tr><td>6</td><td>A-G-H-F-Y-H-L-G-X</td><td>1002.7</td><td>1002.5</td></tr>
<tr><td>8</td><td>A-G-H-F-Y-H-L-G-X'</td><td>984.6</td><td>984.5</td></tr>
<tr><td>9</td><td>PE (on C)
|
L-C-N-A-G-L-S-G-K-$CONH_2$
| (on K)
A-G-H-F-Y-H-L-G-X</td><td>1949.9</td><td>1950.0</td></tr>
</table>

[a]Amino terminal sequences and FAB mass spectra were obtained for the cyanogen bromide peptides of PE-STP-3 (see Figure 1) using the methods described under Experimental Procedures.

[b]Single letter abbreviations are used for the amino acids; cysteine was identified as the pyridylethyl (PE) derivative; X = homoserine and X' = homoserine lactone.

[c]Masses are reported as the monoisotopic mass for the molecular ion ($M+H^+$).

this structure, including the presence of homoserine and a carboxyl terminal amide.

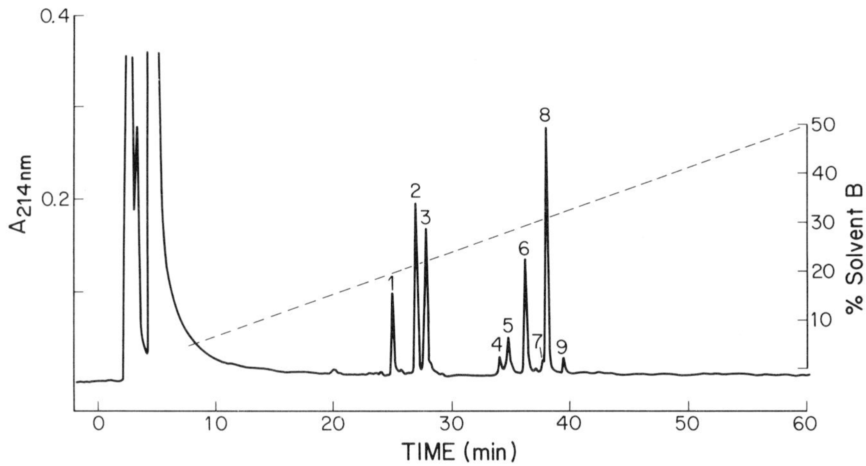

Figure 2. Cyanogen bromide peptide map of PE-STP-3. Cyanogen bromide peptides were prepared and purified by reversed-phase HPLC as described in Experimental Procedures. Each peak was subjected to microsequence and FAB-MS analyses; the results are presented in Table 2.

IV. DISCUSSION

Amino terminal sequence analysis of native STP-3 yielded a double sequence, indicating the presence of two peptide chains. The presence of cystine/cysteine in STP-3 was confirmed by S-alkylating the peptide with 4-vinylpyridine. The product, PE-STP-3, gave the same sequences as native STP-3, and PTH-pyridylethylcysteine was identified on cycles 6 and 11. FAB-MS showed masses of 2777.3 and 2989.4 for STP-3 and PE-STP-3, respectively, thus confirming the presence of two cysteine residues in STP-3. The 212 mass unit difference observed for the peptides also confirmed the presence of a disulfide bond (i.e. cystine) in STP-3. If STP-3 had contained two free cysteine residues, a 210 mass unit increase would have been observed. These initial studies established the presence of a disulfide bond in STP-3; additionally, they indicated that the two peptide chains in STP-3 were linked by some other mechanism.

To assign the amino terminal sequences to their respective chains and to determine the manner in which the two chains were linked, PE-STP-3 was fragmented with cyanogen bromide. Microsequence and FAB-MS analyses of the major peptides led to two proposals for the structure of STP-3 as illustrated below.

A

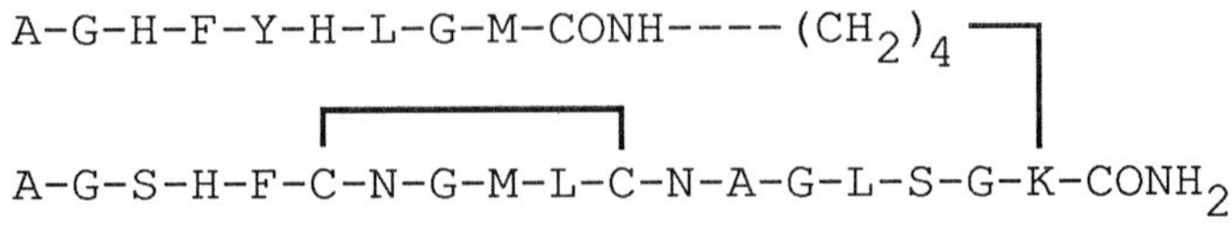

B

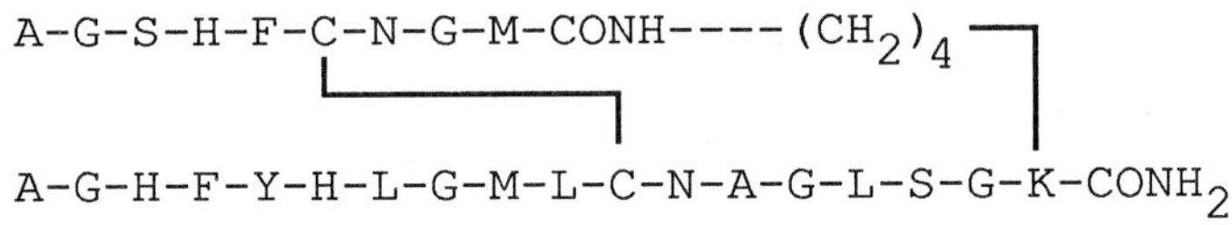

The two peptide chains in each structure are linked by an amide bond formed with the e-amino group of the C-terminal lysine residue on the longer chain and the carboxyl terminus of the shorter chain. The difference between the possible structures involves the relative positions of the shorter chain and the first nine residues of the longer chain. By interchanging these sequences, either **A** with an intrachain disulfide bond or **B** with an interchain disulfide bond can be generated. Because of the location of the methionine residues, the expected cyanogen bromide peptides from **A** and **B** are identical; however, the analysis of two minor cyanogen bromide fragments allowed the structures to be distinguished. As previously indicated, CNBr-5 from PE-STP-3 yielded a single sequence corresponding to the longer chain of **A**, with PE-cysteine on cycles 6 and 11 and homoserine on cycle 9. A second minor peptide, CNBr-9, yielded a double sequence corresponding to the shorter chain of **A**, with homoserine on cycle 9, and the last nine residues of the longer chain. Thus, based on the analysis of CNBr-5 and CNBr-9, structure **A** was determined to be correct for STP-3.

The generation and subsequent analysis of CNBr-5 and CNBr-9 played a key role in this study. The accepted mechanism of cyanogen bromide fragmentation involves the reaction of the methionine sulfur atom with CNBr followed by an intramolecular reaction that expels methyl thiocyanate and generates an iminolactone intermediate. Hydrolysis of the iminolactone usually generates two fragments, one containing a C-terminal homoserine lactone or homoserine and a second containing a free amino terminus. Cyanogen bromide fragmentation of STP-3, for the most part, followed this scheme, although in the case of CNBr-5 and CNBr-9, hydrolysis of the internal iminolactone generated homoserine, without cleavage of the peptide bond. Conversion of methionine to homoserine without peptide bond cleavage has been reported for methionine-threonine and methionine-serine bonds and has been attributed to an intramolecular reaction of the ß-hydroxyl group of threonine and serine with the iminolactone (Schroeder et al., 1969). Similarly, conversion to homoserine without cleavage of methionine-cystine bonds has been explained by an intramolecular reaction between the sulfur atom of the cystine and the iminolactone (Doyen and Lapresle, 1979). A more general mechanism was

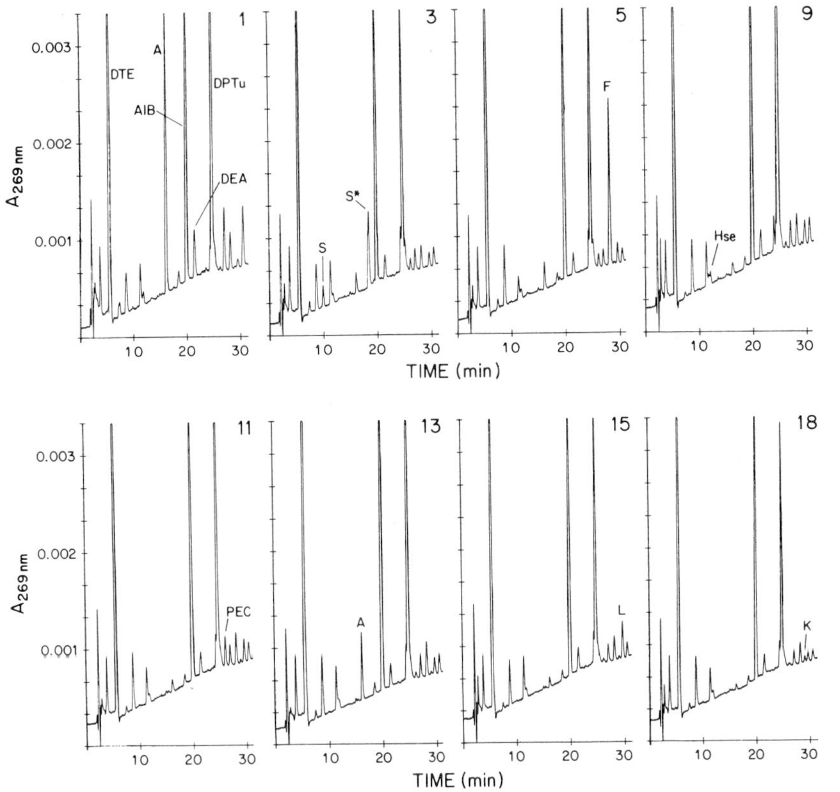

Figure 3. Microsequence analysis of CNBr-5 of PE-STP-3. CNBr-5 was sequenced as described in Experimental Procedures. HPLC chromatograms are shown for representative cycles, including cycle 9 homoserine (Hse) and cycle 11 pyridylethylated cysteine (PEC). Other abbreviations are DTE, dithiothreitol; AIB, PTH-aminoisobutyric acid (HPLC standard); DEA, diethylphenylthiourea; DPTU, diphenylthiourea.

postulated to explain methionine conversion to homoserine without bond cleavage at the methionine-glutamic acid and methionine-isoleucine bonds in cytochrome C (Corradin and Harbury, 1970). In this mechanism, hydrolysis of the iminolactone produces an intermediate that can rearrange to form homoserine lactone with bond cleavage or homoserine without bond cleavage. In the case of STP-3, conversion of methionine to homoserine without bond cleavage at the methionine-leucine and methionine-e-amino lysine bonds was perhaps due to the latter pathway.

V. REFERENCES

Corradin, G., and Harbury, H.A. (1970). Biochim. Biophys. Acta. 221, 489.

Doyen, N., and Lapresle, C. (1979). Biochem. J. 177, 251.

Hawke, D.H., Harris, D.C., and Shively, J.E. (1985). Anal. Biochem. 147, 315.

Shively, J.E., Miller, P., and Ronk, M. (1987). Anal. Biochem. 163, 517.

Schroeder, W.A., Shelton, J.B., and Shelton, J.R. (1969). Arch. Biochem. Biophys. 130, 551.

Xu, Q.-Y., and Shively, J.E. (1988). Anal. Biochem. 170, 19.

55

STRUCTURE OF SYMPOSIUM TEST PEPTIDE-3

James Elliott
Kathryn Stone
William Roberts
Mary LoPresti
Ray De Angelis
Myron Crawford
Janet Kapouch
Eric Jacobsen
Kenneth Williams

School of Medicine Protein and
Nucleic Acid Chemistry Facility

Walter J. McMurray

Comprehensive Cancer Center
Mass Spectrometry Facility

Chin-Kai Meng
Matthias Mann
John Fenn

Department of Chemical Engineering
Yale University
New Haven, Connecticut

I. INTRODUCTION

The STP series of "unknown" peptides distributed in conjunction with the Protein Society Symposia provide one of the few opportunities where success usually depends upon quickly bringing nearly every major area of expertise that is available in a state-of-the-art protein chemistry laboratory to bear on a single problem. The knowledge that only a limited amount of time and peptide are available provides a stimulus to reach the absolute limits of sensitivity of detection and to devise the most ingenious but yet direct strategy to rapidly arrive at a final structure. Since we had previously relied on a tripartite approach based on amino acid sequencing, mass spectrometry and peptide synthesis to solve the structure of STP-2, we approached STP-3 in the same fashion.

II. MATERIALS AND METHODS

A. HPLC Isolation and Amino Acid Sequencing of Cyanogen Bromide Peptides Derived from STP-3.

Upon receipt of the STP-3 aliquot it was dissolved in 200 μl 0.05% trifluoroacetic acid and 20% was taken for hydrolysis followed by ion exchange amino acid analysis as described below. A small scale cyanogen bromide digest was carried out by adding 16 μl of a 1.1 mM solution of cyanogen bromide (175 fold molar excess over methionine) in 88% formic acid to 50 pmol STP-3 that was in a volume of 3 μl 0.05% trifluoroacetic acid. After 24 hrs at room temperature the digest was stopped by adding 190 μl 0.05% trifluoroacetic acid and injecting onto a (2.1 mm x 25 cm) Vydac C-18 column that had been equilibrated on a Waters Associates HPLC System (1) at 0.15 ml/min with 0.05% trifluoroacetic acid. The resulting peptides were eluted by increasing the concentration of buffer B (0.05% trifluoroacetic acid in 80% acetonitrile as follows): 2-37% B, (0-63 min); 37-75% B, (63-95 min); 75-98% B, (95-105 min). A second cyanogen bromide digest was accomplished by adding 20 μl of a 20 mM solution of cyanogen bromide (1000 fold molar excess over methionine) to 400 pmol STP-3 that had been dried in an eppendorf tube. After 24 hrs at room temperature, a 50 pmol aliquot was dried _in vacuo_ and then reduced by incubating for 12 min at 100°C in 20 μl 6 M guanidine hydrochloride, 10 mM dithiothreitol at pH 4.5. Peptides selected for amino acid sequencing were loaded directly onto an Applied Biosystems Model 470A Gas Phase Sequencer equipped with a Model 120 on-line HPLC PTH-analyzer.

B. Mass Spectrometry and Amino Acid Analysis

Electrospray mass spectrometry was carried out on a VG Micromass 1212 Quadrupole Mass Spectrometer (2). The STP-3 solution (0.05 mg/ml in an equal volume mixture of methanol and water) was pumped into the electrospray chamber at 1.0 μl/min and approximately 25 ng (9 pmole) was required for the spectrum. FAB mass spectrometry was carried out on a VG-ZAB-SE Mass Spectrometer using xenon fast atom bombardment (8 kev at 1 mA).

Aliquots of STP-3, 625 pmol for ion exchange and 125 pmol for reverse phase amino acid analysis after derivatization with phenylisothiocyanate, were hydrolyzed in 6 N HCl, 0.1% phenol for 16 hrs at 115°C prior to analysis on either a Beckman Model 6300 or a Waters Associates Pico-Tag HPLC System respectively (3).

C. Peptide Synthesis

Peptides were synthesized using a Model 430A Applied Biosystems Peptide Synthesizer, incorporating amino acids as their symmetrical anhydrides or activated esters using standard protocols from the manufacturer. HF cleavage was performed at 0°C for 45 minutes in a mixture containing 25 ml hydrofluoric acid, 0.8 ml anisole, 0.3 g 2-mercaptopyridine, and 0.3 g dithiothreitol. After cleavage the HF was removed using nitrogen gas and the resin was extracted with 150 ml ethyl acetate followed by 150 ml diethyl ether. The peptide was then solubilized in 7 M guanidine hydrochloride which was then pumped directly onto two (2.5 cm x 25 cm) Vydac C-18 columns connected in series and eluted at a flow rate of 7 ml/min by increasing the concentration of acetonitrile in 0.05% trifluoroacetic acid at the rate of about 0.5% acetonitrile/min.

III RESULTS AND DISCUSSION

A. Amino Acid Sequencing of STP-3 and Its Cyanogen Bromide Fragments

Based on analytical HPLC on a 2.1 mm ID Vydac C-18 column of an aliquot of STP-3 corresponding to 2% of the total sample, STP-3 appeared to be nearly homogeneous. As a result 2.5% of the sample was subjected to automated amino acid sequencing. As shown in Table I, two PTH-amino acids were released at cycles 3,4,5 and 7. In addition, a plot of the log of the total yield of PTH-amino acids versus cycle number (data not shown) indicated at least a 50% drop in overall yield after methionine 9. Taken together, these data suggested that STP-3 consisted of a 9 residue short chain covalently linked to an 18 residue long chain. Although reduction and carboxamidomethylation of 100 pmol STP-3 followed by reverse phase HPLC resulted in nearly a 90% loss of peptide, sufficient carboxamidomethylated STP-3 remained to permit it to be subjected to amino acid sequencing. The results from this analysis (data not shown) confirmed the data shown in Table I, including the cysteine at residue 11 that had previously been postulated only on the basis of the presence of an absorbance peak corresponding to the DTT adduct of PTH-dehydroserine without a corresponding PTH-serine absorbance, and indicated that STP-3 contained a second cysteine/half-cystine at residue 6. In the absence of reduction and carboxamidomethylation no PTH-dehydroserine was seen at cycle 6 during the sequencing of STP-3; it was therefore concluded that STP-3 must contain a disulfide bond. In addition, since reduction and carboxamidomethylation followed by HPLC did not result in separating the two chains, they must be covalently linked at another position that was either in addition to or instead of the postulated disulfide bond. Based on the sequence in Table I, an obvious possibility for this attachment point was the ϵ-NH_2 group of the carboxy-terminal lysine.

Table I. Amino acid sequencing of 78 picomoles STP-3

Cycle	Amino Acids	Yield (pmol)
1	Ala	83.1
2	Gly	36.9
3	His/Ser	5.1/19.0
4	Phe/His	9.0/5.8
5	Tyr/Phe	14.1/16.3
6	His	7.9
7	Leu/Asn	22.3/15.3
8	Gly	23.1
9	Met	22.6
10	Leu	6.7
11	Cys	6.0
12	Asn	6.5
13	Ala	4.0
14	Gly	4.9
15	Leu	3.0
16	Ser	4.7
17	Gly	1.9
18	Lys	0.4

The available data suggested that cyanogen bromide digestion offered the best means of separating the two individual chains of STP-3. As a result, a small scale cyanogen bromide digest was carried out on 50 pmol STP-3 that produced a low yield of a short peptide (peak 1 in Fig. 1) that gave the sequence, X-X-His-Phe-Tyr-X-Leu-Gly, as well as a quartet of peaks that eluted at around 100 min. Since these latter peaks (the first of which appeared to be intact STP-3 based on HPLC co-elution) were also produced following incubation in 70% formic acid without added cyanogen bromide, they may have resulted from re-arrangements about the Asn-Gly bond. Tests with another methionine-containing synthetic peptide demonstrated that the concentrations of the reactants used in this small scale cyanogen bromide digest (5.2 μM Met-X, 0.93 mM cyanogen bromide) were too low to promote significant cleavage. A second digest which was done at higher concentrations (40 μM Met-X, 20 mM cyanogen bromide) gave upon reduction and HPLC the pattern shown in Fig 2. Since peaks 1 and 2 in Fig 2

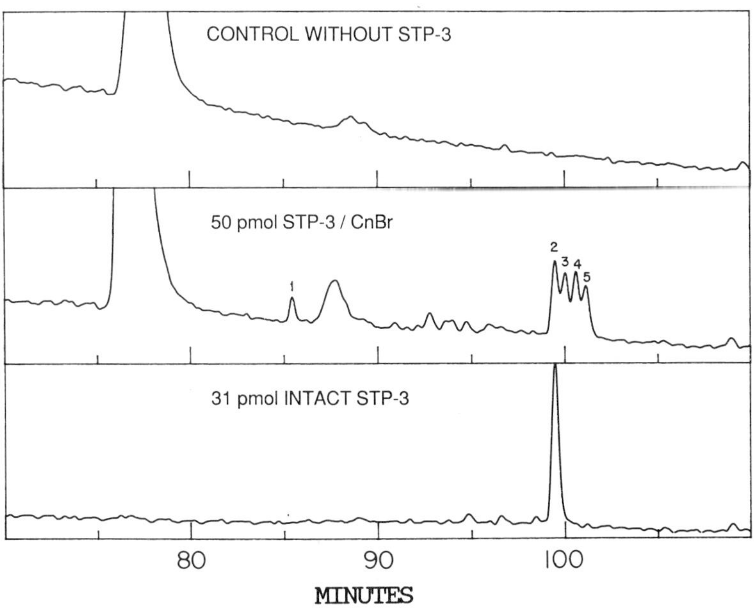

Fig. 1. Reversed-phase HPLC separation of a cyanogen bromide digest of 50 pmol STP-3. In addition to the STP-3 digest shown in the center panel, a control without STP-3 (top panel) as well as 31 pmol intact STP-3 (bottom panel) was also run. The large peak at 77 min in the top two chromatograms is residual formic acid.

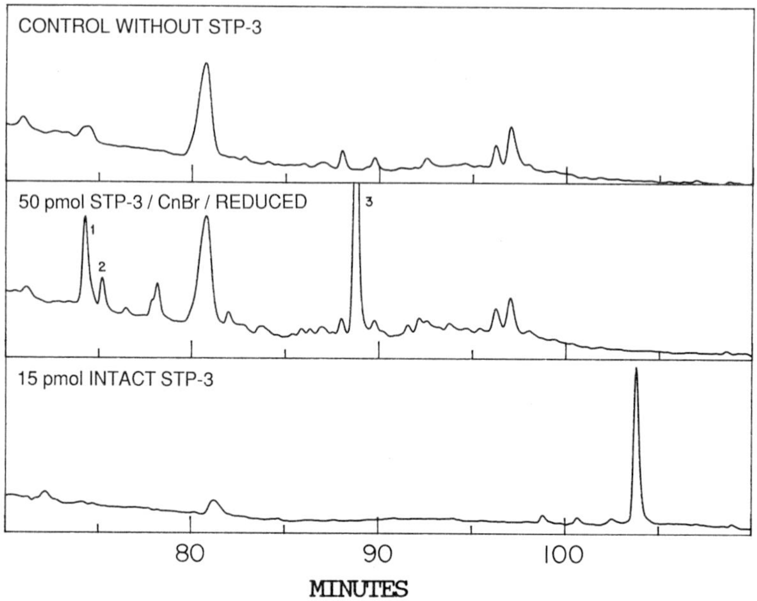

Fig. 2. Reversed-phase HPLC separation of 50 pmol STP-3 following cyanogen bromide cleavage and reduction. An aliquot from a cyanogen bromide digest of 400 pmol STP-3 was reduced and then subjected to HPLC as described in Materials and Methods. In addition to the STP-3 digest shown in the center panel, a control without STP-3 (top panel) as well as 15 pmol intact STP-3 (bottom panel) was also run.

|—— Cn-1 ——→
Ala-Gly-Ser-His-Phe-Cys-Asn-Gly-Met
(1) (5) (9)

|—— Cn-3 ——→ |—— Cn-2 ——→
Ala-Gly-His-Phe-Tyr-His-Leu-Gly-Met-Leu-Cys-Asn-Ala-Gly-Leu-Ser-Gly-Lys
(1) (5) (10) (15) (18)

Fig. 3. Results of amino acid sequencing of cyanogen bromide peptides derived from STP-3. Cn-1, Cn-2 and Cn-3 correspond with the absorbance peaks labelled 1,2 and 3 in Fig. 2.

were only seen following reduction, these two peptides must be covalently linked via a disulfide bond. Amino acid sequencing of the peaks labelled 1,2 and 3 in Fig 2 resulted in the sequences shown in Fig 3.

B. Mass Spectrometry and Amino Acid Analysis of STP-3

Since the data accumulated to this point was consistent with either a 2 or a 4 chain structure for STP-3 we resorted to electrospray mass spectrometry to quickly decide between these two possibilities. This analysis (not shown) indicated an approximate MW of 2779.2 that was only consistent with a 2 chain structure. Now that the approximate MW was known, an FAB/mass spectrometry experiment was carried out (Fig 4) which gave an intense peak at 2777.6 and, based on the average mass of the

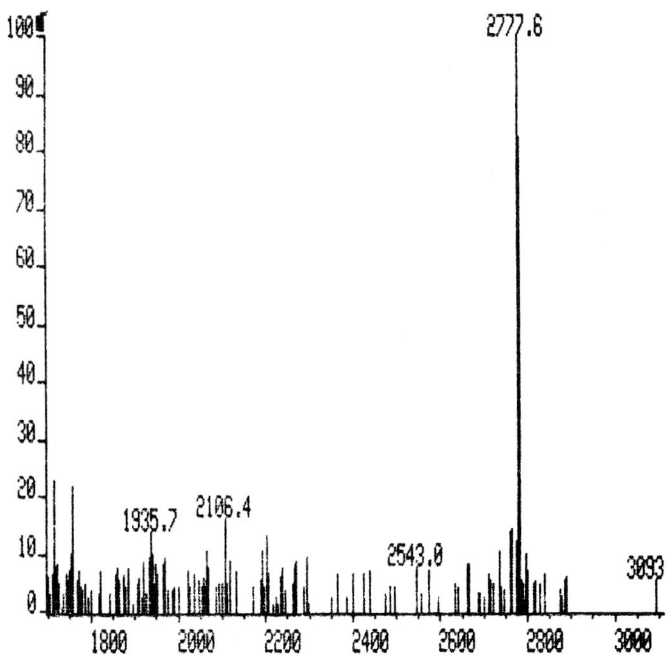

Fig. 4. FAB/mass spectrometry of approximately 150 pmol of STP-3. As indicated above, the most abundant molecular ion cluster occurred at an m/Z of 2777.6. The estimated molecular weight of STP-3, based on the average molecular weight for the major ion cluster shown above is 2778.6.

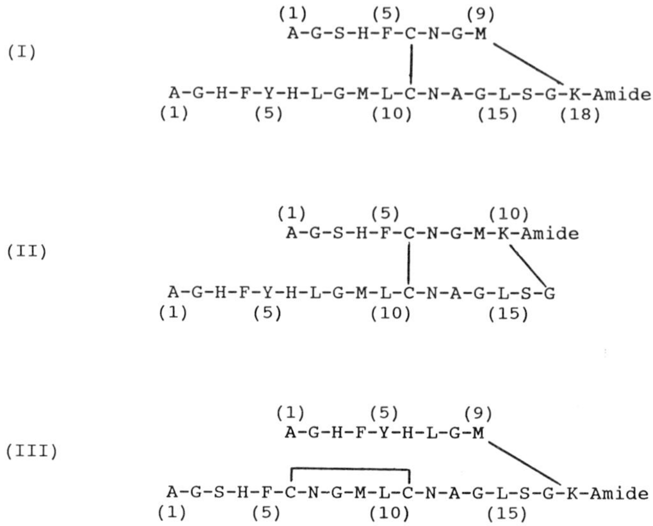

Fig. 5. Three possible isomeric structures for STP-3.

isotopic cluster, an estimated MW of 2778.6. Taken together with the previous amino acid sequencing studies, these data suggested the STP-3 structures shown in Fig 5. The predicted MW for all of these structures is 2778.2 which is within 0.4 atomic mass units of the value that we determined for STP-3. Further confirmation of the proposed COOH-terminal amide was obtained by FAB/mass spectrometry on an aliquot of STP-3 that had been cleaved with cyanogen bromide but not reduced. In this instance, the largest species in the resulting mixture had a measured MW of 1733.8 which exactly corresponds to that predicted for any of the structures in Fig 5 assuming that they ended in COOH-terminal homoserine lactone and that the first 9 amino acids had been removed from the long chain.

Hydrolysis and ion exchange amino acid analysis of STP-3 gave an amino acid composition (data not shown) that agreed within ±4.6% with that predicted for the structures shown in Fig 5. Although our PTC amino acid composition (data not shown) appeared to be less accurate (the overall average error was ±22.6% based on the predicted structures) it nonetheless provided valuable proof that STP-3 did not contain any unusual (acid stable) amino acids.

C. A Synthetic Peptide Approach to Deciding Between Three Isomeric STP-3 Structures

To decide between the first two possible STP-3 structures that we considered ((I) and (II) in Fig 5) we synthesized the COOH-terminal peptides that would result in each case following cyanogen bromide cleavage and reduction of STP-3. In the case of structure (I) at the top of figure 5 this corresponded to a 9 residue peptide ending with an amidated lysine. In the case of the structure (II) in the middle of figure 5 this corresponded to an 8 residue peptide linked via the COOH-terminal glycine to the ϵ-NH_2 group of lysine amide. Following cyanogen bromide digestion and reduction of STP-3, comparative HPLC clearly demonstrated that isomer (II) was incorrect. That is, the synthetic peptide corresponding to the COOH-terminal cyanogen bromide fragment derived from this isomer eluted from a Vydac C-18 column 4 min later than the corresponding cyanogen bromide peptide derived from STP-3. The two remaining possibilities are therefore structures I and III in Fig 5. Although we initially proposed structure (I), none of the available data could differentiate between structure I and III. We elected therefore to synthesize structure (I) using the approach outlined in Fig 6. Following HF cleavage, trypsin digestion (peptide/trypsin (w/w) ratio of 25, 2 M urea, 25 mM NH_4HCO_3, 37° for 24 hrs) and air oxidation, a product was formed that had the correct molecular weight and amino acid sequence. As shown by comparative HPLC in Fig 7, the synthetic peptide

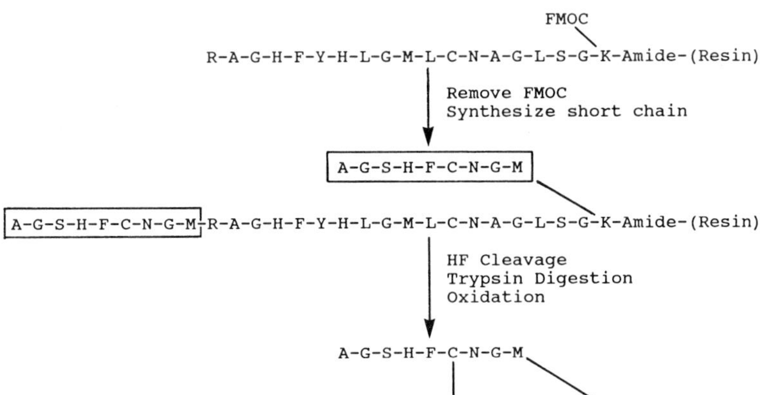

Fig. 6. Solid phase synthesis of the proposed STP-3 (isomer I) structure.

corresponding to structure (I) eluted approximately 1 min prior to STP-3. This result was confirmed (following the Protein Society Symposium) by repeating the HPLC with a synthetic version of structure (I) that was kindly supplied by Jean Rivier. The lack of co-chromatography of STP-3 with a synthetic version of structure (I) (a result which was obtained on the first day of the Protein Society Symposium) demonstrated that our proposed STP-3 structure (structure (I) in Fig 5) was incorrect and suggested that, as demonstrated by 3 other groups (see relevant chapters in this volume), STP-3 has the structure shown at the bottom of Fig 5 (isomer III).

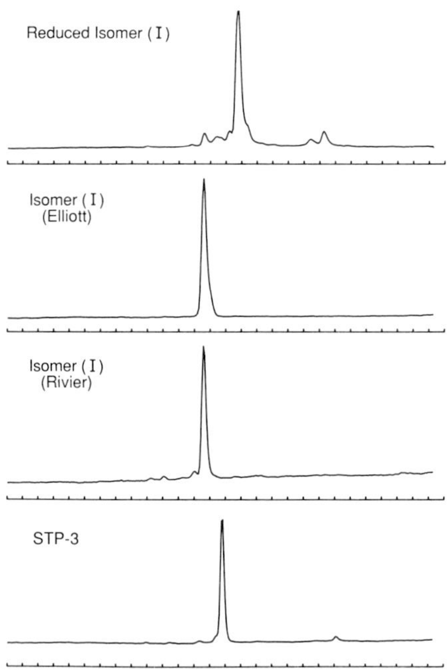

Fig. 7. Reversed-phase HPLC on a 4.6 mm x 25 cm Aquapore C-8 column of synthetic peptides corresponding to the proposed STP-3 structure (isomer I at the top of figure 5). The above chromatograms correspond as indicated to approximately 1 nmol each of STP-3 (isomer III), the proposed STP-3 structure (isomer I) synthesized by J. Rivier and also by J. Elliott and finally, in the top chromatogram, approximately 2 nmol of the reduced isomer I.

IV. REFERENCES

1. K.L. Stone and K.R. Williams (1988) in Macromolecular Sequencing and Synthesis (D.L. Schlesinger, ed) pp. 7-24. Alan R. Liss Inc. New York.

2. C.M. Whitehouse, R.N. Dreyer, M. Yamashita and J.B. Fenn (1985) Analytical Chemistry 57, 675-679.

3. K.L. Stone and K.R. Williams (1986) J. Chrom. 359, 203-212.

56

CHARACTERIZATION OF SYMPOSIUM TEST PEPTIDE-3

Donald F. Hunt
Patrick R. Griffin
John R. Yates, III
Jeffrey Shabanowitz

Department of Chemistry
University of Virginia
Charlottesville, Virginia 22901

Jay W. Fox
Linda K. Beggerly

Department of Microbiology
University of Virginia Medical Center
Charlottesville, Virginia 22908

I. INTRODUCTION

Characterization of Symposium Test Peptide-3 (STP-3) at the University of Virginia involved a collaborative effort between the mass spectrometry group in the Chemistry Department and the protein sequencing facility in the Medical Center. STP-III was shown to consist of a single Lys-amide residue connected through ϵ- and α-amino groups to two peptide chains, 9 and 17 residues in length, respectively. As shown below, the longer of the two chains contains an intramolecular disulfide linkage.

NH_2-A G S H F C N G M L C N A G L S G-NH-CH-$CONH_2$
(disulfide bond between the two C residues; $(CH_2)_4$ links CH to the NH of the second chain)
NH_2-A G H F Y H L G M-NH

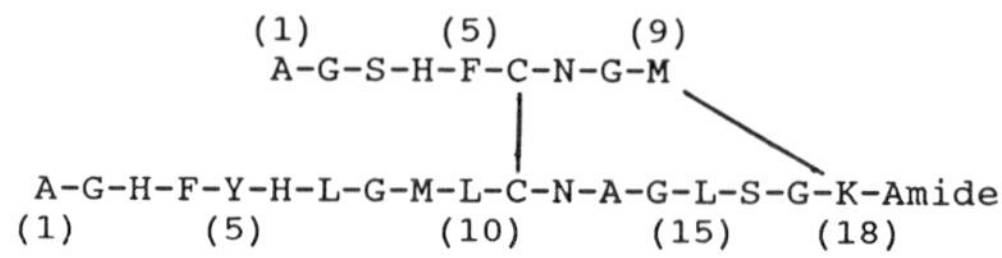

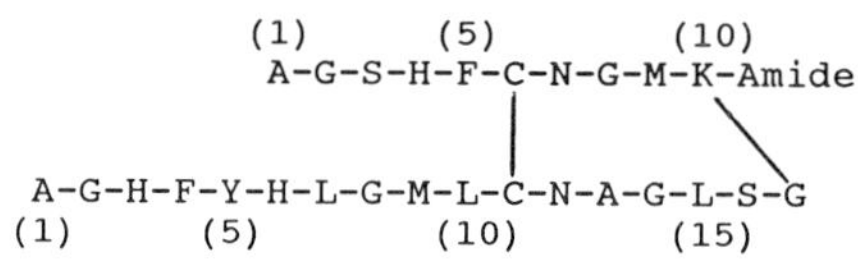

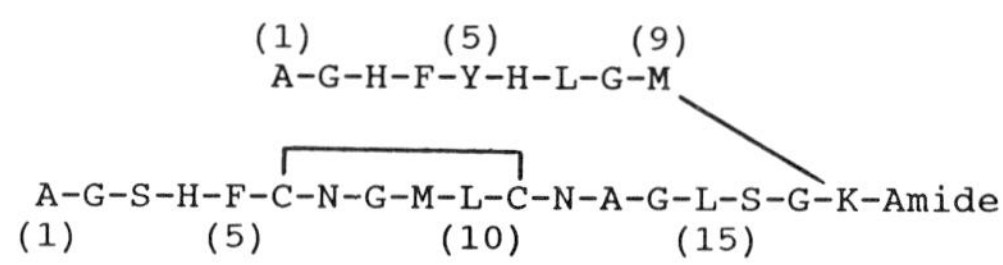

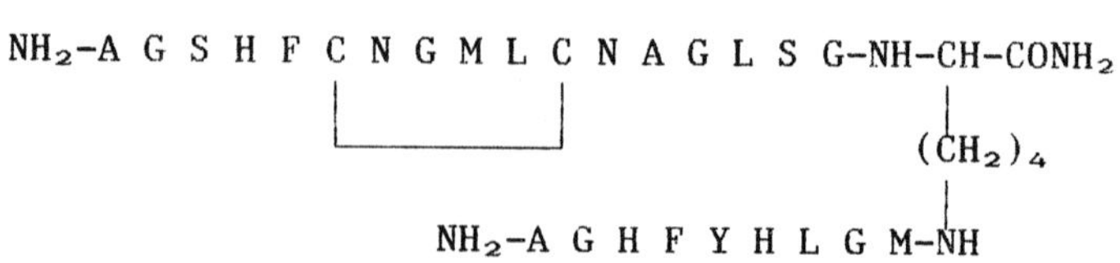

II. EXPERIMENTAL

STP-3, 3-5 nmol, was received as a lyophilized solid in the bottom of a plastic Eppendorf tube. Salts and other contaminants, if present, were removed from the sample by reverse phase HPLC on an Applied Biosystems Model 130A separations system. Sample in 40 ul of 0.1% TFA was injected onto a RP-300 Aquapore column (2.1 mm x 3 cm) and eluted with a 40 min linear gradient of 0% to 60% acetonitrile (0.085% TFA) in 0.1% TFA. Purified sample (200-300 pmol) was analyzed on an Applied Biosystems Gas Phase Sequenator. Trial digests with cyanogen bromide and a variety of enzymes were performed at the 50-100 pmol level and followed by reverse phase HPLC and/or mass spectrometry. Mass spectra were recorded on triple quadrupole and tandem quadrupole Fourier transform instruments. Operation of these instruments for oligopeptide sequence analysis has been described previously (1-3). Collision activated dissociation mass spectra recorded on the triple quadrupole instrument were performed at the 100-300 pmol level. Molecular weight determinations and laser photodissociation experiments were performed on the tandem quadrupole Fourier transform instrument at the 10-60 pmol level.

III. RESULTS AND DISCUSSION

A. Summary of Chemical Experiments and Molecular Weight Determinations.

1. Fig. 1A shows the low resolution mass spectrum (16K transform) recorded on 40-60 pmol HPLC purified STP-3. The unresolved (M+H)+ multiplet appears at a m/z 2778.3. Accordingly, STP-3 was assigned an average molecular weight of 2777.3

2. Acetylation of STP-3 was performed on the solids probe of the mass spectrometer by addition of 0.5 uL of 3/1 MeOH/acetic anhydride and a trace of pyridine to the sample. Mass spectra recorded thirty seconds later showed that the (M+H)+ ion had shifted to higher mass (m/z 2906) by 126 units. Acetylation in the absence of pyridine occurs only on amine groups. In the presence of pyridine, Tyr residues are also acetylated. Introduction of a single acetyl group increases the mass of the molecule by 42 mass units. The shift of 126 mass units observed for the (M+H)+ ion of STP-3 suggested that the molecule contained a total of 3 amino and/or phenolic groups.

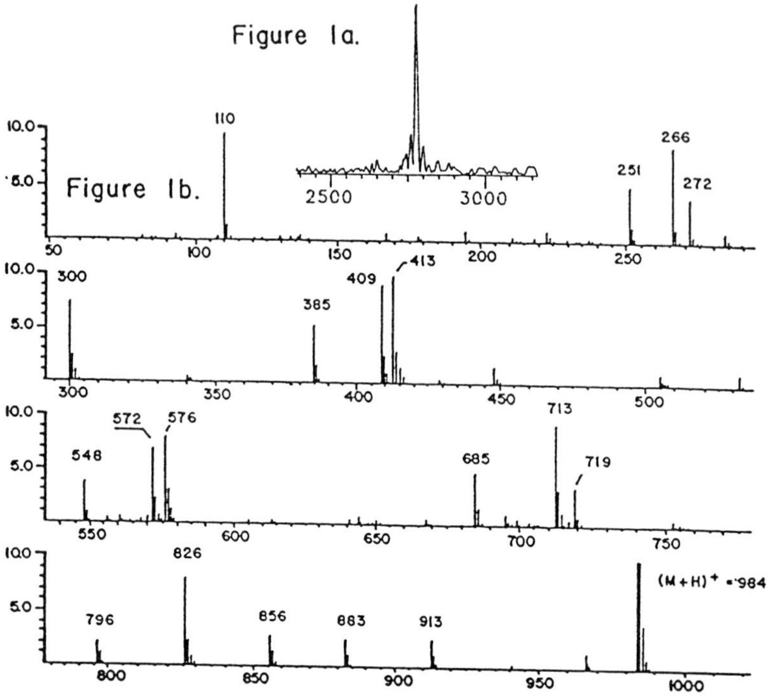

Fig. 1. (a) (M+H)+ ion at m/z 2778.3 in the low resolution Fourier transform mass spectrum recorded on 40-60 pmol of STP-3; (b) collision activated dissociation mass spectrum recorded on the homoserine lactone (M+H)+ ion derived from the CNBr fragment of mol mass 983.

3. Performic acid oxidation of STP-3 was performed on the solids probe of the mass spectrometer by addition of 0.5 uL of standard reagent to the sample matrix. Mass spectra recorded several minutes later showed the presence of new (M+H)+ ions at m/z 2908, 2924, and 2940. The observed mass shifts of 130, 146 and 162 units are consistent with the incorporation of two hydrogens (1 mass unit each) plus 8, 9 and 10 oxygen atoms (16 mass units each), respectively. Oxidation of two Cys residues to cysteic acid and two Met residues to Met sulfoxide and/or sulfone groups would explain the above data.

4. STP-3 (60 pmol) was converted to the corresponding methyl ester on treatment with 2N HCl/MeOH for 1 hr at room temperature. A mass spectrum recorded on the product of this transformation showed a new (M+H)+ ion at m/z 2808. Conversion

of side chain or C-terminal COOH groups to the corresponding methyl esters ($-COOCH_3$) increases the mass of the molecule by 14 units/COOH. Two explanations were considered for the observed mass shift of 30 units: (a) conversion of two acid sensitive Asn or Gln residues to the corresponding methyl esters (15 mass units/$CONH_2$) or (b) oxidation of 1 Met residue to the sulfoxide and conversion of one COOH group to the corresponding methyl ester. Only the first explanation is consistent with the structure proposed for STP-3.

B. Summary of Results Obtained From Chemical and Enzymatic Digests.

1. Treatment of STP-3 at the 100 pmol level with trypsin, Asp-N, Glu-C, and Lys-C failed to cleave the molecule. Mass spectra recorded on an aliquot of each reaction mixtures showed that the (M+H)+ ion remained at m/z 2778.3.

2. Treatment of STP-3 (500 pmol) with cyanogen bromide afforded two fragments. (M+H)+ ions for the two species were observed at m/z 984 and 1735, respectively. Since treatment of the two fragments with 2N HCl in MeOH shifted each (M+H)+ ion to higher mass by 32 units, it was concluded that both molecules contained homoserine lactones and that neither had a free carboxyl group. It was also noted that the molecular weights of the two fragments (containing Met rather than homoserine lactone residues) minus water summed to give the original molecular weight of STP-3. This could only be the case if one of the two Met residues were part of a ring system.

Shown in Fig. 1B is the collision activated dissociation mass spectrum recorded on the (M+H)+ ion at m/z 984. The amino acid sequence deduced from the spectrum is shown below. When a collection of peptide (M+H)+ ions are allowed to collide with argon atoms we expect to see more or less random fragmentation along the peptide backbone. Ions of type B are produced if the extra proton, and therefore the positive charge, remains on the fragment containing the N-terminus of the peptide. Ions of type Y" result if the charge remains with the fragment containing the C-terminus. Fragment ions of type B and Y" predicted for the above sequence are shown above and below the structure. Those observed in the spectrum are underlined. Note that the fragmentation patterns observed on the triple quadrupole mass spectrometer do not allow differentiation of Leu and Ile, two residues of identical mass. Residue 7 in the cyanogen bromide fragment is assigned as Leu because amino acid analysis performed on STP-3 showed that Ile was not present in the molecule.

72	129	$\underline{266}$	$\underline{413}$	$\underline{576}$	$\underline{713}$	$\underline{826}$	$\underline{883}$	$\underline{984}$	Y"
Ala	Gly	His	Phe	Tyr	His	Leu	Gly	hSer	
$\underline{984}$	$\underline{913}$	$\underline{856}$	$\underline{719}$	$\underline{572}$	$\underline{409}$	$\underline{272}$	159	102	B

3. Digestion of STP-3 (500 pmol) with chymotrypsin produced two small peptides the mass spectra of which showed (M+H)+ ions at m/z 728 and 993. Treatment of the smaller peptide with methanolic HCl failed to shift the (M+H)+ to higher mass. On-probe acetylation added two acetyl groups to the peptide and thus shifted the (M+H)+ to higher mass by 84 units. The above results indicated that the smaller chymotryptic fragment contained two amino groups and a blocked C-terminus. These structural features could be accommodated by either (a) a conventional linear peptide containing a Lys residue and a blocked C-terminus or (b) an unconventional peptide containing two chains linked to the α- and ε-amino groups of Lys residue blocked at its C-terminus. Only the latter possibility agreed with the fragmentation pattern obtained in the collision activated dissociation mass spectrum recorded on the (M+H)+ ion at m/z 728 (Fig. 2). The amino acid sequence deduced from this spectrum is shown below. Predicted fragment ions containing His and Ser as the N-termini are shown above and below the structure, respectively. Those observed in the spectrum are underlined.

				$\underline{584}$		
				$CONH_2$		
138	$\underline{251}$	$\underline{308}$	$\underline{439}$	|	$\underline{641}$	$\underline{728}$
NH_2-His	Leu	Gly	Met	Lys	Gly	Ser-NH_2
$\underline{728}$	$\underline{591}$	$\underline{478}$	$\underline{421}$	$\underline{290}$	145	88

Esterification and acetylation of the larger chymotryptic fragment shifted the (M+H)+ ion at m/z 993 to higher mass by 14 and 42 mass units, respectively. These results are those expected for a peptide containing a single amino group and a single carboxyl group. A collision activated dissociation mass spectrum recorded on the (M+H)+ ion for the methyl ester (m/z 1007) provided the necessary fragmentation to deduce the sequence of this peptide. Confirmation of the deduced sequence was obtained by treating the peptide with dithiothreitol and iodoacetic acid. Both Cys residues were reduced and carboxymethylated under the above conditions and the (M+H)+ at m/z 993 shifted to m/z 1111 as expected.

Cys Asn Gly Met Leu Cys Asn Ala Gly Leu

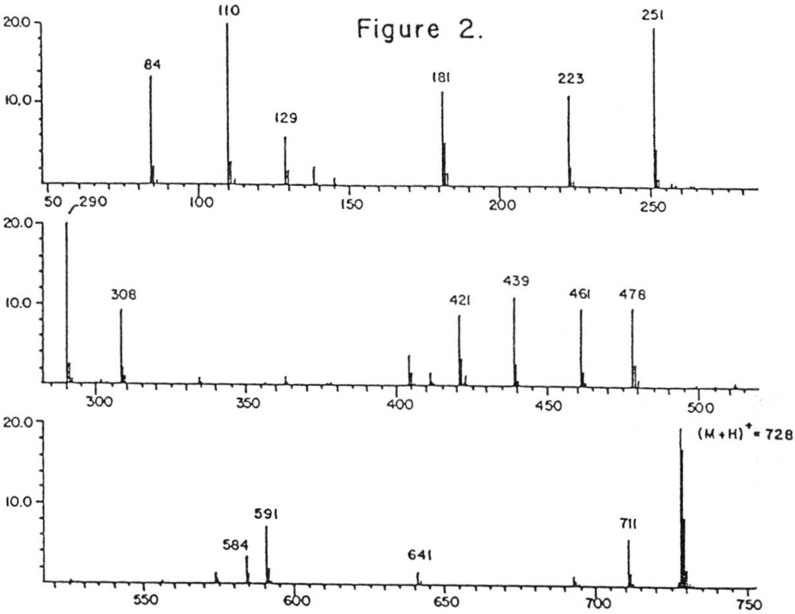

Fig. 2. Collision activated dissociation mass spectrum recorded on the (M+H)+ ion, m/z 728, derived from a chymotryptic fragment of STP-3.

5. Information to align the above sequences within the structure of STP-3 was obtained by analysis of a laser photodissociation mass spectrum recorded on 50 pmol of the intact parent molecule. Photodissociation of the (M+H)+ ion at m/z 2778.3 gave the spectrum shown in Fig. 3. Observed fragment ions are consistent only with attachment of the two chymotryptic peptides and CNBr fragment in the order shown below.

```
┌──────────────────────────────┐
|                              | 1658 1544 1473 1416 1303 1216 1159
Cys Asn Gly Met Leu Cys Asn  Ala  Gly  Leu  Ser  Gly  Lys-CONH2
                                                       |
                          Ala Gly His Phe Tyr His Leu Gly Met
                           72 129  266 413 576 713 826 883 1014
```

6. Substractive Edman degradation performed at the 500 pmol sample level was employed to deduce the order of the missing amino acids at the N-terminus of the larger peptide chain in STP-3. In this experiment mass measurements were made on the (M+H)+ ions observed for the parent molecule before and

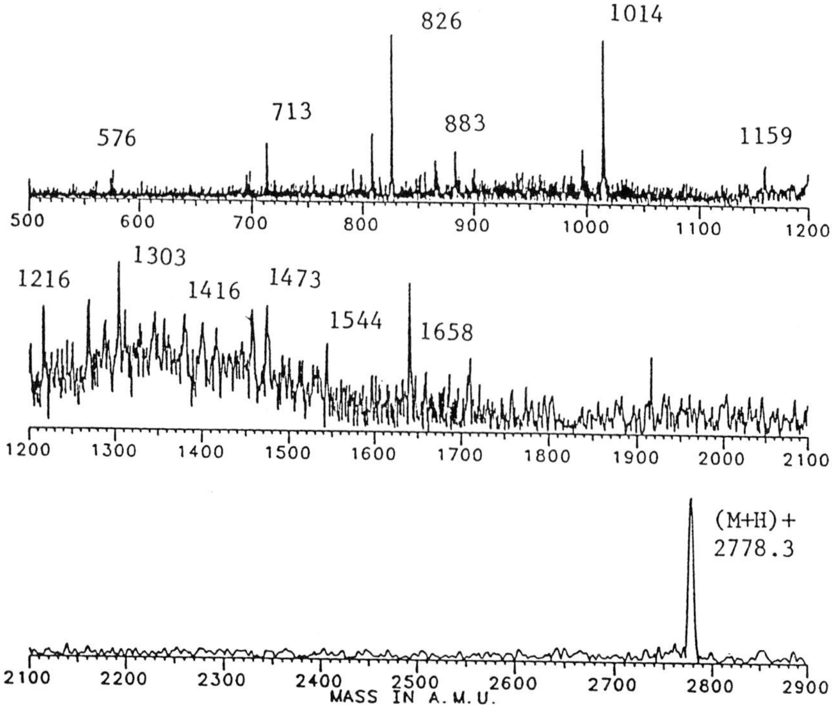

Fig. 3. Laser photodissociation Fourier transform mass spectrum recorded at the 50 pmol sample level on the (M+H)+ ion, m/z 2778.3, of STP-3.

after removal of an amino acid by manual Edman degradation. Results obtained after each of five such cycles of Edman degradation are summarized in Table I. Since STP-3 contains two amino termini, two amino acids were released during each cycle of Edman degradation. As expected, one of the two amino acids removed in each cycle matched that predicted from the sequence of the CNBr fragment discussed above. The other amino acid released in each cycle was assigned to the amino terminus of chain 2. Addition of amino acids, AGSHF-, to the N-terminal side of the first Cys residue brought the mass of the deduced structure up to that measured for the (M+H)+ ion of STP-3.

C. Data Obtained From Automated Gas Phase Edman Degradation Experiments.

Results obtained from an independent analysis of intact STP-3 at the 100-200 pmol level by automated gas phase Edman degradation are summarized in Table I. Quantitation of liberated PTH derivatives by HPLC indicated that two amino acids were removed from STP-3 in 8 of the first 9 cycles. A blank from one chain at cycle 6 and low yield of dhSer at

Table I. Data obtained on intact STP-3 by (a) manual Edman degradation plus mass spectrometry and (b) automated gas phase Edman degradation and PTH analysis by HPLC.

	Manual Edman Degradation and Mass Spectrometry				Automated Gas Phase Edman Degradation	
Cycle No.	Observed (M+H)+	Mass Lost	Amino Acid Identified		PTH Identified	
0	2778	-				
1	2636	142	A	A	A[2]	A[2]
2	2522	114	G	G	G	G
3	2298	224	H	S	H	S
4	2014	284	F	H	F	H
5	1704	310	Y	F	Y	F
6					H	-
7					L	N
8					G	G
9					M	M
10						L
11						dhSer[3]
12						N
13						A
14						G
15						L
16						S
17						G
18						-

[1]Quantitation of PTH derivatives facilitated detection of more than one mole of amino acid residue liberated in cycles 1-9. [2]Products of this cycle had to be analyzed at the end of the run due to mechanical failure of the HPLC pump. [3]Detection of a single His residue at cycle 6 and a weak signal due to dhSer at cycle 11 suggested the presence of two Cys residues linked through a disulfide bond at positions 6 and 11 in the longer of the two peptide chains.

cycle 11 are consistent with the presence of Cys residues at those positions. Derivatization of STP-3 with vinylpyridine was not performed so the two Cys residues were not detected.

D. Linkage Assignment of the Two Peptide Chains.

To determine which peptide chain was linked to the α- and ε-amino groups of Lys, STP-3 (200 pmol) was reduced, carboxymethylated, and cleaved with CNBr. The fragment containing Lys at the C-terminus, $(M+H)^+$ = 919 (see below) was then acetylated under conditions that failed to introduce an acetyl group at the ε-amino group of model peptides (3/1 $MeOH/Ac_2O$ containing 2% TFA). Acetylation of the N-terminus of the model peptides occurred to the extent of 40% under these conditions. Acetylation of the cyanogen bromide fragment from STP-3 with the above reagent incorporated a single acetyl group and shifted the observed $(M+H)^+$ ion from m/z 919 to 961. This result strongly suggested that the cyanogen bromide fragment contained a free ε-amino group on Lys and, therefore, that the longer of the two peptide chains was linked to the α-amino group of Lys.

IV. REFERENCES

1. Hunt, D. F., Yates, J. R., III, Shabanowitz, J, Winston, S., and Hauer, C. R. (1986) Proc. Natl. Acad. Sci., USA **83**, 6233-6237.
2. Hunt, D. F., Shabanowitz, J., Yates, J. R., III, Zhu, N. Z., Russell, D. H., and Castro, M. E. (1987) Proc. Natl. Acad. Sci., USA **84**, 620-623.
3. Brinegar, A. C., Cooper, G., Stevens, A., Hauer, C. R., Shabanowitz, J., Hunt, D. F., and Fox, J. E. (1988) Proc. Natl. Acad. Sci., USA **85**, 5927-5931.

57

CHARACTERIZATION AND SYNTHESIS OF STP-3

Sylvia W. Yuen, Kenneth M. Otteson, Joel C. Colburn, William T. Moore *, Tim D. Schlabach, David R. Dupont, and Robert J. Mattaliano

Research and Development, Applied Biosystems, Inc., Foster City, CA 94404

*Analytical Chemistry Center, University of Texas Health Science Center, Houston, TX 77225

I. INTRODUCTION

The determination of structural information for peptides and proteins requires the application of a variety of analytical techniques in a complementary fashion. Information derived from each technique is used as the basis for a structural model which is consistent with experimental results. Proof of molecular structure rests on comparative physico-chemical analysis of the test molecule versus a molecule specifically synthesized on the basis of a proposed structure.

The STP-1 (1) and STP-2 (2) peptides are examples of samples prepared by Jean Rivier and co-workers (The Salk Institute, La Jolla, CA) to provide an analytical challenge and to demonstrate the application of a variety of modern methods for the analysis of polypeptide structures. This abbreviated text reports the structural analysis of the third in this series of test polypeptides, STP-3.

II. MATERIALS and METHODS

The instrumentation utilized in the structural elucidation of STP-3 for peptide sequencing, amino acid analysis, synthesis and chromatographies was primarily that available from Applied Biosystems, Inc. (ABI). High resolution mass spectra were collected as described previously (2). Upon receipt, STP-3 was dissolved in 0.1% TFA with moderate vortexing. Aliquots were removed from the thawed sample which was stored at -20°C. A more detailed description of the materials and methods used in the characterization and synthesis of STP-3 may be obtained directly from ABI (3).

III. RESULTS and DISCUSSION

A. Preliminary Experiments

The chromatographic and electrophoretic properties of STP-3 were

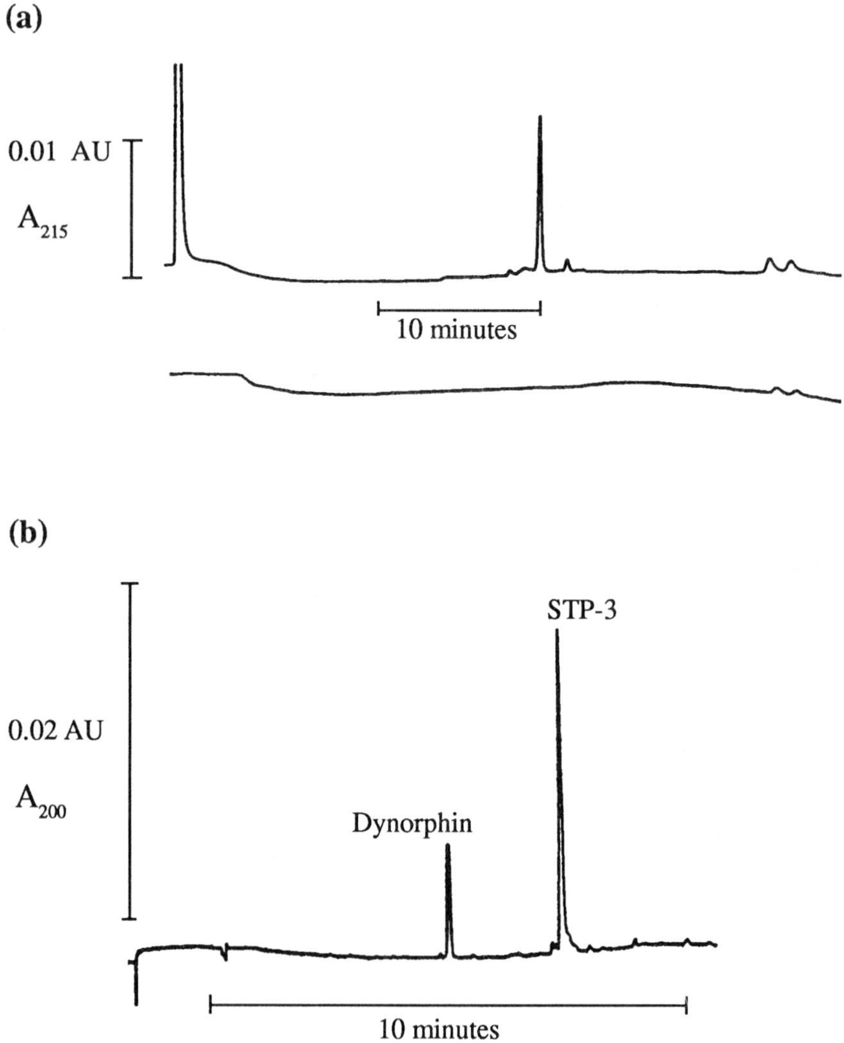

Fig. 1. (a) Narrowbore RP-HPLC of 0.5% of the STP-3 sample as compared to baseline. (b) CE electropherogram of STP-3 mixed with dynorphin.

examined using both reverse phase HPLC (RP-HPLC) and capillary electrophoresis (CE). Figure 1a shows the RP-HPLC trace of STP-3 as compared to that for a baseline control which demonstrates that STP-3 behaved as a nearly homogenous peptide. As a method which complements HPLC, CE consumes an insignificant percentage of sample (several nanoliters) when used analytically. Figure 1b shows the electropherogram of STP-3 mixed with dynorphin, a peptide included as a mobility reference. The electrophoretic behavior of STP-3 during CE also indicated near homogeneity.

Peptide sequence analysis of STP-3 (3.3% of the sample) revealed several interesting structural features (see Table Ia). Single residues were observed during cycles 1, 2, 6 and 8 through 18. However during cycles 3, 4, 5 and 7, two residues were detected. Ser residues were confirmed at cycles 3 (in addition to His) and 16, e.g., authentic PTH-Ser was observed in addition to the DTT adduct of PTH-dehydroalanine (dhA*), which forms by beta-elimination of the side chains of Ser, phospho-Ser and Cys residues. At cycle 6, His was detected along with a trace increase of dhA* and at cycle 11, only dhA* was observed. At cycle 10 (Leu), the pMole yield dropped nearly in half as compared to cycle 9 (Met). The last detectable residue was Lys at cycle 18. The peptide sequence yield / cycle data suggest that the unknown is comprised of two covalently attached peptide chains of 9 and 18 residues, respectively. Each chain contains the same residues at positions 1, 2, 8 and 9 with possibly a disulfide or modified amino acids at positions 6 and 11.

Another aliquot (3%) was subjected to vapor phase acid hydrolysis. The composition of the hydrolysate was determined using the PTC-amino acid derivatives. The results are shown in Table Ib. The composition data indicate that the quantity of peptide recovered from the vial by dissolution in dilute acid was 1.8 nanomoles. The composition of STP-3 is consistent with the sequence data, using Asx=2 for the mole ratio calculation. No unusual PTC-amino acids were observed during the analysis.

The occurrence of a Cys at position 11 could be inferred due to the detection of dhA* in that cycle. Another Cys at position 6 was possible since the yield of His at that cycle was low and a trace of dhA* was also detected. Thus, the presence (or absence) of Cys residues at positions 6 and 11 and the possibility that they were linked via a disulfide bond needed examination. Aliquots (~ 80 pMoles each) of STP-3 were subjected to alkylation with 4-vinylpyridine (4VP) in the presence or absence of reductant. Each peptide was then repurified by RP-HPLC. Sequence analysis of the sample subjected to alkylation in the absence of reductant yielded results identical to those for the native peptide. However, sequence analysis of STP-3 alkylated after reduction with mercaptoethanol (PE-STP-3), confirmed cysteine residues at positions 6 and 11, but otherwise gave results identical to those for the native sample. These results indicate that 1) the Cys residues at positions 6 and 11 are oxidized since they could not be alkylated in the absence of reductant, and 2) the two polypeptide chains are also attached by a non-reducible cross-link.

An accurate peptide mass calculation is possible based upon the composition but must take into account the nature of the cross-link between the two chains. With the exception of the cysteine residues at positions 6 and 11, the

Table I. STP-3 (a) Peptide sequence and (b) Compositional analysis results

(a) Peptide sequence analysis

Cycle	Residue(s)	pMoles		Total pMoles
1	Ala	72.4		72.4
2	Gly	68.2		68.2
3	Ser, dhA* + His	4.3+9.0*	11.6	24.9
4	Phe + His	34.8	11.3	46.1
5	Phe + Tyr	36.0	27.7	63.7
6	(dhA*) + His	(0.2*)	9.1	9.1
7	Asn + Leu	21.1	25.3	46.4
8	Gly	43.4		43.4
9	Met	34.9		34.9
10	Leu	14.2		14.2
11	(dhA*)	(2.4*)		(2.4)
12	Asn	12.9		12.9
13	Ala	11.7		11.7
14	Gly	8.4		8.4
15	Leu	7.4		7.4
16	Ser, dhA*	1.27+1.9*		3.2
17	Gly	4.8		4.8
18	Lys	1.8		1.8
19	- -	- -		- -
20	- -	- -		- -

(b) Compositional analysis

Residue	pMoles	Mole Ratio		Sequence Observed
D	43.76	2.0	(2)	2
E	4.28	0.2		
S	45.75	2.1	(2)	2
G	132.89	6.1	(6)	6
H	60.75	2.8	(3)	3
R	3.79	0.2		
T	4.07	0.2		
A	71.21	3.3	(3)	3
P	4.44	0.2		
Y	21.40	1.0	(1)	1
V	5.68	0.3		
M	11.33	0.5	(1)	2
I	2.77	0.1		
L	59.30	2.7	(3)	3
F	31.40	1.4	(1)	2
K	11.72	0.5	(1)	1

only other residues with reactive side chains capable of covalent linkages are the 9 residue chain C-terminal Met and the 18 residue chain C-terminal Lys. Assuming the presence of an amide cross-link between the Met carboxyl and the Lys epsilon-amino group and the disulfide linkage, the predicted molecular weight of STP-3 would be 2779.9 for the C-terminal acid form and 2778.9 for the C-terminal amide form. A molecular weight of 2777.9 for native STP-3 was obtained by FAB-MS using 10 pMoles of peptide. This value is more consistent with the predicted mass of the simple inter-chain amide bond cross-linked structure ending as a C-terminal amide as described above.

B. Analysis of Structural Detail

```
A G [H F Y H L] G M L C N A G L S G K-x
A G [S H F C N] G M ------------------┘
1         5         10        15
```

STP-3 working model; x = C-terminal acid or amide.

The model shown above (disulfide not included) summarizes the results of previously described experiments. Ambiguities remaining (in brackets) relate to the relative chain positions of each residue between positions 3 and 7, inclusive. To resolve these assignments, 200 pMoles of PE-STP-3 was subjected to CNBr cleavage. The subsequent peptide mapping/sequence analysis results are shown in Figure 2. Sequence analysis of CNBr peptides peaks 2 and 10, each ending in homoserine (hS), allowed the specific sequence assignment of residues on the amino terminal side of the Met residues at position 9. Peak 3 was the fragment derived from residues 10 - 18 of the long chain. However, this cleavage failed to clarify if the 9 residue cysteine containing peptide (AGSHFCNGM) was attached by an amide bond to Leu (10), i.e part of the long chain, or to the side chain amino group of Lys (18).

Since the sequencing results indicated the presence of an N-G bond on one chain at positions 7-8, another type of chemical cleavage was performed using hydroxylamine. The predicted cleavage products would be sufficiently different in the case of the SHFCN segment being part of the long chain versus the short chain as to allow the resolution of this ambiguity. The resultant peptide map and sequence data are summarized in Figure 3. The peptide corresponding to peak 2 would be produced by hydroxylamine cleavage in either case. However, the sequence data for peptides from peaks 8 and 12 can only be accommodated by a model in which the SHFCN segment is actually part of the long chain. Peak 8 peptide is the product of cleavage between the N-A peptide bond on the long chain. Peak 12 peptide is the product of cleavage between the N-G peptide bond on the long chain. The double sequence observed at certain cycles for these two peptides occurs since the resultant fragments are comprised of two chains attached by a cross-link between the C-terminal residues. Peak 14 appears to be uncleaved STP-3. Thus, the complementary use of sequence data from peptides prepared by both CNBr and hydroxylamine cleavage of STP-3 allowed for the solution of the primary structure for each chain.

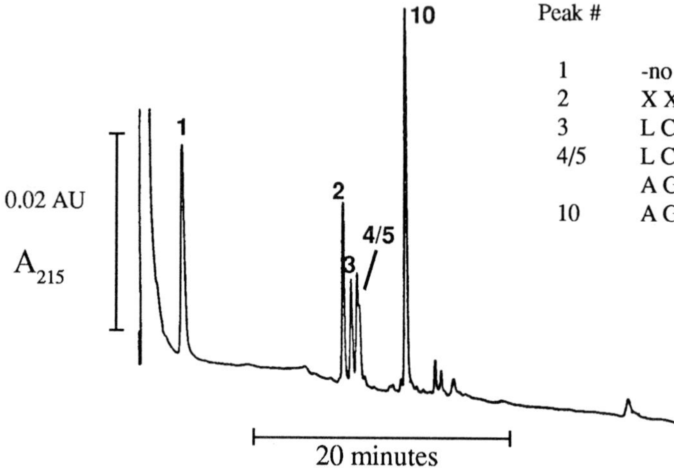

Peak #	
1	-no sequence-
2	X X S H F C N G ^{h}S
3	L C N A G L S G (K)
4/5	L C N A G L S G K
	A G S X F C N G X
10	A G H F Y H L G ^{h}S

Fig 2. RP-HPLC mapping and sequence results for STP-3 CNBr digest peptides.

Peak #	
1	-no sequence-
2	A G S H F C N
8	A G H F Y H L G M
	L S G
12	A G H F Y H L G M
	G M L C N A G L S G K
14	A G H F (Y)(H)(L) G M L X N A
	X X (F)(C)(N)

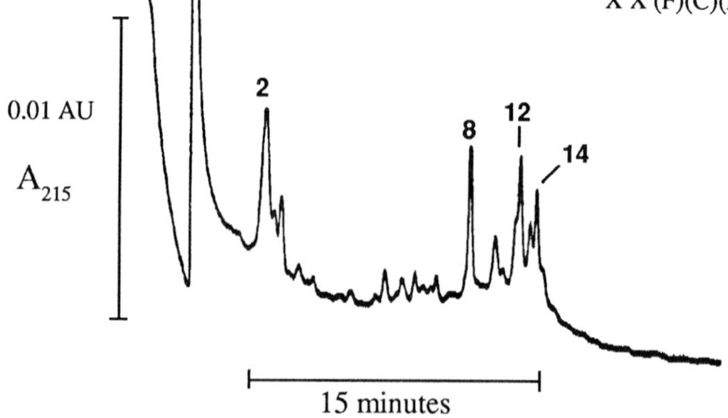

Fig 3. RP-HPLC mapping and sequence results for PE-STP-3 hydroxylamine digest peptides.

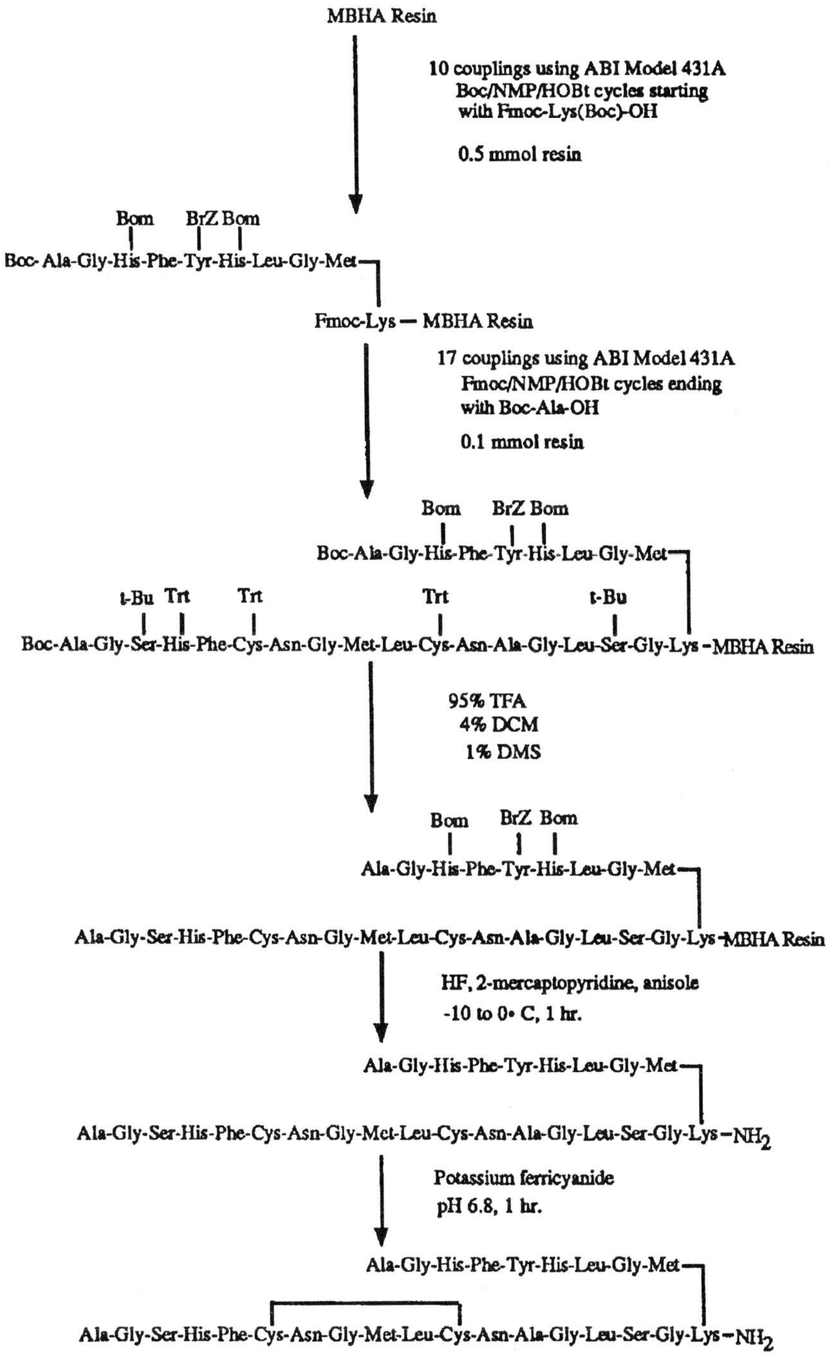

Fig . 4. Synthesis of STP-3 using sequential tBoc and Fmoc chemistries.

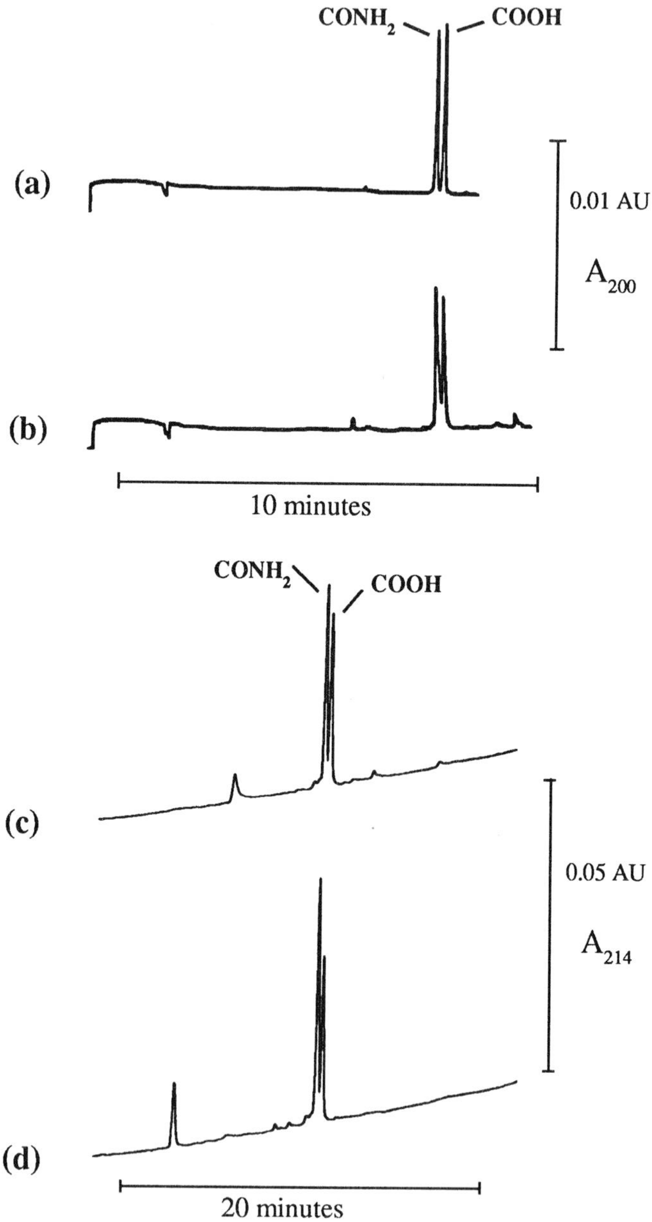

Fig. 5. Comparative CE and RP-HPLC of ABI synthetic peptides versus STP-3. (a) CE, mixture; (b) CE, mixture plus STP-3; (c) RP-HPLC, mixture; (d) RP-HPLC, mixture plus STP-3.

C. *Comparison of STP-3 to Synthetic Analogs*

The mass data suggested that STP-3 contained a C-terminal amide. To confirm this, both the C-terminal acid and C-terminal amide forms were synthesized for comparison to authentic sample. The synthetic scheme for the amide form is outlined in Figure 4. Fmoc-Lys(Boc) was activated to the HOBt ester prior to coupling to the MBHA resin. The acid form was prepared using hydroxymethyl resin to which Fmoc-Lys(Boc) was attached via the symmetric anhydride/DMAP route. Otherwise both approaches are identical and were performed on a synthesizer (431A) capable of both Fmoc and Boc chemistries.

Compositional and sequence analyses of both the C-terminal amide and acid forms of ABI synthetic peptides gave results which were identical to those for STP-3. The chromatographic and electrophoretic mobilities of STP-3 were then compared to the synthetic products. Figure 5 outlines the results of these comparative analyses. Both RP-HPLC and CE were able to resolve the C-terminal acid and amide forms. STP-3 co-eluted with the ABI synthetic amide form during both CE and RP-HPLC. Also, mixing STP-3 together with the ABI synthetic analogs resulted in an increase in the amplitude of C-terminal amide form relative to the acid form using either RP-HPLC or CE. These results confirmed that authentic STP-3 contains a C-terminal amide.

IV. SUMMARY

In conclusion, a variety of analytical methods were used in complementary fashion to design a model based upon experimental observation. Actual proof of STP-3 structure was facilitated by comparison of STP-3 to synthetic analogs. The structure of STP-3 is shown at the end of figure 4. STP-3 is comprised of two peptides of 9 and 18 residues, respectively. The C-terminal Met of the short chain is covalently attached to the C-terminal Lys epsilon-amino group of the long chain via an amide bond. The 18 residue chain contains an intra-chain disulfide bond between residues 6 and 11 and ends as a C-terminal amide.

V. REFERENCES

1. Wilson, K.J. et al, (1985) ABI Protein Sequencer User Bulletin # 15
2. Wilson, K.J. et al, (1987) ABI Protein Sequencer User Bulletin # 33
3. Yuen, S.W. et al, (1988) ABI Protein Sequencer User Bulletin # 38

Acknowledgement: We thank Pau Yuan, Ken Wilson, and Mike Hunkapiller for support during this project and critical reading of the manuscript.

Index

A

F

G

N

O

R

S

T

W